Reliability of Organic Compounds in Microelectronics and Optoelectronics

Willem Dirk van Driel • Maryam Yazdan Mehr
Editors

Reliability of Organic Compounds in Microelectronics and Optoelectronics

From Physics-of-Failure to Physics-of-Degradation

Springer

Editors
Willem Dirk van Driel
Delft University of Technology
Delft, The Netherlands

Maryam Yazdan Mehr
Delft University of Technology
Delft, The Netherlands

ISBN 978-3-030-81575-2 ISBN 978-3-030-81576-9 (eBook)
https://doi.org/10.1007/978-3-030-81576-9

This Springer imprint is published by the registered company Springer Nature Switzerland AG
The registered company address is: Gewerbestrasse 11, 6330 Cham, Switzerland

Preface

Further developments in microelectronics and optoelectronics industries necessitate materials that can withstand harsh and demanding working conditions. Reliability and lifetime assessment of organic compounds in microelectronic and/or optoelectronics devices are two key critical issues when it comes to the development of new products. Reliability is defined as the probability of failure in a system over a given lifetime, with defined operating conditions. When it comes to reliability, two factors come into play: stable performance and predictability of failures within the so-called "service lifetime". Developing accurate reliability models for microelectronic and/or optoelectronic devices requires an in-depth understanding of failure and degradation mechanisms in organic compounds. Here, organic materials relate to moulding compounds, die-attach, underfill, coatings, silicones, polycarbonate and many more. Failures in sensitive microelectronic and optoelectronic devices could have severe consequences on the functioning and profitability of many plants and infrastructures. Understanding the root-cause of failures in electronic devices and how materials degradation can contribute to such failures are therefore extremely important in developing more reliable organic materials and systems. Such knowledge can also be directly used in optimizing service conditions (voltage, temperature, environmental parameters and mechanical stresses). Having a clear picture of a failure scenario is very useful in resolving the complexity of the inter-relations between material properties and service conditions, which in case of an improper functioning result in failure. This book aims to provide a comprehensive reference into the critical subject of failure and degradation in organic materials, used in optoelectronics and microelectronics systems and devices. Key unique features of this book are:

- Documenting and introducing failure mechanisms of organic materials, used in microelectronics and optoelectronics devices
- Inter-relating ageing of organic materials to product failure and how to use available simulation techniques to optimize the design and performance of a device

- Investigating the integration of several stresses (thermal, moisture, light radiation, mechanical damage and more) into the performance of a large-scale system in several industrial domains (lighting, automotive, transport and more)
- Introducing state-of-the-art multi-scale/multi-physics simulation and experimental techniques to study failures of organic compounds in micro/optoelectronic devices

This book contains chapters related to (i) organic materials (silicones and polycarbonate), (ii) degradation mechanisms in microelectronics materials like moulding compounds, (iii) degradation mechanisms in optoelectronics components like OLED and LEDs and (iv) state-of-the art modelling and lifetime assessment techniques.

Parts of the contents in this book are first-hand results from industrial research and development projects. Reading this book, students in different engineering disciplines get an insight and develop an in-depth understanding of different failure and/or degradation mechanisms in organic materials. Also, this book will certainly be useful when it comes to training methodologies of assessing failures, degradations and reliability of different engineering materials for students. Further, students, engineers and technicians in different industrial sectors will certainly find this book interesting and informative.

We would like to thank all the authors for their valuable contributions to the book. The undersigned would also like to make acknowledgements to many of their colleagues in Signify and Delft University of Technology who have contributed to this book in one way or another.

Personal Acknowledgements

Willem van Driel is grateful to his wife Ruth Doomernik, their two sons, Juul and Mats, and their daughter, Lize, for their support during the writing and editing of this book. Besides that, the idea of the book was given during the Corona Davide Challenge (100 bowls of Italian ice cream during the 2020 lockdown), fruitfully accomplished with Silas and Joachim, and many thanks go to their parents Roger Gerritzen and Tineke van den Heuvel. Really amazing ice creams: www.gelateria-davide.nl.

Maryam Yazdan Mehr is grateful to her husband Abbas and their daughter Adrina for their support during the writing and editing of this book.

Xuejun Fan is grateful to his wife, son and parents for their unselfish support and love. G.Q. Zhang is grateful to his wife and their two children.

Delft, The Netherlands — W. D. van Driel
Maryam Yazdan Mehr
Beaumont, TX, USA — Xuejun Fan
Delft, The Netherlands — G. Q. Zhang
May 2021

Contents

Chapter 1
Degradation Mechanisms of Silicones

François de Buyl and Shin Yoshida

1 Introduction to Silicones

1.1 Polydimethylsiloxane

The best known of the commercially available silicones are polydimethylsiloxanes (PDMS), so these will be our main concern. They are available as linear fluids, cyclics, gels, and resins, depending on the degree of cross-linking, or as elastomers, when fillers are incorporated in cross-linked polymers.

Silicone polymers or resins display the unusual combination of an inorganic chain similar to silicates or glass, which is often associated with high surface energy, and side methyl groups that are, conversely, organic by nature, and thus exhibit low surface energy [1]. The Si–O bonds of silicone are strongly polarized and, without protection, should lead to strong intermolecular interactions. Methyl groups weakly interact with each other, so they effectively shield the main molecular chain. This is further facilitated by the absence of side groups at the oxygen atom, which results in the high flexibility of the siloxane backbone.

The bond lengths and angles are also responsible for the properties of silicones (Fig. 1.1). The typical C–C and C–O bond lengths are 1.54 and 1.41 Å, respectively, while the Si–O bond length is much longer with 1.65 Å. The bond angles between C–C and C–O bonds are also different than the Si–O bond angles. A typical C–C–C or C–O–C bond has an angle near 110° (tetrahedral spatial configuration). The Si–O–Si bond angle is 142° and the O–Si–O bond angle is 109° [2].

F. de Buyl (✉)
Dow Silicones Belgium srl, Seneffe, Belgium
e-mail: francois.debuyl@Dow.com

S. Yoshida
Dow Toray Co. Ltd., Tokyo, Japan

W. D. van Driel, M. Yazdan Mehr (eds.), *Reliability of Organic Compounds in Microelectronics and Optoelectronics*,
https://doi.org/10.1007/978-3-030-81576-9_1

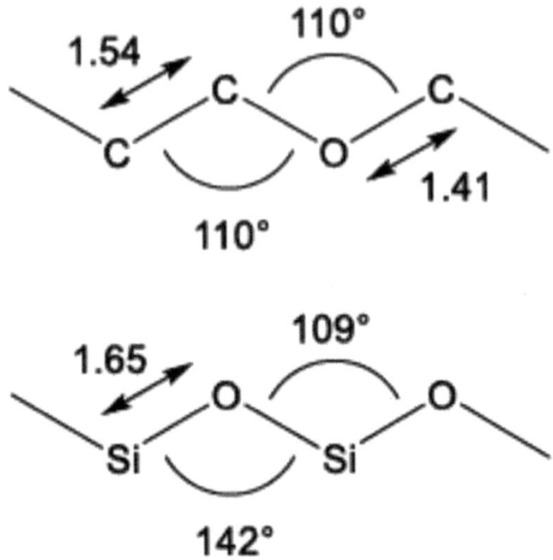

Fig. 1.1 Bond angles and lengths of carbon and silicon-based polymers

While these differences between carbon-based polymers and silicon-based polymers can seem small, the longer bond lengths and the widened bond angles relieve significant barriers to bond rotation. This open molecular structure also explains the lower barrier to rotation around the Si–O bond, of about 2.5 kJ/mol (0.5 kcal/mol), approximately equal to kT at room temperature, compared with barriers of about 17 kJ/mol in alkanes (4 kcal/mol) [3, 4]. Thus, the polydimethylsiloxane chain can adopt many conformations. In more practical terms, this means silicone's polydimethylsiloxane backbone can easily adopt many shapes for easier processing and molding.

Low intermolecular interactions and high free volume of polydimethylsiloxane compared to hydrocarbons result in other physical characteristics, such as a low glass transition temperature (Tg) of 146°K (−127 °C), high flexibility, good vibration absorption characteristics as well as high impact resistance [1]. Despite the high solubility and high diffusion coefficient of gases such as oxygen, nitrogen, or water vapor, silicone resists wetting due to its very low surface tension [1]. Therefore, with their low moisture uptake and ability to withstand harsh environmental conditions, conventional silicones are already frequently used by the electronics industry to protect fragile components against damage and corrosion.

Compared to many organic materials, the chemical backbone of silicones makes them particularly well-suited to manage the increasingly high temperatures of today's and tomorrow's LED lighting systems. The combination of siloxane and methyl has thermal and oxidative stability benefits arising from the polarized Si–O bond that is highly ionic and exhibits a large bond energy, 452 kJ/mole (108 kcal/mol) due to the electro-negativity difference, Δ_{EI} of 1.7, between silicone (1.8) and carbon (2.5). By comparison, the Si–C bond has a bond energy of ca. 318 kJ/mole (76 kcal/mol) – which is slightly lower than a C–C bond with ca. 346 kJ/mol (83 kcal/mol) and C–O bond with ca. 358 kJ/mol (86 kcal/mol) – while the Si-Si bond is weak, 193 kJ/mole (46 kcal/mole) (Fig. 1.2).

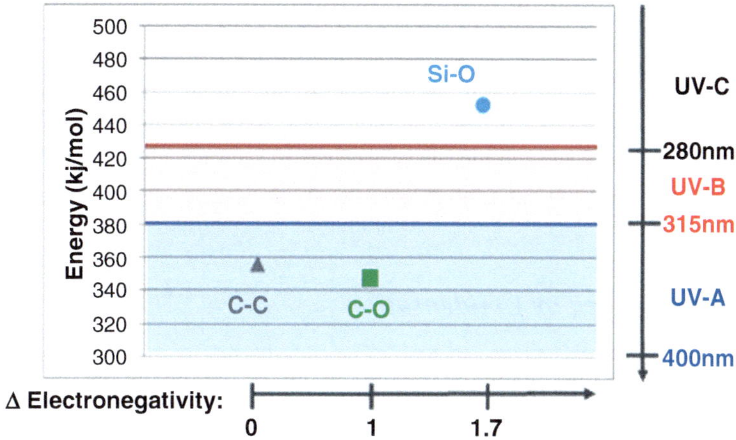

Fig. 1.2 Bonds energy and difference in electronegativity of C–C, C–O, and Si–O bonds against UV radiation scale

Methyl groups on silicon in hexamethyldisiloxane, $Me_3Si–O–SiMe_3$, are reinforcing further the Si–O bond strength, with a dissociation energy of 569 kJ/mol (136 kcal/mol) [5].

These values of bond energies and the quite strong polar character of the Si–O bond, i.e., $Si^{\delta+}–O^{\delta-}$, partially explain the stability of the dimethylsiloxane polymer backbone against homolytic scission (oxidation, thermolysis). On the other hand, heterolytic scissions are more favorable, and will be useful for the preparation of high molecular weight PDMS by re-equilibration polymerizations catalyzed by acids or bases [4, 6].

As illustrated in Fig. 1.2, the value of Si–O bond energy also explains the stability against heat and ultraviolet (UV) of PDMS-based optical components used in LED lamps & luminaires applications [7, 8]. In comparison to hydrocarbon-based polymers used for the manufacture of optical components, e.g., acrylics (PMMA) or polycarbonates (PC), which backbone is made of C–C and C–O bonds, the PDMS is perfectly transparent to UV-A (315–400 nm) and UV-B (280–315 nm) radiations, and in a certain extent to UV-C radiation (200–280 nm). Therefore, UV-A/B radiations are not significantly absorbed by PDMS polymer, and therefore no degradation of physical properties occurs. In the UV-C region, the stability of PDMS polymer or any silicone-based materials formulated and designed for a particular application will depend on the exact UV-C wavelength and irradiance at which the material will be exposed.

For applications in LED packaging devices, however, it was discovered and demonstrated [9] that partially replacing methyl groups by phenyl groups along the polysiloxane backbone offers the benefit of increasing the refractive index of the siloxane polymer or resin [10, 11] and therefore enhances light output from and efficiency of LED devices, which will be further discussed throughout this chapter.

A small percentage of phenyl groups along the chain sufficiently perturbs mobility of the polymer chains and disrupts crystallization, allowing the polymer to remain flexible at very low temperatures; the glass transition temperature of methyl-phenyl-siloxane remains quite negative, with T_g of ca. −25 °C.

2 Crosslinking Mechanism

Silicone polymers are easily transformed into a three-dimensional network and an elastomer via a cross-linking reaction, which causes formation of chemical bonds between adjacent polymer chains. This is achieved according to one of the following reactions: cross-linking by condensation, addition, or radical reaction.

2.1 Crosslinking by Condensation

This method is used in sealants such as those available from Do-it-yourself shops. Silicone sealants are also widely used by professionals in original equipment manufactories for sealing windows and doors for either residential or commercial buildings. Most of these products are ready to use one-component Room-Temperature-Vulcanization (1-RTV) and require no mixing: cross-linking starts when the product is squeezed from the cartridge and meets moisture from the atmosphere and/or present at the surface of the substrate [1]. Two-components RTV silicone elastomers however exist too.

Most silicone sealants are formulated from a reactive polymer prepared from a hydroxy-end blocked-polydimethylsiloxane and a large excess of an organosilane crosslinker, such as, e.g., methyltriacetoxysilane or methyltrimethoxysilane, respectively. Acetic acid or methanol, respectively, is released as a by-product of the cure leading to a slight shrinkage upon cure, ca. 0.5% to 1% linear shrinkage [1].

Mechanism of crosslinking of 1-RTV sealants, and adhesives is essentially dictated by diffusion of moisture into the cross-section of the silicone elastomer layer and was studied extensively [12].

Sealants are designed for use in adhesive joints that undergo significant dimensional changes during their service life. Understanding physics of cure of 1-RTV silicone sealants as function of their curing chemistry (i.e., acetoxy, alkoxy, oxime) [1], as well as the long-term performance of sealant joints is therefore critical to the sealant developer and specifier [13, 14]. For a rapid overview of the important aspects of the science of adhesion, the reader may refer to a reasonably recent publication on the subject, which compares theories and adhesives systems for their respective peculiarities [15].

2.2 Crosslinking by Addition

The shrinkage problem encountered with both 1- and 2-part condensation systems can be eliminated when using an addition reaction to achieve cross-linking. Here, cross-linking is achieved using vinyl end-blocked polymers and reacting them with Si–H groups carried by functional siloxane oligomers [16].

$$\sim\sim\sim OMe_2Si-CH=\mathrm{CH_2}+\mathrm{H}-Si\equiv\xrightarrow{\text{cat}}\sim\sim\sim OMe_2Si-\mathrm{CH_2}-\mathrm{CH_2}-Si\equiv$$

where ≡ represents the remaining valences of the silicon atom.

The addition or hydrosilylation reaction occurs mainly on the terminal carbon (β-addition) and is catalyzed by Pt or Rh metal complexes, preferably organometallic compounds to enhance their compatibility with the medium, such as, e.g., Karstedt catalyst [17]. The following mechanism has been proposed, i.e., oxidative addition of the ≡SiH on the Pt, H transfer on the double bond in the β-position, and reductive elimination of the product [17].

$\equiv Si\text{-}CH{=}CH_2$

•
•
•

Pt <===> $\equiv Si\text{-}CH_2\text{-}CH_2\text{-}Pt\text{-}Si\equiv$ → $\equiv Si\text{-}CH_2\text{-}CH_2\text{-}Si\equiv$

/ \

≡Si H - Pt

where to simplify, other Pt ligands and other Si substituents are omitted.

Differential scanning calorimetry (DSC) is also a powerful technique for studying hydrosilylation reaction, which is associated with a strong exotherm. Dynamic or isotherm modes allow characterization of the cross-linking and optimization of formulations based on the temperature corresponding to the onset of cure or to the maximum cure rate, illustrated in Fig. 1.3.

Handling these two-part products (polymer and Pt catalyst in one component, SiH oligomer in the other), however, requires some precautions: the Pt in the complex is easily complexed to electron-donating substances like amines, urethanes, organotin, and organosulfur compounds to form stable compounds with these poisons, rendering the catalyst inactive.

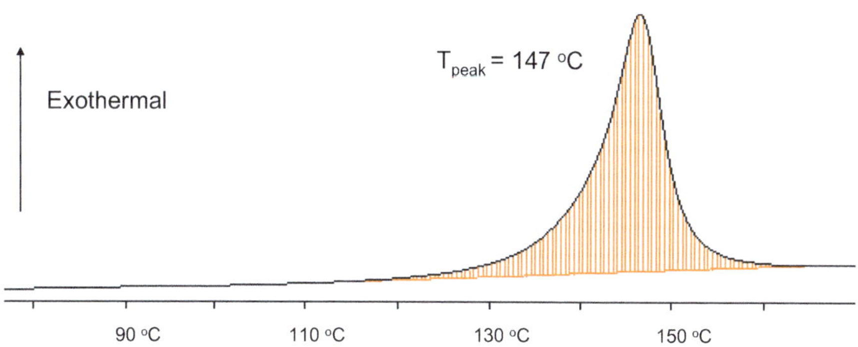

Fig. 1.3 DSC of hydrosilylation reaction between silicone polymers with Si–Vi groups and polymers with Si–H groups in presence of a platinum catalyst [18]

Caution needs also to be taken as the reaction of SiH intermediate with water may lead to highly exothermic process, liberating explosive hydrogen gas.

There is no by-product with this reaction: molded pieces made with a product using this cure mechanism are very accurate, with a minimal shrinkage controlled by the linear coefficient of thermal expansion (CTE) of siloxane polymer-based elastomers.

We will discuss further the mechanism of degradation of silicone elastomers made by crosslinking according to this addition curing mechanism in presence of Pt catalyst and heat. This mechanism of crosslinking is indeed widely used for the fabrication of silicone rubber parts either following an injection-molding process of Liquid-Silicone-Rubber (LSR) [19] or an extrusion-molding process of High-Consistency-Rubber (HCR).

3 Key Properties

Prior to discussing the mechanisms of degradation of silicones in LED lighting applications in detail, we will give a short overview of key properties of silicones, e.g., thermal, mechanical, surface tension, dielectric, permeability, and optical – including key references to papers and textbooks that were published already.

3.1 Thermal Properties

Silicones are showing a quite large range of temperatures (−45 °C to 200 °C) in which there are no thermal transitions, and therefore properties will not change in applications operating in that range of temperatures. PDMS has the lowest glass transition temperature (T_g), measured at −127 °C, and the lowest melting temperature of crystalline phase (T_m) measured at ca. −45 °C, by DSC as illustrated in Fig. 1.4 [18, 20].

A typical temperature frequency sweep by dynamic mechanical analysis (DMA) of any silicone elastomer will show a drop in both storage and loss modulus of more than two orders of magnitude from −60 °C to −35 °C due to the first-order phase transition from crystalline to amorphous state at $T_m = -45$ °C [21].

To further reduce T_g and T_m, methyl groups can be replaced by ethyl groups, as indicated by the values in Table 1.1. The methyl groups mobility is higher in polydiethylsiloxane (PDES) than in PDMS because of the longer distance to the siloxane backbone. Therefore, the overall chain flexibility and side groups mobility increases, and the glass transition temperature is lower. Crosslinking of PDMS to form silicone elastomers networks has little impact on T_g but it can disrupt or even eliminate crystallinity [20].

Phenyl groups dramatically modify siloxane backbone flexibility and mobility, and T_g increases to −25 °C for $–Si(Ph)_2O–$ polymer.

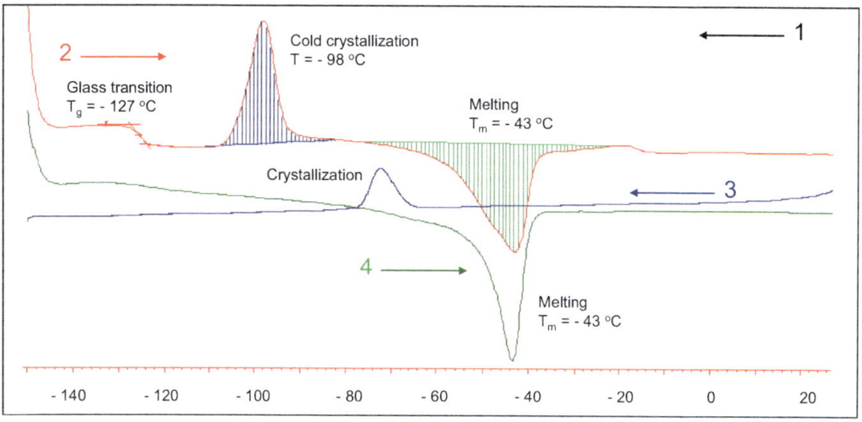

Fig. 1.4 Typical low-temperature DSC analysis of a silicone elastomer: the sample is supercooled at −150 °C (1 – black/no curve), then heated from −150 °C to 25 °C (2 – red curve). The following are detected: glass transition (T_g) at −127 °C, cold crystallization at −98 °C and melting (T_m) at −43 °C. Afterward, the sample is cooled down at a low cooling rate (3 - blue curve) and reheated. A crystallization exothermal peak is observed during the cooling step (3 – blue curve) and only a single melting endothermic peak during the second heating step (4 – green curve) [18]

Table 1.1 Thermal transition in polymers [2, 20, 22]

Polymer	T_g, °C	T_m, °C
–$Si(me)_2O$– (PDMS)	−127	−45
–$Si(et)_2O$– (PDES)	−138	−82
–SiMeEtO–	−128	N.D.
–SiMePhO– (PMPS)	−23	N.D.
–$Si(Ph)_2O$–	−25	191
HDPE	−123	118
Polybutadiene	−109	5
Polyisobutylene	−73	N.D.
Polypropylene (isotactic)	2	186
PET	70	271
Polystyrene (atactic)	100	N.D.
PMMA (isotactic)	105	160

Other unique thermal properties of silicones are shown in Table 1.2. For instance, thermal conductivity (Cp) of PDMS materials is lower than that of HDPE, and thus PDMS-based materials are good thermal insulator unless fillers (e.g., aluminum trihydrates, aluminum oxide, aluminum nitride, boron nitride, zinc oxide, or silica) are added to increase their thermal conductivity [24].

The coefficient of linear thermal expansion (CTE, or Lin K) for PDMS is, however, known to be higher than most other polymeric materials. Practically speaking, an average silicone fluid will increase its volume nearly 9% after increasing its temperature by 100 °C. Interestingly, PDMS and poly-methyl-phenyl-siloxane

Table 1.2 Thermal properties of common polymers [23]

Property	PDMS	HDPE	PMMA	PC
$T_{g,}$ (°C)	−127	−123	114	145
C_p (Wm^{-1} K^{-1})	0.2	0.5	0.2	0.2
CTE, Lin K (K^{-1})	9.3×10^{-4}	12×10^{-5}	7×10^{-5}	6.5×10^{-5}
Specific heat, (Jkg^{-1} K^{-1})	1.6	2.3	1.5	1.2
Continuous high temp limit (°C)	250	120	88	120-130

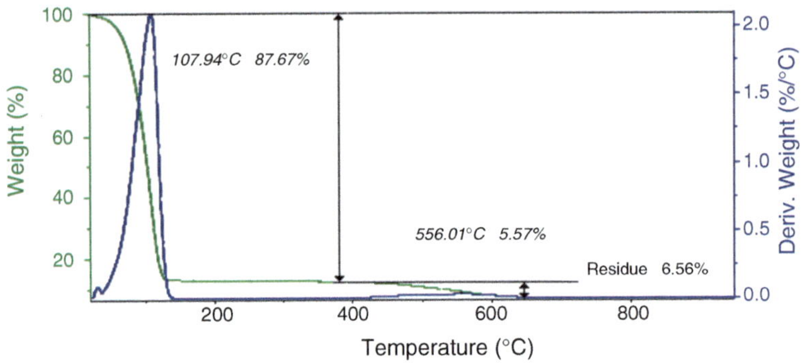

Fig. 1.5 TGA traces under dry nitrogen of a blend of low MW silicone volatile species and cross-linked silicone elastomer (weight loss curve in green, first derivative curve in blue); precise content in volatile species (weight loss up to 150 °C) and in elastomer (second weight loss step between 400 to 700 °C and residue in 950 °C) are indicated [18]

materials are widely used as encapsulants in electronic circuit boards and optoelectronics applications despite this because they are relatively soft and compliant compared to competitive epoxy resins, for instance. These optical silicone encapsulants are materials formed from the cross-linking reaction of either PDMS or methyl-phenyl-siloxane copolymers, which dramatically reduces their CTE as the cross-linking density increases, however.

Concerning the thermal stability of silicones, thermo-gravimetric analysis (TGA) is the most appropriate technique for measuring the onset of weight loss (temperature ramp mode) or the amount of weight loss at a fixed temperature (isotherm mode). It is recommended to run TGA under inert atmosphere as the presence of oxygen will allow, at temperature higher than 300 °C, the partial oxidation of the PDMS backbone into silica and the formation through methyl substituents oxidation into species like carbon monoxide, carbon dioxide, formaldehyde, hydrogen, and water. However, unlike GC-MS techniques, TGA does not give information about the nature of the species volatilized.

Figure 1.5 illustrates a TGA analysis under nitrogen according to a temperature sweep mode, for a blend of low molecular weight volatile siloxane species and a crosslinked silicone elastomer, showing the important weight loss up to 150 °C corresponding to the silicone volatile fraction, followed by the weight loss of the silicone elastomer with an onset degradation temperature at 400 °C. The residue of ca.

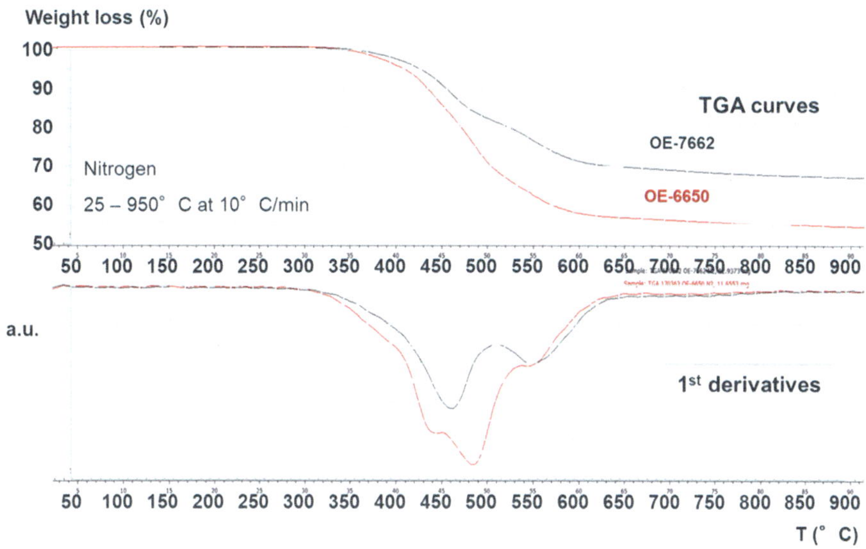

Fig. 1.6 TGA traces under dry nitrogen of phenyl-siloxane optical resins (PMPS, OE-6650 and OE-7662 Optical Encapsulant[1]). Weight loss (above) and first derivatives (bottom) between 25 and 950 °C

6.6% in this case corresponds to the silica ($SiO_{4/2}$) char formed upon decomposition of the silicone elastomer.

TGA traces of representative phenyl-siloxane optical resins (PMPS, OE-6650, and OE-7662 Optical Encapsulant[1]) that are commonly used in optoelectronics applications are shown in Fig. 1.6. It can be clearly observed that these crosslinked optical elastomers are quite stable against high heat, with onset temperature of their first mass loss step in the range 300–350 °C.

These optical elastomers were crosslinked upon heat curing reaction in presence of Karstedt Pt catalyst, which have been shown to give unreliable residue levels at the high decomposition temperatures in the range 550–700 °C.

3.2 *Mechanical Properties*

Cured silicone materials have useful mechanical properties when reinforced by fillers such as fumed silica or siloxane resins made of tetra-(Q) or tri-(T) functional siloxane units (Fig. 1.7). These fillers are added to increase toughness of the elastomer, after curing by either condensation or addition mechanism [1, 23].

[1] Both supplied by Dow Corning Corporation in 2016, now Dow Silicones for the experiments reported in this paper. Currently available from DuPont.

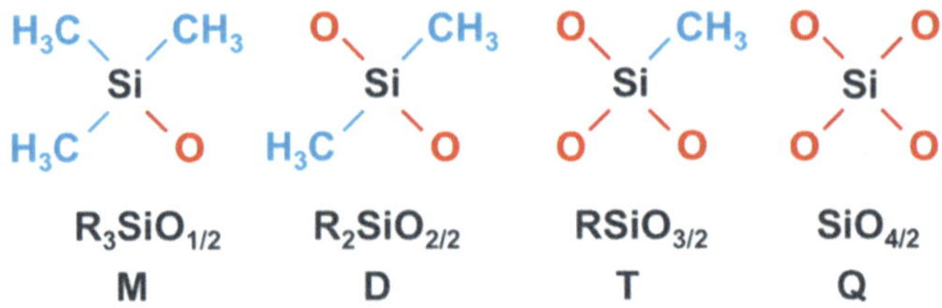

	M	D	T	Q
T_g, °C	–163	–123	63	1200
ρ, g·cm^{-3}	0.76	0.98	1.37	2.20
k, W·m^{-1}·K^{-1}	0.10	0.16	0.24	1.38
G, Pa	$10^{-1.3}$	$10^{5.3}$	$10^{8.8}$	$10^{10.5}$

Fig. 1.7 M, D, T, Q siloxane units and corresponding physical properties

When condensed together, MDTQ units can show extreme physical properties. M units have a low glass transition, while Q units have a high glass transition. Densities of M and D units are below 1, while densities of T and Q units are above 1. Thermal conductivity of Q units is one order of magnitude above thermal conductivities of M and D units. At room temperature, Young's moduli of T and Q units are several orders of magnitude higher than Young's moduli of M and D units. For applications in microelectronics and photonics, optimization of a siloxane-based material relies on art and skill to combine these MDTQ units in such a manner that the resulting material properties meet application requirements and stability throughout the life of the devices under its operating conditions, such as low and high temperatures, exposure to high relative humidity or UV radiation, and combination of them.

After crosslinking into a three-dimensional reinforced network, silicone elastomers are well suited for applications requiring high elasticity because the long SiO bonds and the flat valence angles provide for highly flexible chains. Methylsiloxane-based materials can also range in hardness from a Shore A of 10 up to 85. For comparison, this corresponds to materials ranging in hardness from rubber bands to tire treads. Higher hardness can be reached by replacing methyl- with phenyl- groups along the siloxane backbone [9]. This led to the development and commercialization of a family of optical encapsulants characterized by their high hardness (e.g., Shore D 50-70), high refractive index (e.g., 1.54), and low gas permeability, and are used today in photonics applications, further discussed below in this chapter [9, 11, 25–27].

3.3 *Surface Tension*

Surface tension (γ) or surface energy is the propensity for a substance to attract itself. The higher the surface tension, the stronger the cohesive interactions between itself, relative to the environment – typically ambient air. Another way of stating this

Table 1.3 Surface energies of polymers

Substance	Surface tension (mN/m) (i)	Critical surface tension (mN/m) (ii)	HSP δ ($MPa^{1/2}$)
Water	72.8	–	48.0
PDMS	21.5	24	15.3
–Si(Pr^F)MeO–	26.5 ± 1	–	–
–SiPhMeO– (PMPS)	25.5 ± 3	–	–
HDPE	35.7	31	16.7
PMMA	41.1	39	19.0
PTFE	21.1	19	12.3
PVC	41.1	39	19.6
PET	44.0	43	21.2

(i) Refers to liquids; (ii) refers to solids

is that the surface tension is the amount of energy it takes to increase the surface area of a material. Molecularly, this means breaking intermolecular interactions of molecules to form interactions with the environment, i.e., air [28].

To increase the surface area of a water droplet in air by 1 m^2, for example, it would take 72.8 J of energy. The lower the surface tension, the easier it is to stretch the liquid or the lower the penalty for breaking cohesive interactions. The Hildebrandt Solubility Parameter (HSP) is a measure of these intermolecular forces and trends well with surface energy (Table 1.3).

Silicones have substantially lower intermolecular forces or surface energy compared to most other materials. Thus, when silicone fluids are placed onto most surfaces, it is energetically more favorable for the surface to be wet by the fluid than by air. In the values given in Table 1.3, the surface tension for PDMS (21.5 mN/m) is lower than the critical surface tension (the value for solids) for PMMA (39 mN/m) for instance, and, therefore, PDMS fluid will easily spread on and wet the surface. PDMS will not, however, wet the surface of PTFE for which the critical surface tension is 19 mN/m. The other unique property of silicones is that their surface tension (liquid) is lower than their critical surface tension (solid), meaning they can wet themselves [29].

3.4 Dielectric Properties

Silicone also has excellent dielectrical properties, making it a widely used material in circuit boards and electrical components protection. From Table 1.4, one can see that the dielectric strength (voltage needed to short through the material) and volume resistivity (electrical resistance of a given volume of material) are quite comparable with other materials such as epoxy and PMMA and are good insulators. Some organic polymers like polyethylene and other polyolefins, however, have weaker dipoles and hence a lower dielectric constant whereas others such as

Table 1.4 Dielectric properties of common polymers [23]

Property	PDMS	Epoxy	PMMA	PC
Dielectric constant (1 kHz)	2.86	3.5–5.0	3.0–3.6	2.8–3.0
Dissipation factor (1 kHz)	2.5×10^{-4}	100×10^{-4}	570×10^{-4}	90×10^{-4}
Dielectric strength ($kV.mm^{-1}$)	16–20	12–16	16	34
Volume resistivity (log[Ω cm])	14–15.6	>14	14–15	14

polymethylmethacrylate (PMMA) and polycarbonate have stronger dipoles which result in higher dielectric constants.

Despite strongly polarized Si–O bonds, silicone polymers are non-polar as the Me side groups prevent Si–O dipoles from approaching each other too closely. As a result, the intermolecular forces are weak and mainly composed of London-van der Waals interactions that decrease with the square of the distance between molecules. Due to this ambivalent character of the PDMS polymer chain, the polarizability of the molecule accounts for a relatively high dielectric constant of silicones in comparison to a non-polar polymer like polyethylene.

Although the dielectric properties of silicones are good, they are not exceptional in comparison with organics. The success of silicone-based products is certainly more related to their stability over a wide range of temperature, humidity, and frequency. The reason is that their electrical properties work in concert with the other key silicone properties. For instance, silicones can be made to be relatively soft compared to epoxy and acrylic materials. When used as a coating for a printed circuit board (PCB), this allows the material to move past electronic component as the circuit board heats up, whereas epoxy and acrylic materials can damage components since they are much harder materials. Siloxane polymers where Me groups have been partially substituted with more polarizable phenyl groups lead to increases in dielectric constant and dielectric strength in the material [30].

Given the rapid development in power electronics technology and the trend to design electrical appliances with continuously increasing power density, switching frequency and miniaturization of devices, the need to understand how thermally induced dielectrical breakdown, also referred to "thermal breakdown", was studied for silicone-based dielectric elastomers [31].

Additionally, silicones have low surface energy and can properly wet out a surface to ensure a uniform defect-free coating (Table 1.3), not to mention that silicones are stable at high temperatures (Table 1.2). This combination of properties helps to make silicones the lower cost-in-use material technology compared to other polymers and address the challenges raised by micro- and opto-electronics applications.

3.5 Permeability

The permeability of silicones is also an important and unique property. From the permeability data shown in Table 1.5, it is clear the silicones have some of the highest permeabilities compared to other commercial polymers. Gas permeability is the

Table 1.5 Permeability of polymers against carbon dioxide (CO_2), nitrogen (N_2), and water vapor (H_2O) [32–36]

Polymer	P CO_2	P N2	P H_2O
PDMS	3250	351	36,000
–Si(Pr^F)MeO–	–	76	–
–SiPhMeO– (PMPS)	–	9	–
–Si(n-Oct)MeO-	–	88	–
HDPE	0.37	0.15	9
PMMA	0.34	0.01	3200
PTFE	9.8	1.9	4.8
PVC	–	2.1	275
PET	–	–	130
PVDC (Saran)	0.29	0.0094	0.5

product of diffusion coefficient and gas solubility. While gas solubility of silicone rubber is on the same order of magnitude as organic rubbers, the gas diffusion is outstandingly higher, and therefore silicone polymer-based materials are about 1–2 orders of magnitude more permeable to gas than organic materials at a given temperature.

This is, in part, due to their molecular structure having loosely packed polymer chains (high free volume) and high flexibility, as well as weak intermolecular interactions, and also because PDMS is above its glass and melt transition temperature in most applications. Adding phenyl groups decreases the permeability of the material by nearly two orders of magnitude. Adding fillers or silicone "MQ" resins can also decrease the permeability of silicones. Optimizing the phenyl-siloxane content as well as the Q units proportion in silicone resins to decrease their gas permeability has enabled the broad adoption of optical silicone materials in optoelectronic applications [27].

The high permeability of silicones can make them advantageous in applications. In hydrophobic coatings or treatments, for instance, it is desirable to prevent water absorption, but also to ensure that water vapor can escape. This is an important aspect to consider in the protection of electronic circuitry. The high permeability of silicones to water vapor also enables a faster moisture cure compared to other materials.

3.6 *Optical Properties*

In the middle of the 1960s, both dimethyl- and phenylmethylsiloxane resins were studied for their benefits in already quite diverse optical applications, such as, e.g., (i) optically transparent organosilicone encapsulating resins with the required mechanical strength and flexibility to satisfactory operate over a wide range of temperature conditions and under severe mechanical stresses for electrical insulation systems [37]; (ii) laser communication systems capable of high-efficiency transmission of a laser beam, without being subject to breakage, collapse or corrosion using flexible silicone for the core of the transmission line, and a coating with lower refractive index than the silicone for the clad in order to limit losses in transmission efficiency [38]; and (iii) optical compositions of silicone rubber for contact lenses applications, obtained by radical cross-linking reaction either in presence of peroxide or by irradiation [39].

Later in the 1980s, further developments were investigating the possible applications of silicone resins in the fields of, for instance, (iv) pressure-sensitive optical waveguides that remain optically transparent, showing minimal light attenuation, and easy deformable under relatively low pressure, prepared by addition reaction crosslinking of vinyl-functional PDMS and MQ resins compositions with SiH-functional siloxane crosslinker in presence of Pt catalyst [40], or (v) hydrophilic inherently wettable optical devices such as eye contact or intraocular lenses manufactured from acrylate-functional siloxane resins [41].

Since the discovery of the use of blue LED for illumination in the early 2000s by Nobel price Shuji Nakamura [42], the development and application of optical silicone resins have continued in a variety of fields, such as, e.g., (vi) planar waveguides assembly with at least one silicone core for optical interconnects [43]; (vii) impact resistant waveguides that include a clad surrounding a core and the method of manufacture of the waveguide including injecting the high refractive index (1.46–1.56) curable silicone composition into a clad made of silicone elastomeric tubing of a lower refractive index (ca. 1.41) and thereafter curing the curable silicone composition [44]; (viii) curable optical silicone composition of various refractive index for LED devices encapsulation, with enhanced thermo-optical performances, mechanical strength and adhesion to LED packaging components, superior in reliability against conventional optical epoxy resins [9]; and (ix) curable optical silicone composition for the fabrication by liquid-injection-molding process of free-form optics for controlling light beam in LED lamps and luminaires applications [45], which as now used in numerous LED lighting fixtures and applications, such as, e.g., automotive head- and rear-lamps, and outdoor (e.g., architectural, stadium, street, and parking bays) lighting [7, 8, 46–48].

The primary reason for the continued expansion of silicones in optical and lighting applications is their high transparency throughout the visible and UV wavelengths, as well as their photo-thermal stability (Fig. 1.8). Overtones of C–H of CH_3 groups along the siloxane backbone cause absorption in the near infrared (IR) region above 1050 nm. Absorption in the deep UV region becomes significant at wavelengths below 280 nm and depends on the degree of purity of the siloxane polymers intermediate and balance of MDTQ monomers.

These properties allow siloxanes to transmit light over long path lengths and for long periods of time in harsh environmental conditions without yellowing or distorting the light. Applications such as Adaptive-Driving-Beam (ADB) automotive LED headlamps [49], LED packaging, stadium lighting, and street lighting are just a few examples where silicones excel [46–48].

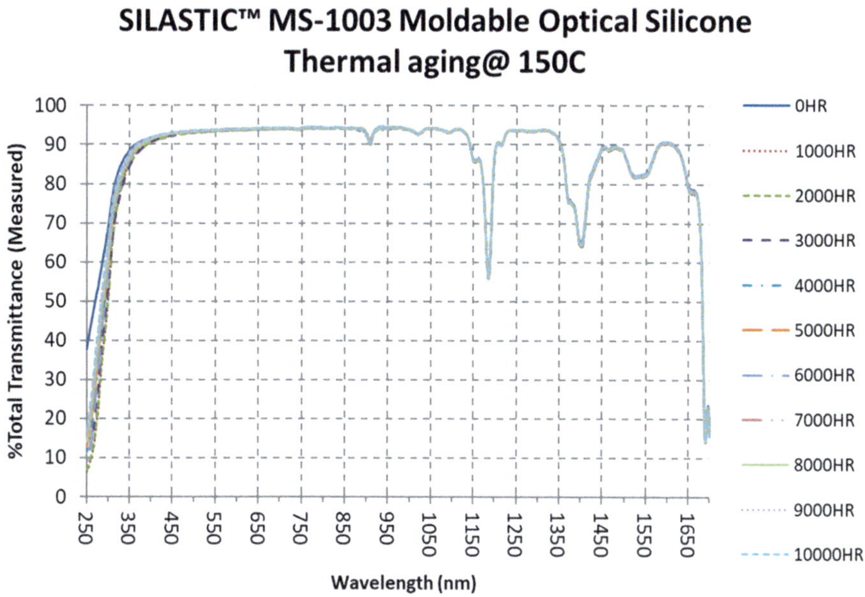

Fig. 1.8 Total transmittance of optical silicone elastomer of 3.2 mm thickness in the UV-VIS-IR wavelengths range, including thermal aging effect up to 10 k h @150 °C

The success of silicone materials in optoelectronic applications is heavily dependent, however, on how the materials are prepared. Features like surface texture, incorporation of fillers, and contaminants such as excessive Pt, vinyl, or others, can cause the material to be sensitive to discoloration, leading to changing optical properties over time.

Compared to other polymers, PDMS materials have lower refractive indices (RI: 1.41–1.43) and higher Abbe numbers (Table 1.6).

In lighting applications where optical silicone elastomers are used for designing total internal reflection (TIR) free forms lenses, on one hand, its lower refractive index compared to PMMA or PC for instance, means that larger and thicker optics is required for obtaining the same angle of refraction of the light beams. On the other hand, the lower refractive index of silicone results in less loss of light by reflection at each air-material interface of the lenses. With an RI of 1.41 for silicone at 700 nm, one single interface reflection loss is calculated to 2.9%, while polycarbonate with an RI of 1.58, one single reflection loss is calculated to 5.1%.

The Abbe number is a measure of optical dispersion or how much the refractive index changes with wavelength. The higher the Abbe number, the more constant the RI is across the visible spectrum and thus, the lower the dispersion of light. From Table 1.6, one can see that PDMS has a relatively high Abbe number giving it low dispersion, comparable to PMMA, but still not as high as pure quartz.

Replacing methyl groups by phenyl on the siloxane polymer backbone (PMPS) increases the light dispersion into the material, explaining why PMPS is almost

Table 1.6 Optical properties of various polymers

Polymer	Refractive Index (700–400 nm)	Abbe Number (V_D)	Photo-thermal stability
PDMS	1.41–1.43	52	Yes
–SiPhMeO– (PMPS)	1.55–1.56	35	Limited
Quartz (SiO_2)	1.47–1.46	68	Yes
PC	1.58–1.62	30	No
PMMA	1.49–1.50	57	No
PS	1.59–1.60	30	No
Epoxy	1.50–1.57	34	No

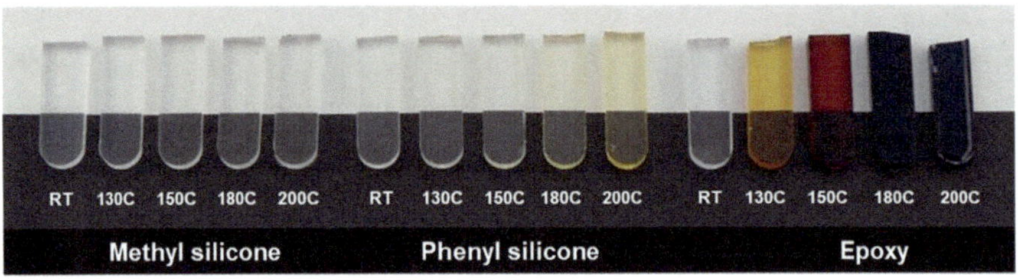

Fig. 1.9 10 k h accelerated heat aging of methyl-siloxane resin (OE-6351 Optical Encapsulant), phenyl-siloxane resin (OE-6630 Optical Encapsulant), and epoxy optical encapsulant materials used in LED packaging devices. (Picture by Toshiki Nakata from Dow Toray Co. Ltd)

exclusively used as a high hardness, low gas permeability optical resin for LED die encapsulation in phosphor-layer conversion LED devices, as well for dispersing phosphors in volume conversion LED devices. Refractive index of PMPS is also higher, a consequence of the presence of phenyl groups incorporated along the siloxane backbone. In LED packaging design, the encapsulation of the LED die with higher refractive index silicone (RI 1.55) enables ca. 7% more light extraction in comparison to normal refractive index (RI 1.41) [27]. Silicones containing phenyl groups are not as photo-thermally stable as purely methyl silicones, however (Table 1.6), which will be the subject of further discussions in the paragraphs below.

From a heat stability perspective, optical transparency of methyl-siloxane resin (PDMS, OE-6531 Optical Encapsulant[2]), phenyl-siloxane resin (PMPS, OE-6630 Optical Encapsulant[2]), and epoxy resins exhibit significant differences, as illustrated in Fig. 1.9. Methyl-siloxane is optically stable up to 200 °C; phenyl-siloxane shows a moderate yellowish discoloration starting at 150 °C, becoming more severe at 200 °C; optical transparency of epoxy-based formulations begins to degrade by 130 °C. Most blue LED packaging with phosphor conversion into white light are now made from either PDMS or PMPS because of their photo-thermal stability at LED operating temperatures. Possible degradation mechanisms do exist, however, and are discussed later in this chapter.

[2] Both supplied by Dow Corning Corporation in 2016, now Dow Silicones for the experiments reported in this paper. Currently available from DuPont.

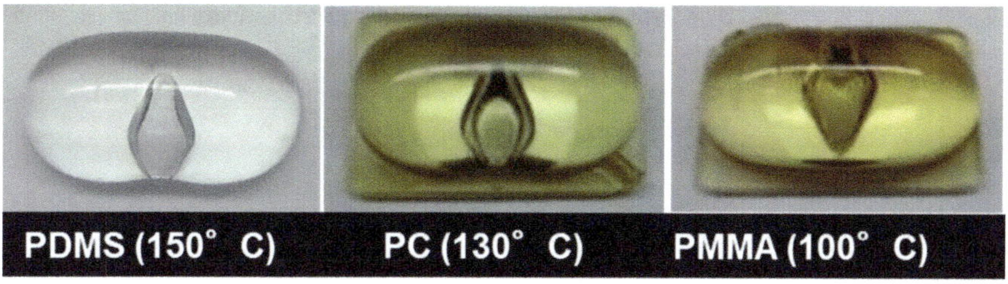

Fig. 1.10 6 k h accelerated heat aging of lenses made of PDMS (SILASTIC™ MS-1002), PC (non-stabilized resin) and PMMA (non-stabilized resin). (Picture by François de Buyl from Dow Silicones Belgium srl)

For LED-based lighting fixtures applications, secondary optics materials used for the design and fabrication of free-form lenses are typically made of either thermoplastic resins (PC, PMMA) or silicone (PDMS). From a high heat stability standpoint, if the thermoplastic resins are not protected against heat and/or sunlight UV radiations, photothermal degradation causes significant yellowish discoloration with time (Fig. 1.10), reducing light output efficacy of the lighting fixture [46]. Lenses made by injection molding process of optical LSR, such as, e.g., SILASTIC™ MS-1002 have been tested up to 6000 h, showing stable optical properties at use temperatures up to 150 °C, as well as after exposure to accelerated sunlight UV-A/B radiations.

On the other hand, several studies have been published to understand degradation mechanisms of optical materials used in solid-state lighting fixtures, with particular focus on comparing polycarbonates, acrylics and silicones [50–52]. These papers aim at describing the influences of chemical structure and service conditions on the degradation mechanisms of these thermoplastics or silicones optical materials used to control and collimate light beam emitted by the LED light soucres. Degradation of thermoplastics (e.g., PC, PMMA) is often observable as yellowish discoloration, which lead to lumen depreciation, and color shift of the LED light output from the fixture. The contributions of different degradation mechanisms of optical and package materials in LED-based products to the lumen depreciation and color shift were methodically reviewed. In most recent years, UV LED [53]-based lamps have started to replace mercury lamps in multiple industrial applications, such as printers, 3D printers, production lines for curing of adhesives, inks and coatings, water and air disinfection and marine antifouling.

As illustrated by the UV spectral transmittance plots of Fig. 1.11, the good transmittance of PDMS polymer-based silicone resins down to UV radiation [54] has enabled the development of cost-effective water purification devices using UV-C LED sources [55]. An efficient and reliable water purification system was designed by selecting the LED emitting at the appropriate wavelength and intensity in the deep UV region (275 nm, 8 mW power), as well as selecting the appropriate optical silicone material to fabricate the lens capable of sustaining the exposure to UV-C radiation up to 10 k h operating conditions [56–59].

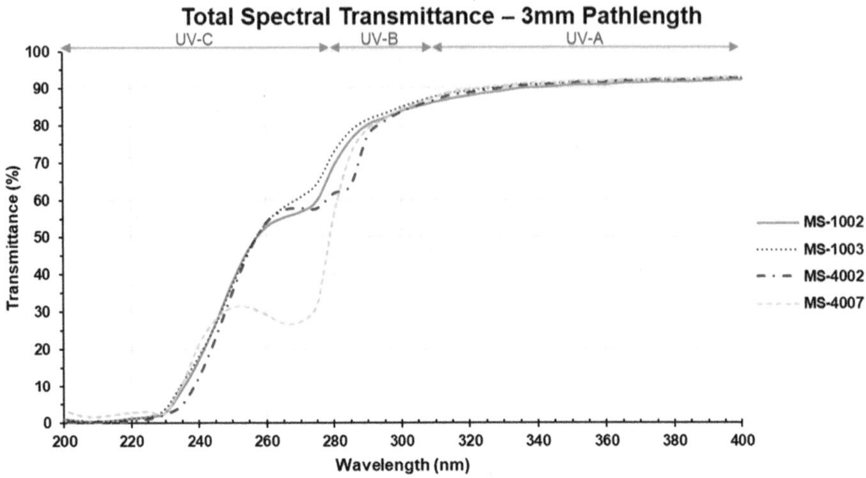

Fig. 1.11 Transmittance of optical grades silicone rubbers in the UV-A/B/C wavelengths. MS-1002, MS-1003, MS-4002, and MS-4007 are Dow SILASTIC™ Moldable Silicones designed for secondary optics applications

In the field of UV-curing processing, the wavelengths of 365 and 395 nm in the UV-A region are commonly used for curing inks, coatings, or adhesives, with irradiances varying from 1 to 3 W/cm^2 range up to 8 to 16 W/cm^2 range [60]. The results and conclusions reached from a study where SILASTIC™ MS-4007 optical grade LSR was exposed to UV-365 and UV-395 LED source show stable optical and mechanical properties upon UV radiation exposures at 365 and 395 nm wavelengths up to 3–4 W/cm^2 irradiance, with moderate hardness increase (up to 87 Shore A at 365 nm and 82 Shore A at 395 nm, against initial 75 Shore A). At higher irradiance of 8 W/cm^2 at 395 nm, embrittlement of MS-4007 specimens was observed, with hardness increasing up to 91 Shore A. The loss of light transmission at 365 and 395 nm was ca. 4% (down to ca. 92% from initial 96% on a 2 mm path length) after 280 days exposure to these UV wavelengths, which is acceptable for the foreseen applications.

3.7 *Optical Siloxane Resin Synthesis*

Silicones have a wide variety of structural constituents based on M ($R_3SiO_{1/2}$), D ($R_2SiO_{2/2}$), T ($RSiO_{3/2}$), and Q ($SiO_{4/2}$) units as previously explained and illustrated in Fig. 1.7. The selection of the substituent, R, is also a variable that enables the material designer some control of the physical properties of the final material. Methyl groups are the most common silicon substituent (R groups) in industrial applications of silicones, followed by phenyl. Silicone oils and elastomers are the most common industrial products and consist mainly of linear polymers based on the methyl-substituted D unit and are referred to as polydimethylsiloxane (PDMS).

In contrast, silicone *resins* are 3D-network polymers primarily built from T [$RSiO_{3/2}$] and Q [$SiO_{4/2}$] units. Silicones resins made only from T unit are also called silsesquioxanes [27] that have been developed for applications in LED packaging encapsulation [9, 61].

Polysiloxanes are synthesized by hydrolytic polycondensation of chlorosilanes or alkoxysilanes of corresponding M, D, T, and Q combinations. The polysiloxanes are often cured to provide a stable, solid form for a given application. As introduced previously, curing is typically achieved by condensation of residual silanol or alkoxy groups or by an addition cure, hydrosilylation reaction between Si-Vinyl and Si-H functionalities.

In order to vary the refractive index (RI) of the siloxane resin, it is possible to adjust the relative molar ratio of methyl against phenyl groups, increasing therefore RI from ca. 1.41 for a pure dimethyl-siloxane material, up to ca. 1.57 for a high content of phenyl-siloxane species (Table 1.6 and Fig. 1.12) [10].

The control of the refractive index of silicone resin encapsulants in optoelectronics contributes to the quality of the resulting LED light, and the light extraction efficiency. The refractive indices (RI) of an LED chip, phosphors, and air are ~2.5, ~1.8, and 1.0, respectively, while those of methyl and phenyl silicones are ~1.41 and 1.53–1.55, typically. A greater amount of light was extracted using a lower RI encapsulant for blue light without phosphor, while white LED with phosphors conversion layer showed higher brightness using silicone encapsulants with a higher RI [27].

Another benefit of a high phenyl content in the silicone resins used in LED packaging is the reduced gas permeability, as high gas permeation of encapsulants can reduce durability. For instance, corrosion of the silver reflector by sulfur gas will reduce the light intensity by damaging the reflectivity at the bottom of the device.

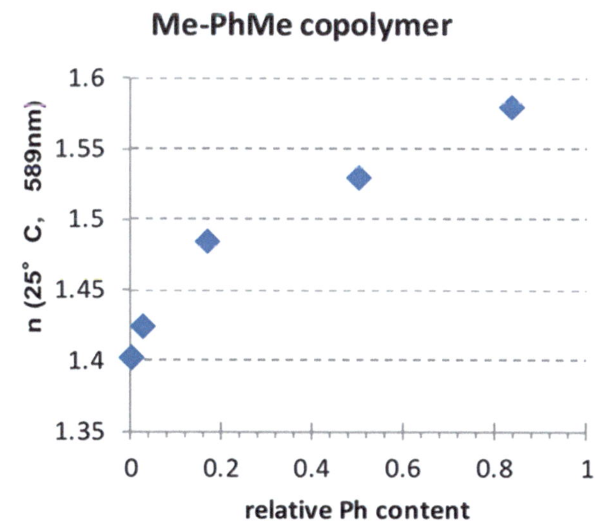

Fig. 1.12 Refractive index of siloxane resins as function of phenyl (Ph) group substitution on the poly-dimethyl-siloxane chain. From reference [10], with permission

Methyl silicones show high gas permeation with water vapor permeation rate (WVTR) of ~100 g/m^2/day for 1 mm thickness at 40 °C/90% relative humidity. But this value decreases to 10 ~ 15 g/m^2/day for phenyl silicones, which is one key reason why phenyl silicones are preferred for this application [27].

4 Physical Degradation

4.1 Thermal Effect

The resistance to oxidation of the silicone backbone when exposed to elevated temperatures is superior to most organic polymers, and thus, polysiloxanes are often used for high-temperature protective coatings. Degradation processes of polysiloxanes are observed, however, when polysiloxanes become highly stressed by extreme conditions. It has been reported that thermal stresses cause yellowing and cracking of PDMS when temperatures become excessive. As introduced in the earlier paragraph on thermal properties of silicones, TGA (Fig. 1.6) of polysiloxanes shows an onset temperature of degradation in the range 300–350 °C under air or nitrogen. Degradation is caused by an oxidative depolymerization of PDMS or PMPS polymer backbone initiated above 340 °C, resulting in the formation of linear and cyclic oligomers [62]. Since oxidation of polysiloxanes with long aliphatic hydrocarbon chains occurs more readily than for dimethyl or phenyl-methyl-based polysiloxanes, these two later silicone resin families are preferably used for LED packaging applications.

Heat aging at moderate temperatures, below 300 °C, leads on the other hand to condensation reaction of residual silanol groups, increasing crosslinking density and formation of Si–O–Si bonds between PDMS polymer chains, and therefore hardening of the cured silicone material, resulting in increased internal mechanical stress and crack [63].

4.2 Blue Light Absorption: Photo-Thermal and Stokes Shift Effect

There are two primary root causes for degradation of LED silicone encapsulants exposed to high heat and intense blue LED radiation (450 nm, 50 W/cm^2) radiation or UV radiation with wavelength higher than 300 nm (up to ca. 0.15 W/cm^2) [64, 65].

The first root cause of significant yellowish discoloration and darkening of the LED silicone encapsulant on its own (without phosphors) is caused by the degradation of epoxy-functional adhesion promoter additives contained in the silicone LED encapsulants upon moderate heat exposure already. Also, epoxy resins used in white reflectors of LED packaging, die-attach, and solder paste, can leave small amount of epoxy resin residues inside silicone composition. These *organic* impurities can accumulate at the interface between LED die and silicone optical encapsulant, resulting in a darkening of the LED die surface, like darkening illustrated of epoxy resin in Fig. 1.9.

Cured phenyl-methyl-siloxane specimens were tested against their thermal degradation, showing no discoloration up to 190 °C for 226 h, but when the temperature was increased up to 300 °C, significant yellowish discoloration was observed [66].

The second root cause of significant yellowish discoloration and darkening of the LED silicone encapsulant is when heat is generated by Stokes shift effects issued from the blue radiation conversion into white light in presence of phosphors (i.e., phosphorescent chemical components) dispersed into the silicone optical encapsulant. In these conditions, although silicone materials are thermally more stable than organic materials, irradiation by intense blue light and UV light accelerates discoloration of LED silicone encapsulant.

The Stoke shift effect was well illustrated in a study with LED packages containing phosphor particles dispersed in silicone encapsulant [67]. As temperature of the phosphor particles rises to 300 °C, blue light radiation of the die is converted into yellow-red component of the white light (depending on the exact nature of the phosphor). During the phenomenon of phosphor quenching, local temperatures can gradually increase up to higher than 500 °C, and therefore causing darkening of the silicone encapsulant as well as hardening until it becomes possibly brittle and generate cracks at the interface of the silicone encapsulant and the phosphor conversion layer.

As illustrated in Fig. 1.13, upon normal operating conditions of the LED package, no degradation is observed (Fig. 1.13a), while at temperatures of 450 and 500 °C, dark spots appeared at the silicone-phosphor interface, starting to grow from the center of the package.

To discriminate phosphor degradation from silicone encapsulant degradation phenomena, pure silicone cuboids were cured without phosphor particles, showing that wrinkles at the silicone cuboid surface started to appear at 450 °C, and severe carbonization (darkening) at 500 °C as illustrated by the pictures of Fig. 1.14.

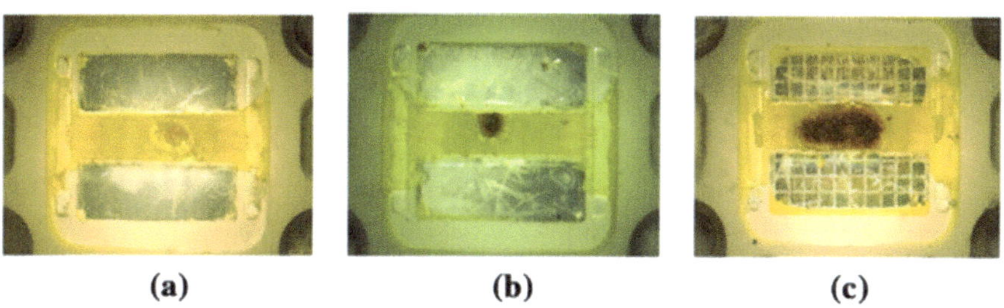

Fig. 1.13 Images of silicone-phosphor LED packages in different temperatures: (**a**) ambient temperature, (**b**) 450 °C, and (**c**) 500 °C. (From reference [67], with permission)

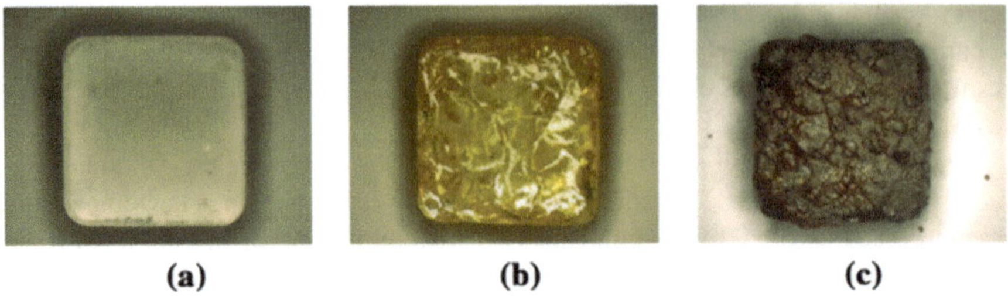

Fig. 1.14 Images of cured pure silicone cuboid (without phosphor particle) in different temperatures: (**a**) ambient temperature, (**b**) 450 °C, and (**c**) 500 °C. (From reference [67], with permission)

5 Chemical Degradation

5.1 *Thermal Effect*

Thermal degradation of PDMS by oxidative process has been studied [68, 69]. Two competing mechanisms, Si–O bond scission depolymerization and crosslinking reactions, have been reported as the two key, thermally induced degradation mechanisms of PDMS.

As already introduced in a prior paragraph on thermal properties of silicone, PDMS thermally decomposes to cyclic oligomers through Si–O bond scission.

Degradation in nitrogen gas atmosphere results in depolymerization and production of cyclic DMS oligomers [70, 71]. The mechanism for formation of the smallest cyclic DMS (hexamethylcyclotrisiloxane) is illustrated in Fig. 1.15 [69].

Another degradation mechanism, a radical mechanism, occurs by Si–CH_3 bonds scission. Scission of Si–CH_3 bonds results in cross-linking of the PDMS macro-radicals, decreasing of the flexibility of the PDMS chain and hindering cyclic DMS oligomer generation.

In photo-ageing conditions, crosslinking reactions are more important than cyclic formation. Depolymerization chain scission mechanism and cross-linking mechanism at 100 °C have been compared for hydroxy-terminated-PDMS (Mw g/mol = 110,000) by GPC using toluene as solvent (Fig. 1.16). An increase in molecular weight was observed at 100 °C after 10 h. However, molecular weight was not changed significantly at 100 °C for 20 to 75 h. This result suggests that crosslinking occurs in the first 10 h and chain scission mainly occurred after 20 h [62].

5.2 *Pt Nano Clustering Effect*

As introduced in prior paragraph, hydrosilylation reactions are widely used in the silicone industry, including silicone LED encapsulant applications. A silicon–hydrogen (Si–H) bond is added across an unsaturated carbon–carbon double bond

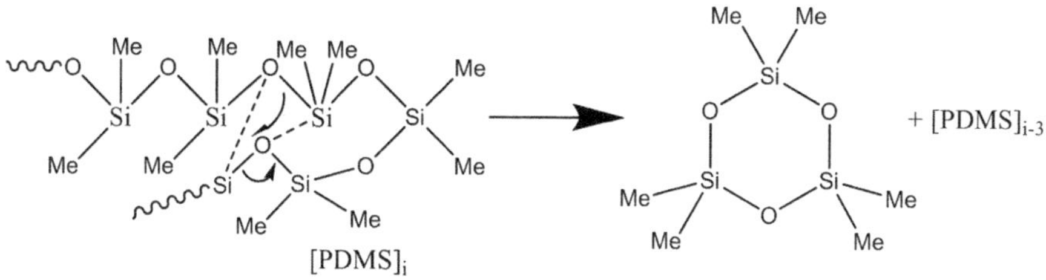

Fig. 1.15 PDMS molecular depolymerization mechanism. (From reference [69], with permission)

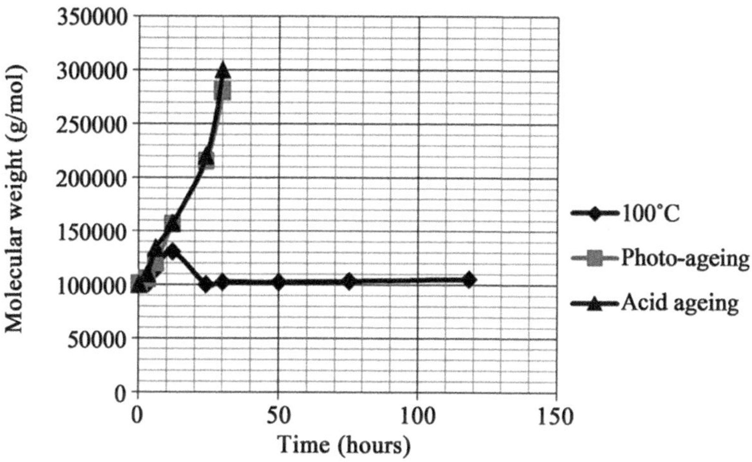

Fig. 1.16 GPC results using toluene as solvent. Comparison of changes in molecular weight (Mw) upon various aging of hydroxy-terminated PDMS. (From reference [62], with permission)

(C=C) resulting in the formation of a silicone–carbon (Si–C) bond. Hydrosilylation reactions are catalyzed by platinum catalyst such as the Karstedt's catalyst. Pt catalysts have been widely studied [72]. Karstedt's catalyst is prepared by the reaction of chloroplatinic acid, H_2PtCl_6, with vinyl-silicon containing compounds such as divinyltetramethyldisiloxane $M^{Vi}M^{Vi}$. At the end of a hydrosilylation reaction, Pt species are observed, depending on the ratio of Si–H to Si-vinyl and the nature of the olefin (hydrocarbon or silicon-vinyl). Specie **8** is formed at high vinyl concentration and **9** is formed at more stoichiometric conditions (Fig. 1.17).

The platinum end product, however, is a function of the stoichiometry of the reactants. At excess Si-vinyl concentration, the platinum end product contains only platinum-carbon bonds, whereas at high Si–H concentration, the platinum end product is multinuclear and also contains platinum-silicon bonds [73].

Platinum nanoparticles formed after the hydrosilylation reaction have been studied. UV-Vis spectra of platinum nanoparticles are brown in color and absorb UV light. In the final steps of the hydrosilylation reaction, all systems show a dark-brown colloidal solution, indicating the formation of Pt nanoparticles. In light of the

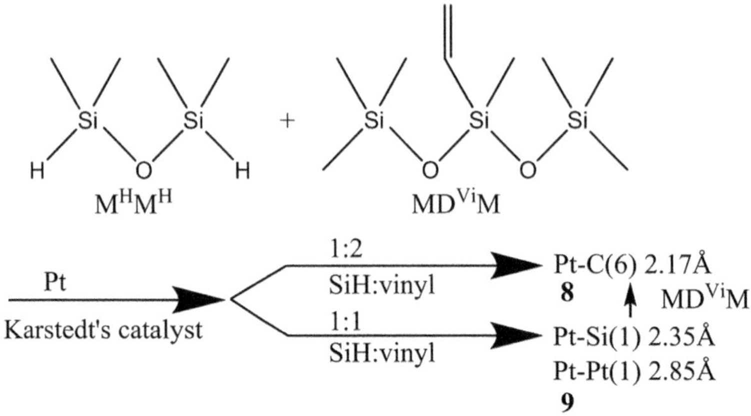

Fig. 1.17 Pt catalyst species formed during hydrosilylation reaction. (From reference [72], with permission)

Lewis mechanism of hydrosilylation reactions, the reaction solutions of different colors indicate the formation of different Pt complexes, the Pt nanoparticles that are derived from the Pt–Si–H complex contain Pt–Si bonds, and the Pt nanoparticles that are derived from the Pt–Si–vinyl complex contain Pt–C bonds [74].

Since reactivity of Pt nanoparticles can be a root cause of photo-degradation of silicone materials, effects to the thermal/UV stability by residual Pt catalysts or Pt nanoparticles have been studied. Improvement of thermal stability of cross-linked PDMS from 405 to 480 °C was reported [75].

5.3 Si–Me and Dimethylene Group Scission

The dimethylene group (Si–CH_2–CH_2–Si) is commonly observed in silicone materials as the product of hydrosilylation reaction between a silicon hydride (Si–H) and a silicon vinyl (Si-vinyl) group, as the hydrosilylation reaction is one of the most prevalent curing reactions in silicone elastomers used in micro- and optoelectronics protection applications. Thus, photo-oxidation of PDMS containing dimethylene groups has been analyzed and reported [76]. The dimethylene group is comprised of a carbon–carbon bond which is unstable under UV irradiation as it has a lower bond energy than the Si–O–Si bond of the PDMS backbone. It was reported that unspecific radicals (R · or ROO ·) can be formed, including the formation of macroradical, by hydrogen abstraction. The dimethylene bond is cleaved and the generated alkyl radical leads to the formation of carbon-hydroperoxide via a peroxy radical. Carbon-hydroperoxide can be decomposed to produce silyl radicals and a new Si–O–Si bond (Fig. 1.18).

The dimethylene group scission may occur in both methyl silicone materials and phenyl silicone materials formed from hydrosilylation reaction and used for applications in optoelectronics.

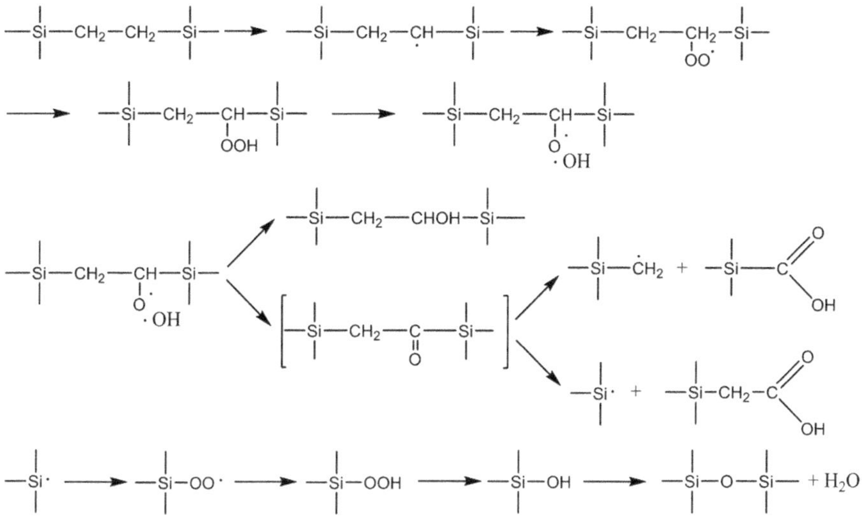

Fig. 1.18 Mechanism of oxidation of ethylene linkage via alkyl, carbon-hydroperoxide and peroxy radicals formation. (From reference [76], with permission)

5.4 Phenyl Oxidation

The major mass loss of PDMS is due to the formation of cyclic siloxane trimers as illustrated in previous section. Incorporation of phenyl groups into polymer chain may change the dominant degradation reaction, however.

Photooxidation of polysiloxane-containing phenyl groups has been examined [63].

FTIR observation indicates, (i) a decrease of aromatic group by phenyl scission, (ii) formation of carbonyl groups and esters by radical oxidation, and (iii) decrease of initial silanol by cross-linking. Since the oxidation products represent only 20–30% of the phenyl loss, phenyl ring scission reaction is accounting for the complementary degradation mechanism. Fluorescence spectroscopy suggests the presence of phenol group after aging. Deep UV-C light irradiation (254 nm) increases yellowing of phenyl siloxane. Like for PDMS, longer wavelengths UV radiation (>300 nm) may cause yellowing in phenyl siloxane optical encapsulants due to chromophore impurities, such as, e.g., epoxy residues or other degradation products from organic components (die attached and solder paste), not associated to phenyl siloxane groups.

As phenyl groups are incorporated in the polysiloxane backbone, the pre-exponential factor of the Arrhenius plot for polymer degradation is reduced as the polymer chain mobility is reduced due to steric hindrance of phenyl groups [77]. The reduction of pre-exponential factor indicates therefore that phenyl siloxane polymer chains are more rigid and more difficult to attain the necessary configuration for being degraded by cyclic siloxane formation like in case of PDMS like illustrated in Fig. 1.15. The small mass loss of phenyl siloxane polymer at temperature inferior to 250 °C was assigned to the de-benzylation reaction rather than cyclic siloxane trimer generation. The mass loss due to cyclic siloxane trimer generation at higher temperatures could be reduced by incorporating phenyl groups into polysiloxane.

Therefore, even though stability of phenyl siloxane against heat and blue LED light is enhanced because less prone to degradation into cyclic formation, phenyl siloxane remains susceptible yellowish discoloration and darkening due to extrinsic chromophore impurities, such as, e.g., epoxy residues from white reflector, die-attach, and solder paste used in LED packages.

5.5 *Radical Crosslinking*

Oxy-luminescence, or chemiluminescence has been investigated as a means of studying the oxidative degradation of organic polymers including polytetrafluoroethylene, polystyrene, polyurethane, polymethyl methacrylate, Nylon 66, and polypropylene [78]. The luminous intensity can decrease or increase as the atmosphere surrounding the material is changing from inert to oxidative. The luminescence was observed from polypropylene sample at 150 °C. The substitution of inert gas such as nitrogen, or argon causes a very rapid loss of luminous intensity of the polypropylene sample. The intensities of light emission by polymers heated in oxygen reflect their ease of initial oxygen absorption or overall oxidative stability.

The authors of this book chapter have demonstrated that chemiluminescence of methyl-siloxane resins (OE-6351 Optical Encapsulent[3]) and phenyl-siloxane resins (OE-6630 Optical Encapsulent[3]) was a valuable mean to evaluate these optical resins stability against deep UV-C radiation exposure at 254 nm (Fig. 1.19).

Before UV aging, chemiluminescence of phenyl silicone material (OE-6630) was observed at 160 °C and higher temperatures. Chemiluminescence of phenyl silicone material (OE-6630) significantly increased after UV irradiation, which indicates radical degradation reaction relating to phenyl functionalities. Chemiluminescence intensity of methyl silicone material (OE-6351) after UV irradiation was less than that of phenyl silicone (OE-6630).

5.6 *Environmental Degradation*

Many studies have been conducted to evaluate the fate and effects of silicones in the environment throughout their life cycle [79, 80]. Cured silicone-based materials such as, e.g., conformal coating, die-attach adhesives, and optical encapsulants used

---

[3] Both supplied by Dow Corning Corporation in 2016, now Dow Silicones for the experiments reported in this paper. Currently available from DuPont.

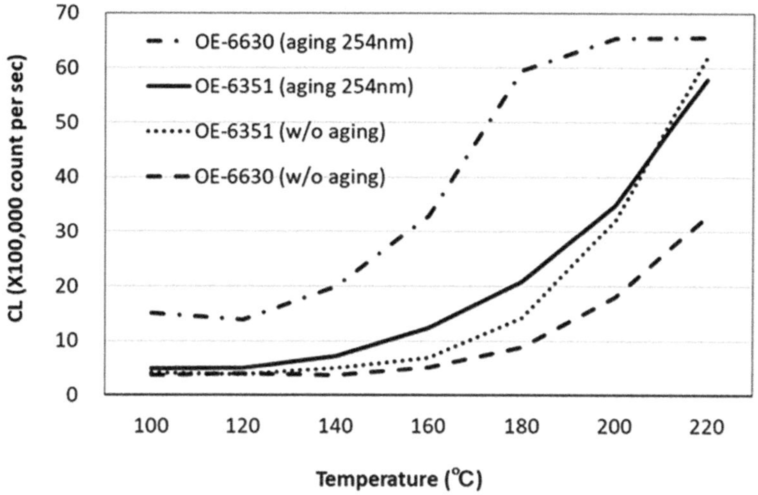

Fig. 1.19 Chemiluminescence (CL) of hydrosilylation reaction products. Comparison of phenyl-siloxane resin (OE-6630) and methyl-siloxane resin (OE-6351) optical encapsulants exposed to increasing temperature between 100 and 220 °C, without light irradiation, against with light irradiation of 254 nm for 12 h, respectively

in optoelectronics may however enter the environment as a component of domestic or industrial waste, that will possibly be either land filled or incinerated. In the latter case, they are converted back to inorganic ingredients, amorphous silica, carbon dioxide, and water vapor [81].

In a specific study where silicone fluid was spreading on photocatalytic self-cleaning glasses used for window glazing [82], the mechanism of degradation of the PDMS species was studied as a function of exposure to outdoor weathering conditions. In these self-cleaning glass applications, one of the most popular photocatalyst materials used is titanium dioxide (TiO_2) [83]. The use of anatase crystalline form of TiO_2 of an appropriate thickness is suitable to achieve good hydrophilicity and a high activity coefficient under sunlight UV irradiation. The detailed mechanism of degradation of PDMS oil on such a self-cleaning TiO_2 coated glass was studied by surface analysis techniques (XPS, ToF SIMS, and water contact angle), and demonstrated that silicon oxide, SiO_2, was formed on the surface of the glass, originated from the breakdown reaction of PDMS oil upon the photo-catalytic action of the anatase TiO_2 layer in presence of sunlight UV radiation and water.

Acknowledgments The authors are grateful to Ryan Baumgartner and Mike Ferritto of Dow Corporation for their inspiring contribution to the paragraph on key properties of silicones; to Rachel Rademacher and Jake Steinbrecher for their contribution to optical characterization of SILASTIC™ Moldable Optical Silicone; and lastly to Anne Dupont for her contribution and discussion of thermal measurements of silicones by DSC and TGA.

References

1. F. de Buyl, Silicones sealants and structural adhesives. Int. J. Adh. Adhes. **21**, 411–422 (2001)
2. J.E. Mark, *Polymer Data Handbook*, 2nd edn. (Oxford University Press, 2009)
3. F. Weinhold, R. West, The nature of the silicon-oxygen bond. Organometallics **30**(21), 5815–5824 (2011)
4. S.J. Clarson, J.A. Semlyen, *Siloxane Polymers*, Polymer Science and Technology Series, ISBN 0-13-816315-4 (Elice Horwood and Prentice Hall, Englewood Cliffs, 1993)
5. R. Walsh, in *Bond Dissociation Energies in Organosilicon Compounds: Silanes and Silicones*, ed. by B. Arkles, G. Larson, 2nd edn., (Gelest Inc, Morrisville, 2008), pp. 200–207
6. W. Noll, *Chemistry and Technology of Silicones* (Academic, New York, 1968)
7. H. Da Silva, Moldable optical silicones enable LED lamps and luminaires designs. LEDs Mag (October 2012)
8. F. de Buyl et al., How silicones are evolving to meet the growing needs of the LED lighting industry? LED Prof. **36**, 40 (March/April 2013)
9. M. Yoshitake, et al., Light emitting device encapsulated with silicones and curable silicone compositions for preparing the silicones. US 8,258,502 (2012)
10. A. Lee Smith, *The Analytical Chemistry of Silicones* (Wiley, New York, 1991)
11. A. Norris et al., High reliability of silicone materials for use as polymer waveguides, in *Proceedings of the SPIE Linear and Nonlinear Optics of Organic Materials III*, ed. by M. Kuzyk, M. Eich, R. Norwood, vol. 5212, (SPIE, Bellingham, 2003), pp. 76–82
12. F. de Buyl, J. Comyn, N.E. Shephard, N.P. Subramaniam, Kinetics of cure, crosslink density and adhesion of water-reactive alkoxysilicone sealants. J. Adh. Sci. Technol. **16**(8), 1055–1071 (2002)
13. J.M. Klosowski, *Sealants in Construction* (Marcel Dekker, New York and Basel, 1989)
14. N.E. Shephard, J.M. Klosowski, Effects of degradation factors on sealant adhesion, in *Durability of Building Sealants, State-of-the-Art Report #21 of RILEM Technical Committee 139-DBS*, ed. by A. T. Wolf, (RILEM Publication S.A.R.L, Cachan, 1999), pp. 107–135
15. J. Comyn, *Adhesion Science*, RSC Paperbacks (Royal Society of Chemistry, Cambridge, 1997)
16. F.O. Stark, J.R. Falender, A.P. Wright, Silicones, in *Comprehensive Organometallic Chemistry*, ed. by G. Wilkinson, F. G. A. Stone, E. W. Abel, vol. 2, (Pergamon Press, Oxford, 1982), pp. 305–363
17. J. Stein, L.N. Lewis, Y. Gao, R.A. Scott, In situ determination of the active catalyst in hydrosilylation reactions using highly reactive Pt(0) catalyst precursors. J. Am. Chem. Soc. **121**, 3693–3703 (1999)
18. A. Dupont, Chapter 2: Characterization of silicones, in *Silicones in Industrial Applications*, "Inorganic Polymers" Encyclopedia, ed. by R. De Jaeger, M. Gleria, (Nova Sciences, New York, 2008)
19. M. Bont, C. Barry, S. Johnston, A review of liquid silicone rubber injection molding: Process variables and process modeling. Polym. Eng. Sci., 1–17 (2021). https://doi.org/10.1002/pen.25618
20. A. Ghanbari-Siahkali et al., Investigation of the hydrothermal stability of cross-linked liquid silicone rubber (LSR). Polym. Degrad. Stab. **90**, 471–480 (2005)
21. C.E. Weir, W.H. Leser, L.A. Wood, Crystallization and second-order transitions in silicone rubbers. J. Res. Natl. Bur. Stand. **44**, 367–372 (1950)
22. T.R. Crompton, *Polymer Reference Book* (Smithers Rapra Technology Limited, Shrewsbury, 2006), p. 219
23. F. Rodriguez, *Principles of Polymer Systems* (Hemisphere Publishing Corporation, Washington, DC, 1982)
24. H. Chen, V.V. Ginzburg, J. Yang, Y. Yang, W. Liu, Y. Huang, L. Du, B. Chen, Thermal conductivity of polymer-based composites: Fundamentals and applications. Prog. Sci. **59**, 41–85 (2016)

25. A. Hermann et al., The influence of phosphor particles on the water transport in optical silicones for LEDs. Opt. Mater. **6**, 100047 (2020). Available online https://doi.org/10.1016/j.omx.2020.100047
26. J.V. Degroot, Chapter 12: Silicones in industrial applications, in *Inorganic Polymers, an Advanced Research Book*, ed. by R. DeJaeger, M. Gleria, (Nova Science Publishers, New York, 2008), pp. 56–59
27. M. Itoh, M. Suto, T. Iimura, J.H. Chae, R.G. Schmidt, Silicone encapsulants for high brightness LED. RadTech Asia Proceedings, Tokyo, October 24–27 (2016)
28. S. Wu, *Polymer Interface and Adhesion* (Marcel Dekker, New York, 1982)
29. M.J. Owen, P.R. Dvornic, Silicone Surface Science, in *Advances in Silicon Science*, vol. 4, (Springer, Cham, 2012) . ISBN 978-94-007-3876-8
30. J.W. Swanson, F.C. Dall, On the dielectric strength of synthetic electrical insulating materials, in *IEEE International Symposium on Electrical Insulation*, (IEEE, Piscataway, 1976)
31. L. Yu, S. Vudayagiri, L.A. Jensen, A.L. Skov, Temperature dependence of dielectric breakdown of silicone-based dielectric elastomers. Int. J. Smart Nano Mater. **11**(2), 129–146 (2020). https://doi.org/10.1080/19475411.2020.1768605
32. M.E. Cifuentes, C.L. Lee, H.L. Chapman, K.M. Lee, K.L. Ulman, Investigation of structure-permeability relationships of silicone polymers. International Symposium on Synthetic Membranes, Michigan Molecular Institute, Midland, MI, August 20–24 (1984)
33. C.L. Lee, H.L. Chapman, M.E. Cifuentes, K.M. Lee, L.D. Merril, K.L. Ulman, K. Venkataraman, Silicone polymer membranes for air separation. 4th BOC Priestly Conference on Membranes in Gas Separation and Enrichment, University of Leeds, UK, September 16–18 (1986)
34. S.A. Stern, Structure-permeability relationships in silicone polymer membranes. 4th BOC Priestly Conference on Membranes in Gas Separation and Enrichment, University of Leeds, UK, September 16–18 (1986)
35. J. Brandrup, E. H. Immergut (eds.), *Polymer Handbook*, 3rd edn. (Wiley, New York, 1989)
36. P.M. Bhadha, How weld hose materials affect shielding gas quality. Weld. J. **8**(7), 35–42 (1999)
37. M.E. Nelson, M.E. Nelson, Organosiloxane encapsulating resins, US 3,284,406 (1966)
38. E.L. Kern, Laser communication system using flexible silicone transmission line, US 3,434,776 (1969)
39. K.E. Polmanteer, Optical compositions of silicone rubber, US 3,518,324 (1970)
40. L.A. Kroupa, Liquid curable polyorganosiloxane compositions, US 4,529,789 (1985)
41. D.D. Mitchell, Wettable silicone resin optical devices and curable compositions therefor, US 4,487,905 (1985)
42. B. Johnstone, *Brilliant. Shuji Nakamura and the Revolution in Lighting Technology* (Prometheus Books, Amherst, 2007)
43. G. B. Gardner and R. G. Schmidt, Planar optical waveguide assembly and method of preparing same, US 6,905,904 (2005)
44. J. DeGroot et al., Impact resistant optical waveguide and method of manufacture thereof, US 7,551,830 (2009)
45. M. Bahadur, R. Nelson, M. Strong, Silicone composition for producing transparent silicone materials and optical devices, WO 2010/138221 (2009)
46. F. de Buyl et al., Dow corning moldable silicone leading innovation in LED light fixtures. J. Soc. Silicon Chem. Jpn **31**, 23–38 (2014)
47. Moldable optical silicone elastomers spark creativity in LED lighting. Laser Focus World (September 30, 2019). Published online on https://www.laserfocusworld.com/optics/article/14040094/moldable-optical-silicone-elastomers-spark-creativity-in-led-lighting
48. G. Nolan, ENGEL, ACH-solution and Dow silicones to display LED lenses made of LSR at Fakuma. Br. Plast. Rubber Mag. (September 10, 2018). Published online on https://www.britishplastics.co.uk/industry-events/Fakuma2018/engel-ach-solution-and-dow-silicones-to-display-led-lenses-m/
49. M. Maier, J. Moisel, F. Herold, Multibeam headlights in the Mercedes-Benz CLS-class. ATZ World **117**, 4–9 (2015). https://doi.org/10.1007/s38311-015-0156-0

50. M. Yazdan Mehr, W.D. van Driel, K.M.B. Jansen, P. Deeben, M. Boutelje, G.Q. Zhang, Photodegradation of bisphenol A polycarbonate under blue light radiation and its effect on optical properties, Optical Materials, **35**, 504–508 (2013)
51. M. Yazdan Mehr, A. Bahrami, W.D. van Driel, X.J. Fan, J.L. Davies, G.Q. Zhang, Degradation of optical materials in solid-state lighting systems, Int. Materials Review, 1743–2804 (2019), https://doi.org/10.1080/09506608.2019.1565716
52. M. Yazdan Mehr, W.D. van Driel, F. de Buyl, G.Q. Zhang, "Study on the Degradation of Optical Silicone Exposed to Harsh Environments", Materials, **11**, 1305 (2018)
53. UV LED – Technology, Manufacturing and Application Trends. Yole Dévelopement – EPIC, Market and Technology Report (2016)
54. J.V. DeGroot, A. Norris, S.O. Glover, T.V. Clapp, Highly transparent silicone materials. Proceedings of the SPIE 5517, Linear and Nonlinear Optics of Organic Materials IV, Vol. 5517, pp. 116–123 (2004)
55. F. de Buyl et al., Innovation in water purification systems with UV-C LEDs and silicone optics. Proceedings of the 7th Light Professional Symposium, pp. 346–352 (2017)
56. Philips Natural Trust UVC LED Module for Clean Water. Technical bulleting, www.philips.com/uvpurification (2018)
57. F. de Buyl et al., Innovation in water purification systems with UV-C LEDs and silicone optics. Presented at 7th Light Professional Symposium, Bregenz, Austria September 26–28 (2017)
58. Dow and Philips lighting collaborate to change the future of UV purification. Dow case study, Ref. No. 11-3847-01A (2018)
59. M.A.P. Janssen, Device useful for purifying UV-C fluid comprises UV-C LED modules for administering UV-C light to fluid, each module comprising printed circuit board (PCB) having bottom and top surfaces, UV-C LED, and UV-C-transmissive window elements. Patent Application WO2017202508A1, November 30, 2017 (Derwent World Patent Index 201781175G)
60. UWAVE, UV LED curing lamps, on http://www.uwave.fr/en/uv-led-curing-lamps.html, consulted on October 2, 2020
61. C. Hasegawa, M. Yoshitake, H. Akitomo, Curable silicone composition that provides a highly transparent cured silicone material. Dow Corning Toray Co Ltd., US 8,859,693, January 29, 2010
62. N.S. Tomer et al., Oxidation, chain scission and cross-linking studies of polysiloxanes upon ageings. Open J. Org. Polym. Mater. **2**, 13–22 (2012)
63. Y. Israeli, J. Lacoste, J. Cavezzan, J. Lemaire, Photo-oxidation of polydimethylsiloxane oils and resins. Part IV – Effect of phenyl groups. Polym. Degrad. Stab. **47**, 357–362 (1995)
64. N.S. Tomer et al., Oxidation, chain scission and cross-linking studies of polysiloxanes upon aging. Open J. Org. Polym. Mater. **2**, 13–22 (2012)
65. S. Yoshida, T. Iimura, Silicone material, curable silicone composition, and optical device. Dow Corning Toray Coo Ltd., US 9,550,916 B2 (March 24, 2016)
66. J. Huang et al., Rapid degradation of mid-power white-light LEDs in saturated moisture conditions. IEEE Trans. Device Mater. Reliab. **15**(4), 478–485 (2015)
67. X. Luo, F. Xing, F. Chen, H. Zheng, Phosphor self-heating in phosphor converted light emitting diode packaging. Int. J. Heat Mass Transf. **58**, 276–281 (2013)
68. G. Camino et al., PDMS thermal degradation part 1. Kinetic aspects. Polymer **42**(6), 2395–2402 (2001)
69. G. Camino et al., PDMS thermal degradation part 2 the degradation mechanism. Polymer **43**(7), 2011–2015 (2002)
70. R.N. Lewis, Methylphenylpolysiloxanes. J. Am. Chem. Soc. **70**(3), 1115–1117 (1948)
71. N. Grassie, I.G. MacFarlane, The thermal degradation of polysiloxanes I: Poly(dimethysiloxane). Eur. Polym. J. **14**, 875–884 (1978)
72. L.N. Lewis et al., Platinum catalysts used in the silicones industry. Platin. Met. Rev. **41**(2), 66–75 (1997)
73. J. Stein et al., In situ determination of the active catalyst in hydrosilylation reactions using highly reactive Pt(0) catalyst precursors. J. Am. Chem. Soc. **121**(15), 3693–3703 (1999)

74. J. Huang et al., Platinum nanoparticles from the hydrosilylation reaction: Capping agents, physical characterizations, and electrochemical properties. Langmuir **21**(2), 699–704 (2005)
75. M. Planes et al., Rhodium and platinum hydrosilylation catalysts for increasing UV stability of poly(dimethylsiloxane) in geostationary environment. Polym. Degrad. Stab. **142**, 111–116 (2017)
76. Y. Israeli, J. Lacoste, J. Cavezzan, J. Lemaire, Photo-oxidation of polydimethylsiloxane oils. Part III – Effect of dimethylene groups. Polym. Degrad. Stab. **42**, 267–279 (1993)
77. G. Deshpande et al., Kinetic aspects of the thermal degradation of poly(dimethyl siloxane) and poly(dimethyl diphenyl siloxane). Polym. Degrad. Stab. **76**, 17–24 (2002)
78. M.P. Schard et al., Oxyluminescence of polymers. I. General behavior of polymers. J. Appl. Polym. Sci. **8**, 985–995 (1964)
79. European Centre for Ecotoxicology and Toxicology of Chemical: Linear Polydimethylsiloxanes (viscosity 10-100,000 centistokes), ECETOC Joint Assessment of Commodity Chemicals N°. 26., September 1994
80. C. Stevens, Environmental degradation pathways for the breakdown of polydimethylsiloxanes. J. Inorg. Biochem. **69**, 203–207 (1998)
81. C. Stevens, Chapter 22: Silicones and their impact on the environment, in *Silicones in Industrial Applications*, "Inorganic Polymers" Encyclopedia, ed. by R. De Jaeger, M. Gleria, (Nova Sciences, New York, 2008)
82. J.-P. Hautekeer, F. de Buyl, L. Gallez, Best sealing practice for easy-to-clean and self-cleaning glasses. Int. Glass Rev. **2**, 52–55 (2003)
83. A. Fujishima, K. Hashimoto, T. Watanabe, *TiO2 Photo-catalysis, Fundamentals and Applications*, 1st edn. (BKC, Inc, Tokyo, May 1999) ISBN 4-939051-03-X

Chapter 2
Degradation Mechanisms of Aromatic Polycarbonate

T. M. Eggenhuisen and T. L. Hoeks

1 Introduction

Bisphenol-A polycarbonate is an amorphous thermoplast patented virtually simultaneously by Dr. Schnell and Dr. Fox in the mid-1950s [1]. Its toughness and high-temperature resistance made it stand out, and later also the transparent appearance and high Limiting Oxygen Index (LOI) flammability index were added to the recognized unique set of properties. Bisphenol-A polycarbonate, usually referred to as simply polycarbonate (PC), is a polymer consisting of the monomer bisphenol-A (Bisphenol Acetone, BPA) linked by carbonate bonds (see Fig. 2.1). It has found use in diverse industries combining 2 or 3 of those mentioned main properties, with applications in, e.g., optical media, construction (sheet materials), electrical and electronics (covers), automotive (headlamp lenses), water bottles, health care (dialysis equipment), etc. Its use as a homopolymer or in blends with other engineering thermoplasts, such as ABS, PBT, or PET, has brought a wide range of materials to the industry that allowed replacement of metals or glass providing lightweight solutions, freedom of design by using shaping techniques such as injection molding, and improved safety. Global production of polycarbonate exceeded 5 million MT in 2016 [2]. The polymer is the polycondensation product of BPA with a carbonate carrier, typically phosgene, or is produced by trans-esterification with di-alkyl or di-aryl carbonates (diphenylcarbonate, dimethylcarbonate).

The 4 main properties of high heat performance, impact, transparency, and low flammability translate into now well-defined and well-developed material characteristics fitting the wide range of applications described above. High-temperature resistance results from the high glass transition temperature of the polymer at 145 °C, which is measured using differential scanning calorimetry. The high glass

T. M. Eggenhuisen (✉) · T. L. Hoeks
SABIC, Bergen op Zoom, The Netherlands
e-mail: Tamara.Eggenhuisen@sabic.com

W. D. van Driel, M. Yazdan Mehr (eds.), *Reliability of Organic Compounds in Microelectronics and Optoelectronics*,
https://doi.org/10.1007/978-3-030-81576-9_2

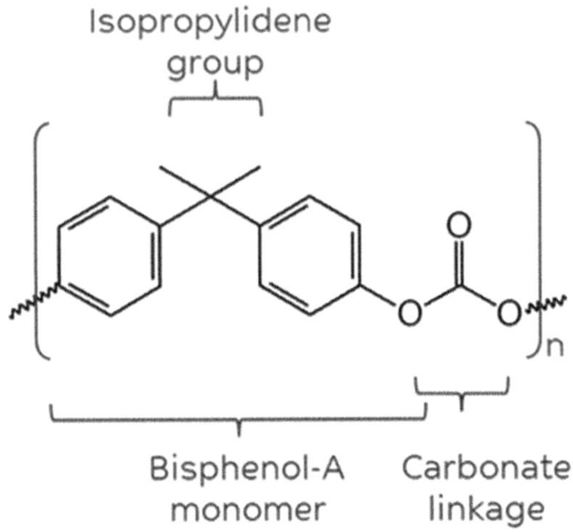

Fig. 2.1 Chemical structure of bisphenol-A polycarbonate

transition temperature is related to the stiffness of the polymer chain and its packing, requiring a relatively high temperature to allow polymer chain mobility. Alternatively, the heat deflection temperature (HDT) or VICAT temperature indicates when the polymer starts to deform under a certain load (ISO75 and ISO306 respectively). These are predictors for structural integrity of a part under use conditions. The impact performance is typically expressed as the material being ductile or brittle at a certain temperature and at a certain part thickness. When hit with a projectile with a certain energy, the material can absorb the energy in different ways, or shatter into pieces. This property relates to the molecular weight of the polymer and can vary for different polycarbonate blends and copolymers. The toughest polycarbonates (homopolymer) show ductile impact performance at −30 °C, meaning that the material absorbs the impact energy without shattering (ISO 180/1A). Polycarbonates or polycarbonate blends are used as safety windows in the forestry industry, motorcycle helmets, and other heavy-duty safety applications. The light output through a polycarbonate part is largely determined by the refractive index, which is slightly above that of glass and of PMMA ($n_{PC} = 1.58$, $n_{glass} = 1.54$, $n_{PMMA} = 1.49$) [3]. As will be discussed in more detail below, the aromatic structure of PC absorbs UV light, with a small band of absorption extending into the visible light. Hence, the PC has a yellow appearance (typically indicated by the Yellowness Index (ASTME313)), but often pigments or dyes are added to make the part look blue. The light absorbance by the polycarbonate and the added dyes or pigments affect the total transmission, as well as the appearance of the light passing through a polycarbonate part. Finally, the low flammability of polycarbonate is related, partly, to its molecular structure. Having a low C/H ratio in the aromatic rings gives it a relatively high limiting oxygen index (LOI), which is the minimum level of oxygen to sustain a flame. Flame retardant additives make it even more flame resistant.

Table 2.1 Degradation mechanisms for polycarbonate

External factor	Degradation mode
Heat, no oxygen	Thermal degradation
Heat and oxygen	Thermal oxidation, heat aging
Light (UV) and heat	Photo-oxidation, UV weathering
Light (blue LED) irradiation (and heat)	Photo-thermal degradation, LED aging
Chemicals	Hydrolysis, chemical degradation, stress cracking
Water (and heat)	Hydrolysis, hydrothermal aging
Sterilization by irradiation	γ-Ray degradation

The aromatic ring structure of BPA brings forth many of the PC's selling properties, but combined with the carbonate linkages, they make the molecular structure of polycarbonate sensitive toward degradation under influence of different external factors. In Table 2.1, degradation modes of polycarbonate are listed in relation to the external factors causing the polymer to degrade. During production of the polycarbonate, the polycarbonate compound, the part, or the assembled device, and during use of the device a combination of these effects may occur and need to be controlled to ensure proper lifetime of the assembly. In general, all factors can cause polymer chain scission or changes in the chemical structure leading to loss of mechanical properties (part breakage) and poor appearance (yellowing, haze, crazing).

Being an amorphous polymer, polycarbonate is prone to chemical attack by a large variety of chemicals, including water. The random packing of the polymer chains allows for diffusion of small molecules, causing weakening of the part integrity or chemical attack. Even small amounts of chemicals coming from, e.g., oils present on screws used for assembly, or chemicals coming from glues or sunscreen lotion can have negative effects on the integrity of a part or cause haze and poor aesthetics. The carbonate condensation bond is relatively easy to break, leading to chain scission. The sensitivity toward these attacks also relates to the specific part's stress level (residual tension in the part from cooling the molten plastic into the part mold, or stress applied by assembly), the exposure time and temperature as well as the chemical concentration. Water by itself can readily hydrolyze the carbonate bond, and this is accelerated by temperature and catalyzed by acids, bases, or metal impurities in the water [4]. Chemical attack can lead to visual effects such as haze by surface erosion, bubble formation, or crazing (large number of small cracks in the surface, yielding a shimmering effect), and in severe cases, or when combined with high part stress levels, lead to part breakage. Higher molecular weight polycarbonate is more resistant toward chemical attack. After chain scission, the polymer fragments are long enough to retain mechanical integrity as long as the residual molecular weight is higher than the entanglement molecular weight. However, longer exposure times will still yield part failure.

Polycarbonate has good resistance against ionizing radiation, such as γ-rays, which are used for sterilization. The polymer integrity remains at exposure to doses at 2.5–5.0 Mrad, however a yellow color appears [5]. The exact chemical origin of the color formation remains unproven, but color compensation gives the product a blue appearance. Obviously, steam sterilization poses a challenge to polycarbonate, given its sensitivity toward water, leading to the carbonate bond breakage.

In the paragraphs below, the degradation of PC under different combinations of exposure to heat and light are discussed in more detail. Typical exposure modes are described, followed by an explanation of the underlying chemistry. The effect of the degradation on the physical appearance and performance of polycarbonate is explained, as well as common methods to measure the degradation and methods of predicting the lifetime. Finally, current solutions are shared which are used to help avoid or slow down the degradation. Overall, as with all organic polymers, it is not a question of whether the polymer will degrade, but when.

2 Effect of Processing on Polycarbonate Degradation

In the process of producing a polycarbonate compound and part, thermal exposure already occurs, starting the process of potential polymer degradation. The polycarbonate resin is molten in an extruder and mixed with additives or other polymers to produce the compound with the right properties needed for the application. Shaping the polycarbonate compound into a part starts with, again, melting the polycarbonate, for example, in an injection molding machine to fill a mold creating the part's shape. Alternatively, the polycarbonate can be extruded into a flat surface with various thicknesses (from 50 μm, yielding a film, up to 12 mm forming a thick solid sheet). In these processes, the polycarbonate endures temperatures around 250 up to 300 °C, well above its glass transition temperature at 145 °C to obtain a molten polymer that has high enough flow for the process. Additional heat may be generated in the process of filling the part by the shear induced on the polymer during injection through a narrow mold entrance (gate) into the mold cavity. Oxygen can be present in the extruder, or at the mouth of the molding machine screw. While the extrusion process is usually fast (30 s), during injection molding the polycarbonate may be kept molten for several minutes.

Chemical bond strengths between atoms are indicative of a molecule's thermal stability. In the BPA-PC structure, the weakest bond was calculated to be the C-C bond linking the methyl group in the isopropylidene bridge [6]. Thermal energy will cause the bond to break and leads to the formation of a radical. In the absence of oxygen this will lead to chain scission and further reactions of the radical such as branching, recombination, or formation of small organic molecules (see Scheme 2.1) [6]. In presence of oxygen, the radical may be converted into a peroxide radical, followed by similar degradation mechanisms yielding chain scission and a variety of side products, referred to as auto-oxidation [7, 8]. The second weakest bond in polycarbonate is the C(=O)-O carbonate bond. It is especially sensitive to hydrolysis or alcoholysis and cleavage occurs readily in the presence of, e.g., water under thermal conditions. It can also undergo rearrangements. These reaction pathways have been identified based on structures formed in subsequent reactions such as

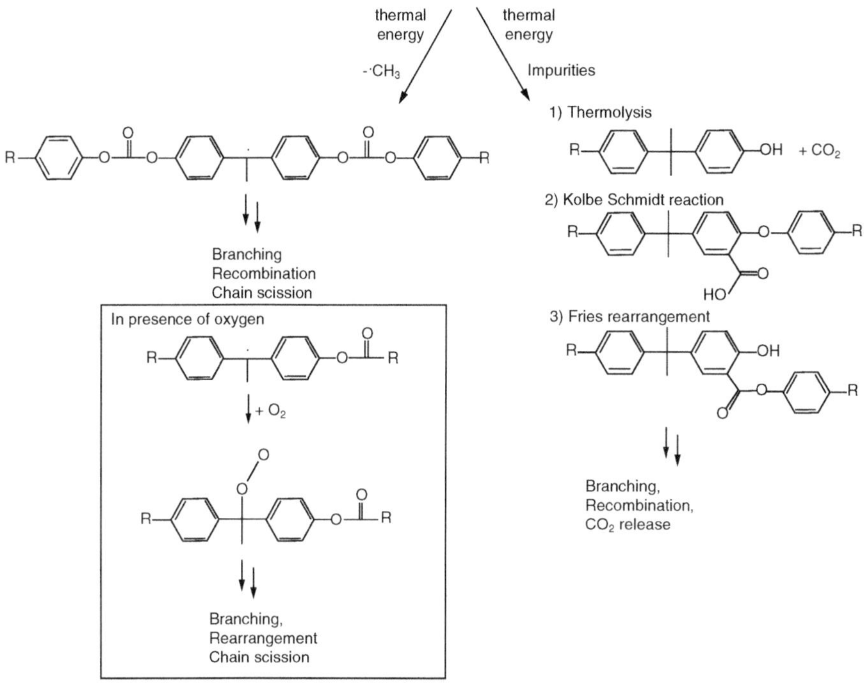

Scheme 2.1 Thermal degradation mechanism of BPA-PC, based on [4, with permission]

recombination, branching, chain scission, etc. (a.o. carboxylated phenol derivatives (called Fries structures), hydroxyl acetophenone, biphenyls, isopropenylphenol), and are proceeding in parallel to the radical degradation pathways and are catalyzed by impurities such as acids, bases, or alcohols [4]. The diverse chemistry and low concentration of degradation products in the polymer make it challenging to single out the most destructive pathway.

Many different chemical structures have been identified to be formed after thermal degradation of PC, and all of them together cause a visible physical effect [4, 7, 8]. Chain scission leads to molecular weight loss and can cause negative effects in, e.g., the parts mechanical performance. Among all the different chemical structures, so-called color bodies are formed: oxidized aromatic structures that absorb visible light. This gives the part a yellow appearance (if the part is produced from a transparent or white compound). Small organic molecules (referred to as volatiles), CO_2, and H_2O that are formed as side products can cause marks on the surface of the part where the gas bubble is locked between the molten plastic and the cavity surface.

Thermal degradation will occur when the organic polymer is heated to such an extent that atom bonds can start to break. Impurities, such as metals or salts, can catalyze the degradation. Also, the presence of water during the processing can cause harm, leading to hydrolysis of the carbonate bond. Therefore, to reduce the level of degradation, a clean compound (with low impurity content), that is dried well to remove the absorbed moisture, is needed. The auto-oxidation process generates radicals, and to interrupt this process stabilizers can be added. Primary antioxidants are free radical scavengers that react with the formed radical and form stable compounds to prevent further propagation. Typical examples of these are hindered phenols. Secondary antioxidants react with hydroperoxides to also prevent further degradation by radical formation [9]. Phosphorous compounds are typically used to stabilize polycarbonate during processing, such as phosphines, phosphites, or divalent phosphonites. They can reduce the formed (hydro)peroxides to generate an ether or alcohol group, yielding the oxidized phosphite (see Scheme 2.2). The oxidation reaction may release another radical, which via a continued pathway of oxidation and radical recombination can be neutralized. Alternatively, the phosphite can hydrolyze, releasing an alcohol, which may act as a primary antioxidant, scavenging directly the formed radicals. In addition, the phosphorous compounds can form complexes with multivalent metals, preventing further reaction of these impurities with formed hydroperoxides or other bonds in the polymer structure. Hydrolysis of the phosphorous compounds leads to the formation of phosphorous or phosphonic acids, which can actually lead to degradation of polycarbonate. Hence, balance of the hydrolytic stability of the stabilizer and its preventive oxidation activity are key to achieve the right level of polycarbonate stabilization.

3 Polycarbonate Stability Under Thermal Oxidative Conditions

One of the selection criteria often leading designers to select polycarbonate is the requirement of a certain level of heat resistance in case the part is exposed to elevated temperature during operation of the assembly. As explained in the

$$\mathbf{R_{pol}\text{-}O\text{-}OR} + (R_pO)_3P \rightarrow (R_pO)_3P{=}O + \mathbf{R_{pol}\text{-}OR}$$

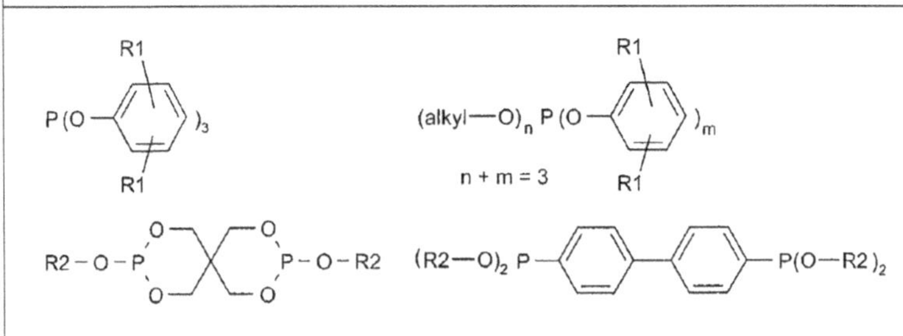

Scheme 2.2 Reaction of phosphite stabilizers with a peroxide. R_{pol} represents the polymer chain, R_p represents the phosphite side group (can be alkyl, phenol, or hydrogen), R represents the side chain that the peroxide reacted with. Typical chemical structures of phosphites and phosphonites [9, with permission]

introduction, the temperature at which deformation of a polycarbonate part starts under a certain load is defined as the HDT temperature and is related to the Tg of the material [ISO75]. For polycarbonate, the HDT temperature typically lies around 130–140 °C, while for, e.g., PMMA it can range from ~90 to 110 °C depending on the polymer backbone orientation (tacticity). For opaque parts, crystalline blends of thermoplastics with a higher VICAT may provide suitable alternatives in case more heat resistance is required. At temperatures below the deformation temperatures, a polycarbonate part can also lose its impact strength due to physical aging, which is a phenomenon generally observed with glassy polymers [10]. Annealing a part after molding can remove residual strains and stresses and help bring the polymer to its equilibrium state. The operating temperature of the assembly also influences the polymer's stability and lifetime, as under atmospheric conditions, in the presence of oxygen and water degradation may occur.

The mechanisms of degradation under thermal oxidative conditions at operating temperatures are found to be similar to what was discussed in the previous paragraph. Most mechanisms are studied under temperatures exceeding 300 °C, but chemical analysis of polycarbonate parts aged at temperatures below the glass transition temperature shows similar chemical degradation products [4, 7, 8]. The mechanisms of thermal and oxidative degradation attacking the PC backbone are complemented by chemistry occurring with impurities. BPA produced on an industrial scale used for PC production will have a p,p-BPA purity level exceeding 99 wt%. However, this means that there can be up to 10.000 ppm other organic structures, some of which have shown to be more sensitive to thermal oxidative conditions, causing faster, visually observable degradation [11]. In addition, polycarbonate polymer chains are typically end-capped by a monophenol, which stops the chain growth during the polymerization reaction. However, this process is not 100% effective and a small amount of free BPA-OH end-groups will remain in the polymeric mixture. These are prone to oxidation and formation of ketones, which are highly colored.

The radical initiated degradation process resulting in the formation of peroxides or leading to chain breakage and rearrangements gives rise to oxidation products which are colored, such as quinones or isoprene and its derivatives [4]. This leads to increased light absorption in the UV and visible range, giving the transparent polycarbonate a yellow appearance. Figure 2.2 shows typical progression of loss of transmission in the blue wavelength range and the resulting effect on total transmission (%T) and increase of yellowness index over time. In addition, chain breakage may affect the mechanical performance. Exposure to higher T in air leads to burning and char formation.

The increased light absorption of the polycarbonate can readily be measured as a function of time and temperature and the kinetics can be analyzed using the Arrhenius equation (see Eq. 2.1). Using the calculated activation energy and acceleration factor, a lifetime prediction can be made to reach the limits of the level of transmission needed for good operation. Typically, a molded part with thickness representative for the application is exposed to temperatures at 110 °C up to 140 °C,

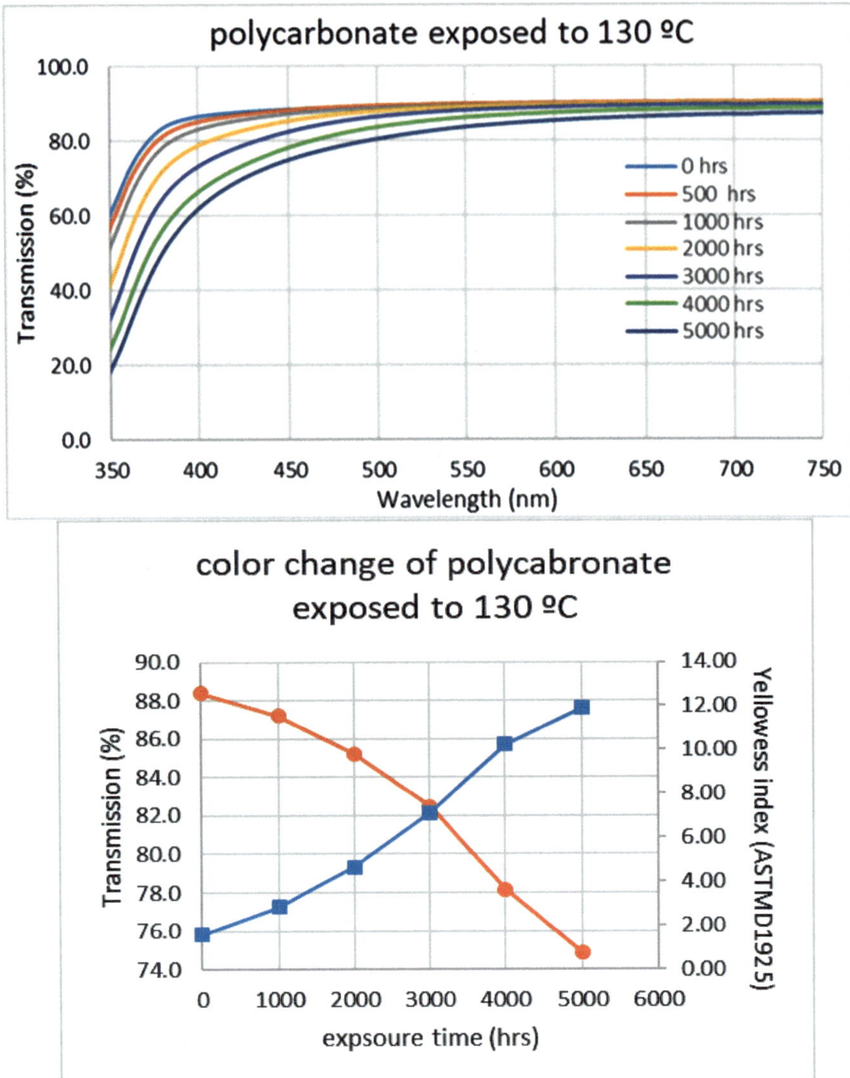

Fig. 2.2 Transmission spectra measured on a 2.5 mm polycarbonate plaque exposed to 130 °C for up to 5000 h (top) and evolution of total transmission (circles) and yellowness index (squares) over time (Bottom)

while following the shift in the desired property (i.e., %T at a specified wavelength, or yellowness index (YI)). Figure 2.3 shows that the increase in YI of polycarbonate exposed to different temperatures follows the expected kinetic behavior. An activation energy of 77 kJ/mol, or 18.3 kcal/mol, was calculated [4, 12, 13]. Note that at temperatures below 110 °C, the increase in YI or loss in transmission at certain wavelengths is slow, and long exposures are needed to obtain color shifts that can be

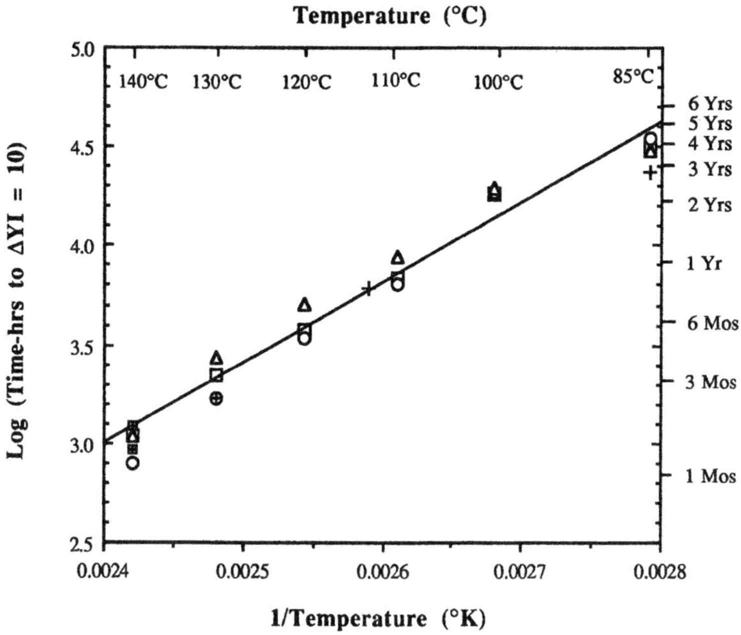

Fig. 2.3 Time-temperature analysis of yellowing of polycarbonate under thermal oxidative conditions [4, with permission]

properly measured. To reach the delta YI of 10, as used in the graph below, an exposure of 1 year was needed, making these analyses slow. At the same time, 140 °C is the upper limit of exposure, as at this temperature, parts can already start to show deformation preventing a good color measurement.

$$AF = \exp\left[\frac{E_{act}}{k_B}\left(\frac{1}{T_{use}}\right) - \frac{1}{T_{stress}}\right] \tag{2.1}$$

With: AF = acceleration factor, Eact = activation energy (eV), k_B = Boltzman's constant, T_{use} = Application temperature (K), T_{stress} = test temperature.

In a similar manner to the color, the retention of other properties, such as electrical, mechanical impact, and tensile are tested at varying temperatures. The relative temperature index (RTI) indicates the temperature at which 50% of the indicated property is retained after 50.000 h (5.7 years) exposure [14]. For polycarbonate typical values for retention of electrical properties are 130 °C, mechanical impact 125 °C, and mechanical strength 125 °C. To obtain these values, the relevant property is measured after exposure of a representative part at different temperatures and time. Plotting the log (time-to-50% retention) versus the different exposure temperatures will give a linear fit. In practice, the line fit obtained for shorter exposure times is compared to a control sample with similar chemical structure that was already fully analyzed.

Reducing the operating temperature of a device by 10 °C will slow down the degradation of the polycarbonate by a factor of 2–3. Hence, if discoloration is affecting the lifetime of a particular assembly, lowering the temperature by allowing ventilation, increasing the distance between the polycarbonate part and the heat source, or lowering the temperature of the heat source will have a benefit on the total lifetime. Polycarbonate compounds can contain a phenolic antioxidant, which complements the stabilizing function of the phosphite stabilizer that prevents degradation during processing. The antioxidants can scavenge radicals and thereby prevent the degradation pathway [9]. In addition, the organic purity of the BPA in the polymer backbone was shown to affect the color stability over time, as well as the presence of residual monomer, endcap, and oligomers [11].

4 Polycarbonate Stability Under Photo-Oxidation

Polycarbonate is mostly transparent for the visible and IR wavelengths, but the aromatic structure of polycarbonate absorbs light in the UV region (see Fig. 2.4) [5]. While the overlap of the light absorbance of polycarbonate and the solar spectrum is minimal, it is sufficient to cause visible degradation effects within the lifespan of typical outdoor polycarbonate applications in, e.g., sheets (10–20 years) or covers for forward automotive lighting (5–15 years) and industrial lighting. In addition to UV irradiation, other environmental influences, such as rain, heat, algae growth and

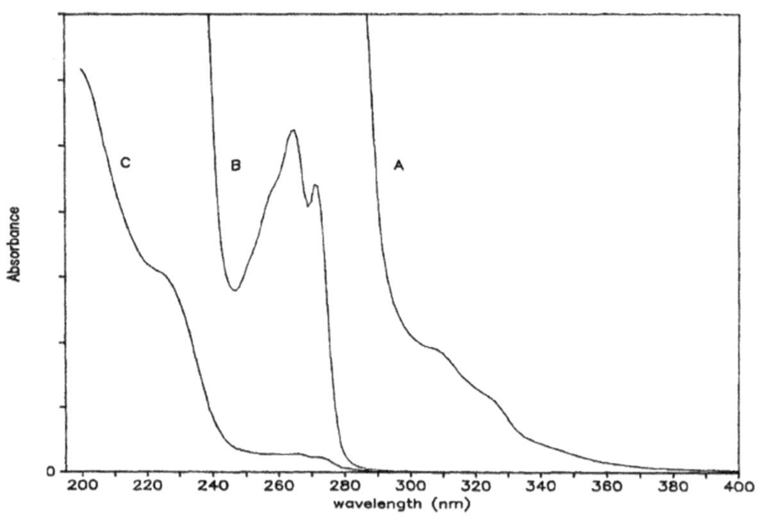

Fig. 2.4 UV absorption spectrum of bisphenol A polycarbonate (A ~ 380 μm film, B ~ 2 μm film, C. ~0.15 μm film) [5, with permission]

other eco-attack can lead to initiation of degradation. Even indoor applications of polycarbonate suffer some yellowing over time, due to a low amount of UV light being scattered. Outdoor exposure combining UV, rain, and heat poses quite a challenge for polycarbonate lifetime.

A photon with the right energy (wavelength) can be absorbed by the polycarbonate backbone structure leading to an excited state. The energy gained by the photon absorption needs to be released, which can occur via different chemical rearrangements which then lead to the formation of various degradation products. Direct excitation of the aromatic chromophoric group at 254 nm (the maximum in the absorbance spectrum) is known to lead to a chemical pathway called the photo-Fries rearrangement (see Scheme 2.3) [15]. The salicylate intermediate and the benzophenone structure can be detected using infrared spectroscopy, and their formation rate can be followed. However, as observed in the solar spectrum depicted in Fig. 2.5, the solar spectrum contains virtually no UV light with wavelengths below 295 nm. Hence, the direct excitation pathway is not likely to occur in standard outdoor use conditions. Literature shows different observations, finding salicylate, and benzophenone products in PC exposed to light resembling the solar spectrum, and discussion remains about the mechanism.

Upon degradation under UV light, many different products were detected, such as biphenols, hydroxyphenyl ethers, quinones, and other oxidized products [4]. Different quantum mechanical states of polycarbonate show absorbance of wavelengths below 300 nm, but several of the transitions were calculated to have tails of

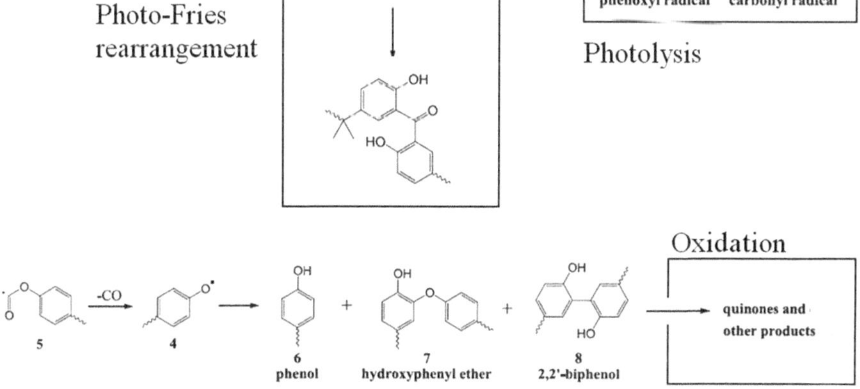

Scheme 2.3 Degradation pathways under UV light exposure [16, adapted from 17, with permission]

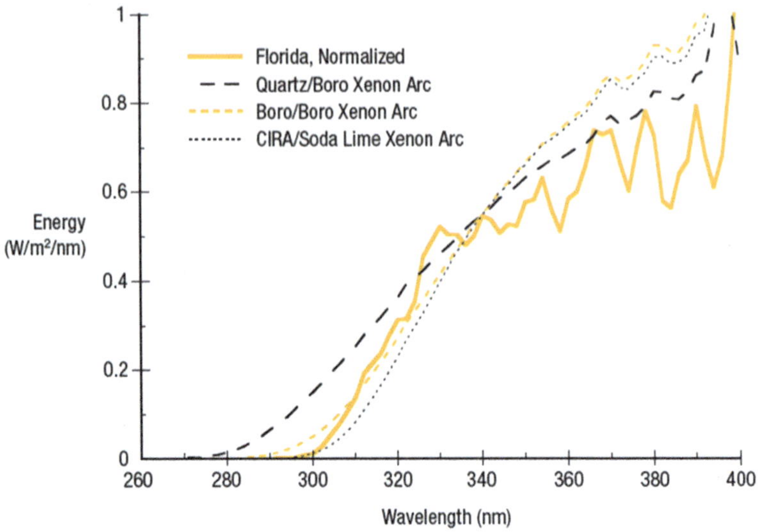

Fig. 2.5 Spectral power distribution of sunlight and three common xenon arc filer combinations, all normalized to 0.55 W/m^2/nm at 340 nm [20, with permission]

absorbance toward longer wavelengths, albeit with low extinction coefficients and being forbidden transitions [11]. Hence, the energy of the UV photons in natural light is too low to directly generate excited states, which lead to the observed chemical rearrangements, chemistry still occurs. Possibly, the tailing absorbance toward lower energy of some particular transitions supported by, e.g., impurities causing the slightest changes in local energy states are at the start of UV-induced degradation.

Overall, the exposure of polycarbonate to natural light leads to different chemical pathways: Fries-rearrangement, photolysis (or chain scission), photooxidation (radical initiated oxidation), and ring oxidation. As depicted in Scheme 2.3, Fries-rearrangement is still observed, even though the wavelength does not meet the energy required for the direct excitation. Photolysis leads to the formation of two radicals, breaking the carbonate bond. Those radicals undergo rearrangements to form structures that are prone to oxidation and form quinones and other highly colored oxidized species.

The formation of colored chemical structures causes absorption of the visible light in the region of 380–500 nm and gives a transparent or white material a yellow appearance. Colored materials may suffer from color shift by the combination of yellowing of the polycarbonate with possible bleaching of the dies or pigments. Chain scission leads to the formation of small molecules, which can sublime, or be removed from the surface causing erosion and leading to diminished aesthetics of the part [18]. A typical erosion rate for an unprotected white polycarbonate was found to be 2.5 μm/(MJ/m^2/nm@340 nm) or 0.83 μm/month of outdoor exposure in Florida. The increased surface roughness causes a haze effect and reduced transmission for a transparent polycarbonate, while for opaque black parts this is observed as loss of gloss and a dull appearance. Crazing is another observed effect, where small cracks in the surface cause a shimmering light effect or lighter surface appearance. Most of the degradation of polycarbonate under UV light occurs in the top 3–30 μm [19]. Penetration of the UV light into the polycarbonate goes deeper, but once color absorbing bodies are formed, UV penetration is limited to the top layer. Therefore, the effect of UV weathering on the mechanical properties is limited. For tensile strength, the bulk fraction dominates the performance, and remains relatively unharmed, although effects from continued exposure to heat (depending on the outdoor conditions) may cause a negative effect. The brittle surface layer may affect the impact behavior, acting as a notch and causing crack propagation with brittle failure as a result [18]. On the other hand, when the surface layer is too brittle, it will not act as notch initiator in impact tests, retaining ductile impact behavior.

To measure the effect of UV exposure on the performance of polycarbonate, parts can be placed in the outdoor environment. However, depending on the geographical location, different results will be achieved as the light dose (in MJ/m^2) varies [pickett05]. Other external factors, such as rainfall, rain acidity, or temperature have minor effects. In addition, microbial growth can occur locally. Typically, weathering sites in Florida and Arizona are used as two reference sites with high light dosage and posing opposites on the spectrum of humidity. However, as degradation is slow, and testing location may not be readily accessible, accelerated lab test methods have been developed. Critical to these so-called weatherometers is the light spectrum of the light bulb. In set ups commonly used nowadays, the solar light is simulated by using a Xenon arc lamp, combined with specific light filters. As seen in the graph in Fig. 2.5, there are small details in how the irradiance (in $W/m^2/nm$) varies with the wavelength in the critical range at 280–400 nm. For example, a Xenon arc light combined with a borosilicate filter still exposes the part to a small amount of light with wavelengths 290–300 nm, which is not present in natural sunlight. Given the discussion above regarding the delicate balance between photon energy (wavelength) and polycarbonate excitation states, these differences cause an effect in how well the accelerated tests predict real outdoor exposure lifetime. The better the match of filter spectral distribution with solar light, the better the prediction.

Accelerated weathering tests are correlated to outdoor exposure time by the light dose in MJ/m^2 at 340 nm. An exposure to 3500 kJ/m^2 at 340 nm is found to represent 1 year of exposure in Florida [20]. For a part exposed to a light source emitting at a typical value of 0.55 W/m^2 at 340 nm, this means 1768 hrs of exposure, or 10.5 weeks. In addition to the total light exposure and the spectral light distribution, light and dark cycles, as well as rain and temperature are controlled in weatherometers to simulate as best as possible real outdoor conditions. For an unprotected polycarbonate, effects may be observed already after 1 year of outdoor exposure, while protected sheets will have a guaranteed lifetime of e.g., 10 years. The activation energy of yellowness increase for a white PC was determined to be 5 kcal/mol [21].

The use of UV additives is common practice to expand the lifetime of polycarbonate when the materials are intended for outdoor applications or may be exposed to low levels of UV in an indoor environment. Commonly used UV additives are benzotriazoles or triazines, molecules that can release the energy of absorbed photons by internal hydrogen transfers [22]. Further lifetime extension of parts that are exposed to direct sunlight, such as sheet or headlamp lens covers, a protective top layer is typically applied. This layer can be either a silicone or acrylate hard coat, a PMMA layer, or a UV protecting cap layer. Preventing the UV to reach the bulk of the polycarbonate by a 8–50 μm layer is sufficient, and performance can be guaranteed up to 5 years for yellowing, and 10 years for breakage.

5 Polycarbonate Stability Under LED Light and Thermal Conditions

The development of high-power LED lights has revolutionized the automotive lighting designs and has led **to** great advances in other lighting applications. Applications of polycarbonate endure less heat as compared to more conventional light sources, such as halogen bulbs in, e.g., automotive headlamp lenses. However, locally high temperatures still occur, and depending on the assembly design, a polycarbonate optic, light bar, or cover in close proximity of the LED might still be exposed to temperatures which affect the lifetime performance. The spectrum of a blue light LED does not lead to thinking the light will interact with the polycarbonate in a harmful way, with wavelengths in the visible region and with low emission at wavelengths close to 400 nm. These wavelengths are well above the harmful wavelengths discussed earlier, when analyzing the degradation of polycarbonate under UV light. Still, excessive degradation of polycarbonate in LED assemblies can be observed. The nature of this unexpected degradation process has recently caught attention and has been described in several papers but is not yet as well understood as, for example, UV weathering [12, 23, 25]. The first investigations of this new phenomenon point toward the combination of the LED light exposure combined with elevated temperatures, hence it has been referred to as LED aging, photodegradation under blue LED light, or photo-thermal aging under blue LED light.

At first sight, the degradation of PC exposed to LED light and elevated temperature appears similar to thermal oxidation. Indeed, in the graph in Fig. 2.6, the yellowness index change over time follows the same pattern as can be observed for thermal oxidation without light exposure, but at a higher temperature. The effect of the light exposure may be just a faster thermal oxidization. However, as the pictures show in Fig. 2.7, after a certain amount of time and linear increase in the yellowness, the polycarbonate burns, referred to as catastrophic failure. Knowing at which point in time this occurs is essential to be able to predict the lifetime of the assembly.

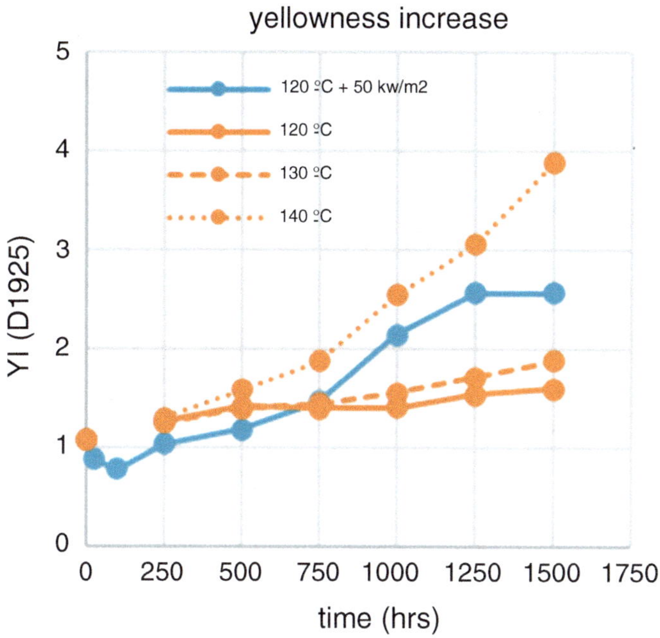

Fig. 2.6 Increase in yellowness of PC exposed to LED light at 120 °C, compared with yellowness increase of PC under thermal oxidation at various temperatures

Fig. 2.7 Catastrophic failure of PC after exposure to LED light, time increase from left to right [12, with permission]

The transmission spectra of the polycarbonate exposed to 140 °C in dark conditions and 120 °C in combination with a blue LED light with irradiance at 50 kW/m^2 shows that the decrease in transmission at wavelengths below 400 nm is much larger for the LED exposed sample than the dark thermal aged sample (see Fig. 2.8). As the calculations of color parameters, such as the yellowness index are often based on the visible wavelength region only (400–750 nm), this effect is not revealed when looking at the yellowness index only. Indeed, the absorbance at 360 nm

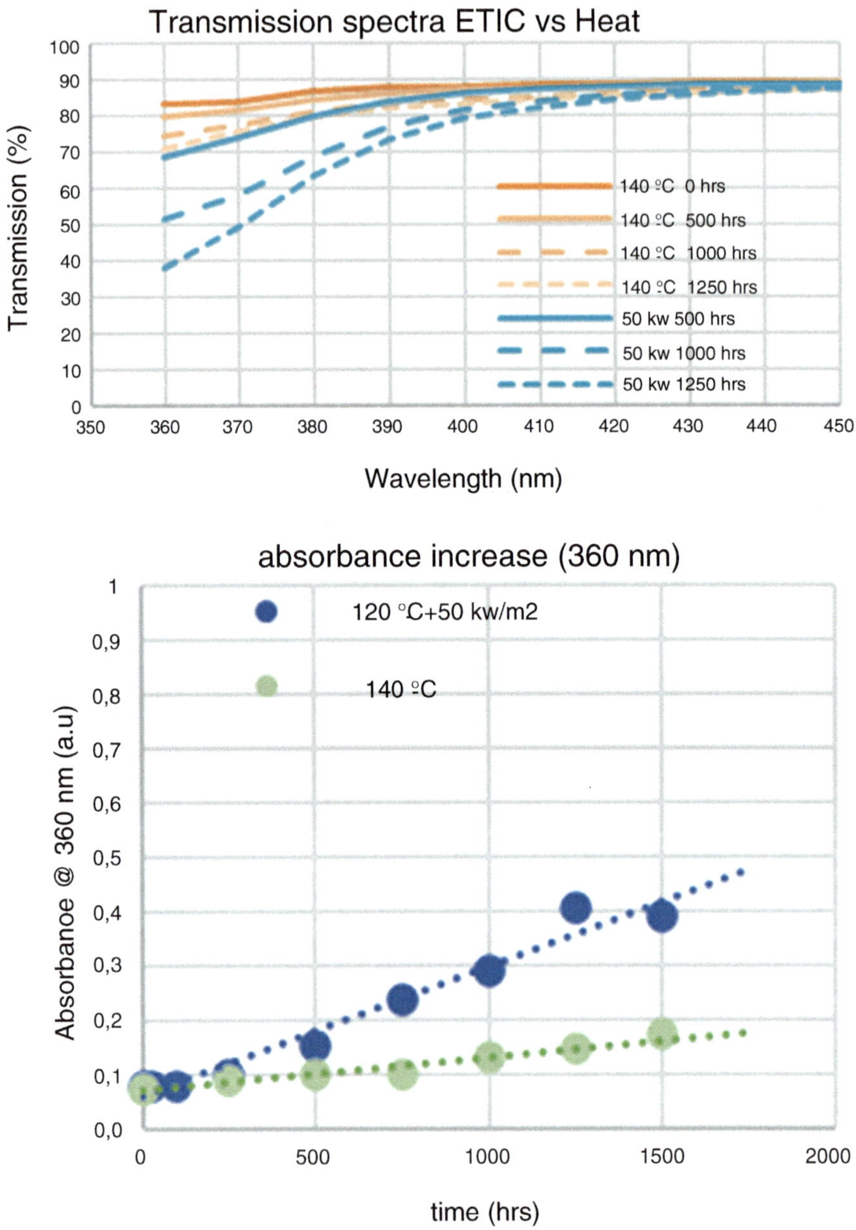

Fig. 2.8 A polycarbonate sample exposed to LED light at 450 nm and 120 °C as compared to a sample exposed to 140 °C in dark conditions. TOP: Reduction of transmission at wavelengths below 450 nm for, BOTTOM: Increase of absorbance at 360 nm as a measure for degradation

increases at a much faster rate under LED exposure than under dark conditions, indicating differences in the chemical changes that are occurring. Initial studies using FTIR and UV-Vis showed the presence of species similar to those observed for photodegradation, such as Fries-rearrangement and photo-oxidation products (see Scheme 2.3 above) [23]. However, the wavelength emitted by the LED does not match the absorbance of polycarbonate and the mechanism and energetic pathways are still to be understood.

Set ups have been designed to perform accelerated testing to predict the lifetime performance. Different published methods use blue LED lights with different irradiance levels in a thermally controlled environment, referred to as Highly Accelerated Stress Test [24, or Chap. 9 in this book], the Elevated Temperature Irradiance Chamber [12]., or the Monitored Liquid Thermostatted Irradiation Setup [25]. These allow testing of the effect of LED at different light intensities in combination with the environment temperature. In the HAST and ETIC set ups, the part temperature is monitored by a thermocouple, while convection heating is used to raise the chambers' temperatures. Similar to thermal oxidation or UV weathering, the activation energy of this photo-thermal degradation process is measured by following the change in physical parameters. In the work of Gandhi et al., the absorbance at 360 nm was found to show linear behavior in the initial stage of the degradation process and showed reciprocity for irradiance levels ranging from 5 to 75 kW/m^2 (see graphs in Fig. 2.9) [12]. The increase in absorbance at different temperatures followed the Arrhenius kinetics, and activation energies between 30 and 48 kJ/mol were calculated for polycarbonates with different types of stabilizer, above the typical activation energy for UV weathering, but below that of thermal oxidation. Using a different set up, Mehr et al. reported an activation energy of 30 kJ/mol based on the time to 30% lumen depreciation [24]. However, in these experiments, there was a nonlinear response to the light intensity, but the range of irradiance was below and at the lower end of the work covered by Gandhi et al. (0.825–13.2 kW/m^2).

As the exact mechanism of degradation is not fully understood, there are not yet specific methods developed to avoid polycarbonate degradation by blue LED light in order to extend the assembly lifetime. However, general concepts, such as reducing the operating temperature by improved ventilation, cooling of the LED, or increasing the distance between the LED and the polycarbonate are applicable. Interestingly, different stabilizers added to polycarbonate yielded different activation energies [12]. This can indicate that impurities may have an accelerating factor, similar to what was observed for thermal oxidation. At the same time, it also provides potential as to develop the right stabilization package that can interfere with the degradation pathway and improve the overall lifetime performance.

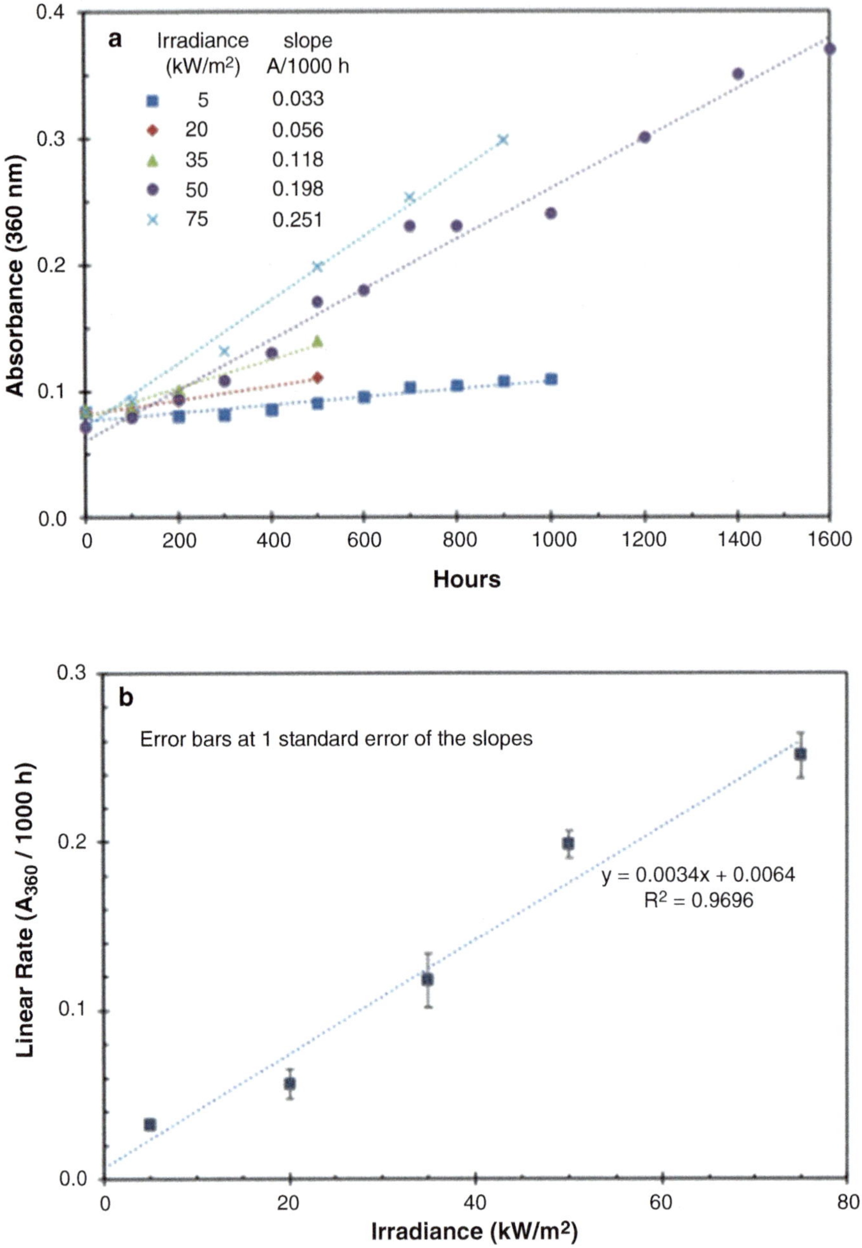

Fig. 2.9 Increase in absorbance at 360 nm for PC-1 exposed with irradiance 50 kW/m2 at 90, 100, 110, and 120 °C. (**b**) Data with time axis shifted to cause superposition on the 120 °C curve [12, with permission]

6 Conclusions

In this chapter, we have presented the current state-of-the-art on degradation mechanisms in bisphenol-A polycarbonate. The main mechanisms are presented, but it is noted that several details are still unknown. Further research is needed to unveil the full detail of these degradation mechanisms under external loadings.

References

1. W.F. Christopher, D.W. Fox, Polycarbonates, in *Chemistry and Physics of Polycarbonates*, ed. by H. Schnell, (Interscience publishers – John Wiley & Sons, 1964)
2. https://www.statista.com/statistics/720476/polycarbonate-global-production-capacity-distribution-by-producer/, Sep 23 2020
3. D.G. Legrand, Optical properties of Polycarbonates, in *Handbook of Polycarbonate Science and Technology*, ed. by J. T. Bendler, 1st edn., (CRC Press, 1999)
4. A. Factor, Mechanisms of Thermal and Photodegradations of Bisphenol A Polycarbonate, Polymer Durability, Chapter 5. Adv. Chem. **249**, 59–76 (1996). https://doi.org/10.1021/ba-1996-0249.ch005
5. A. Factor, Degradation of Bisphenol A Polycarbonate by light and γ-Ray irradiation, in *Handbook of Polycarbonate Science and Technology*, ed. by J. T. Bendler, 1st edn., (CRC Press, 1999)
6. W. Jang, Polym. Degrad. Stab. **86**, 419–430 (2004)
7. Thermochim. Acta **426**, 73–84 (2005)
8. Montaudo et al., Macromol **35**, 4297–4305 (2002)
9. W.D. Habicher, I. Bauer, J. Pospisil, Macromol. Sympos. **225**, 147–164 (2005)
10. D. Legrand, Annealing of polycarbonates, in *Handbook of Polycarbonate Science and Technology*, ed. by J. T. Bendler, 1st edn., (CRC Press, 1999)
11. H. de Brouwer, J. van den Bogerd, J. Hoover, Color stability of polycarbonate for optical applications. Eur. Polym. J. **71**, 558–566 (2015)
12. K. Gandhi, C.L. Hein, R. van Heerbeek, J.E. Pickett, Acceleration parameters for polycarbonate under blue LED photo-thermal aging conditions. Polym. Degrad. Stab. **164**, 69–74 (2019)
13. J Appl Polym Sci **13**, 2119 (1969)
14. Standard for Polymeric materials – Long term property evaluations, https://standardscatalog.ul.com/ProductDetail.aspx?productId=UL746b, Standard 746B, Edition 5. Edition Date: August 15, 2018
15. A. Rivaton, Recent advances in bisphenol-A polycarbonate photodegradation. Polym. Degrad. Stab. **49**, 163–179 (1995)
16. J.E. Pickett, Influence of photo-Fries reaction products on the photodegradation of bisphenol-A polycarbonate. Polym. Degrad. Stab. **96**, 2253–2265 (2011)
17. D. Bellus, P. Hrdlovic, Z. Manasek, Photoinitiated rearrangements in poly[2,2-propanebis(4-phenyl carbonate)]. J. Polym. Sci. Part C: Polym. Lett. **4**, 1e5 (1966) photofries
18. O. Kuvshinnikova, G. Boven, J.E. Picket, Weathering of ar-omatic engineering thermoplastics: Comparison of outdoor and xenon arc exposures. Polym. Degrad. Stab. **160**, 177–194 (2019)
19. Polym. Degrad. Stab. **2**, 203–223 (1980)
20. J.E. Pickett, Highly Predictive Accelerated Weathering of Engineering Thermoplastics. Mater. Test. Prod. Technol. News **35**(73) (2005)
21. J.E. Pickett, D.A. Gibson, S.T. Rice, M.M. Gardner, Effects of temperature on the weathering of engineering thermoplastics. Polym. Degrad. Stab. **93**, 684–691 (2008)

22. F. Gugumus, Light Stabilizers, in *Plastics Additives Handbook*, ed. by H. Zweifel, R. Maier, M. Schiller, 6th edn., (Hanser Publishers, 2009), pp. 139–424
23. M.Y. Mehr, W.D. van Driel, K.M.B. Jansen, P. Deeben, M. Boutelje, G.Q. Zhang, Photodegradation of bisphenol A polycarbonate under blue light radiation and its effect on optical properties. Optical Mater. **35**, 504–508 (2013)
24. M.Y. Mehr, W.D. van Driel, G.Q. Zhang, Reliability and life-time prediction of remote phosphor plates in solid-state lighting applications using accelerated degradation testing. J. Electron. Mater. **45**(1) (2016)
25. J. Meyer, F. Walther, Advanced test setup for accelerated aging of plastics by visible LED Radiation Moritz Hemmerich. Materials **13**, 4261 (2020)

Chapter 3
EMC Oxidation Under High-Temperature Aging

A. Inamdar, P. Gromala, A. Prisacaru, A. Kabakchiev, Y. Yang, and B. Han

1 Thermal Aging of Epoxy Molding Compound

Epoxy molding compounds (EMC) are commonly used for housing the semiconductor chip and other internal circuitry of electronic packages. EMC protects electronic components and systems from various environmental conditions such as mechanical shock, chemical loads, heat, and moisture [1]. EMC is a composite material consisting of epoxy resin as the matrix, silica as the filler material, hardeners, and other additives [2]. The cross-linking reaction between epoxy resin and hardeners forms the thermosetting polymer.

A typical EMC contains very high content of silica filler, which helps in attaining desired thermomechanical properties. Hardeners contribute to the heat resistance and storage stability. Additives such as cure-promoter and flame-retardants provide increased reaction between epoxy and hardener and lower risk of flammability, respectively. The constituent materials of EMC contribute to its high adhesion strength, low shrinkage, excellent chemical resistance, and heat resistance [3]. The ability of EMC to be molded, its mechanical and chemical properties, and a relatively low cost compared to traditional ceramic packaging make it a better alternative as an encapsulating material. In the automotive industry, EMC is often used to protect not only single integrated-circuit (IC) packages but also entire electronic control units (ECU). Compared to the consumer electronics, automotive electronics

A. Inamdar (✉)
Delft University of Technology, Delft, The Netherlands
e-mail: A.S.Inamdar-1@tudelft.nl

P. Gromala · A. Prisacaru · A. Kabakchiev
Robert Bosch GmbH, Automotive Electronics, Reutlingen, Germany

Y. Yang · B. Han
University of Maryland, College Park, MD, USA

W. D. van Driel, M. Yazdan Mehr (eds.), *Reliability of Organic Compounds in Microelectronics and Optoelectronics*,
https://doi.org/10.1007/978-3-030-81576-9_3

is exposed to harsher environments such as moisture and temperature variation, which cause aging of EMC.

Aging can be defined as the long-term, irreversible changes that occur in a material due to its exposure to ambient conditions. The changes can be in the structure, composition, and morphology of the material [4]. In general, aging may result from a combination of mechanical loading and chemical changes. In case of an electronic package, aging refers to the changes occurring in its constituent materials during its operation-lifetime. Figure 3.1 shows a schematic diagram of a typical electronic package of the type quad flat no-leads (QFN). Among all of its components, the one most exposed to the atmosphere is the EMC. Therefore, the most important aging effect is usually the oxidation of EMC through the contact with atmospheric oxygen [4]. This type of aging is governed by the exposure to elevated temperatures, and hence it is referred as thermal aging of EMC.

The operating temperature around the automotive electronics located near a combustion engine or transmission can be as high as 150 °C [5]. Such exposure to high temperatures causes oxidation of EMC. This can significantly change its mechanical properties [6–8] and thus can affect the reliability of electronic components. Yet, only limited information can be found in the literature, and it is the motivation of the present study. This study focuses on the following key aspects of EMC oxidation –

- Experimental study of the time and temperature dependent growth of EMC oxidation layer, followed by developing an activation-energy-based mathematical model;
- Characterization of the thermomechanical properties of fully oxidized EMC specimens using standard experimental tests;
- Moiré interferometry-based comparison of the thermomechanical behavior of a thermally aged and pristine molded package during a temperature cycle;
- Finite element simulation-based solder-fatigue analysis of thermally aged molded package to determine its effect on solder joint reliability.

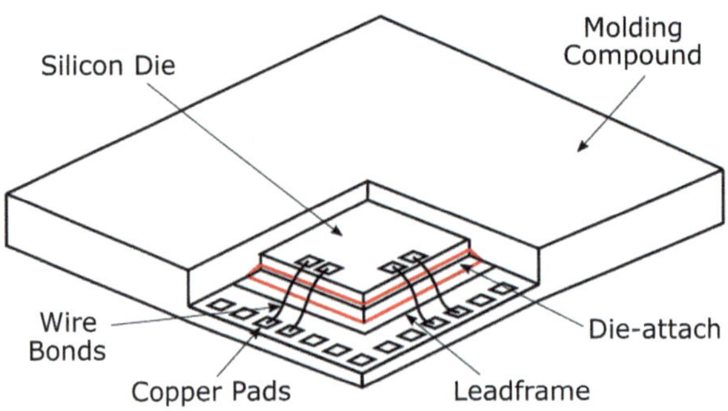

Fig. 3.1 Construction of a typical QFN package

2 Characterization of Oxidation Growth During Thermal Aging

EMC oxidation is a diffusion-based process, and thus the oxidation rate is time and temperature-dependent [9]. For complete characterization, the oxidation growth in EMC is measured as a function of time at various temperatures. The results are discussed to provide an insight into the oxidation process. The experimental data are utilized to develop an empirical model for oxidation growth.

2.1 Measurement of Oxidation Layer

EMC oxidation is time and temperature-dependent. It has been established that the oxidation layer grows with the increase in the ambient temperature, as well as the time-duration of its exposure to EMC [10]. Based on this qualitative information, a design of experiments was developed for quantifying the time-temperature dependency of the oxidation growth. A total of 24 specimens of EMC with a rectangular cross section (10 mm × 4 mm) were prepared for the test. Batches of eight specimens were subjected to three different high-temperature storage (HTS) conditions at 170 °C, 200 °C, and 230 °C, respectively. A single specimen was removed from each batch at 2, 4, 8, 24, 100, 500, 1000, and 1500 hours, and the thickness of the oxidation layer was measured.

After being removed from the HTS chamber, the specimen was sectioned and polished. Thickness measurements were performed using a fluorescence microscope equipped with ultraviolet illumination [11]. It is worth mentioning that an initial attempt was made using a digital microscope, but it was difficult to identify the oxide layer boundary clearly from the microscope images. Figure 3.2 shows representative fluorescence microscopy images, obtained from specimens aged for 1000 h at 170 °C, 200 °C, and 230 °C, respectively.

Under the fluorescence microscope, the pristine EMC was seen as green, while the oxide layer as light to dark brown. During aging at 230 °C, the oxide layer becomes dark brown, which is clearly distinguishable from pristine EMC. However, the specimens kept at lower temperatures show a prominent color gradient from brown to green, indicating the gradual transition from oxide layer to pristine EMC. In such cases, the gradient was closely observed, and the boundary was defined at the region where the brown portion of gradient ends and the green region begins.

For each specimen, 9 independent measurements of oxidation thickness were made and averaged to minimize the uncertainties associated with the boundary detection. A typical standard deviation was 1.1 μm, which established some guideline of thickness measurement accuracy.

The results are shown in Fig. 3.3. It is clear that the oxidation growth rate and thus the total thickness of the oxidized layer increases as the aging temperature

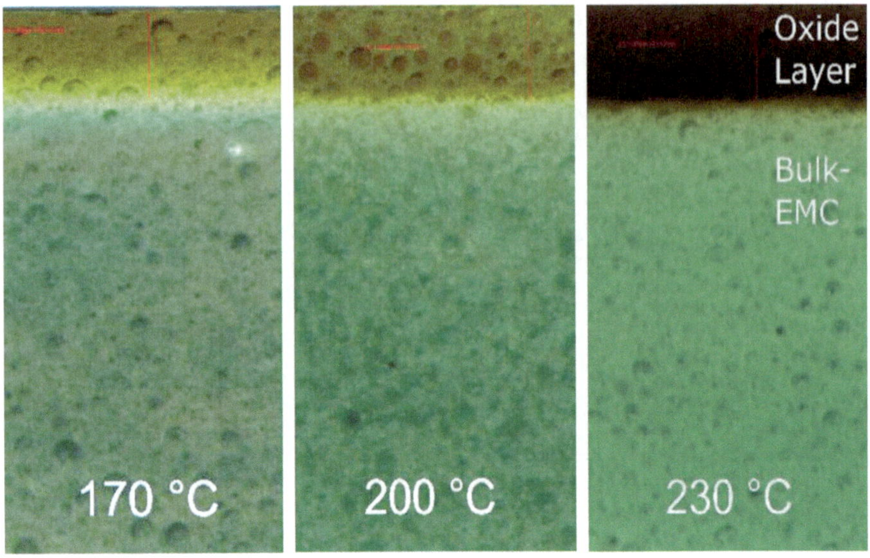

Fig. 3.2 Representative fluorescence microscope images obtained after 1000 h of thermal aging

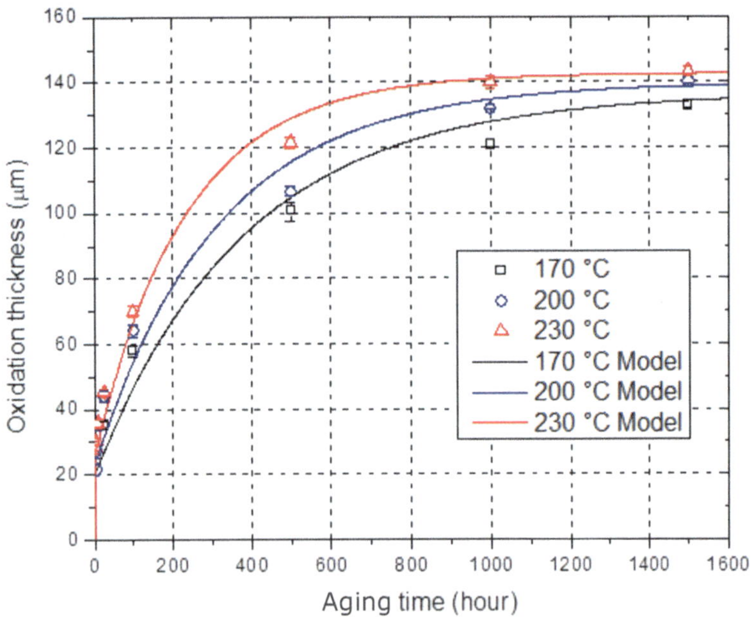

Fig. 3.3 Oxidation layer thickness as a function of aging time

increases. The thicknesses of the oxidized layer after 1500 h of aging are 133.0, 140.0, and 143.3 μm for the aging temperatures of 170 °C, 200 °C, and 230 °C, respectively.

It is important to note that the oxidized layer thickness increased rapidly during the first 30 h of aging. The thicknesses of the oxidized layer only after 24 h of aging were 35.5, 43.7, and 45.4 μm for 170 °C, 200 °C, and 230 °C, respectively. This is nearly 30% of the thickness of the oxidized layer after 1500 h.

2.2 *Behavior of Oxidation Growth*

Oxidation of EMC results from a combination of two processes – diffusion of oxygen and reaction of oxygen with EMC [8, 9]. Atmospheric oxygen diffuses into the porous network of EMC through its exposed surfaces. The diffused oxygen reacts with the matrix resin in EMC to form an oxide compound. As a result, a layer of oxidized EMC is formed along the exposed surface of the encapsulant. Both processes (i.e., diffusion and reaction) occur simultaneously, and the slower of these two determines the effective rate of oxidation.

When the reaction is slower than the diffusion, oxygen can diffuse into EMC beyond the oxide layer. As a result, the oxidation layer does not have a distinct separation from the pristine EMC, which is seen clearly in Fig. 3.2 in the fluorescence microscope image obtained at 170 °C. For the same reason, the diffused but unreacted oxygen could be found into the bulk of EMC [10]. On the other hand, when the reaction is faster than the diffusion, the diffused oxygen is consumed rapidly by the reaction. Furthermore, the oxidized layer hinders the diffusion, and thus the oxygen cannot penetrate beyond the oxide layer. One possible reason for this is the smaller diffusivity of the oxidized molding compound [12]. As a result, a more distinct oxide layer boundary can be observed. This can be seen clearly in Fig. 3.2 in the fluorescence microscope image obtained at 230 °C.

This can also be used to explain the rapid initial growth of the oxidized layer observed at all three aging temperatures. For a short period of time after the pristine EMC is placed inside the aging chamber, oxygen in the air diffuses freely into the pristine EMC, and is consumed simultaneously to produce the oxidized layer. As mentioned earlier, the oxidized layer slows down the oxygen diffusion process, which subsequently stabilizes the oxidation growth rate.

2.3 *Empirical Model for Oxidation Growth*

The measurements of oxidation layer thickness clearly show a trend of rapid initial growth, especially within the first 30 h of thermal aging. In addition, the growth rates decrease significantly after 500 h of storage. An empirical model, which can reproduce the aforementioned features, is proposed to predict the oxidation thickness at different values of temperature and aging time. The proposed model can be expressed as:

$$h(t) = A(T) + B\left(1 - e^{-C(T)t}\right) \tag{3.1}$$

$$A(T) = a_1 T + a_2 \tag{3.2}$$

$$C(T) = C_0\, e^{-\left(\frac{E_C}{RT}\right)} \tag{3.3}$$

Equation (3.1) defines the oxidation thickness, h, as a function of time t. In the equation, $A(T)$ is a temperature-dependent constant that describes the rapid initial growth of oxidation, and it can be expressed as a linear function of T in Kevin (Eq. (3.2)), where a_1 and a_2 are constants. B is a temperature-independent constant, while the temperature-dependent constant $C(T)$ is defined by the Arrhenius relationship (Eq. (3.3)), where C_0 is a constant, E_C is the activation energy, and R is the ideal gas constant.

The parameters A, B, and C must be determined in order to completely define the empirical model. Model parameters were determined using the measured data and standard nonlinear regression techniques. The values are summarized in Table 3.1, and the oxidation growth curves predicted by the model are shown in Fig. 3.3.

3 Effect of EMC Oxidation on Thermomechanical Properties

Polymeric materials such as epoxy molding compounds typically show different mechanical behavior at different temperatures. There exists a certain characteristic temperature beyond which they transform from a glassy state to a rubbery state, and thus, they exhibit different thermomechanical properties. Figure 3.4 shows a typical variation of thermomechanical properties against temperature.

The transformation from glassy to rubbery state can be easily understood by the rapid fall in the value of elastic (storage) modulus with respect to increase in temperature, as indicated in Fig. 3.4a. Similarly, the coefficient of thermal expansions

Table 3.1 Parameters of empirical model

Parameter	Value
a_1 (μm/K)	0.1
a_2 (μm)	−24.3
B (μm)	116.58
C_0 (1/hour)	0.1662
E_C (J/mol)	1.5426×10^4

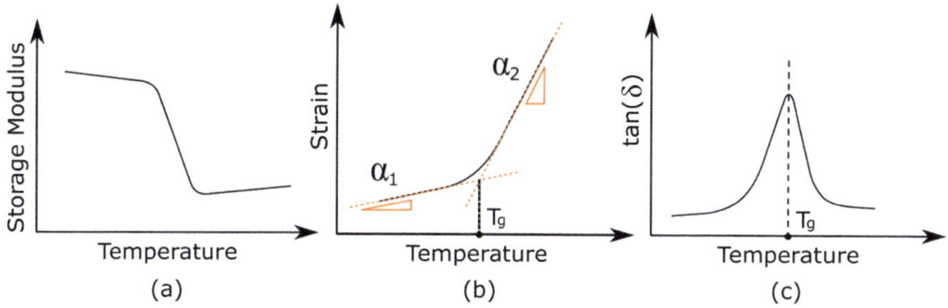

Fig. 3.4 Glass transition of polymeric materials – plots of (**a**) storage modulus, (**b**) thermal strain, and (**c**) tangent delta against temperature

(CTE), which is evaluated as the slope of thermal strain versus temperature plot as shown in Fig. 3.4b, also has 2 different values below and above the characteristic temperature. This phenomenon is known as glass transition, and the characteristic temperature is called as glass transition temperature T_g.

Three crucial thermomechanical properties of oxidized EMC are characterized – elastic (storage) modulus, E, glass transition temperature T_g, and coefficient of thermal expansion CTE. The standard experimental tests used for this characterization are – dynamic mechanical analysis (DMA) and thermomechanical analysis (TMA).

3.1 *Study of Partially Oxidized EMC Specimens*

Experimental study of the effect of thermal aging on thermomechanical properties of EMC can be found in the literature. In [10], EMC specimens were subjected to various sets of standard thermal aging processes, e.g., sold reflow, thermal cycling, high-temperature storage, etc. Similarly, in [12, 13], thermomechanical properties were characterized using EMC specimens, which were thermally aged (high-temperature storage) for different time duration.

Experimental characterization using these specimens shows anomalies. For example, instead of a typical single peak in the tan(δ) versus temperature plot (Fig. 3.4c), two local maxima were observed for the aged specimens. The second peak clearly indicates the presence of a second phase in the material, i.e., the oxide layer [13]. In other words, this suggests that the oxidation layer should also exhibit different thermomechanical behavior than the bulk of EMC.

It is important to note that these experimental tests were carried out on partially oxidized EMC specimens. In [10], purely analytical methods have been described for evaluating effective material properties of oxidized EMC on the basis of experimental data of the partially oxidized EMC specimens. In this study, fully oxidized specimens were prepared and tested to avoid any ambiguity in measured properties.

3.2 *Preparation of Fully Oxidized EMC Specimens*

Thin specimens (≈ 300 μm thickness) were fabricated by grinding molded parts (≈ 2 mm thickness) that were prepared for mechanical testing. The thin specimens are shown with the regular specimens in Fig. 3.5. The thin specimens were then stored at a constant temperature of 150 °C for 2000 h. Based on the data shown in Fig. 3.3, it was expected that complete oxidation occurred through the specimen, as oxidation occurred from both top and bottom surfaces of the specimen. This was later confirmed by observing the cross sections under the fluorescence microscope. The image of the cross section is shown in Fig. 3.6, where complete oxidation over the thickness is evident.

3.3 *Experimental Characterization of Thermomechanical Properties*

Thermomechanical properties of the fully oxidized EMC specimens are compared with those of the pristine EMC specimens.

3.3.1 Storage Modulus

The storage modulus was obtained using dynamic mechanical analysis (DMA) under three-point bending mode, which is illustrated schematically in Fig. 3.7. Storage modulus is proportional to the maximum elastic energy stored during one

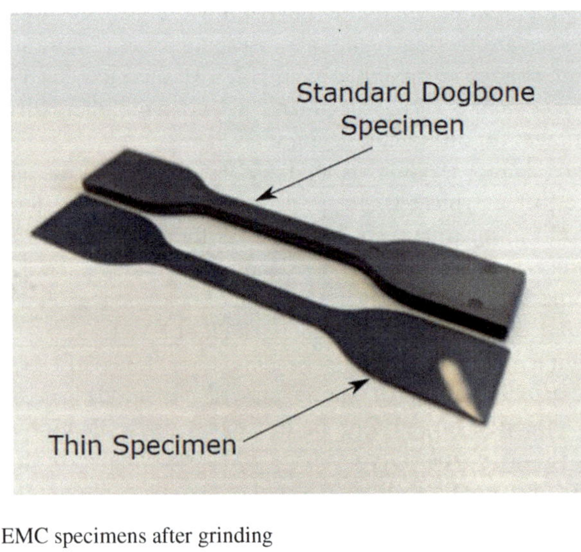

Fig. 3.5 Thin EMC specimens after grinding

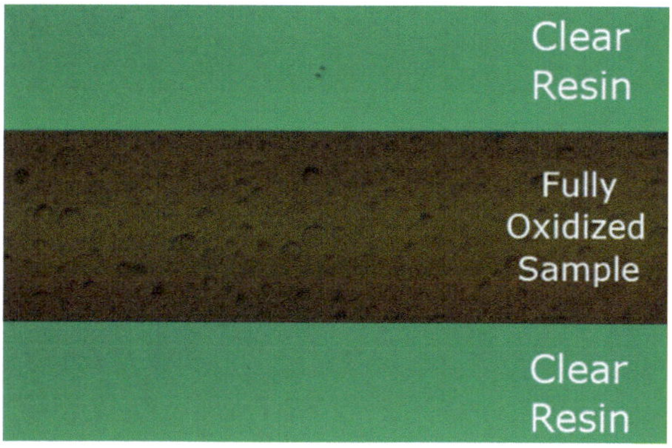

Fig. 3.6 Cross section of fully oxidized thin specimen

Fig. 3.7 Schematic diagram of DMA test in three-point bending mode

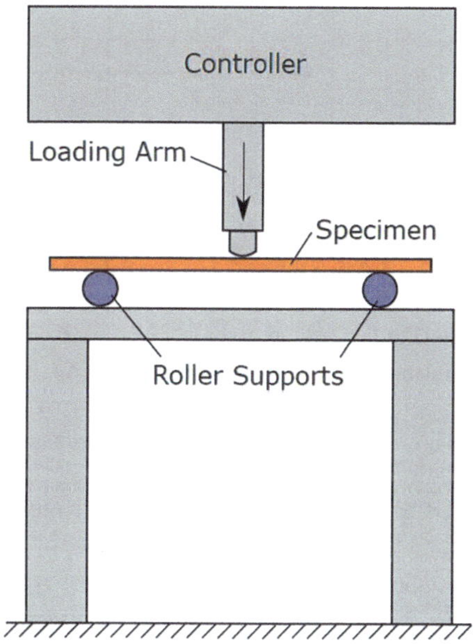

load period, and thus the plot of storage modulus versus temperature obtained from DMA is used for assessing the elastic properties of a material.

Figure 3.8 shows the temperature-dependent storage moduli of pristine and fully oxidized EMC specimens, obtained at a frequency of 10 Hz. The plot clearly indicates that the modulus of EMC increased significantly for the entire range of temperatures after oxidization. For example, the modulus of oxidized EMC at room temperature (20 °C) increased by nearly 20%. It is important to mention that the beginning of the transition zone of oxidized EMC is shifted by nearly 80 °C, and the transition zone of oxidized EMC is widened as well.

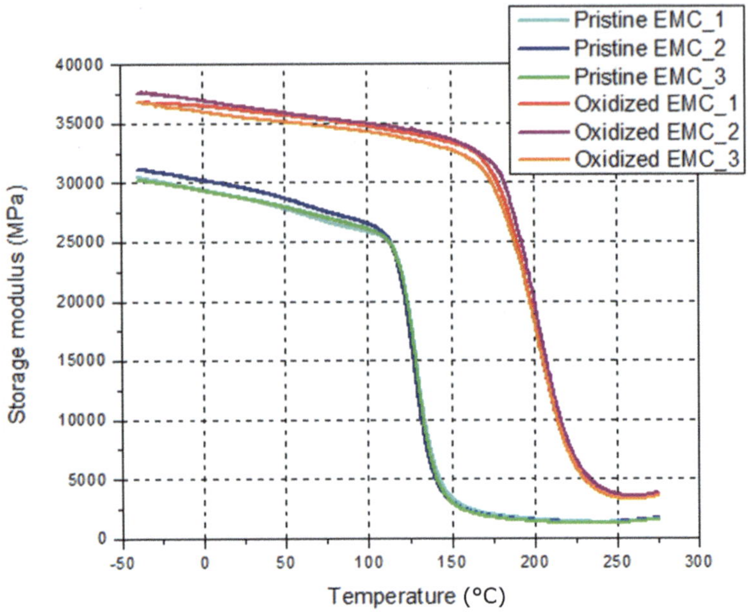

Fig. 3.8 Comparison of DMA results for pristine and oxidized EMC

3.3.2 Coefficient of Thermal Expansion

Coefficient of thermal expansion (CTE) can be measured routinely by Thermomechanical Analyzer (TMA). The specimens used for thermomechanical property characterizations are, however, too thin (≈ 300 μm) to be measured by TMA. A full-field displacement measurement technique called digital image correlation (DIC) was employed instead to measure the CTE of fully oxidized and pristine EMC specimens. The experimental setup of DIC is shown schematically in Fig. 3.9. An integrated heating plate provides an accurate temperature control, and two cameras record the images of a specimen surface. The procedure to measure CTE using the DIC technique is straightforward. Two images of a specimen are captured before and after temperature change, and then a DIC analysis proceeds with those images to determine thermally induced displacements [14], from which thermal strains, and thus CTE are determined.

Figure 3.10 shows a representative deformation field obtained at 120 °C, where the horizontal lines in black color are the paths used for strain calculations. Thermal strains obtained from DIC tests are shown in Fig. 3.11. The glass transition temperature, T_g, and the CTEs below and above T_g (α_1 and α_2, respectively) were determined from the strains, and the results are summarized in Table 3.2.

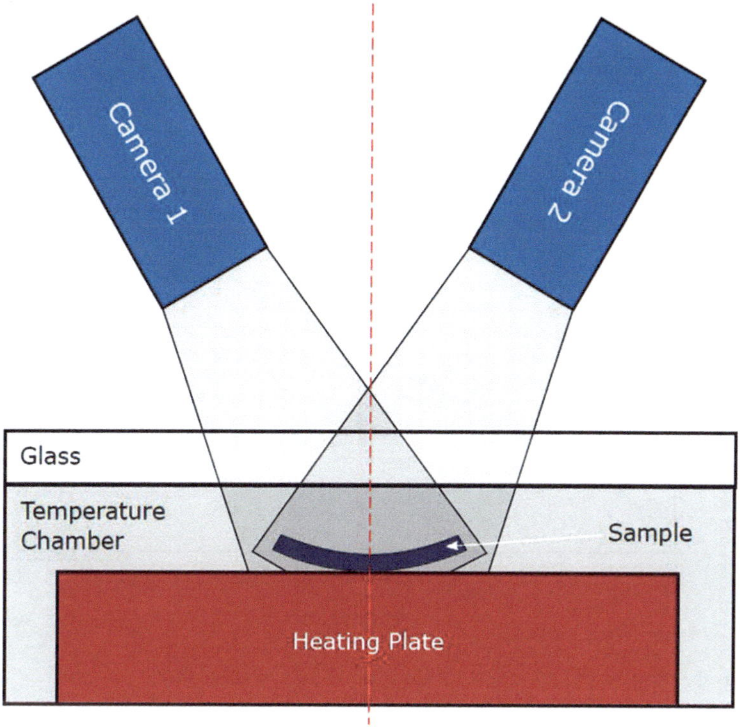

Fig. 3.9 Schematic illustration of the DIC setup

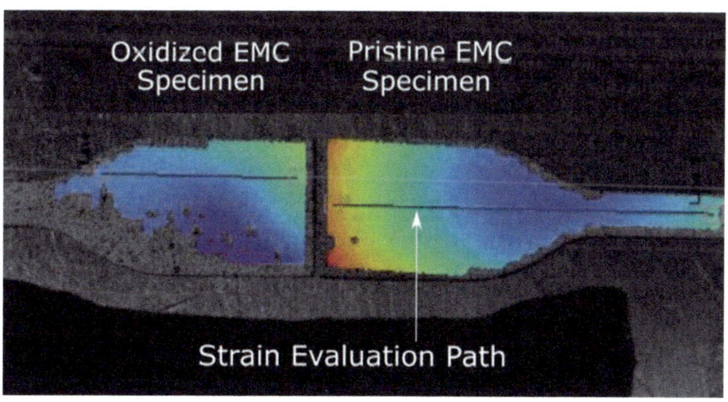

Fig. 3.10 Evaluation of thermal strain at 120 °C along a path using DIC software

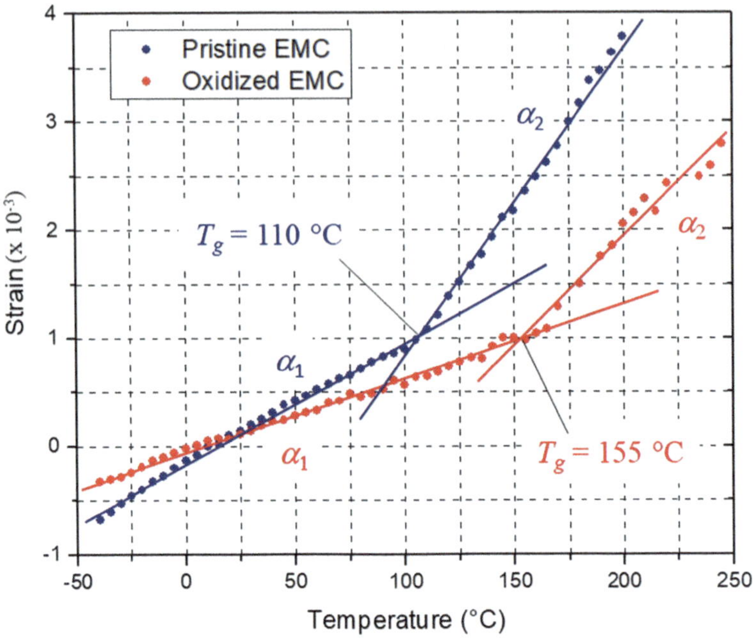

Fig. 3.11 Comparison of DIC results for pristine and fully oxidized EMC

Table 3.2 CTE values for pristine and fully oxidized EMC

	T_g (°C)	α_1 (ppm/°C)	α_2 (ppm/°C)
Pristine EMC	110	11.4	30.3
Oxidized EMC	155	6.5	20.2

As expected from the DMA results, oxidized EMC shows a significantly higher value of T_g (an increase of 45 °C). The CTE values of oxidized EMC are much lower than the corresponding values of pristine EMC for the entire range of temperatures; the CTEs of oxidized EMC have become 57% and 66% of those of pristine EMC below and above T_g, respectively. It is worth noting that the largest CTE difference (23 ppm/°C) occurs in the temperature range between $T_g^{pristine}$ and $T_g^{oxidized}$.

These results highlight that the thermomechanical properties of oxidized EMC are significantly different from those of pristine EMC, and the properties of oxidized EMC must be considered when reliability of devices subject to a long period of high temperature operating conditions is to be assessed. A next section is devoted to an investigation of the effect of the oxidation layer on the thermomechanical behavior of a package.

4 Effect of EMC Oxidation on Thermomechanical Behavior of Package

An electronic package typically consists of multiple layers of different materials, molding compound being the outermost one. Due to the differences in the thermomechanical properties of different materials involved in the multi-layered structure of a package, temperature changes can lead to mismatch in the thermal expansion of individual components. This causes warping of the entire structure and stresses are generated within as well as along the interfaces of different components. These stresses can lead to different failure mechanisms, and thus their quantification is an important aspect of reliability assessment of an electronic system.

Thermal aging of an electronic package causes oxidation of EMC along its exposed surfaces. Due to the significantly different thermomechanical properties of oxidized EMC, the behavior of the entire electronic package can also be affected significantly. The effect of oxidation layer on the thermomechanical behavior of an actual package is studied by an optical technique called moiré interferometry.

4.1 Preparation of Thermally Aged Package Specimens

Quad flat no-leads (QFN) packages were used for testing. The dimension of the package is 12 mm × 12 mm × 0.85 mm. The internal structure of the QFN package is schematically shown in Fig. 3.12a. For an automobile with a combustion engine, the operating temperature of electronic components located near an engine can be as high as 150 °C [5]. In real world usage, a consumer car is used for about 10 h a week [15, 16]. Thus, in just 4 years, the components would be subjected to a high operating temperature for more than 2000 h. Therefore, the QFN package specimens were thermally aged at 150 °C for 2000 h.

After thermal aging process, cross sections of a few specimens were prepared using the same procedure described before for the measurement of oxidation layer thickness. Figure 3.12b shows the cross-section observed under the microscope. The presence of a thick oxide layer as well as a gradient from fully oxidized to pristine EMC is clearly visible. The measured thickness is ≈ 150 μm, which matches reasonably well the value calculated using the empirical model (132 μm).

Two thermally aged specimens as well as two pristine QFN packages were analyzed by moiré interferometry. The packages were cut in half along one of the planes of symmetry, and ground flat for specimen grating replication.

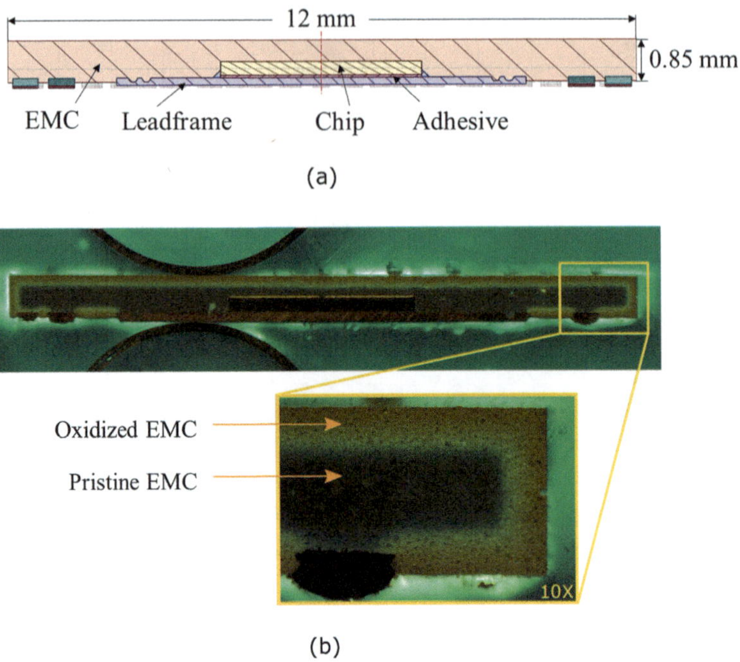

Fig. 3.12 Cross section of QFN package specimen – (**a**) internal structure (**b**) oxidation layer due to thermal aging at 150 °C for 2000 h

4.2 Experimental Analysis Using Moiré Interferometry

Real-time moiré interferometry was employed to investigate the effect of the oxidized layer on the package deformations. Moiré interferometry is a full-field optical technique to measure the in-plane deformations with high sensitivity [17], and it has been used widely for electronic packaging design and reliability assessment [18]. The outputs are the contour maps of in-plane displacements, which are related to the fringe orders by:

$$U(x,y) = \frac{1}{f} N_x(x,y) \quad V(x,y) = \frac{1}{f} N_y(x,y) \tag{3.4}$$

where N is the fringe order and f is the frequency of the virtual reference grating. In routine practice, a virtual reference grating with a frequency of 2400 lines/mm is used, which provides a contour interval of 417 nm per fringe order.

The optical/mechanical configuration used in the study is illustrated in Fig. 3.13. It consists of (1) a portable engineering moiré interferometer that provides two sets of virtual reference gratings and (2) a conduction chamber built on a

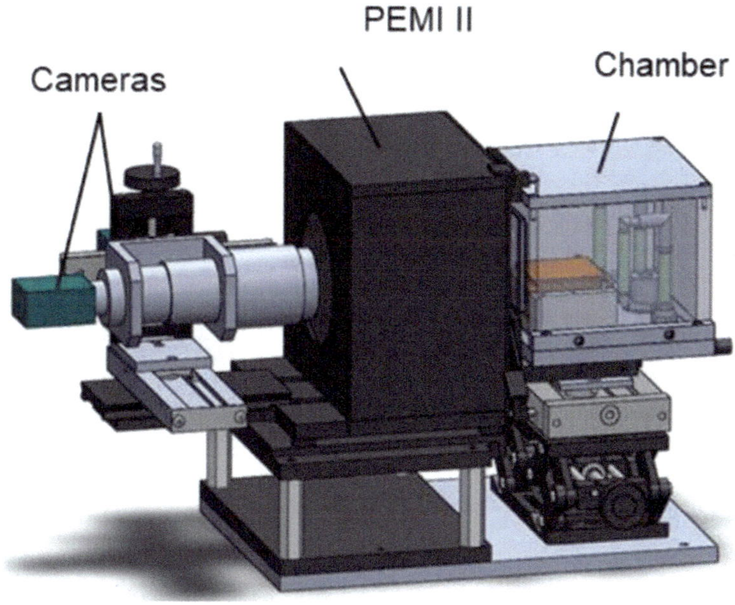

Fig. 3.13 Schematic illustration of the optical/mechanical configuration for moiré interferometry

high-performance thermo-electric cooler that provides accurate temperature control. The thermal conduction chamber is mounted on an *x*-*y*-*z* translation stage, which allows positioning as well as focusing the specimen. More details of the system can be found in Ref. [19].

The specimen gratings with a frequency of 1200 lines/mm were replicated on the cross sections at room temperature (20 °C). The details about the grating and replication procedures used in this study can be found in [17, 20]. The packages with the gratings were then heated to 85 °C and 150 °C, and the thermal deformations were documented by the moiré system. It is to be noted that the first temperature (85 °C) is below the T_g values of both pristine and oxidized EMCs, while the second one (150 °C) lies between the T_g values of the pristine and oxidized EMC.

The *U* and *V* fringe patterns of the right half of the package, representing the in-plane displacements in the *x* and *y* directions, respectively, are shown in Fig. 3.14, where the contour interval is 417 nm per fringe. The displacements were evaluated along the path defined in the top inset of Fig. 3.15 to investigate the effect of oxidized EMC on the package behavior. The results are shown in Fig. 3.15a, b for 85 °C and 150 °C, respectively.

As expected, the results at 85 °C show similar behaviors. At 150 °C, however, two packages exhibit radically different behaviors. The *U* displacement of the pristine package is much larger than that of the oxidized package, while the *V* displacement of the pristine package is significantly smaller than that of the oxidized package. The larger *U* displacement of the pristine package at 150 °C is attributed to the fact that the testing temperature is higher than T_g of the pristine EMC but

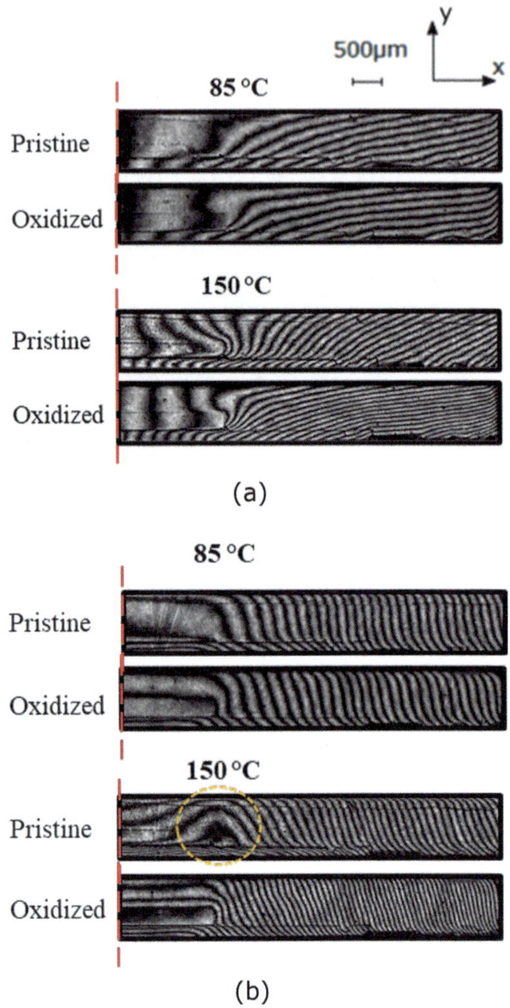

Fig. 3.14 (**a**) U-field and (**b**) V-field displacements of the right half of the package, obtained at 85 °C and 150 °C, where the contour interval is 417 nm per fringe

below T_g of the oxidized EMC. The relatively low CTE of the oxidized EMC also contributed to the large difference.

The most peculiar difference was seen in the *V* displacements at 150 °C. The *V* or bending displacements are caused mainly by the CTE mismatch between the EMC and the leadframe. The CTE of the leadframe is 16.2 ppm/°C, which is much larger than α_1 (the CTE below T_g) of both pristine and oxidized EMCs. As a result, heating from room to 85 °C produced upward bending (∪). The pristine EMC passes T_g (110 °C) during the second stage of heating to 150 °C. The CTE above T_g, α_2, is 30.3 ppm/°C, which is larger than the leadframe CTE. Consequently, the second stage of heating produced downward bending (∩), and thus the net bending displacement of the pristine package was reduced at 150 °C (from 11 μm to 9 μm). The

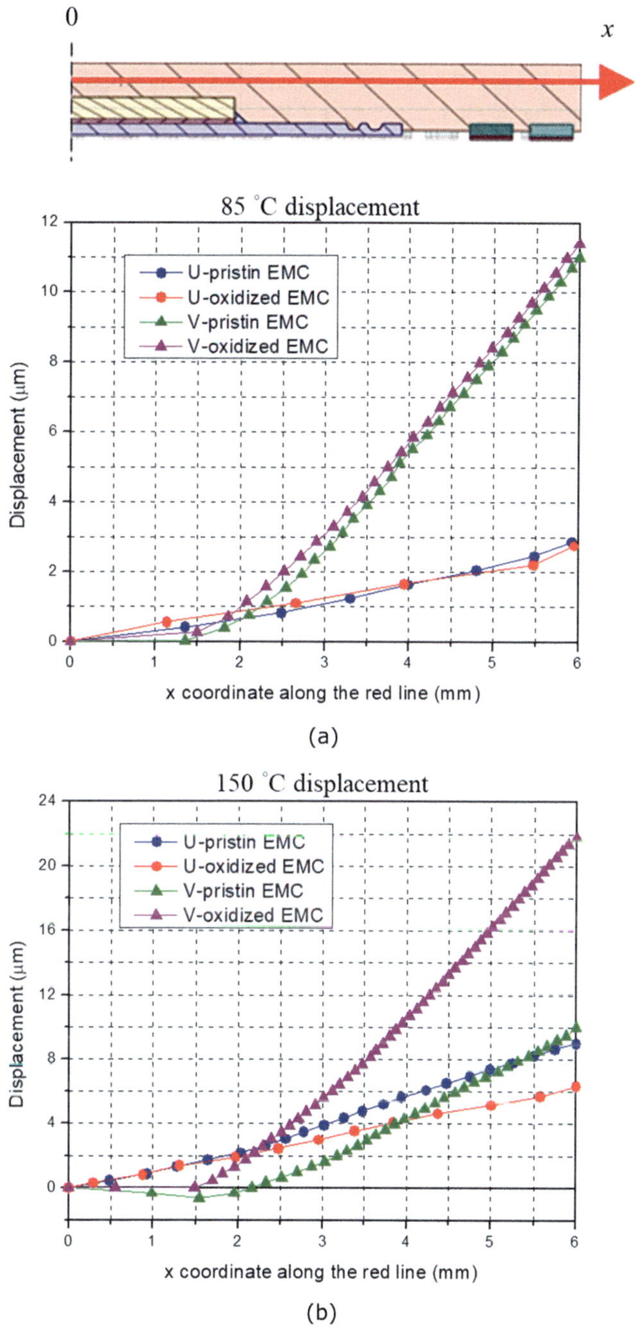

Fig. 3.15 U- and V-displacement at (**a**) 85 °C and (**b**) 150 °C along the red line

unusual bending displacement near the chip, marked by a dotted circle in Fig. 3.14b, was caused by this complex loading condition.

In the oxidized package, the package behavior was controlled by the thick oxide layer surrounding the package, as shown Fig. 3.12b due to its high modulus. Although a large amount of pristine EMC was still present inside, the large modulus of the oxidized EMC did not allow the pristine EMC to deform freely. In addition, the oxidized EMC remained below T_g (155 °C) during the second stage of heating, i.e., the CTE remained as α_1. The net result was that the bending displacements were proportional to ΔT. The bending displacement at 85 °C was 11 μm, which was caused by ΔT of 65 °C (= 85–20) and the bending displacement at 150 °C was 22 μm, which was caused by ΔT of 130 °C (= 150–20). This stiffer behavior must be considered when long-term board-level reliability is to be assessed.

5 Effect of EMC Oxidation on Reliability of Electronic Components

Electronic components have a wide range of failure modes, viz. packaging failure, contact failure, PCB failure, and semiconductor failure. Causes of failure range from a sudden change in temperature, excess current or voltage, mechanical shock, stress or impact, etc., to the fatigue caused by prolonged exposure of these parameters. Statistics show that 70% of the failures are related to packaging and assembly, and the primary reason behind them is the solder joint failure [21]. Moreover, about 45% of the failures are caused by the exposure of ambient thermal conditions [22]. Thus the failure of solder joint due to the temperature variation is one of the most important modes of failure.

5.1 Fatigue Failure of Solder Joints

Solder fatigue is the mechanical degradation of solder joint due to its deformation under cyclic loading. As a result of repeated temperature fluctuations, mechanical vibrations, or mechanical loads, fatigue failure occurs often at stress levels below the yield stress of the solder material. For this study, the effect of temperature cycles is considered.

An electronic package can experience various temperature cycles during its multi-step fabrication and assembly process as well as during its operation lifetime. Temperature variation causes warping of the package due to the CTE-mismatch among its constituent components. This leads to cyclic stresses within the package as well as within the solder joints. Standard temperature cycles from −40 °C and 150 °C were considered for the reliability assessment under solder-fatigue. There are several models for analyzing thermal fatigue failure of solder joints. They can be classified as – strain-based, stress-based, and energy-based models [22]. For this study, a strain-based model was considered.

EMC oxidation results into a much stiffer behavior of the QFN package under a temperature cycle. This change in behavior can significantly affect the solder fatigue. A finite-element simulation-based analysis was carried out to study the effect of oxidation layer on the solder joint reliability. A calibrated model of the QFN package including its geometry and material behavior was utilized. The procedure of modeling and calibration is discussed in this section.

5.2 *Modelling of Package Geometry*

An accurate solid model of the QFN package was prepared based on extensive measurements carried out using X-ray imaging and digital microscope. Cross sections of the package-PCB assembly at different locations were prepared. Most important dimensions such as chip dimensions, die-attach meniscus, EMC height, copper-pads dimensions, solder standoff height, etc., were measured carefully. Important features such as the skew shape of the solder joints, resulting from an offset between copper pads on the QFN package and on the PCB, were precisely replicated in the model. An average value of solder stand-off height was determined for the model by fitting a plane to the spatial plot of the solder-height data.

Subsequently, the solid model was further divided into multiple sub-components based on the requirements of stress-evaluation and finite-element mesh. In the ANSYS Workbench environment, a finite-element mesh was generated using a combination of 20-node homogeneous (hexahedral) structural solid and 10-node tetrahedral structural solid 3D elements.

5.3 *Modeling of Material Properties*

5.3.1 QFN Package

The QFN package consists of following materials – silicon, adhesive, copper, and molding compound. Two well-known materials – copper and silicon – were modeled as temperature-dependent linear elastic isotropic and orthotropic materials, respectively. Material behaviors of EMC and adhesive materials are highly viscoelastic. Based on the complexity of the simulation-based analysis and the available experimental data, an equivalent linear elastic model was proposed for each material. Experimental results discussed in the Sect. 3 of this chapter were utilized for preparing models of the pristine and oxidized EMC. The equivalent material models were calibrated using the thermomechanical behavior of QFN packages, in order to reproduce the effective viscoelastic behavior of these materials.

5.3.2 Calibration Using Moiré Interferometry

Experimental results from real-time moiré interferometry were utilized for the calibration. The models were modified over several iterations of finite-element simulations. Due to the symmetry, a quarter model of the QFN package with appropriate constraints replicating the experimental conditions was utilized. Calibration was carried out at 85 °C, 125 °C, and 150 °C in the same order. Figure 3.16 shows the simulation results of *U*-field and *V*-field displacement at 85 °C using the calibrated models. Further calibration was carried out at 125 °C and subsequently at 150 °C. The simulation results at 150 °C using the finalized model are shown in Fig. 3.17. Note that two different sets of models for adhesive, pristine-EMC, and oxidized-EMC were created for the case of pristine-package and aged-package, respectively.

During calibration, the elastic modulus of adhesive was significantly reduced (from 0.17 GPa to 0.01 GPa) after its T_g in both sets of models for replicating an effective viscoelastic relaxation. Similarly, the elastic modulus of pristine-EMC was significantly reduced to 0.15 GPa after its T_g, but only in the case of pristine-package. This allows reproducing the W-shaped deformation curve, as indicated in Fig. 3.17.

In contrast, for the case of aged-package, the much stiffer outer layer of the oxidized-EMC did not allow the pristine-EMC in the bulk to relax. Therefore, the corresponding model of pristine-EMC was left unchanged. Material models of the oxidized-EMC for both sets did not require any modification.

5.3.3 Solder and PCB

Solder joint was modeled as a viscoplastic material, and accumulated plastic strains after each thermal cycle were evaluated. Solder alloys can have different compositions, which can lead to different microstructures (i.e., variations in phases, grain size, etc.) [23]. For simplicity, the effect of microstructures was not considered. Instead, the bulk behavior of solder joint was calculated by a rate-dependent inelastic model with kinematic hardening.

Lastly, considering the construction of the PCB, it was modeled as a single substrate material with equivalent linear elastic orthotropic material properties. All material models were assigned to their respective components in ANSYS environment using an APDL script.

5.4 *Solder Fatigue Analysis*

5.4.1 Simulation Settings

Finite-element simulation was carried out using a quarter geometry of the package-PCB assembly with appropriate symmetry boundary conditions. Figure 3.18 shows the quarter geometry model of the QFN package. The oxidation layer within the

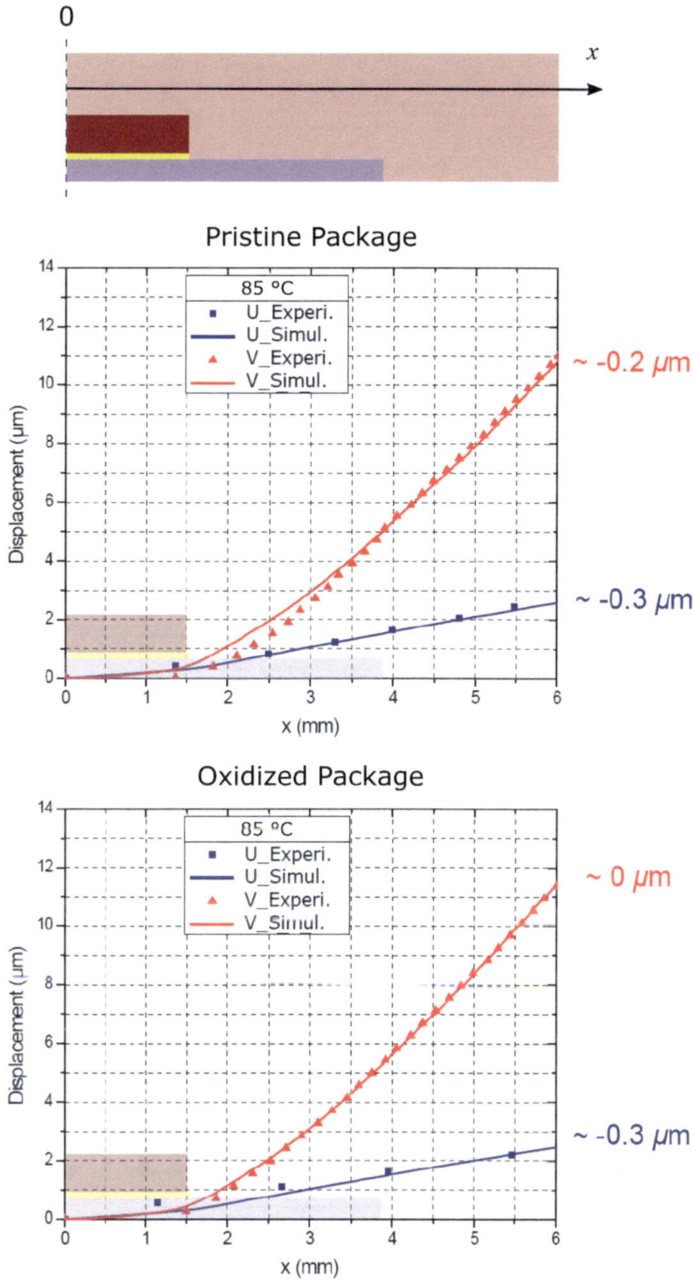

Fig. 3.16 Calibration of model at 85 °C using moiré interferometry simulation

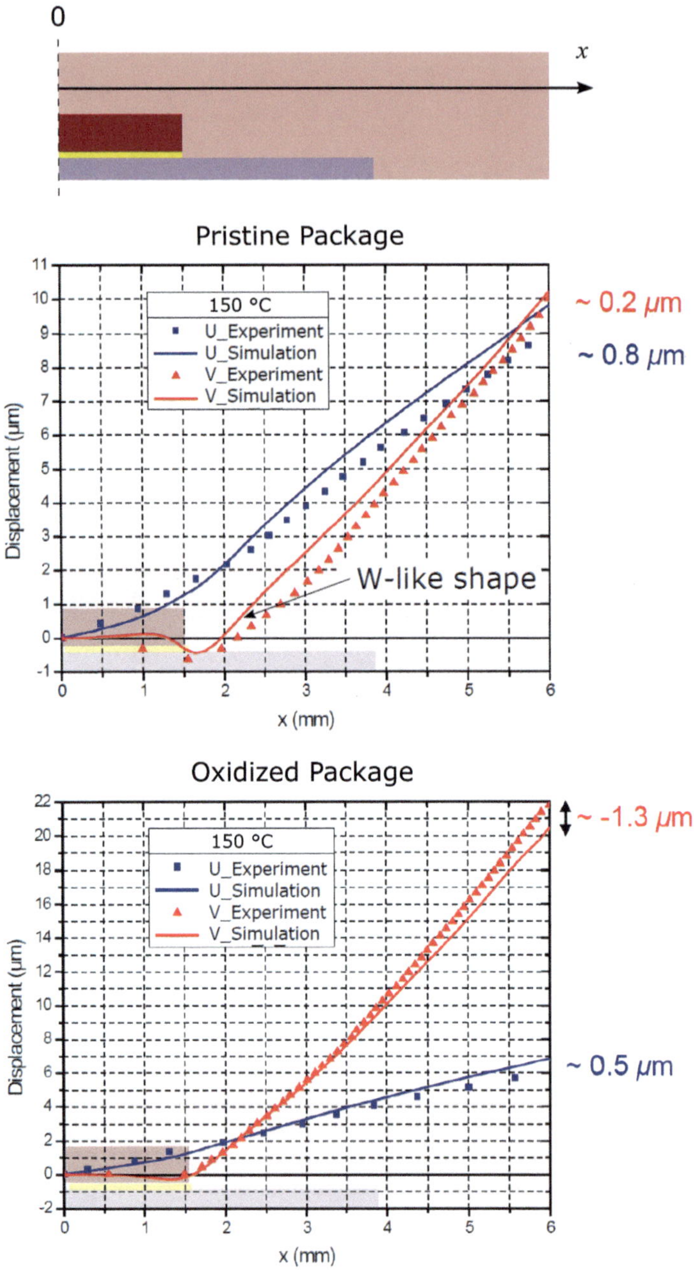

Fig. 3.17 Simulation results at 150 °C using finalized model

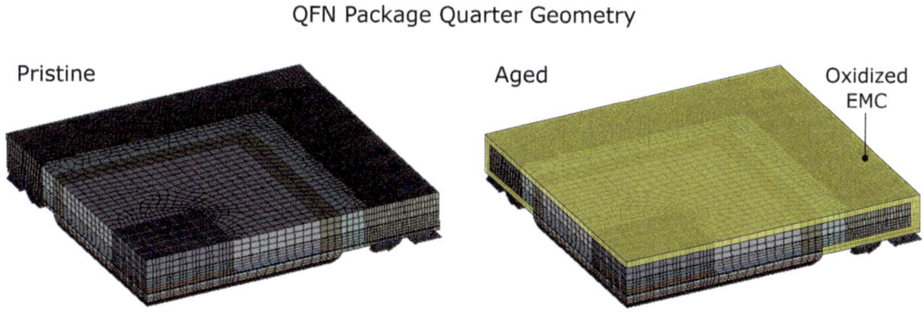

Fig. 3.18 Oxidation layer within the finite-element model of QFN package

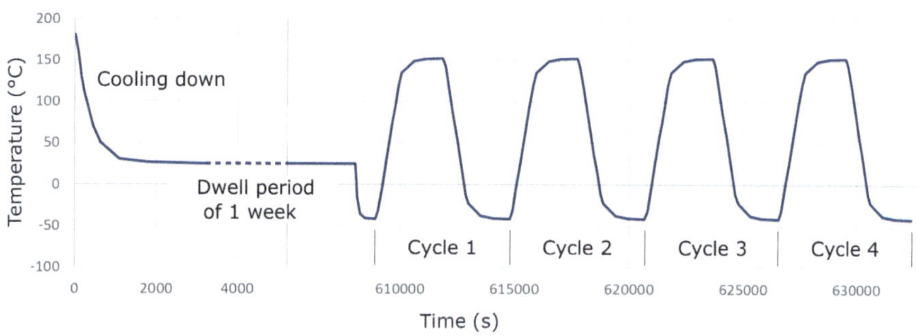

Fig. 3.19 Thermal condition for solder fatigue analysis

thermally aged QFN package was represented with a separate (150 μm thick) layer of elements within the EMC.

Figure 3.19 shows thermal loading conditions used in the analysis. It consists of three parts – (1) cooling down from 180 °C to 25 °C after the reflow soldering process, (2) dwell period of 1 week corresponding to the actual storage of components, (3) thermal cycles (−40 °C to 150 °C) used for standard reliability tests. It is important to note that the first two parts are necessary to replicate the real-world conditions, which introduce residual stresses within the solder joints and stress relaxation due to the prolonged storage, respectively.

5.4.2 Simulation Results

Fatigue failure of the solder material is caused by crack initiation and propagation, which is quantified as the accumulated fatigue damage, in this case – the accumulated plastic strain (APS) over several thermal cycles. A damage parameter for a solder joint is evaluated based on the APS of all of its elements. It is a volume-based weighted average of the APS, and thus represents a single value for the entire solder joint (SJ).

For this study, a damage parameter "DP20" was utilized, which is based on the finite elements with highest value of APS and constituting 20% of the SJ volume. Figure 3.20a shows an example of a solder joint, highlighting the elements selected for the calculation of DP20. Among all solder joints, indicated with numbers in Fig. 3.20b, SJ-9 showed the highest DP20 values in each temperature cycle, and thus it was monitored closely.

Convergence of the strain energy density accumulated during each cycle was studied in Ref. [24]. It was observed that regardless of the type of solder (with or without lead), at least 3 thermal cycles are required for the convergence of the averaged strain energy density. Therefore, more than 4 cycles were considered for this study. Figure 3.21 shows the DP20 value for SJ-9 after each thermal cycle.

Values of DP20 for both pristine and aged packages converged well after sixth cycle. Up to 8 cycles were studied to confirm this, and the values corresponding to eighth cycle were utilized for a further analysis. Damage parameter of SJ-9 in the pristine-package starts with a much higher value, but converges to a lower value than the aged-package. The comparison of converged values shows that the aged-package has about 0.06% more plastic strain accumulated per cycle than the pristine-package. Preliminary estimation of the fatigue life was carried out using the Coffin-Manson relation and the DP20 values of SJ-9 after eighth thermal cycle.

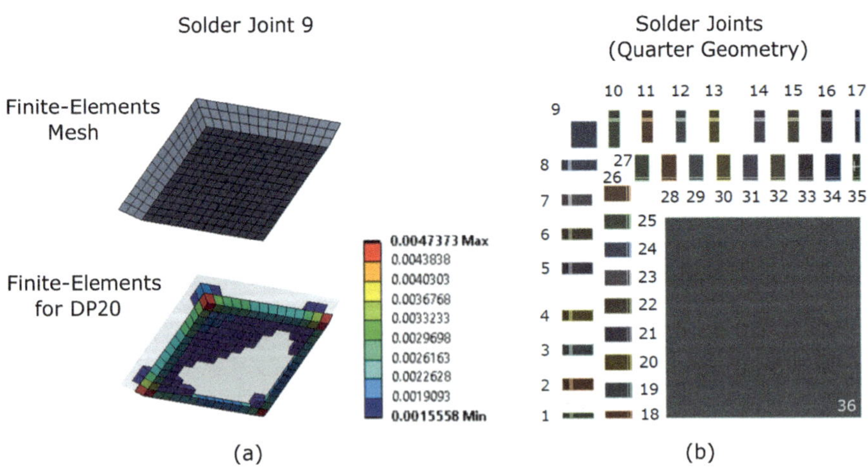

Fig. 3.20 Finite-element simulation – (**a**) evaluation of damage parameter for SJ-9 and (**b**) layout of solder joints for the quarter geometry of QFN package

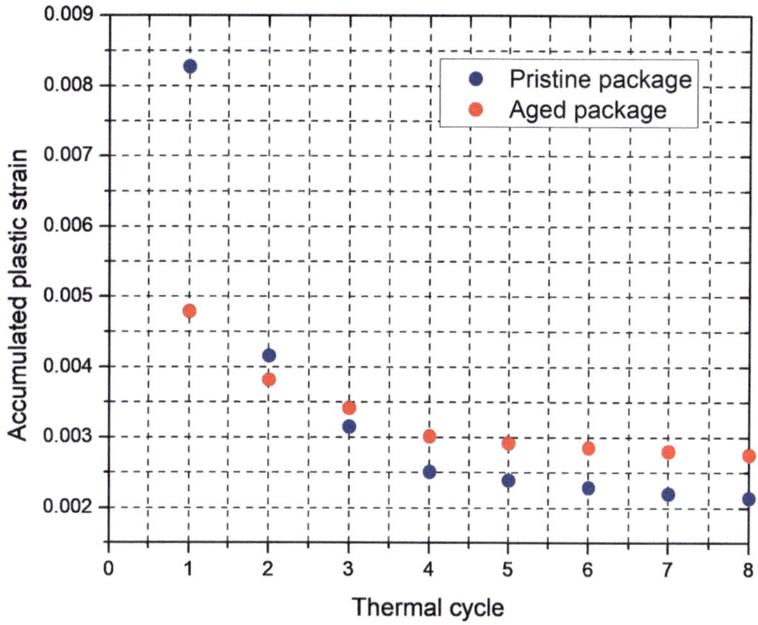

Fig. 3.21 Convergence of accumulated plastic strain values for SJ-9

Coffin-Manson model describes the relationship between the fatigue-lifetime of a material undergoing cyclic loading, and it is expressed as:

$$\left(\Delta\varepsilon^{pl}/2\right) - \varepsilon_f\left(2N\right)^C \tag{3.5}$$

where $\Delta\varepsilon^{pl}$ is accumulated plastic strain in one cycle; N is the number of cycles before failure (fatigue-lifetime); ε_f is fatigue ductility coefficient, which roughly corresponds to the true elongation at breaking point in a quasi-static tensile test; and C indicates fatigue ductility, which has value in the range [−0.7, −0.5] for ductile materials [25]. For the sake of the first order comparison, the value of C is assumed to be −0.6 for the solder material. Thus, the ratio of the fatigue-lifetime of solder joints between the pristine and aged packages can be given as:

$$\frac{N_{aged}}{N_{pristine}} = \left(\frac{\Delta\varepsilon^{pl}_{aged}}{\Delta\varepsilon^{pl}_{pristine}}\right)^{1/C} \approx \frac{2}{3} \tag{3.6}$$

Due to the same solder material used in pristine and aged package, corresponding ε_f values are the same. Substituting $\Delta\varepsilon^{pl}_{aged}$ and $\Delta\varepsilon^{pl}_{pristine}$ with the corresponding values of DP20 of SJ-9 after 8th thermal cycle, the ratio is approximately $\left(2/3\right)$.

The result implies that the oxidation layer would cause 33% reduction in the lifetime of the solder joints, which is significant.

6 Conclusion

Thermal aging of epoxy molding compound (EMC) leads to its oxidation. The process of oxidation and its effects on the material and package-level thermomechanical behavior as well as on the solder joint reliability was studied. The growth of an oxidation layer was first determined experimentally by subjecting EMC specimens to various aging temperatures. The results clearly showed the temperature-dependent rate increase. It was also observed that the thickness of the oxidation layer increased rapidly at the early stage of aging ($\approx$ 30 h). An activation energy-based empirical model was proposed for predicting oxidation layer thickness as a function of aging time and temperature, and the model constants were determined using the experimental results.

In the second task, thin EMC specimens were prepared to produce fully-oxidized test specimens, and they were tested to obtain three crucial thermomechanical properties of oxidized EMC. The results clearly showed significant changes in properties after oxidization, specifically (1) an increase of the elastic modulus, (2) a decrease of CTE, and (3) an increase of T_g. As the third task, the effect of oxidation layer on the thermomechanical behavior of an actual package encapsulated by EMC was investigated by an optical technique called moiré interferometry. A thermally aged package exhibited a much stiffer behavior under a thermal cycle, compared with a pristine package. This behavior change should be considered for accurate reliability assessment.

Lastly, the effect of EMC oxidation on thermal fatigue of the solder joints was studied using finite-element analyses. A set of calibrated models for the behavior of a pristine and an aged QFN package were utilized. The ratio of the number of cycles for fatigue-failure between aged and pristine QFN was evaluated based on the converged accumulated plastic strain using the Coffin-Manson model. The results showed that the oxidation layer could cause as much as 33% reduction in the solder joint life.

Acknowledgments A part of the work has been performed in the project ArchitectECA2030 under grant agreement No. 877539. The project is co-funded by grants from Germany, Netherlands, Czech Republic, Austria, Norway and Electronic Component Systems for European Leadership Joint Undertaking (ECSEL JU).

A part of the work was funded by the Federal Government of Germany, Framework Program for Research and Innovation 2016-2020 Microelectronics – Innovation Drivers of Digitalization (Rahmenprogramm der Bundesregierung für Forschung und Innovation 2016-2020 Mikroelektronik aus Deutschland – Innovationstreiber der Digitalisierung) – Project smartSTAR (No. 16ES0965). This funding is gratefully appreciated.

References

1. K.M.B. Jansen, J. de Vreugd, L.J. Ernst, C. Bohm, Thermal aging of molding compounds, 11th International Conference on Electronic Packaging Technology & High Density Packaging 2010, (August 2010), pp. 778–780
2. E. Nguegang, J. Franz, A. Kretschmann, H. Sandmaier, Aging effects of epoxy moulding compound on the long-term stability of plastic package, 11th International Conference on Thermal Mechanical and Multiphysics Simulation and Experiments in Micro-Electronics and Micro-Systems EuroSimE 2010, (April 2010), pp. 1–6
3. M. Linec, B. Mušič, The Effects of Silica-Based Fillers on the Properties of Epoxy Molding Compounds. Materials **12**(11), 1811 (2019)
4. C.J.G. Plummer, Testing of Polymeric Materials, in *Comprehensive Materials Processing*, (Elsevier Publication, 2014), pp. 35–70
5. E.H. Amalu, N.N. Ekere, R.S. Bhatti, High temperature electronics: R&D challenges and trends in materials, packaging and interconnection technology, 2nd International Conference on Adaptive Science & Technology (ICAST), (December 2009), pp. 146–153,
6. J. de Vreugd, K.M.B. Jansen, L.J. Ernst, C. Bohm, R. Pufall, High temperature storage influence on molding compound properties, 11th International Conference on Thermal Mechanical and Multiphysics Simulation and Experiments in Micro-Electronics and Micro-Systems EuroSimE 2010, (April 2010), pp. 1–6
7. J. de Vreugd, A. Sanchez Monforte, K.M.B. Jansen, L.J. Ernst, C. Bohm et al., Effect of postcure and thermal aging on molding compound properties", 11th Electronics Packaging Technology Conference 2009, (December 2009), pp. 342–347
8. R. Li, D. Yang, P. Zhang, F. Niu, M. Cai, G. Zhang, Effects of high-temperature storage on the elasticity modulus of an epoxy molding compound. Materials **12**(4), 684 (2019)
9. A. Quintana, M.C. Celina, Overview of DLO modeling and approaches to predict heterogeneous oxidative polymer degradation. Polym. Degrad. Stab. **149**, 173–191 (2018)
10. E.N. Ngnetiwe, Experimental and Simulation-based Investigations on the Influence of Thermal Aging and Humidity on the Warpage of Molded Plastic Packages, PhD Thesis, Institute for Micro Integration, University of Stuttgart, (2013)
11. F. Rost, Fluorescence Microscopy, Applications, in *Encyclopedia of Spectroscopy and Spectrometry* (Third Edition, 2017), pp. 627–631
12. B. Zhang, M. Johlitz, A. Lion, L. Ernst, K.M.B. Jansen, D.K. Vu, L. Weiss, Aging of Epoxy Moulding Compound - Thermomechanical Properties during High Temperature Storage, 17th International Conference on Thermal, Mechanical and Multi-Physics Simulation and Experiments in Microelectronics and Microsystems (EuroSimE), (April 2016), pp. 1–6
13. J. de Vreugd, The Effect of Aging on Molding Compound Properties, PhD Thesis, Delft University of Technology (TU Delft), (2011)
14. B. Pan, K. Qian, H. Xie, A. Asundi, Two-dimensional digital image correlation for in-plane displacement and strain measurement: a review. Measurement Sci. Technol. **20**(6) (2009)
15. A. Stetzer, Americans Spend More Time Behind the Wheel, Despite Reduction in Daily Mobility, According to Ketchum's Daily Ride Index, Ketchum. Retrieved from https://www.prnewswire.com/news-releases/americans-spend-more-time-behind-the-wheel-despite-reduction-in-daily-mobility-according-to-ketchums-daily-ride-index-300863559.html, 06 June 2019
16. G. Pasaoglu, D. Fiorello, A. Martino, G. Scarcella, A. Alemanno, A. Zubaryeva, C. Thiel, Driving and Parking Patterns of European Car Drivers - a Mobility Survey, Scientific and Technical Research Reports, European Commission, Publications Office of the European Union, (2012)
17. D. Post, B. Han, P. Ifju, High Sensitivity Moire: Experimental Analysis for Mechanics and Materials, Mechanical Engineering Series, Springer, 1994. (Student edition, 1997)
18. B. Han, Thermal stresses in microelectronics subassemblies: quantitative characterization using photomechanics methods. J. Therm. Stress. **26**(6), 583–613 (2003)

19. B. Wu, B. Han, Advanced mechanical/optical configuration of real-time moire interferometry for thermal deformation analysis of fan-out wafer level package. IEEE Trans. Comp. Packaging Manuf. Technol. **8**(5) (2018)
20. B. Han, Recent advancements of moire and microscopic moire interferometry for thermal deformation analyses of microelectronics devices. Exp. Mech. **38**(4), 278–288 (1998)
21. R. Tilgner, Physics of failure for interconnect structures: An essay. Microsyst. Technol. **15**(1), 129–138 (2009)
22. X. Li, R. Sun, Y. Wang, A Review of Typical Thermal Fatigue Failure Models for Solder Joints of Electronic Components, IOP Conference Series: Materials Science and Engineering, Volume 242, 2017 and 3rd International Conference on Applied Materials and Manufacturing Technology (ICAMMT 2017) 23–25, (June 2017)
23. W. Engelmaier, *Solder Joints in Electronics: Design for Reliability* (Workshop Notes, Engelmaier Associates, Inc., 1999)
24. C. Han, B. Han, Board level reliability analysis of chip resistor assemblies under thermal cycling: A comparison study between SnPb and SnAgCu. J. Mech. Sci. Technol. **28**, 879–886 (2014)
25. M. Rocotta, Simple expressions to estimate the Manson-Coffin curves of ductile cast irons. Int. J. Fatigue **78**, 38–45 (2015)

Chapter 4
Peridynamic Modeling of Thermo-oxidative Degradation in Polymers

P. Roy, D. Behera, E. Madenci, and S. Oterkus

1 Introduction

Extensive and critical applications of microelectronics require accurate failure prediction of encapsulating materials for improved service life under harsh environments. The encapsulation protects the components of the microcircuit from external mechanical loads such as impact, chemical and ultraviolet ray exposure and heat, etc. Thus, the encapsulating material plays an important role in maintaining the integrity and functioning of the microcircuit during its service. Degradation of encapsulation may result in malfunctioning of the device; thus, leading to severe consequences in many critical applications in aerospace and medical fields. One of the common encapsulation materials is epoxy molding compounds (EMC) due to its affordability and processing efficiency. EMC is also preferred due to its high adhesion strength, high resistance to chemical and moisture exposure, high heat resistance, and superior electrical properties. During service, the encapsulation material may experience degradation due to high temperature, moisture absorption, and oxygen diffusion. Significant changes in material properties of EMC may occur under high-temperature storage condition in the presence of oxygen. Oxidation of EMC leads to scission and crosslinking of molecular chains in microscale which results in increase in elastic modulus, shrinkage of the material, and material embrittlement.

P. Roy · D. Behera · E. Madenci (✉)
Department of Aerospace and mechanical Engineering, University of Arizona, Tucson, AZ, USA
e-mail: madenci@arizona.edu

S. Oterkus
Department of Naval Architecture, Ocean and Marine Engineering, University of Strathclyde, Glasgow, UK

W. D. van Driel, M. Yazdan Mehr (eds.), *Reliability of Organic Compounds in Microelectronics and Optoelectronics*,
https://doi.org/10.1007/978-3-030-81576-9_4

Therefore, design with polymers for long-term service under high temperature requires consideration of factors such as physical aging, chemical changes, and thermo-oxidative degradation. Also, the coefficient of thermal expansion, glass transition temperature, and fracture properties experience change such as decrease in fracture strain. Such changes in the material properties cause residual stress in the oxidized layer resulting in damage in the form of cracks.

Hence, understanding the long-term reliability of EMCs under extreme environments through experimental characterization and numerical simulations is essential. Several numerical and experimental studies have been carried out to understand the oxidation process and its effect on the mechanical response of materials. Cui et al. [1] analyzed thermo-oxidation diffusion of EMC using 2D finite element (FE) simulations of the oxidation diffusion. They calculated the distribution of elastic modulus in the oxidized layer for 150 °C aging temperature and compared it with the experimental observations. Li et al. [3] performed dynamic mechanical analysis (DMA) analysis of EMCs under high-temperature storage (HTS) condition to measure change of elasticity modulus with time. EMCs with two different filling contents are studied for various temperatures and aging times. They observed an increase in the elasticity modulus with thermal aging. Molecular dynamics (MD) simulations are carried out by Fan et al. [2] to determine the interfacial energy between EMC and the copper substrate under thermal cycling. The MD simulations predicted that the cuprous oxide content in the copper substrate affects the interfacial adhesion between EMC and copper substantially which agrees with the experimental results. Mei et al. [15] used Dinger-Funk equation to determine the optimal packing of spherical silica fillers used in EMC. The effect of oxidation on viscoelastic properties of a commercial molding compound is studied by Jansen et al. [5]. It is observed that the storage modulus and the glass transition temperature increase with the aging time. They also investigated the shrinkage due to the oxidation. Experimental study on the mechanical properties of molding compounds under thermal aging is performed by Yang and Cui [8]. They measured the variation of the elastic modulus, coefficient of thermal expansion, and the glass transition temperature with different temperature and aging times using DMA and thermomechanical analysis (TMA). Through a 3D finite element analysis, they also analyzed the effect of aging on the stress-strain response of the electronic package. Zhang et al. [11] proposed an equivalent layer model which offers a simplified and efficient method to determine the effect of thermal aging process of EMCs on the stress state of the package. Also, they measured the shrinkage due to aging. Li et al. [3] investigated how glass transition temperature of EMC with fillers is affected by thermal oxidation under high-temperature air storage conditions. They observed two glass transition temperatures for unoxidized core and completely oxidized surface material which depends on the temperature and aging time. Yang et al. [6] analyzed how high-temperature aging affects the reliability of automotive electronics. They carried out experiments on aging effect in the thermomechanical properties and microstructural changes in the aged materials using DMA, TMA, and TGA methods. Significant effects on the rubbery modulus, glass transition temperature, and the strength are observed during aging process associated with shrinkage. They also performed a 2D finite element analysis and determined the effect of aging on the stress-strain response of the electronic package. Inamdar et al. [14] carried out experiments on oxidation growth rate and mechanical properties of thermally oxidized EMC. They measured the thickness of the oxidized layers for various temperatures and storage times using fluorescent microscopy and computed the activation energy for the oxidation growth rate. Also, they measured the coefficient of thermal expansion (CTE), glass transition temperature, and modulus of elasticity of thin samples under HTS using digital image correlation (DIC) and dynamic mechanical analysis (DMA) techniques. Vreugd et al. [9] studied how post-cure and thermal aging affect the mechanical properties of molding compounds through experiments. They found increase in the glass transition temperature, rubbery modulus, and shrinkage of the molding compound due to oxidation. Zhang et al. [12] performed experiments on a bimaterial sample made of molding compound and copper substrate under high-temperature storage (HTS) and temperature cycling (TC) conditions. They also carried out 3D FE simulations and validated their modeling assumptions. Manoharan et al. [13] studied the effect of aging on elastic modulus, coefficient of thermal expansion, and glass transition temperature for epoxy "green" mold compounds used in commercial devices through nanoindentation and TMA techniques, and computed the activation energy of the oxidized layer. Zou et al. [7] used real-time Moire interferometry technique to measure coefficient of thermal expansion (CTE) and glass transfer temperature of a molding compound using a vacuum thermal chamber and a computer imaging analysis system. They also investigated thermomechanical deformation of a biomaterial consisting of silicon and molding compound with interface delamination and furnished associated fringe pattern using this technique. Hölck et al. [10] used MDS to estimate the material properties of 3D networked epoxy molding compounds. They developed a 3D crosslinking scheme and analyzed interfaces of crosslinked epoxies to an amorphous silicon-dioxide surfaces. Zhang et al. [4] stored thick and thin EMC samples in the oven at 175 °C for several weeks and measured the thickness of the oxidation layer of partly and fully aged samples through cross-sectioning and fluorescence microscopy methods. The viscoelastic material parameters are measured using Dynamic Mechanic Analysis (DMA) tests. They measured the CTE as function of aging time for unaged and partly aged EMC samples and measured the shrinkage due to aging. They carried out FE simulations and observed how aging process affects the distribution of stress inside a package.

Pochiraju and Tandon [20] developed a model for high-temperature aging of carbon fiber-reinforced PMR-15 composites while considering the interaction between oxidation and damage. They observed that oxidation in fiber-reinforced composites is orthotropic with more oxidation growth in the fiber direction. They specifically analyzed oxidation growth in the transverse direction and oxidation in the presence of a discrete crack. Tandon et al. [19] simulated thermo-oxidative aging of neat resin specimens of high-temperature polyimide PMR-15 through a diffusion-reaction model that includes temperature, oxygen concentration, and weight loss effects. They compared one-dimensional simulation results on the evolution of the oxidative layer with the experimental observations. They also inferred the diffusivity of the oxidized material by comparing simulation results with the experimental observations. Tandon et al. [17] simulated thermo-oxidative aging of neat resin using a diffusion-reaction model also considering temperature, oxygen concentration, and weight loss effects. They considered anisotropy in the thermo-oxidative behavior through a unit cell analysis using the 3D FE model. A few parametric studies were also carried out and discussed how fiber and interface diffusivity may affect oxidation growth. Yu and Pochiraju [18] proposed a model for diffusion of moisture, oxygen, and nitrogen in polymer matrix composites considering the effect of fiber-matrix interphase on diffusivity. They also performed numerical simulations and compared the results with experimental observations. An et al. [16] studied the effect of metal coating and coating-polymer interface on the oxidative degradation of BMI resin and its unidirectional composites and observed that coating decelerate the oxidation.

The previous models are based on classical continuum mechanics (CCM) and solved using finite element methods for oxygen concentration, deformation, and failure prediction. Accurate prediction of thermo-oxidative response of EMC requires the consideration of multiple length scales, nonlinear material response, and emergence of cracks. Failure modeling within the CCM framework is cumbersome as it requires smoothness of the field variables and needs an external criterion for initiation and propagation of cracks. A nonlocal continuum theory, bond-based peridynamics (BB-PD), introduced by Silling [21] removes the smoothness requirement of the field variables through integro-differential equations. The BB-PD assumes finite distance interaction among material points through bonds. The extent of the interaction for a material point is known as horizon which also serves as an internal length parameter. Damage in BB-PD is introduced through breakage of bonds using a bond-breaking criterion based on critical stretch which allows for spontaneous emergence and propagation of cracks. Modeling of damage under thermo-oxidation is challenging due to the multi-physics and multi-scale nature of the problem. Recently, Madenci and Oterkus [22] developed a PD model for oxidative diffusion. BB-PD offers an efficient computational framework to analyze mechanics of deformation and oxygen concentration in a single framework.

This chapter presents a BB-PD model for coupled analysis of oxidation-diffusion and mechanical deformation of epoxy resins under isothermal condition. The critical stretch criteria is used to break the PD bonds for crack initiation and propagation. The capability of the model is demonstrated through numerical simulations on isothermal oxidation-diffusion in 977–2 epoxy resin, thermo-oxidation and shrinkage in PMR-15 resin strip with and without pre-cracks, and crack propagation in a 977–2 epoxy resin and copper bimaterial configuration under oxidation.

2 Thermal Oxidation

Oxidation of epoxy resin at high temperature is an irreversible process leading to changes in micro and macroscale properties. Reaction of oxygen with epoxy causes chain scission and oxidative cross-linking at macromolecular scale, thus, making the material denser. On the other hand, the oxidation and volatile compounds create voids within the material leading to weight loss. These effects result in irreversible volume shrinkage, increase in elastic modulus, decrease in coefficient of thermal expansion, glass transition temperature, and failure strain. As a consequence, these factors along with physical restraints induce residual stresses within the material.

2.1 Diffusion-Reaction Model

The rate of oxidation in epoxy resins is governed by two competing phenomena – oxygen diffusion in bulk material and chemical reaction of oxygen with epoxy. The diffusion-reaction equation for a material body Ω with boundary $\partial\Omega$ can be given as [17]

$$\frac{\partial C}{\partial t} = D(\phi,T)\nabla^2 C - R(\phi,T,C) \tag{4.1}$$

where $C(\boldsymbol{x}, t)$ is the molar concentration of oxygen, D the diffusivity, R the oxidation reaction rate of oxygen with epoxy, ϕ the oxidation state parameter, and T the absolute temperature. The initial and boundary conditions can be specified as

$$C(\boldsymbol{x},0) = C_0(\boldsymbol{x}) \tag{4.2a}$$

$$C(\boldsymbol{x},t) = C_s, \boldsymbol{x} \in \partial\Omega_D \text{ for boundaries exposed to oxygen} \tag{4.2b}$$

$$\nabla C \cdot \boldsymbol{n} = 0, \boldsymbol{x} \in \partial\Omega_N \text{ for symmetric and no flux boundaries} \tag{4.2c}$$

where $\partial\Omega = \partial\Omega_D \cup \partial\Omega_N$ with $C_0(\boldsymbol{x})$ being the initial molar concentration field of oxygen and C_s the oxygen concentration in exposed boundary, $\partial\Omega_D$.

2.2 Reaction Model

As introduced by Tandon et al. [17], the state of oxidation is denoted with a variable, ϕ. The value $\phi = 1$ represents the unoxidized state, $\phi = \phi_{ox}$ denotes the completely oxidized state, and $\phi \in (\phi_{ox}, 1)$ designates the intermediate state of oxidation as shown in Fig. 4.1.

The oxidation reaction rate, R models the reaction of oxygen with epoxy. It depends on the temperature, T, the oxygen concentration, C, and the oxidation state, ϕ as

$$R_c(\phi,T,C) = \begin{cases} R(T,C) & \text{for } \phi > \phi_{ox} \\ 0 & \text{for } \phi = \phi_{ox} \end{cases} \tag{4.3}$$

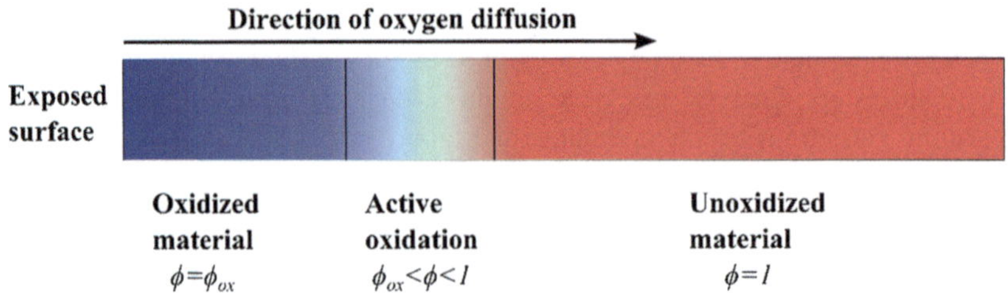

Fig. 4.1 Schematic view of state of oxidation in polymers

in which $R(T, C)$ represents the dependence of reaction rate on temperature, T, and the oxygen concentration, C. It is worth noting that a material point is completely oxidized when $\phi = \phi_{ox}$, and the reaction rate vanishes. The expression for $R(T, C)$ is expressed as

$$R(T,C) = R_0(T)\frac{2\beta C}{(1+\beta C)}\left[1-\frac{\beta C}{2(1+\beta C)}\right] \tag{4.4}$$

where, $R_0(T)$ is the maximum or saturation reaction rate in the presence of oxygen and the parameter β facilitates a dimensionless form of C. The temperature dependence of R_0 is modeled through an Arrhenius relationship as

$$R_0(T) = R_0^{'} \exp\left(-\frac{R_a}{R_u T}\right) \tag{4.5}$$

where $R_0^{'}$ is the rate constant, R_a is the activation parameter, and R_u is the universal gas constant.

2.3 Reaction Termination Model

The oxidation state parameter is a function of the weight loss due to oxidation, and its evolution can be expressed as

$$\frac{d\phi}{dt} = -\gamma(t)R(C,T) \tag{4.6}$$

where $\gamma(t)$ is the weight loss parameter and it is represented as

$$\gamma = \begin{cases} \gamma_{\max} - (\gamma_{\max} - \gamma_{\min})\dfrac{t}{t_{cr}}, & t < t_{cr} \\ \gamma_{\min}, & t > t_{cr} \end{cases} \tag{4.7}$$

where t_{cr} denotes the time after which $\gamma(t)$ remains constant. The values γ_{max}, γ_{min}, and t_{cr} can be determined from experiments.

2.4 Oxidation-Induced Chemical Strain and Stress

Shrinkage due to oxidation of epoxy causes strain on the material. The change in shrinkage strain can be quantified in terms of unoxidized (α^{un}) and oxidized (α^{ox}) shrinkage coefficients as

$$\Delta\boldsymbol{\varepsilon}_{sh} = \left[\alpha^{un}\left(\frac{\phi-\phi^{ox}}{1-\phi^{ox}}\right)+\alpha^{ox}\left(\frac{1-\phi}{1-\phi^{ox}}\right)\right]\Delta t\boldsymbol{I} \tag{4.8}$$

where a linear interpolation exists between α^{un} and α^{ox} to calculate effective shrinkage coefficient. It is worth noting that the effective shrinkage coefficient is zero in the absence of oxidation. The induced stress can be evaluated as

$$\sigma = \boldsymbol{C} : (\varepsilon - \varepsilon_{sh}) \tag{4.9}$$

where **C** is the fourth-order constitutive tensor.

2.5 Change in Material Properties

As experimentally observed [19, 23], the thermo-oxidative aging of epoxy resin changes the chemical composition and physical properties of the material. Specifically, the changes in elastic modulus and diffusivity can be quantified as

$$E(\phi) = E^{un}\left(\frac{\phi-\phi^{ox}}{1-\phi^{ox}}\right)+E^{ox}\left(\frac{1-\phi}{1-\phi^{ox}}\right) \tag{4.10}$$

and

$$D(\phi) = D^{un}\left(\frac{\phi-\phi^{ox}}{1-\phi^{ox}}\right)+D^{ox}\left(\frac{1-\phi}{1-\phi^{ox}}\right) \tag{4.11}$$

3 Bond-Based Peridynamics

The peridynamic (PD) theory introduced by Silling [11] is a non-local continuum approach, and it remains valid even in the presence of a discontinuity in the form of a crack. As shown in Fig. 4.2, material point $\boldsymbol{x}$ interacts with other material points $\boldsymbol{x}'$ in its horizon (domain of interaction), $H_{\boldsymbol{x}}$. The force density vector, $\boldsymbol{f}$ arising from the interaction between material points $\boldsymbol{x}$ and $\boldsymbol{x}'$ is aligned with the bond between these points. The force density vector depends on the bond stretch, s, bond constant, c, and the status parameter, μ. The stretch is defined as

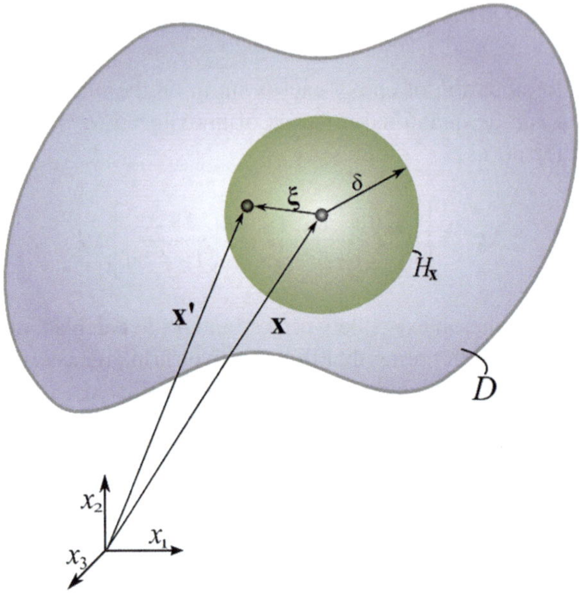

Fig. 4.2 PD material points with its horizon

$$s = \frac{|\boldsymbol{\xi} + \boldsymbol{\eta}| - |\boldsymbol{\xi}|}{|\boldsymbol{\xi}|} \tag{4.12}$$

where $\boldsymbol{\xi} = \boldsymbol{x}' - \boldsymbol{x}$ is the PD bond in the reference configuration, $\boldsymbol{\eta} = \boldsymbol{u}' - \boldsymbol{u}$ is the relative displacement of the PD bond, and the operator || denotes Euclidean norm.

The PD bond constant, c can be determined through comparing the variation of the strain energy densities from CCM and PD for simple loading conditions. For isotropic materials, under plane stress condition, it is of the form

$$c = \frac{9E_{bond}}{\pi h \delta^3} \tag{4.13}$$

in which h is the thickness and δ the radius of the horizon (extent of the interaction domain). The effective Young's modulus (E_{bond}) of the bond, $\boldsymbol{\xi}$ can be evaluated as

$$E_{bond} = \frac{2E_x E_{x'}}{E_x + E_{x'}} \tag{4.14}$$

where $E_x = E(\phi(\boldsymbol{x}))$ and $E_{x'} = E(\phi(\boldsymbol{x}'))$ are the Young's modulus at points $\boldsymbol{x}$ and $\boldsymbol{x}'$ respectively. It is worth noting that Poisson's ratio is restricted to 1/3 in bond-based PD for plain stress condition.

The variable μ is a history-dependent scalar-valued function that represents the status of the bond, i.e., whether the bond is intact or failed. The criterion for failure of the bond can be given as

$$\mu(\boldsymbol{x}-\boldsymbol{x}',t)=\begin{cases}1\,\text{if}\; s(\boldsymbol{x}'-\boldsymbol{x},t')<s_c\;\;\text{for all}\,0<t'<t\\ 0\;\text{otherwise}\end{cases} \tag{4.15}$$

where s_c represents the critical stretch of a bond. The damage at a point is defined by the ratio of number of broken bonds to the total number of bonds of a material point in its horizon as

$$\varphi(\boldsymbol{x},t)=1-\frac{\int_{H_x}\mu(\boldsymbol{x}'-\boldsymbol{x},t)\,dV'}{\int_{H_x}dV'} \tag{4.16}$$

3.1 Coupled Deformation and Oxidation

The equation of motion in BB-PD proposed by Silling [21] is given as

$$\rho(\boldsymbol{x})\ddot{\boldsymbol{u}}(\boldsymbol{x},t)=\int_{H_x}\boldsymbol{f}(\boldsymbol{x}'-\boldsymbol{x},\boldsymbol{u}'-\boldsymbol{u})\,dV'+\boldsymbol{b}(\boldsymbol{x},t) \tag{4.17}$$

where ρ is the mass density and $\boldsymbol{b}$ the body force vector. The force density vector can be expressed as

$$\boldsymbol{f}(\boldsymbol{u}'-\boldsymbol{u},\boldsymbol{x}'-\boldsymbol{x},t)=\mu c(s-s_{sh})\frac{\boldsymbol{\xi}+\boldsymbol{\eta}}{|\boldsymbol{\xi}+\boldsymbol{\eta}|} \tag{4.18}$$

in which s_{sh} is the shrinkage stretch due to oxidation. As proposed by Madenci and Oterkus [22], the rate of shrinkage stretch s_{sh} can be given as

$$\dot{s}_{sh}=\alpha^{un}\left(\frac{\phi-\phi^{ox}}{1-\phi^{ox}}\right)+\alpha^{ox}\left(\frac{1-\phi}{1-\phi^{ox}}\right) \tag{4.19}$$

3.2 PD Diffusion-Reaction Equation

As proposed by Madenci and Oterkus [22], the PD form of diffusion-reaction equation can be expressed as

$$\frac{\partial C}{\partial t}=\int_{H}\mu\left(\lambda\frac{C'-C}{|\boldsymbol{x}'-\boldsymbol{x}|}\right)dV'-R_c(\phi,T,C) \tag{4.20}$$

where the oxidation reaction rate, R_c can be evaluated using Eq. (4.3). Similar to PD elasticity parameters, the PD parameter for diffusion λ can be determined for 2D analysis as

$$\lambda = \frac{6D_{bond}}{\pi\delta^3 h} \tag{4.21}$$

where D_{bond} is the effective diffusivity of the bond. For a bond, $\boldsymbol{\xi}$, it is expressed as

$$D_{bond} = \frac{2D_x D_{x'}}{D_x + D_{x'}} \tag{4.22}$$

where $D_x = D(\phi(\boldsymbol{x}))$ and $D_{x'} = D(\phi(\boldsymbol{x}'))$ are the diffusivities at points $\boldsymbol{x}$ and $\boldsymbol{x}'$.

3.3 *Boundary Conditions*

The boundary conditions are prescribed using fictitious boundary layers as suggested by Madenci and Oterkus [24]. The width of the fictitious boundary layer can be taken as the radius of the horizon [25].

3.3.1 Dirichlet Boundary Condition

The condition of $C(x^*, y^*, t) = C_s$ is imposed on the boundaries exposed to oxygen. This is achieved by applying the following constraint condition on the fictitious region $\mathcal{R}_f$ along the boundary of the surface, $\mathcal{S}$ (Fig. 4.3) as

$$C_f\left(x_f, y_f, t+\Delta t\right) = 2C_s\left(x^*, y^*, t+\Delta t\right) - C\left(x, y, t\right) \tag{4.23}$$

with $\left(x^*, y^*\right) \in \mathcal{S}, \left(x_f, y_f\right) \in \mathcal{R}_f, \left(x, y\right) \in \mathcal{R}$.

Similarly, the displacement boundary conditions in both directions can be imposed as

$$u_f\left(x_f, y_f, t+\Delta t\right) = 2U^*\left(x^*, y^*, t+\Delta t\right) - u\left(x, y, t\right) \tag{4.24a}$$

and

$$v_f\left(x_f, y_f, t+\Delta t\right) = 2V^*\left(x^*, y^*, t+\Delta t\right) - v\left(x, y, t\right) \tag{4.24b}$$

For boundaries on roller support where the displacement constraint is applied only in the horizontal direction, and is allowed to move freely in vertical direction, the boundary conditions are employed as

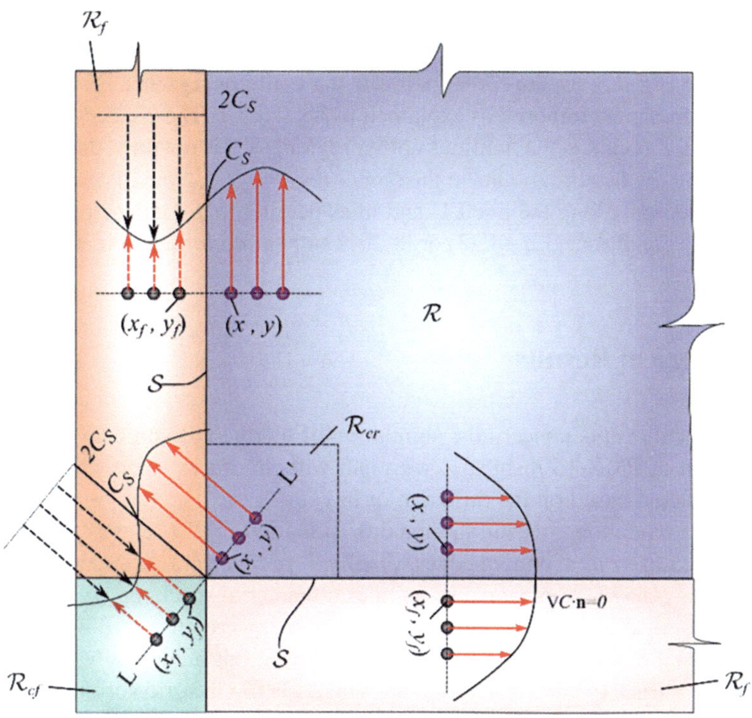

Fig. 4.3 Imposition of Dirichlet and no flux boundary conditions

$$u_f\left(x_f,y_f,t+\Delta t\right)=2U^*\left(\mathrm{x}^*,y^*,t+\Delta t\right)-u\left(x,y,t\right), \tag{4.25a}$$

and

$$v_f\left(x_f,y_f,t+\Delta t\right)=v\left(x,y,t\right) \tag{4.25b}$$

Similar approach can be used if displacements in vertical direction are constrained and in horizontal direction are free.

3.3.2 Zero Flux Boundary Conditions

Zero flux condition for oxygen concentration, i.e., $\nabla C \cdot \boldsymbol{n} = 0$ is enforced by applying constraint equations on material points in the fictitious region, $\mathcal{R}_f$ as.

$$C_f\left(x_f,y_f,t+\Delta t\right)=C\left(x,y,t\right) \tag{4.26}$$

3.3.3 Treatment of Corners

As shown in Fig. 4.3, for the corner points in the fictitious region, a Dirichlet boundary condition can be imposed as explained in Sect. 3.3.1. The concentration values of the material points in the fictitious corner region, $\mathcal{R}_{cf}$ is obtained through extrapolation using the material points in the corner region, $\mathcal{R}_{cr}$. Note that, the extrapolation is carried out along the line LL' and lines parallel to it. Depending on the type of boundary conditions, Eq. (4.23) or (4.26) can be used for extrapolation.

4 Numerical Results

The numerical results concern the simulation of thermo-oxidation and mechanical deformation of PMR-15 resin strip with and without cracks. Also, crack propagation is predicted based on the diffusion of oxygen in the material. Subsequently, it presents the oxidation concentration and deformation in 977–2 epoxy resin specimen exposed to oxygen on two edges. Finally, it presents crack propagation due to oxidation in 2D 977–2 epoxy resin – copper biomaterial specimen.

The numerical results are achieved based on a uniform discretization of the 2D computational domain. Spatial integration is carried out numerically by assigning an area $A_{(k)} = (\Delta x)^2$ to each material point with Δx being the grid spacing. A circular horizon of radius δ is used to construct family members associated with material point $\boldsymbol{x}_{(k)}$ which satisfy: $\|\boldsymbol{x}_{(j)} - \boldsymbol{x}_{(k)}\| < \delta$, where $\boldsymbol{x}_{(j)}$ represents the neighboring points. The area of the horizon of a material point, $\boldsymbol{x}_{(k)}$ is $H_{(k)} = N_{(k)}A_{(k)}$ where $N_{(k)}$ is the total number of material points inside the horizon. As the bond constants are derived in terms of classical parameters based on the assumption of circular horizon, a volume correction factor is required to be applied [26] due to discretization. In this study, the horizon size is taken as $\delta = 3.015\Delta x$. A forward Euler scheme is used for the time integration of the PD diffusion equation and an adaptive dynamic relaxation (ADR) procedure [27] is employed to solve for displacements under plane stress assumption. The ADR time step is considered to be unity. However, the real-time step for solving diffusion-reaction equation is chosen based on numerical stability condition. The bond-breaking criterion is based on critical stretch, i.e., when mechanical stretch exceeds the value of critical stretch, the interaction (bond) is removed.

4.1 Thermo-Oxidation in PMR-15 Resin Strip

A PMR-15 resin strip has length $L = 200\mu m$ and width $W = 20\mu m$. The specimen is exposed to oxygen along the left and right edges by specifying, $C = C_s$. No flux condition is prescribed on the top and bottom edges by imposing, $\nabla C.\,\boldsymbol{n} = 0$ where $\boldsymbol{n}$ is the outward normal vector to the boundary. For deformation analysis, top and bottom edges are considered to be fixed with left and right boundaries traction free

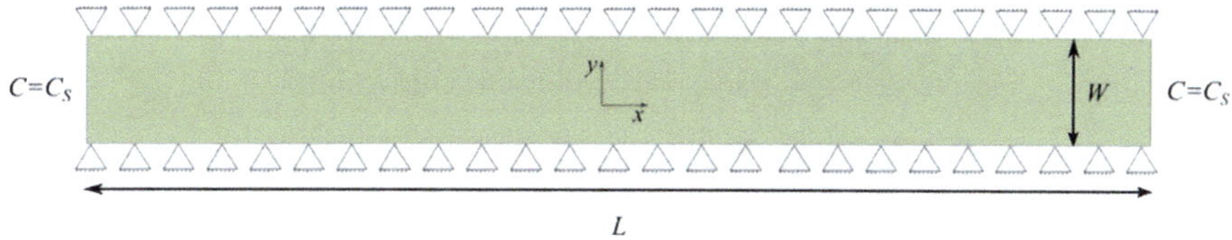

Fig. 4.4 PMR-15 specimen under thermo-oxidation

Table 4.1 Parameters of PMR-15 resin at 288°*C*

$D^{un}\left(\frac{mm^2}{h}\right)$	$D^{ox}\left(\frac{mm^2}{h}\right)$	$E_a\left(\frac{J}{mol}\right)$	$R\left(\frac{J}{molK}\right)$	$R_o'\left(\frac{mol}{m^3 \min}\right)$	
0.8×10^{-3}	1.17×10^{-3}	19700	8.31447	536.94	
$R_a\left(\frac{J}{mol}\right)$	$\beta(m^3/mol)$	φ_{ox}	γ_{max}	γ_{min}	$t_{cr}(h)$
23379.63	0.92	0.18	0.01	0.0033	40

as shown in Fig. 4.4. For numerical simulation, the domain is uniformly discretized into 4000 material points. The time step chosen for explicit scheme is $\Delta t = 0.005min$.

The parameters for diffusion-reaction equation are provided in Table 4.1 [17]. The strip is exposed to oxygen concentration of $C_s = 2\text{mol/m}^3$ under isothermal condition at 288°*C*. The unoxidized and oxidized Young's modulus PMR-15 resin are specified as E^{un} = 7.64GPa and E^{ox} = 8.74GPa, respectively. Shrinkage coefficients of unoxidized and oxidized material are $\alpha^{un} = -1.89\times10^{-6}\left(\frac{1}{h}\right)$ and $\alpha^{ox} = -8.29\times10^{-6}\left(\frac{1}{h}\right)$.

The oxygen concentration field at times 0.625*h*, 12.5*h*, and 125*h* are shown in Fig. 4.5. As expected, the oxygen diffuses into the material body as time progresses.

As the oxygen diffuses into the material, it reacts with epoxy. At a given time, three regimes – completely oxidized, active oxidation zone, and unoxidized can be identified from the oxidation state parameter field as shown in Fig. 4.6.

The horizontal and vertical displacement predictions at time 125*h* due to oxidation-induced shrinkage are shown in Figs. 4.7 and 4.8, respectively.

4.2 *Thermo-Oxidation in PMR-15 Resin Strip with Cracks*

Two symmetrical pre-existing cracks of length a = 25μm are introduced in the PMR-15 specimen considered in the previous section (Fig. 4.9). The crack surfaces are also exposed to oxygen. The oxidation-induced shrinkage due to the fixed top and bottom edge leads to tensile stretches in the bonds near the crack tip which results in propagation of crack. In this study, the critical stretch is considered as 1×10^{-6}. For numerical simulation, the domain is uniformly discretized into 4000

Fig. 4.5 Concentration predictions at times: (a) 0.625h, (b) 12.5h, and (c) 125h

Fig. 4.6 Oxidation predictions at times: (a) 0.625h, (b) 12.5h, and (c) 125h

Fig. 4.7 Horizontal displacement (m) prediction at 125h

Fig. 4.8 Vertical displacement (m) prediction at 125h

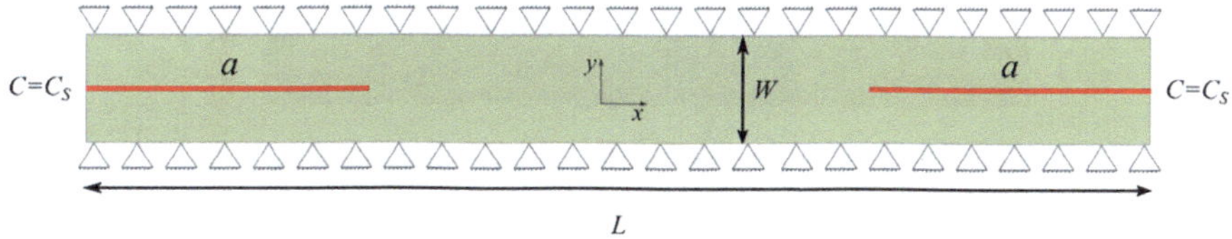

Fig. 4.9 PMR-15 specimen with symmetric cracks under thermo-oxidation

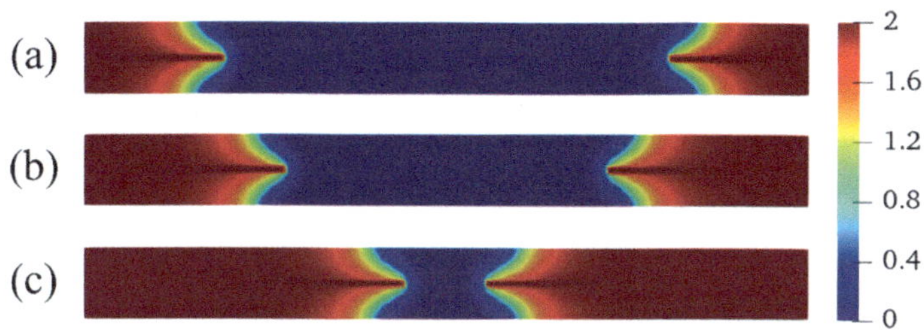

Fig. 4.10 Concentration predictions in presence of cracks at times: (a) 3.125h, (b) 6.25h and (c) 12.5h

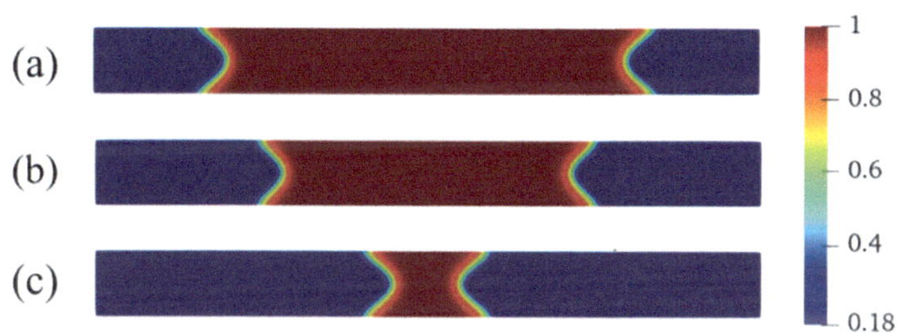

Fig. 4.11 Oxidation state predictions in presence of cracks at times: (a) 3.125h, (b) 6.25h and (c) 12.5h

material points. For numerical stability, the time step used to solve the diffusion-reaction equation is chosen as $\Delta t = 0.0025min$.

The oxygen diffuses into the material from left and right edges as well as from the crack surfaces. As the crack propagates, new surfaces are exposed to oxygen. In the numerical simulation, this is achieved by specifying the oxygen concentration to be $C_s = 2\text{mol/m}^3$ when damage at a point is greater than 0.4. Figure 4.10 shows the oxygen concentration at material points at times 3.125h, 6.25h and 12.5h.

The progress of oxidation state variable is shown in Fig. 4.11. It is observed that oxidation happens faster in cracked specimen compared to uncracked specimen which can be attributed to diffusion of oxygen from crack surfaces into the material body. Figure 4.12 shows the progression of oxidation-induced damage in the specimen. Oxidation leads to damage propagation through successive breaking of the bonds.

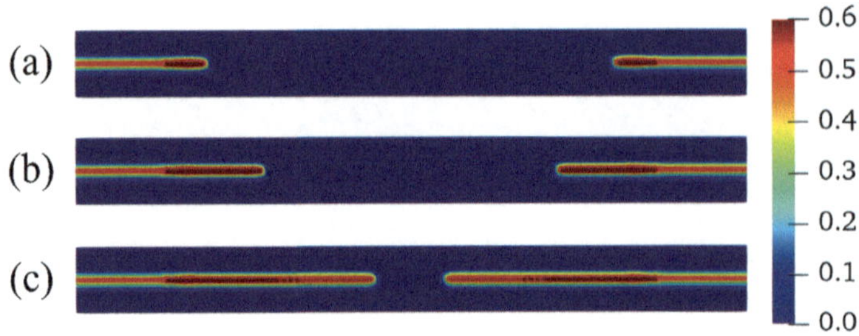

Fig. 4.12 Damage progression due to oxidation at times: (a) 3.125h, (b) 6.25h and (c) 12.5h

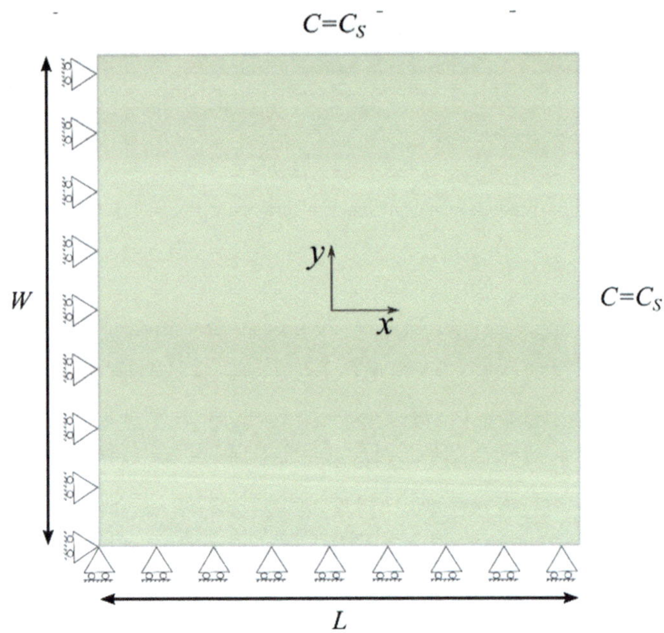

Fig. 4.13 Epoxy resin specimen exposed to oxygen on free edges

4.3 Thermo-Oxidation in 977–2 Epoxy Resin

As shown in Fig. 4.13, a square 977–2 epoxy resin specimen with $L = W = 500\mu m$ is subjected to oxidation on top and right edges. For left and bottom edges, $\nabla C. \boldsymbol{n} = 0$ is specified. The left and bottom edges are on roller support and the top and bottom edges are traction-free for displacement predictions.

Table 4.2 [1] shows diffusion-reaction parameters of 977–2 epoxy resin. The diffusivities of oxidized and un-oxidized resin are assumed to be the same, i.e., $D^{un} = D^{ox} = D$. Oxygen concentration of $C_s = 3.3\text{mol/m}^3$ is specified along the top and right boundaries. Oxidation is assumed to occur at $150^o C$. The Young's modulus of unoxidized and oxidized 977–2 epoxy resin are given as $E^{un} = 4.6\text{GPa}$ and $E^{ox} = 5.5\text{GPa}$, respectively. The shrinkage coefficients of unoxidized and oxidized

Table 4.2 Parameters of 977–2 epoxy resin at $150^{\circ}C$

$D\left(\frac{m^2}{s}\right)$	$E_a\left(\frac{J}{\text{mol}}\right)$	$R\left(\frac{J}{\text{molK}}\right)$	$R_o'\left(\frac{\text{mol}}{m^3\,\text{min}}\right)$
1.3×10^{-12}	19700	8.31447	536.94

$R_a\left(\frac{J}{\text{mol}}\right)$	$\beta(m^3/\text{mol})$	φ_{ox}	γ_{max}	γ_{min}	$t_{cr}(h)$
23379.63	0.15	0.18	0.01	0.0033	40

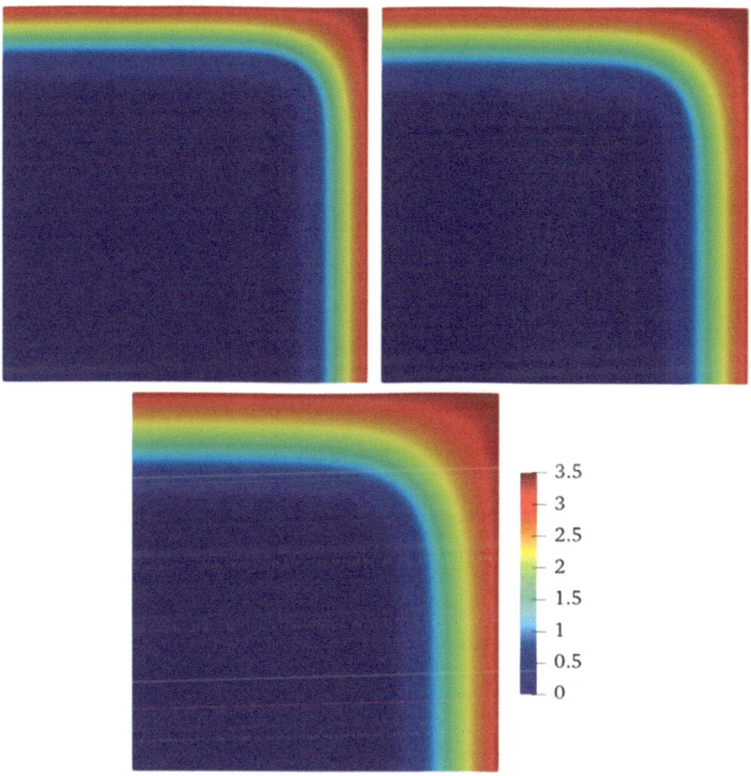

Fig. 4.14 PD predictions of concentration field at times: (a) 25h, (b) 50h and (c) 100h

resin are considered to be the same, i.e., $\alpha^{un} = \alpha^{ox} = -1.5\times10^{-5}\left(\frac{1}{h}\right)$. For numerical simulation, the domain is uniformly discretized into 10,000 material points. The time integration of diffusion-reaction equation is carried out with a time step, $\Delta t = 0.01min$.

Figure 4.14 shows the oxygen concentration at material points at times $25h$, $50h$ and $100h$. Note that as the time progresses, the oxygen seeps into the material

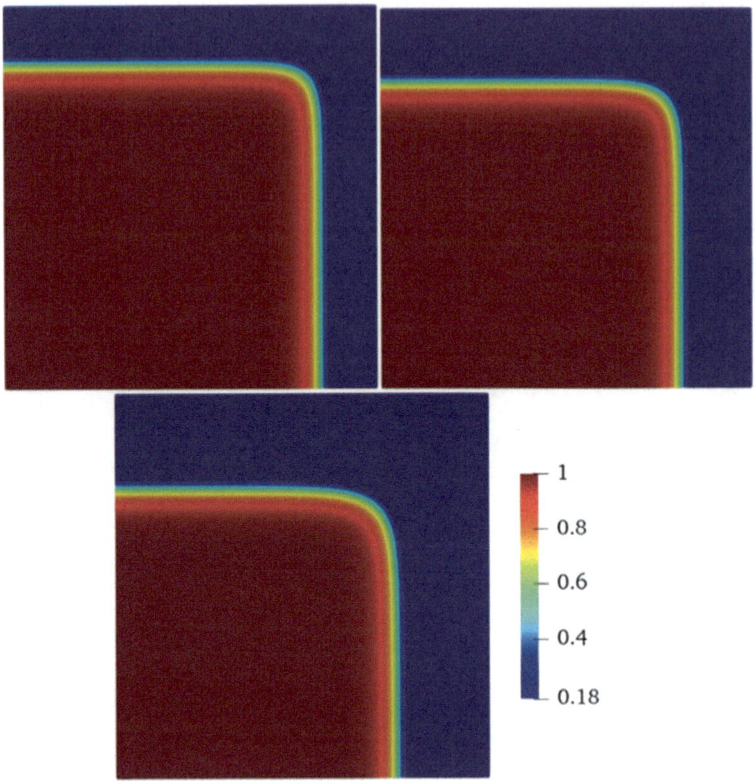

Fig. 4.15 PD predictions of oxidation state field at times: (a) 25h, (b) 50h and (c) 100h

resulting in oxidation of the material. The oxidation state variable representing the degree of oxidation of the material is presented in Fig. 4.15. It is observed that the oxidation progresses uniformly from two sides. The horizontal and vertical displacement fields at time $100h$ are presented in Fig. 4.16. The oxidation-induced shrinkage in the specimen is evident.

4.4 Thermo-Oxidation in 977–2 Epoxy Resin-Copper Bimaterial

As shown in the Fig. 4.17, a square 977–2 epoxy resin-copper bimaterial specimen with dimensions $L = W = 500\mu m$ is considered. The dimension of copper is $500\mu m \times 100\mu m$. An initial crack of length $a = 31.25\mu m$ is introduced at the interface. The left edge of the specimen is exposed to oxygen concentration $C_s = 3.3$mol/m^3. No flux condition, i.e., $\nabla C.\boldsymbol{n} = 0$ is specified at the top, right, and bottom edges. The mechanical boundary conditions are specified as: bottom edge is fixed and

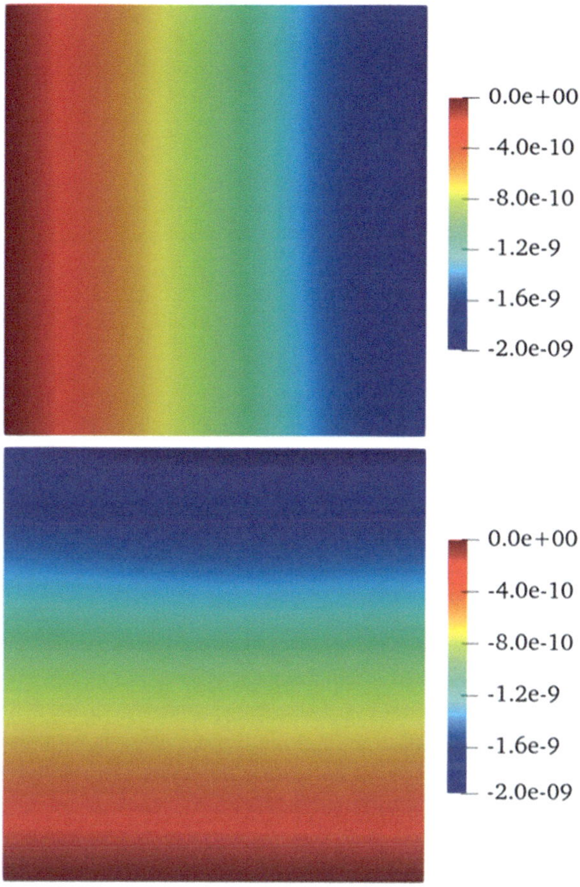

Fig. 4.16 Shrinkage due to thermo-oxidation: (a) horizontal and (b) vertical displacements (m) at time 100h

other edges traction free. A uniform mesh of grid size $\Delta x = 5$ μm is used for numerical simulation. For numerical time integration of diffusion-reaction equation is performed with time size, $\Delta t = 0.005min$.

The material parameters for 977–2 epoxy resin are provided in the preceding section. No diffusion of oxygen is allowed into copper. The Young's modulus of the copper is $E_{cu} = 117$GPa. The bonds at the interface of epoxy and copper are treated in same manner as mentioned in Sect. 3. For bond breaking, the critical stretch used is 1×10^{-5} for epoxy and interface bonds whereas a large value is considered for copper bonds.

Oxygen concentration field at times $1.25h$, $3.125h$, and $6.25h$ is presented in Fig. 4.18. The oxygen diffuses into the material through the left edge and the crack surface. When new surfaces are formed due to crack propagation, they are also exposed to oxygen concentration of 3.3mol/m^3. This allows oxygen to penetrate

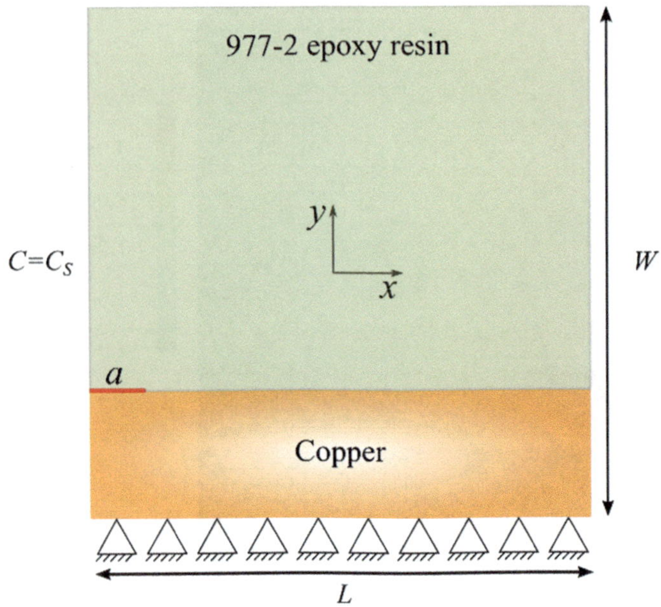

Fig. 4.17 Epoxy resin-copper bimaterial specimen exposed to thermo-oxidation

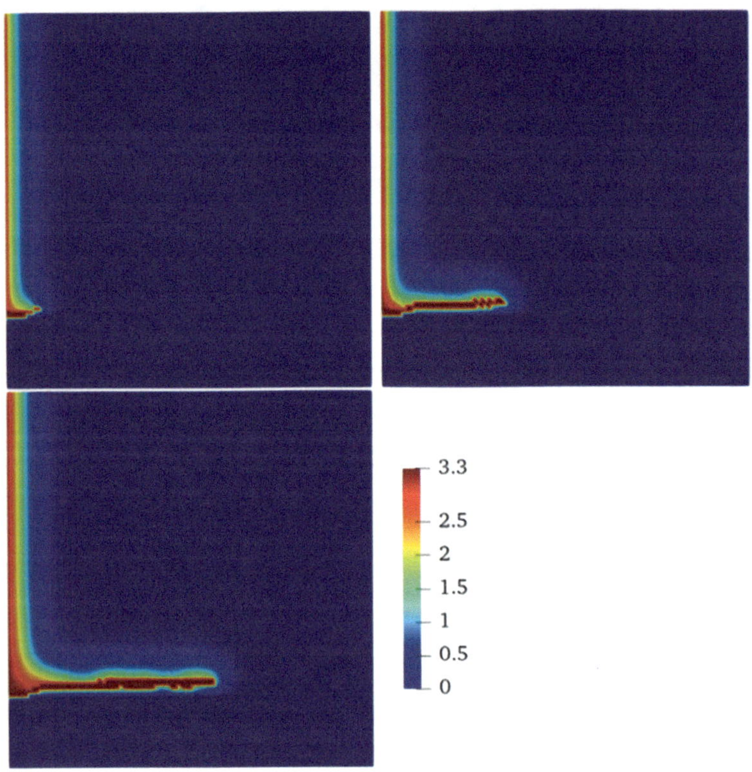

Fig. 4.18 Oxygen concentrations at times: (a) 1.25h, (b) 3.125h, and (c) 6.25h

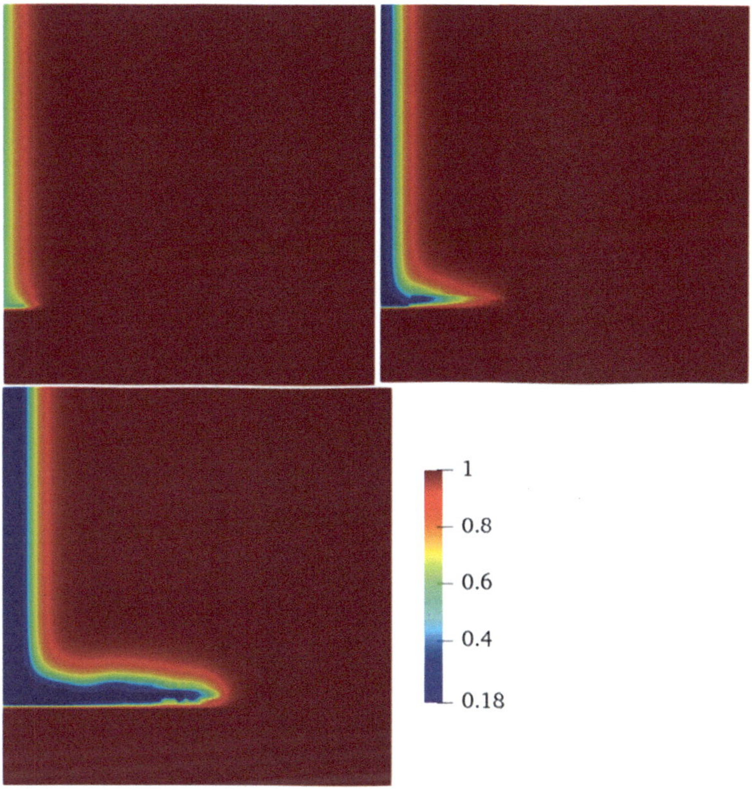

Fig. 4.19 Oxidation state predictions at times: (a) 1.25h, (b) 3.125h, and (c) 6.25h

further into the material. The oxidation state variable showing three regimes of oxidation processes is furnished in Fig. 4.19. The shrinkage-induced mechanical stretch due to oxidation leads to opening of the crack and subsequent propagation. Figure 4.20 shows the progression of damage.

5 Conclusions

This chapter presents a bond-based peridynamic modeling for coupled oxidation-diffusion and mechanical deformation of polymers under isothermal condition while considering the diffusion reaction and weight loss effects. The nonlocal peridynamic theory is based on the interaction among material points through bonds. Also, it is free of the smoothness requirement of the field variables. It successfully captures the effect oxidation-diffusion, thermo-oxidation and shrinkage polymers, and crack propagation due to oxidation. The peridynamic predictions agree well with the previous studies.

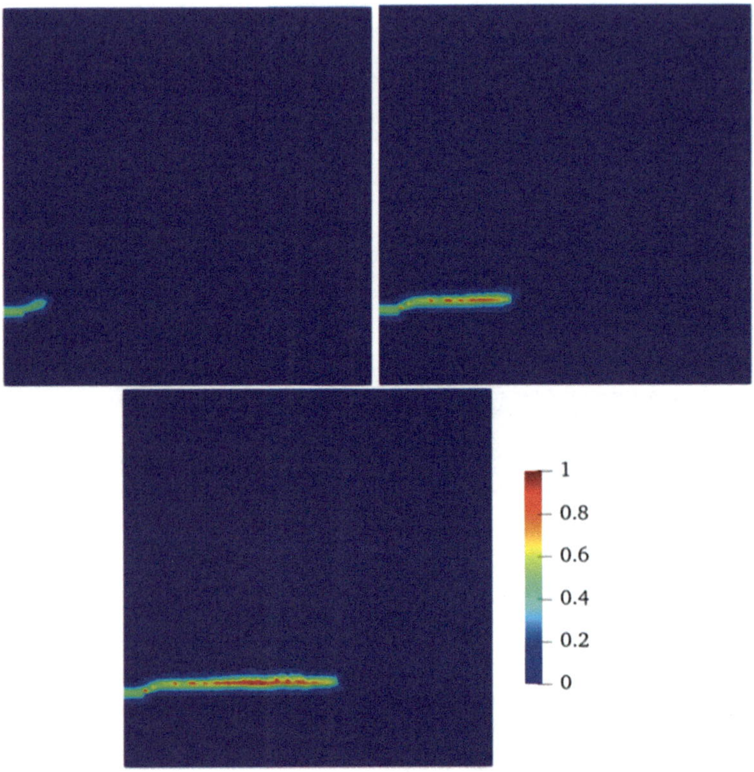

Fig. 4.20 Progressive damage predictions at times: (a) 1.25h, (b) 3.125h, and (c) 6.25h

References

1. Z. Cui, M. Cai, R. Li, P. Zhang, X. Chen, D. Yang, A numerical procedure for simulating thermal oxidation diffusion of epoxy molding compounds. Microelectron. Reliab. **55**(9–10), 1877–1881 (2015)
2. H. B. Fan, E. K. Chan, C. K. Wong, M. M. Yuen, Molecular dynamics simulation of thermal cycling test in electronic packaging (2007)
3. R. Li, D. Yang, P. Zhang, F. Niu, M. Cai, G. Zhang, Effects of high-temperature storage on the elasticity Modulus of an epoxy molding compound. Materials **12**(4), 684 (2019)
4. B. Zhang, M. Johlitz, A. Lion, L. Ernst, K. M. B. Jansen, D. K. Vu, & L. Weiss, Aging of epoxy moulding compound—thermomechanical properties during high temperature storage. in 2016 17th International Conference on Thermal, Mechanical and Multi-Physics Simulation and Experiments in Microelectronics and Microsystems (EuroSimE), pp. 1–6 (2016, April): IEEE
5. K. M. B. Jansen, J. de Vreugd, L. J. Ernst, C. Bohm, Thermal aging of molding compounds. in 2010 11th International Conference on Electronic Packaging Technology & High Density Packaging, pp. 778–780 (2010, August): IEEE
6. D.G. Yang, F.F. Wan, Z.Y. Shou, W.D. van Driel, H. Scholten, L. Goumans, R. Faria, Effect of high temperature aging on reliability of automotive electronics. Microelectron. Reliab. **51**(9–11), 1938–1942 (2011)

7. D. Zou, J. Wang, W. Yang, S. Liu, CTE-measurement and delamination growth by a real time Moire technique. In 1997 Proceedings 47th Electronic Components and Technology Conference, pp. 1124–1127 (1997, May): IEEE
8. D. Yang, Z. Cui, The effect of thermal aging on the mechanical properties of molding compounds and the reliability of electronic packages. In 2011 12th International Conference on Electronic Packaging Technology and High Density Packaging, pp. 1–4 (2011, August): IEEE
9. J. De Vreugd, K. M. B. Jansen, L. J. Ernst, C. Bohm, R. Pufall, High temperature storage influence on molding compound properties. In 2010 11th International Thermal, Mechanical & Multi-Physics Simulation, and Experiments in Microelectronics and Microsystems (EuroSimE), pp. 1–6 (2010, April): IEEE
10. O. Hölck, E. Dermitzaki, B. Wunderle, J. Bauer, B. Michel, H. Reichl, Molecular modeling of a 3D-crosslinked epoxy resin and its interface to native SiO 2—property prediction in microelectronic packaging. In 2010 Proceedings 60th Electronic Components and Technology Conference (ECTC), pp. 135–143 (2010, June): IEEE
11. B. Zhang, A. Lion, M. Johlitz, L. Ernst, K. M. B. Jansen, D. K. Vu, L. Weiss, Modelling of thermal aging of Moulding Compound by using an equivalent layer assumption. In 2017 18th International Conference on Thermal, Mechanical and Multi-Physics Simulation and Experiments in Microelectronics and Microsystems (EuroSimE), pp. 1–6 (2017, April): IEEE
12. B. Zhang, M. Johlitz, A. Lion, L. J. Ernst, K. M. B. Jansen, D. K. Vu, L. Weiss, Thermal aging modeling of molding compound under high-temperature storage and temperature cycling conditions. In 2018 19th International Conference on Thermal, Mechanical and Multi-Physics Simulation and Experiments in Microelectronics and Microsystems (EuroSimE), pp. 1–5 (2018, April): IEEE
13. S. Manoharan, C. Patel, S. Dunford, C. Morillo, P. McCluskey, Aging characteristics of green mold compound for use in encapsulation of microelectronic devices. In 2018 IEEE 68th Electronic Components and Technology Conference (ECTC), pp. 1768–1773 (2018, May): IEEE
14. A. Inamdar, A. Prisacaru, M. Fleischman, E. Franieck, P. Gromala, A. Veres, B. Han, Study of Thermal Aging Behavior of Epoxy Molding Compound for Applications in Harsh Environments. In 2019 IEEE 69th Electronic Components and Technology Conference (ECTC), pp. 811–818 (2019, May): IEEE
15. H. Mei, X. Du, L. Li, W. Tan, Optimal packing research of spherical silica fillers used in epoxy molding compound. In 2009 International Conference on Electronic Packaging Technology & High Density Packaging, pp. 623–626 (2009, August): IEEE
16. N. An, G.P. Tandon, K.V. Pochiraju, Thermo-oxidative performance of metal-coated polymers and composites. Surf. Coat. Technol. **232**, 166–172 (2013)
17. G.P. Tandon, K.V. Pochiraju, G.A. Schoeppner, Thermo-oxidative behavior of high-temperature PMR 15 resin and composites. Mater. Sci. Eng. A **498**(1–2), 150–161 (2008)
18. Y.T. Yu, K. Pochiraju, Modeling long-term degradation due to moisture and oxygen in polymeric matrix composites. Mater. Sci. Eng. A **498**(1–2), 162–165 (2008)
19. G.P. Tandon, K.V. Pochiraju, G.A. Schoeppner, Modeling of oxidative development in PMR-15 resin. Polym. Degrad. Stab. **91**(8), 1861–1869 (2006)
20. K. Pochiraju, G.P. Tandon, Interaction of oxidation and damage in high temperature polymeric matrix composites. Compos. A: Appl. Sci. Manuf. **40**(12), 1931–1940 (2009)
21. S.A. Silling, Reformulation of elasticity theory for discontinuities and long-range forces. J. Mech. Phys. Solids **48**(1), 175–209 (2000)
22. E. Madenci, S. Oterkus, Peridynamic modeling of thermo-oxidative damage evolution in a composite lamina. In 58th AIAA/ASCE/AHS/ASC Structures, Structural Dynamics, and Materials Conference, p. 0197 (2017)
23. J. Liang, K.V. Pochiraju, A methodology for the estimation of transverse failure strength of an oxidized lamina from isothermal aging studies. Compos. A: Appl. Sci. Manuf. **73**, 101–108 (2015)

24. E. Madenci, S. Oterkus, Ordinary state-based peridynamics for plastic deformation according to von Mises yield criteria with isotropic hardening. J. Mech. Phys. Solids **86**, 192–219 (2016)
25. R.W. Macek, S.A. Silling, Peridynamics via finite element analysis. Finite Elem. Anal. Des. **43**(15), 1169–1178 (2007)
26. E. Madenci, E. Oterkus, *Peridynamic Theory and Its Applications* (Springer, Boston, 2014) 978-1-4614-8464-6
27. B. Kilic, E. Madenci, An adaptive dynamic relaxation method for quasi-static simulations using the peridynamic theory. Theor. Appl. Fract. Mech. **53**(3), 194–204 (2010)

Chapter 5
Molecular Modeling for Reliability Issues

Nancy E. Iwamoto

1 Introduction

To understand where molecular modeling fits in the grand scheme of reliability, first consider the hierarchical scale of materials. Simple length scale observations show that different phenomena belong at different scales of physical observation (Fig. 5.1 top). Today, no one modeling tool contains all of the scales, so the types of modeling employed will be different depending upon scale (Fig. 5.1 bottom). As Fig. 5.1 shows, molecular modeling falls in the very small size and time domains. Because of its small nanoscale focus, molecular modeling has often been considered irrelevant for material reliability and not normally engaged to understand issues that are measured on the macroscale level like that of reliability.

However, all reliability issues have to start somewhere via a defect: a crack, a void, a density change, or even a disruption in composition. It is these kinds of "defects" that are most often too small to economically measure experimentally, but their effects can be probed using a molecular model, as these defects or compositional disruptions are potential failure initiation points. The job of the molecular model is then to identify whether such defects are disruptive enough to create a problem and either how the disruption formed and or how problems can be overcome. That is, the molecular scales can be thought of as the first step in understanding the macroscale and entail three different levels: quantum, classic molecular, and classic mesoscale.

There are generally three reasons to engage in molecular modeling for reliability analysis in electronic materials: (a) to specifically simulate the molecular interactions within the composition and define which interactions may either lead to failure or strengthening, (b) to understand how the specific compositions or disruptions originated, and (c) to extrapolate to better materials or processes. The first area can

N. E. Iwamoto (✉)
Honeywell Performance Materials and Technology/Advanced Materials, Ramona, CA, USA
e-mail: niwamoto@prodigy.net

W. D. van Driel, M. Yazdan Mehr (eds.), *Reliability of Organic Compounds in Microelectronics and Optoelectronics*,
https://doi.org/10.1007/978-3-030-81576-9_5

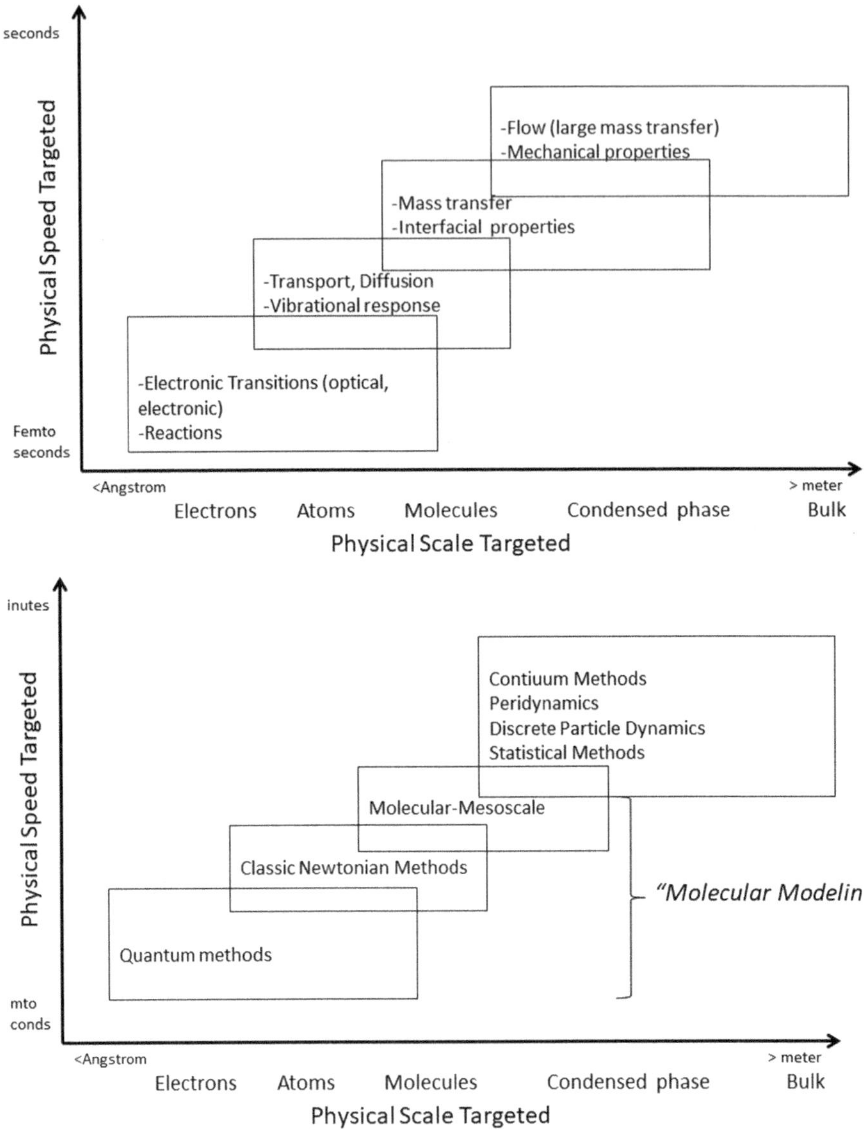

Fig. 5.1 The basic hierarchical size perspective. Top: hierarchy from the physical response perspective. Bottom: hierarchy from the model perspective

describe a molecular-scale mechanical model and involves several steps of identifying the specific molecular structures, modeling the energetics of their interactions, and determining the impact of their interactions. The second area involves determining how the specific molecular structures or the compositions came to be, or evolved. This area is very important when analyzing the impact of processes such as temperature and stress cycling, and diffusion events such as internal diffusion (moisture conditioning) or surface diffusion. Through the understanding of how molecular interactions evolve under stress, specific structures and interactions can then be evaluated as to their ability to survive.

In general, for reliability modeling the usual perspective is to emulate how the material fails compared to experimental tests, and emulation can also be accomplished by molecular models. For instance, the mechanical test of a block of material can be approached by deforming a nanoscale block of material constructed within an infinite periodic cell (infinitely repeating in all three dimensions). However, this perspective can limit the chemists' visualization of the failing interface as the block of material can have both molecular adhesive and cohesive contributions to the response. So, the perspectives covered in this chapter will concentrate more on smaller molecule interactions in which the interface is specifically defined from the molecular level in order to identify how much the specific interactions contribute to the failure under investigation. How these interactions contribute to larger periodic cell model strategies and scaling to mesoscale models of failure will also be discussed.

Most of the molecular modeling methods and examples that will be discussed have already been previously reviewed [1–4], but the small-molecule (cluster) interface strategies were first advanced in the early 1990s within DOD, with computer platforms that were too limited to allow large bulk system simulation on an atomistic scale [5–12] in a timely manner. It was found that performance trends could be easily predicted using small pairwise interactions of molecular models. In particular, for mechanical properties, it was found that modulus could be captured on this scale. For other properties, relative trends were found to be extractable in a meaningful manner to help materials development.

It was also realized at that time that because processes and material history played a large part in how the material failed, and process history could affect the molecular structures and interactions that contribute to performance; even large periodic cell models would not easily identify which interactions were weaker (or stronger). However, if the goal was to search for the best material, the focus could be switched from trying to emulate the failure to looking for best interactions and differences between materials. The best interactions then defined the best the material could do theoretically, regardless of process. This is when the small interface models became an important part of the reliability picture, as the interactions found at this scale could inform large periodic cells models or large continuum models, by identifying whether there should be compositional (or architectural) biases in the multi-interactional environment.

Early work using these smaller cluster interaction models [5–12] which examined dynamic oligomer response to strain leads to the observation that most polymer repeat units had their own specific contact footprint of interaction (both energetically and structurally) that helped the interface resistance to strain. This early observation supported the basic small model strategy to assess relative performance by considering the "best" interaction (or interactions) present. Early work also depended upon having credible experimental benchmarks with known molecular structures to validate model trends. That is, since comparative analysis is a very important function of using molecular models, molecular models should always first be viewed in terms of trends, validated by experimental trends. And, the more the quantitative need (for instance, a property), the higher the importance for experimental validation. Scaling to mesoscale methods can enable simulations equivalent to millions of atoms, and it was found that these same pairwise molecular interactions could be used as the basis to jump length scales between atomistic to large mesoscale beads, which will be discussed in Sect. 3 [13–17].

The molecular interface generously can be categorized by scope of interactions targeted and the type of model. For instance, the interface interactions can be either adhesive (with two different molecular compositions at the interface) or cohesive (with the same molecular composition at the interface). A third category can be loosely defined as mixed interfaces in which the random combination of both the adhesive and cohesive molecular interactions is sampled at the same time such as in miscibility issues, or where multiple interfaces may be involved during the course of the simulation such as in internal diffusion. These can be done by very large cluster models or using 3D periodic cells. (However, it has been found diffusion simulation using large cluster models is slower to optimize due to size.)

To understand individual molecular sources of reliability in electronics, ideally all molecular interface combinations are modeled and compared, including any modifying molecule contribution (such as moisture). However, part of the modeling strategy must always be a critical evaluation of what is most likely present. If the composition cannot be specifically identified, an initial estimate can also be done by modeling (i.e., a reaction model, a diffusion model, or a miscibility model). That is, defining what is present is important to define what models that must be run. This is another use of the small model strategy, as they can quickly sample compositional space as part of the modeling strategy in order to inform a best-guess composition.

This chapter is also meant to illustrate how the chemistry and physical-chemical origins of reliability can be understood from the most basic (molecular structure) level, from fairly small and fast models. By no means is this chapter meant to cover all possible molecular modeling methods and tools, as software developers have been embedding better and better methods, within current commercial software to access large-scale property predictions, using 3D periodic cell techniques. Also, tools such as group contribution methods and those which require large training data sets such as QSAR/QSPR (quantitative structure activity/property relation) and smart AI/deep learning methods will not be discussed here. Instead the important emphasis is how the molecular structure, architecture, and modeling method can be combined in a strategy to obtain the information needed to predict reliability.

2 Molecular Modeling Tools

On the simplest level, molecular modeling can be separated into two tool areas based upon the underlying assumptions used: (1) quantum mechanical models and (2) classic Newtonian molecular models which include molecular or atomistic models (mechanics and dynamics), as well as mesoscale models (sometimes referred to as molecular-mesoscale, in order to distinguish the size of the models possible today from larger mesoscale perspectives). The most sophisticated modeling software tools contain mixtures of these methods (such as embedded atom models) or advanced models (such as reactive force field molecular dynamics), but they will not be covered here. Also, as mentioned previously the statistical methods based upon data training will not be covered.

Quantum methods are generally used when the chemistry or electronic (optical or electrical) response of the material is needed. Because the chemistry can be sampled, they are useful to see if a change in interface is occurring, but the drawback is that they are very time- and computer-intensive, so that atom count is usually limited. On a positive note, most of the periodic table is available using these methods, so are usually the default option when force fields that are used in the classic Newtonian methods are not available.

Classic Newtonian methods allow many more atoms to be involved in simulation. They are much faster than quantum methods and so are especially useful for dynamic responses that are based upon temperature, pressure, and time variation like mechanical reliability issues and are also very useful for random distribution simulations like diffusion and miscibility. However, the classic methods require specific force fields to describe how atoms interact, so as mentioned previously, if a force field is not available to describe the molecular system to be simulated, then quantum methods are recommended.

Regardless of which molecular modeling scale is engaged, the model is comprised of either an isolated 3D molecular cluster or a 3D periodic cell (which is an infinitely repeating unit cell in all 3 dimensions), depending upon the simulation need. The simulation to be engaged will always be dependent upon the investigation, but the basic progression will always be construction of the model, energy optimization search, and then addition of temperature in the dynamics step. Without going into detail, the build step will depend upon the molecular structures defined relevant to the issue being investigated; the optimize step is a search of local space for an energy minimum and will take the form of larger step or finer step procedures; and dynamics usually contains most of the simulation protocols in which stress is added to the system in a time-based manner using temperature, pressure, or specific structural stresses. Temperature dynamics can be classified depending upon the environmental variable that is held constant during equilibration to the target temperature. For example, NVT dynamics will fix the number of molecules or atoms (N), volume (V), and temperature (T) allowing the pressure to vary; NPT dynamics will fix the number of molecules or atoms (N), pressure (P), and temperature (T) allowing the volume to vary. Other stressors can be added such as a forcing potential that will move atoms out of their original position. All stressors can be added in an equilibrium way or nonequilibrium way, depending upon the simulation need.

2.1 Quantum Mechanics

Quantum mechanics revolves around solving wave equations for representing the electron distribution around the nuclei of atoms and the corresponding energies [18]. The basic equation is the well-known Schrodinger's equation. Since exact solutions are difficult, approximations have evolved to better solve Schrodinger's equation (such as Hartree-Fock and DFT), which increased the utility of quantum mechanics. These solutions represent how atomic orbitals combine to describe bonds (valence bond theory) and how the electrons distribute themselves under the influence of the nuclei to represent the electronic structure of a molecule (Molecular Orbital Theory). Hartree-Fock methods use a determinant of single-electron wave functions to represent the many-electron case; whereas DFT methods assume that the many-electron case is represented by a functional of the electron density. For DFT methods functionals are also used which employ approximations for the electron density. All methods used in quantum chemistry require basis set equations which describe the electronic wave functions used to solve the Schrodinger equations and are usually either based upon combinations of atomic orbitals or plane wave descriptions. Approximations to better define non-bond interactions are part of the evolving toolkit of quantum mechanics.

Quantum mechanics (QM) is well suited to understand reactions in the materials being studied and has been particularly useful in cure studies and in determining if side reactions are contributing to property loss. In addition, quantum techniques have been found to be useful in cases where classical methods have lacked adequate force fields, such as in metal diffusion studies. It has also allowed calculations of near-space reactivity and physical-chemical interactions of surfaces, as well as extrapolation of electronic activity into the vacuum space above surfaces. The electronic structure aspect of material development is only mentioned as part of a larger reliability world and will not be covered in this chapter, but these characteristics were found especially useful in our labs for understanding changes in the electronic structure of surfaces responsible for the performance reliability of an optical chip [19, 20]. In general, quantum mechanical models with their high compute requirements end up to be much smaller in size than classical methods (for time turn-around requirements).

2.2 Classic Methods

The classic methods of molecular models are all particle-based, in which the energies are calculated based upon force field relationships that describe the particle interactions. The basic difference between the molecular and mesoscale models is the definition of the particle. The molecular model particle is based upon atoms, and the mesoscale particle is based upon groups of atoms called "beads." Regardless of whether the particles are atoms or beads, classic molecular mechanics and dynamics rely heavily on parameterized interatomic potentials contained in different force fields. So each force field has embedded equations which describes both bond and non-bond interactions. Different force fields are available depending upon how the different interaction energies were defined and the different atoms of the periodic table and their interactions to other atoms. Classic modeling tools are attractive because the energy, deformation, time, and temperature relationships are then predefined and so simpler to calculate compared to quantum mechanics. These are the molecular model's constitutive equations, so that energies and deformations can be tracked during a reliability simulation.

2.2.1 Classic Molecular Models

The force field is distinguished by the different energy components used. All force fields have to have bond and non-bond energies. The simple bond energy is usually defined as a function of distance and force constant, with additional simple angle and dihedral contributions terms. As the force field becomes more sophisticated more cross terms are added, such as bond-angle and bond-dihedral interactions, angle-angle, angle-dihedral, and out-of-plane motion. The non-bond energies are defined by both Van der Waals functions and electrostatic functions (charge distance-based functions). Early force fields were parameterized from analysis of spectra, but today they are commonly parameterized by calculating various deformation (bond, angle, dihedral, etc.) energies using quantum mechanics. The total potential energy of a molecular system is then described as a simple summation of its parts. Kinetic energy is added to describe temperature, using methods to control both the temperature and pressure. For instance, there is always an option to quickly come to temperature using velocity scaling, but more subtle methods are usually available such as Nose and Anderson methods controlled by collision probabilities or Berendsen methods using a defined decay constant. If the model is in a periodic cell, pressures can also be controlled.

However, each force field is different, and one model cannot use multiple force fields in the same model. In order to be sure that a force field is useable for the specific material application studied, the user should first examine the force field file to be sure all of the elements needed are represented. They should also examine the force field to be sure all the important interatomic interactions are present. For instance, if a force field contains a C-O bond representation, but a Cu-O bond is not present, then it cannot be used for an organic-copper oxide interface. For ionic crystalline structures where conditions make it unlikely that the crystalline structure would change, work-arounds are available by constructing the crystalline unit cell and manually defining each atomic charge. The force field will then handle the structure predominantly by the electrostatics of the interactions. Other work-arounds may not be possible, so the only option is to develop your own force field or choose another one that handles all the atoms, bonds, and non-bonds of your model. If you are unfamiliar with the force field choices, the (suggested) best option is to run benchmark tests. For mechanical reliability, one simple property that is convenient to benchmark is modulus.

2.2.2 Classic Mesoscale Models

As described before, the major difference between molecular models and mesoscale models derived from the atomic or molecular scale (or the "molecular-mesoscale" to distinguish this scale from larger mesoscale particle models such as discrete element models) is that the former uses atoms as its basic particle and mesoscale models use groups of atoms that are combined to form a generic bead that represents that grouping of atoms. And just like the molecular model, all force fields are defined based upon the energies of the interacting beads (bond and non-bond energies).

Several common force fields are available, including the Martini [21–23] and the Shinoda [24] force fields.

The same considerations for force field, as discussed previously for molecular models, should be used for mesoscale models. However, one cautionary note is that because the force fields usually have been defined around a certain grouping, the size of the bead bonds, inter-bead distances, and the magnitude of the bond forces defined should be monitored to ensure they are applicable to your situation. Our studies have found it more convenient to develop our own force fields for reliability studies, as the spatial distribution and magnitude of the force field components are specifically designed around our choice of bead definition (atomic grouping choice for the bead) and the energies are then known from the atomistic level. We have found if the force field is adequately defined, there is a continuity of scale, as we progress from molecular models to the mesoscale [13–16], or trends can be approximated from the coarse-grained results [17]. As before with molecular force fields, if using available force fields, it is important to run benchmarks to be sure they are applicable to your material.

2.2.3 General Considerations

In all of these examples, choices of model must be made. The first consideration is the composition of the model. Usually, if metals and inorganic materials are being modeled, to use a classic tool, the force field needs critical examination to ensure that all of the atomic interactions desired are correctly described. For reliability issues in electronics, there is often a mixed interface in which both metal and polymer parameters must be defined within the same force field. If there are missing parameters, then a quantum mechanics tool is usually chosen (but even in this case, care must be taken to be sure the basis set and underlying assumptions will take care of the atom).

The next consideration is the size of the model. While this chapter will concentrate more on using small size models for most of the cases discussed (typically less than a thousand atoms), larger molecular models can also be constructed by increasing the atom count from which macroscale issues, such as voiding, free volume, melting, and glass transitions, can be extracted. Then it is up to the computational resources available to decide whether they can be done in a timely manner. That is, for material developers looking at root cause mechanical reliability issues and searching for interactions, or strongest or weakest contributors, the small modeling strategy is attractive because of quick turn-around. For those looking for specific qualitatively architectural changes (voiding, phase transition, as mentioned previously), then larger models are needed such as large periodic cell molecular models and mesoscale models (which can also be done either in clusters or in periodic cells). Periodic cells are often convenient because an infinitely bonded polymer chain can be represented. If small-scale models are made small enough, both classic and quantum methods can be used for the same interface; but for the large models, usually only classic techniques can be used in a practical timeframe (unless large computer resources are available that can run the larger models with quantum methods).

The next consideration is the kind of information that is needed from the model. If detailed thermodynamic information is needed, usually it is best to design small quantum mechanics simulations. If less detailed thermodynamic comparisons are needed than those usually derived from basic quantum calculations, then classic methods will give the correct energy trends directly from the model without resorting to additional approximations to derive the thermodynamic breakdown [46, 47]. Both quantum and classic methods will yield interaction (dimensional) responses of the interfaces under investigation. Then, it is the geometric analysis tools available that are important to have in your graphical interface tool in order to adequately measure dimensional changes of the molecular system simulated.

It is probably worthwhile to mention before we delve any further that, for practical reliability (molecular) modeling, the molecular structures first have to be identified and built. Although investigators are looking into models that automatically determine the most likely bonding from a composition mixture, at this point such software is not yet available, so the modeler has to start off with a reasonable estimate of the molecular structures that are to be simulated for reliability. Builders in most software packages will do some automated things like build polymers based upon the known monomer reactivity ratios and stoichiometries and automated cross-linking, but the bonds that are to be linked are pre-defined, rather than letting the software decide which bonds are the most likely to form. This has been a disappointment to many engineers who expect the software to build new molecular structures automatically from the molecules they feed the program (a known need that is being researched today). For new materials or materials of unknown structure, the chemists' knowledge of bonding and their chemical reaction knowledge is needed so that relevant structures are used in the simulations, rather than trying to simulate all reaction combinations.

The next sections will try to demonstrate general molecular modeling concepts that can be used for failure and reliability, with examples that have been instructive in the past. Section 3 will concentrate on mechanical properties, Section 4 will cover diffusion examples, and Sect. 5 will highlight indirect analysis in order to predict material weaknesses.

3 Mechanical Reliability on the Molecular Level

3.1 Comparative Stress Models and Size or "How Low Can You Go"

The alarm to most reliability engineers looking into molecular modeling for the first time is the size differential to reality. For example, Fig. 5.2, right, shows a Cu_2O-epoxy interface where the polymer is built as a single strand of ~3500atoms using Epon 828-DETA (1:1), and the copper oxide is at the bottom of the figure. This is a reasonably sized periodic cell model of the interface, with the width at the oxide interface of about 0.0035 microns. However, micrographs of experimental copper oxide grown on a copper surface [25] have shown both globular Cu_2O and feathery CuO crystals with the CuO feather thicknesses appearing smaller than 0.07 μm and the Cu_2O faces appearing smaller than 0.7 μm. The features also exhibit orientations in different directions. So when the molecular model is compared to the micrograph, it is clear that the size of real copper oxide crystal features is much larger than a molecular model. That is, not only is the size of the epoxy polymer layer in the molecular model a mere fraction of any of the actual copper oxide crystal features (demonstrating a high size differential), but the molecular model misses the different copper oxide orientations and morphologies acting at the interface.

This size differential means that a single molecular model like that of Fig. 5.2 would need to be repeated to accumulate an array of oxide deformation directions (and types of oxide) that would be represented in real life, when the individual crystal surfaces are considered. Because of this size differential, smaller, practical molecular models that investigate the chemical structural impact can be used to break issues down into sizes that can address the impact of how the molecular interface deforms in reference to the macroscale. As mentioned previously this will allow identification of which combination of structure deformation is responsible for initiating a larger disruptive event that will cascade into a failure. From this respect, the molecular modeling comes down to determining the impact from an array of different molecular interfacial interactions that could exist and include all adhesive as well as cohesive interactions. The molecular modeling strategy reduces to adequate judgment of which interactions (and molecular structures) are the most

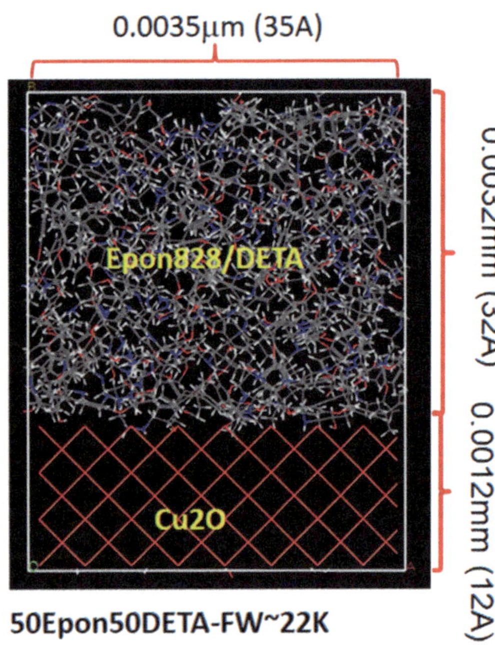

Fig. 5.2 Example of a 3D periodic cell molecular model representing an epoxy-Cu_2O) interface. The molecular model is composed of a 50/50 mole ratio Epon 828-DETA (diethylenetriamine) polymer (top layer) with a molecular weight of ~22,000, on top of crystalline Cu_2O

significant to include in the simulation, the size of the model that is appropriate, and what type of stresses must be included (and how to represent these stresses).

Periodic cell (infinitely repeating 3D unit cells) molecular models represented in Fig. 5.2 can be stressed in shear or tension or even with a twist by deforming the unit cell. These are instructive in looking at more bulk effects from the polymer, such as cross-linking and resin-curative stoichiometry differences [26–28], and for predicting properties averaged over multiple interactions (such as CTE, Tg). Most molecular modeling software available today has embedded methods that will contort periodic cells to the needed deformation, with embedded tools to extract properties.

As mentioned previously, periodic cells tend to have multiple interface response. To demonstrate this concept, Figs. 5.3 and 5.4 show the same unit cell as in Fig. 5.2, before and after deformation is applied to the unit cell (shear in Fig. 5.3 and tension in Fig. 5.4). The relative positions of the copper oxide have been fixed, with the assumption that the polymer is the more pliant of the two materials. The epoxy unit cell has been built using only one continuous polymer chain in the model. For the shear case in Fig. 5.3, the light blue and dark blue atoms represent those atoms that end up at the surface after shear, and so do not represent separate chains or connectivity (i.e., the colors merely represent atom proximity to the surface, and all bonding is maintained by the original continuous polymer chain). Tracking backward (Fig. 5.3 left) in time to the start of the molecular simulation, you can see that the shear action has pulled atoms that are more in the interior of the layer to the interface. The tension example in Fig. 5.4 is interesting in that the deformation could be done either by deforming the full unit cell in the tensile direction or by building the unit cell with a vacuum space that the copper oxide can be moved into and analyzing for the response of the following polymer layer. The former has been performed for Fig. 5.4 and is colored by the proximity of atoms in layers that, when tracked forward in time, illustrate where these atoms end up after the tensile deformation. It is found that the atom layers are mixing after deformation and contact is being lost with atomic scale voiding developed. As no bonding is broken, these void areas

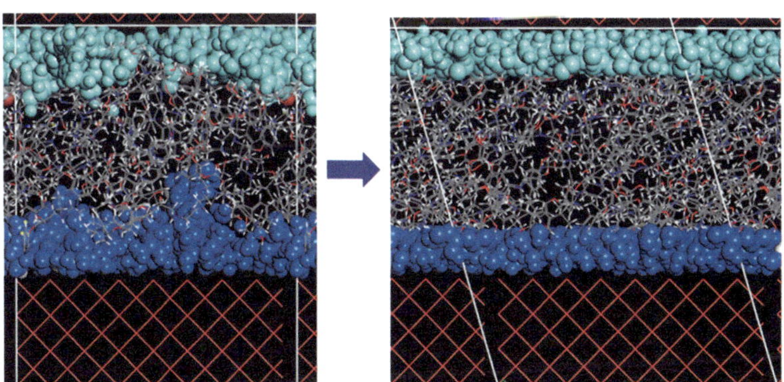

Fig. 5.3 3D periodic cell of a copper oxide-epoxy interface before (left) and after (right) shearing. Atoms are colored by the closest atoms to the interface after deformation

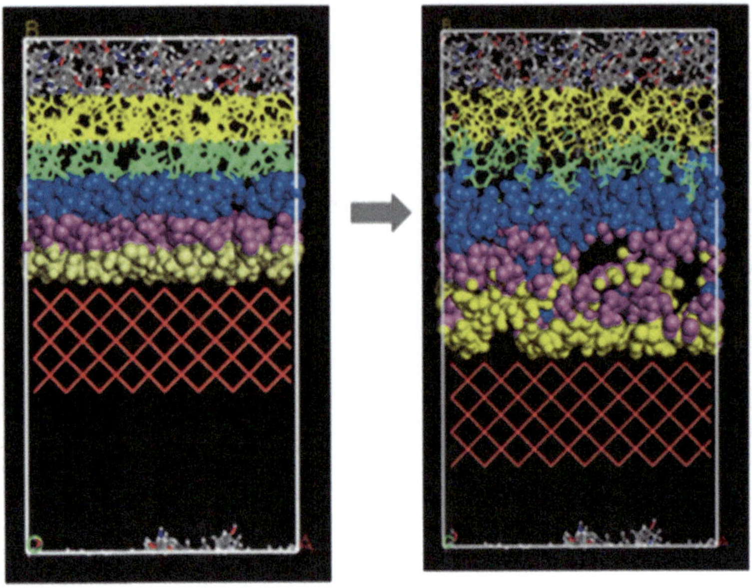

Fig. 5.4 3D periodic unit cell models of a copper oxide-epoxy interface in a vacuum cell. Atoms are colored by layers before the deformation and are illustrative to show where atoms may end up

represent changes in the local packing in the polymer layer on both the adhesive interface and in the cohesive interface, and represent ways in which interaction is lost and the beginnings of how a layer can lose strength. From the macroscale standpoint, the voids found in the molecular model probably could not be directly measured as they are too small (less than approximately 4A in width) and are more indicative of how density can rearrange in a polymer, rather than demonstrate a measurable void. However, on the molecular modeling scale, we see rearranged molecular-scale voids. These molecular-scale voids can be significant when small components of the formulation are tracked or tracking high molecular (or atomic) interactions which contribute to stability.

Both of these simple illustrations show how both polymer cohesion and interface adhesion are acting to pull at interior polymer atoms (cohesion), to represent the total polymer response. These types of models actually represent mixed mode deformations, from our initial perspective of isolating contributing molecular interactions. However, the total energies do not necessarily isolate the specific polymer-polymer interactions from the polymer-surface interactions in order to get a specific ranking of interface contribution that would be important to the engineer or material developer trying to pinpoint causes from (molecular) component level.

In trying to isolate (and rank) molecular interactions, it was found that simple cluster models can generate much of the same information but more quickly, and with more certainty of the specific functional group interaction, than the larger unit cell models. For instance, early models taught us [5–12, 29–34] to obtain a measure of modulus in a polymer system, and we can scale down to a very small oligomer scale, as long as the oligomer piece is large enough to represent the immediate molecular interface bonding. The size of this small piece of polymer then becomes important in considering how small is reasonable. Generally, enough of the monomer-monomer bonding should be present to encompass enough bond rotation during the deformation simulation (generally less than a decamer is usually needed). For example, when a polymer is sheared from the interface, both the bond-bond elongation and bond-rotation will elongate the oligomer in response to the forced deformation and the non-bond forces that contribute to resisting the movement. This literal "unwinding" of the random polymer structure is part of the deformation response, but since we deal with the immediate atomic interface, we can see how the resistance to deformation from the interfacial adhesion is translated to directly to the polymer. (Interestingly other examination of the response of the trailing end has hinted that there may be a chaotic dynamic involved in failure [11].) It was concluded that the basic underlying atomistic interactions that drive the bulk performance are reflected in the small-molecule response. For instance, it was generally found that low strain events such as modulus [34] are quantitative and validated by experiment. Later mesoscale studies [13–16] showed modulus remains quantitative through the scaling process. The quantitative modulus response suggests the origin of the small-strain response lies in important molecular interactions. Outside of the linear elastic response, like strength (taken at the energy barrier), only qualitative trends are obtained from this analysis, but the small model does not adequately treat the actual physical rupture of an interface. So it is the early strain trends that are most instructive when comparing materials in a survey manner using this philosophy on this scale of model. Trend wise, the impact of the treatment history can also be investigated by using the small model approach [32, 33, 35].

Building the polymer interface on this small cluster perspective level is simple. First the most basic composition of the polymer must be determined and built for a random polymer. If the composition is a block copolymer, one strategy is to construct a model of each block separately, as well as a polymer with half of each block so that when analyzed, each block contribution can be isolated. For a cohesive interface, as a starting point with maximum interaction, two chains are built, copied, and offset and then optimized and equilibrated to room temperature. If the chains have enough through-space interaction (Van der Waals and coulombic), then the chains will hug one another; if there is little attraction, then the chains will separate and coil independent of the other chain. The same is done for an adhesive interface, except the substrate is usually built as a planar structure, especially if the substrate is a metal or an inorganic that uses periodic ordered structures; in the case of the periodic ordered substrate structure and for ease in the interface deformation step, the substrate can be spatially fixed. If the substrate is another polymer, it can be built as a planar structure or as a chain (like the cohesive case).

The deformation step of the small cluster molecule model is very simple (Fig. 5.5). Either a forcing potential or a distance change is defined at the ends for the oligomers in order to shear the interfaces apart, in an initial parallel-to-chain-axis motion (or initial parallel to a planar surface if an oligomer is modeled next to a surface). The deformation strain for these small models is usually measured by the distance the leading atom is moved away from its' initial position. The origin of this strategy [6] stems from the observation that when a perpendicular deformation (with respect to the oligomer chains) was attempted from middle of the cluster, the resistance to deformation seemed to originate mostly from shearing against one another until enough separation and enough contact was lost so that the chains began to move uniformly in the tensile strain direction. However, this strategy could change if the molecules are highly rigid, in which case tensile deformation may suffice.

The deformation is performed at several different rates in order to determine the maximum resistance to the imposed deformation. That is, the energy curves that result from the strain imposed will be rate dependent. It was found that usually the initial slopes and the maximum energies increase with an increase in strain rate, as shown in the schematic of Fig. 5.6, to a point where the rate response optimizes. At too high of strain rate level, the maximum energies can decrease (Fig. 5.6, blue line. Suggesting a molecular strain embrittlement), so it's the optimum energy that needs to be defined for the resistance of the interface to deformation. At too slow of deformation rate, the energy response is low, albeit very noisy (not pictured), to the point where the response is more like an equilibrium creep. (The slow rate responses will be important in the diffusion section for mesoscale parameterization, as the response becomes very noisy, like what might be expected more random, unforced, movement.) This rate response is important as it was found to be the most consistent way to compare molecular structures at their very best, especially since we have no idea what kind of molecular rates are actually present. That is, we are discovering the best response of the particular interfaces for direct comparison of different materials.

Static methods can also be used to evaluate the qualitative strength of a molecular interface where the static method is simply the difference between the optimized and equilibrated initial and final state. However, the static results are not necessarily

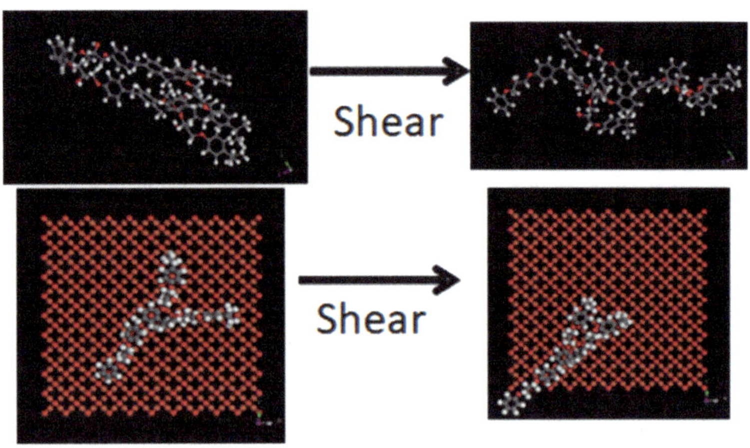

Fig. 5.5 The simplest cluster scale using small oligomer interactions next to itself (top: cohesive) and next to a surface (bottom: adhesive)

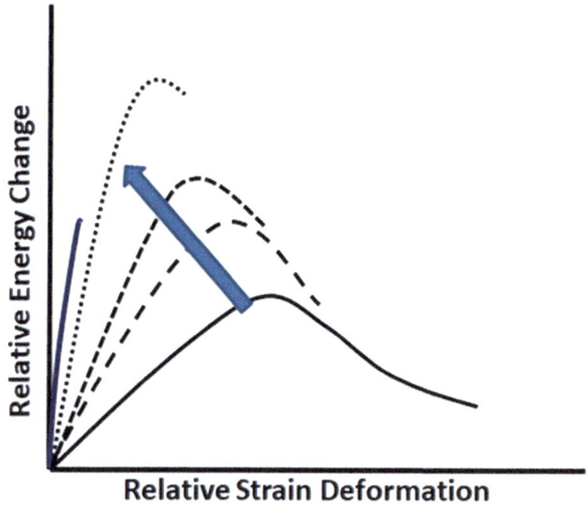

Fig. 5.6 Schematic of rate changes to the energy to strain deformation curves in a molecular model. The arrow indicates the direction of increasing strain rate

equivalent to the dynamics ones. Since the dynamic methods are able to track pathways, the complexity of the evolving interface is captured that will be missing in the static analysis. For example, using dynamics the resistance barrier to deformation can be captured, which the static methods cannot. However, the static methods will tend to capture very gross behavior trends that are usually the same as the dynamic trends. These general static trends can also be expected from surface theory, where the simple Young Dupree concept relates the work or cohesion or adhesion to the surface energy contributions (Eq. 5.1).

$$\mathrm{W}(\mathrm{cohe}) \sim 2 \times SE;$$

$$\mathrm{W}(\mathrm{adhe}) \sim SE_1 + SE_2 - \mathrm{IE}). \tag{5.1}$$

where W = work, SE = surface energy, and IE = interfacial surface energy.

So where static methods leave off, the dynamic adhesion fills in detail.

A good example of using this small-scale model was found in the use of molecular dynamic models to evaluate formulation choices [29–34] in order to develop a silver flake composite via fill that does not crack under HAST (highly accelerated stress test) conditions based on cyanate ester-epoxy copolymers. To demonstrate how small this model was, the polymer used was merely a trimer of the resin-curative combination. Interfaces were both cohesive (i.e., another chain of the trimer) and flat metal surfaces. The oligomer strand was laid out on the surface and a forcing potential applied to one end to shear it from the interface. On this scale, a pure tensile that lifts the polymer chain orthogonally from the surface was thought unlikely due to chain flexure, and when polymer-involved interfaces specifically fail, they probably do so first in a sliding motion (shear), until enough contact is lost and adhesion (or cohesion) is lost. As we are specifically looking at the immediate (adhesive or cohesive) interface, we are not concerned about the polymers surrounding the target oligomer, only in how that immediate contact loses energy with the deformation.

Since the energy and the deformation distance (shear distance) were tracked, an energy deformation curve (basically a stress-strain curve) could be constructed, the initial slope could be used for relative modulus, and the barrier to deformation could be used as the rough yield energy. The cross section of the model is also important, as it is used to determine the resistance force derived from the energy barriers found in simulation and can be measured by the projected area of the chains onto a perpendicular plane. For example, with two maximally interacting polymer strands (where the strands are next to one another in an optimized extended configuration to maximize all possible interactions) when the strain axis used is parallel to the respective chain axes, the cross section is always perpendicular to the strain axis used. For a polymer interacting with a substrate, the strain axis used is parallel to the chain axis, and the cross section includes the surface of the substrate immediately next to the chain. The strain axis employed tends to be the one with the smallest cross section to yield the highest interactions energies. This cross section is an important consideration, because it has always been found that if molecules strongly interact, the cross section tends to be small, whereas if there is weak interaction, the cross section will increase as non-bond distances increase and random coiling of the polymer increase. It was also found that every time water was added to the interface, the modulus dropped as the cross sections increased and energy responses dropped. When evaluating adhesion on substrates, it was clear that this simple method also showed adhesion tendencies where the adhesion increased as the surface energy of the substrate increased, which is expected from Young-Dupree.

For the via fill study, each cured structure from the cyanate ester resin + epoxy resin formulation was modeled. For cyanate esters, there are basically four different kinds of cure structures: the epoxy homopolymer, the triazine structure (from the cyclo-trimerization of the cyanate ester), the isocyanurate (from the cyanate-ester-epoxy), and the oxazolidones (rearranged isocyanurate) that were evaluated. Even more significantly, we found that we could determine the best cure by developing a molecular-scale Coffin-Manson cycling simulation [31–33], establishing a link between the molecular scale to the macroscale Coffin-Manson reliability relationship (Eq. 5.2).

$$\varepsilon = M N^{z} \tag{5.2}$$

where ε = strain, N = #cycles, and z is an experimentally defined material constant.

To accomplish the cycling, the oligomer was sheared from its interface to a specific strain; the forcing potential was then removed, and the interface was allowed to re-equilibrate; and then the strain shear was repeated. This cycling was done until physical separation of the interface, and initial resistance to the shearing potential

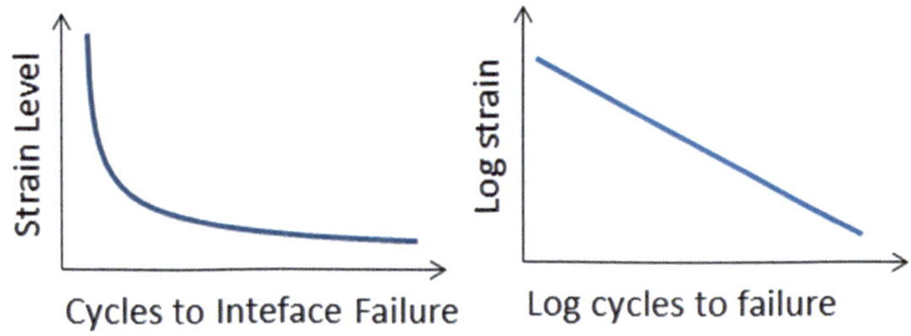

Fig. 5.7 General schematic of a strain-cycling result using a cyanate ester-epoxy polymer

was lost. The procedure was repeated at different strain levels, until enough points were obtained to construct a Coffin-Manson curve (Fig. 5.7 schematic). The most interesting discovery was finding that the strain/cycle plot followed the basic Coffin-Manson equation (which says the number of cycles is related to the strain by exponential factor), so that a log-log plot produced a linear fit between strain and number of cycles. This demonstrated that there is a Coffin-Manson response even at molecular scales that was unique for each oligomer type.

The strain to cycle, Coffin-Manson, curves were used to evaluate resins, curatives, and catalysts (which generate specific structures in the polymers) and to also evaluate the effect of water. This was done by evaluating oligomers comprised of the target ingredients, evaluating the resulting oligomer under molecular stress cycling both with and without water present, and then picking those ingredients that sustained the lowest strain at 1000 cycles (1000 cycles chosen arbitrarily as the target just for modeling purposes) on the plotted modeling data. Using these comparisons, we were able to develop trends on which cyanate ester resin-epoxy resin combination and their cured structure could sustain the highest strain at 1000 cycles. Most significantly, when the effects of these structures were determined, we were able to catalytically shift the chemistry in the appropriate directions as part of the formulation optimization process. The via fill made from the polymer ingredients targeted by the stress-cycling models yielded a commercial via fill which passed experimental HAST cycling tests better than the competitor materials [31–33]. The log-log plots were instructive to distinguish between different additives. For instance, when elastomers were added as modifier components, the zero-strain extrapolation on the log-log plot showed that these interactions had higher strain and the slopes were flatter, both trends consistent with the added compliancy of the rubber.

In another example, small oligomers were used to evaluate the difference between alkyl acrylate and cyanoacrylate responses used as adhesives on metal surfaces in flexible electronics [36]. An example is shown in Figs. 5.8, 5.9, 5.10, and 5.11 for both adhesive cases and cohesive cases. The polymer cohesive cases are particularly interesting and show that water affects the cyanoacrylate more than the acrylate, due to higher polarity functional groups, with the polarity high enough to compete for water attraction. That is, as the alkyl acrylate chains are separated during dynamic shear (Fig. 5.8), water adhesion is weaker than the cohesion of water, so at the end

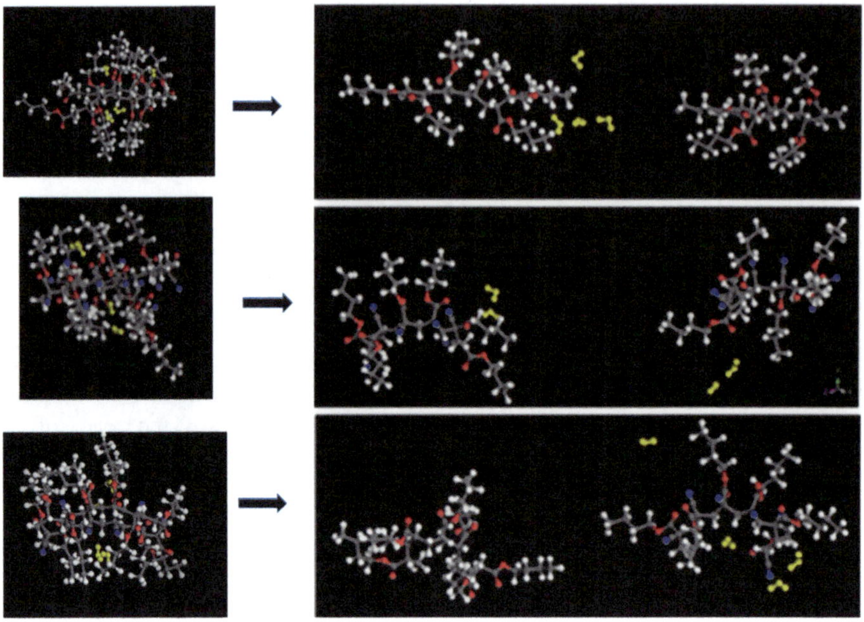

Fig. 5.8 Polymer cohesion simulations of an acrylate hexamer oligomer (top) and a cyanoacrylate hexamer oligomer (middle) and a polymer adhesive case using two (alkyl acrylate+cyanoacrylate) oligomer strands (bottom). Grey = carbon; red = oxygen; white = hydrogen; blue = nitrogen; modifying water molecules are identified in yellow

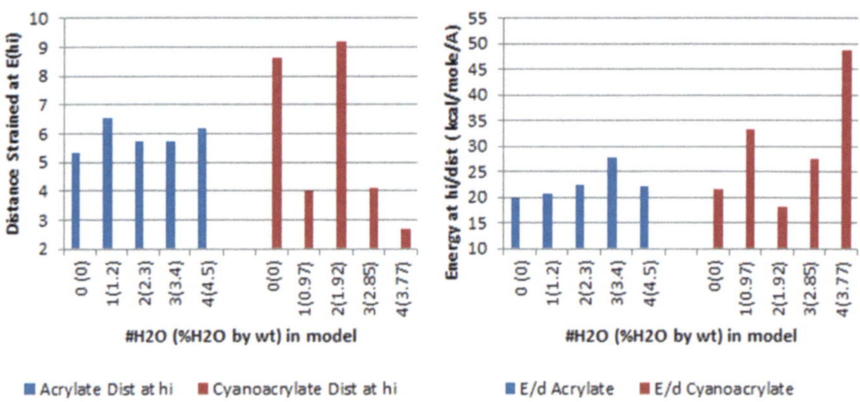

Fig. 5.9 Analytical results from models in Fig. 5.8. Left, relative strain distance over the trajectory for different water levels in the model; right, relative energy barriers to strain

of the trajectory, most of the water has stayed together. However, for the cyanoacrylate the polarity of this polymer backbone competes for water, and at the end of the trajectory, the cluster of water has had to separate, with half attracted to one chain and the other half attracted to the other chain. But note that the final water position

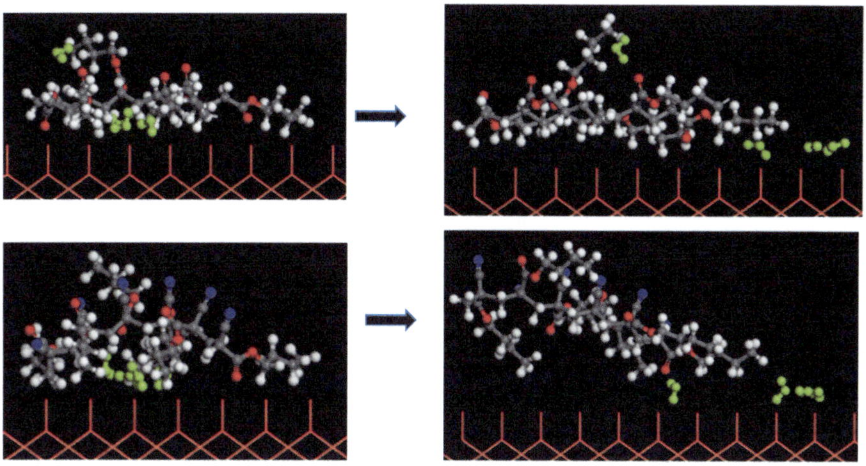

Fig. 5.10 Cu_2O adhesive models before and after applying force. Top, acrylate; bottom, cyanoacrylate. Light green = water molecules

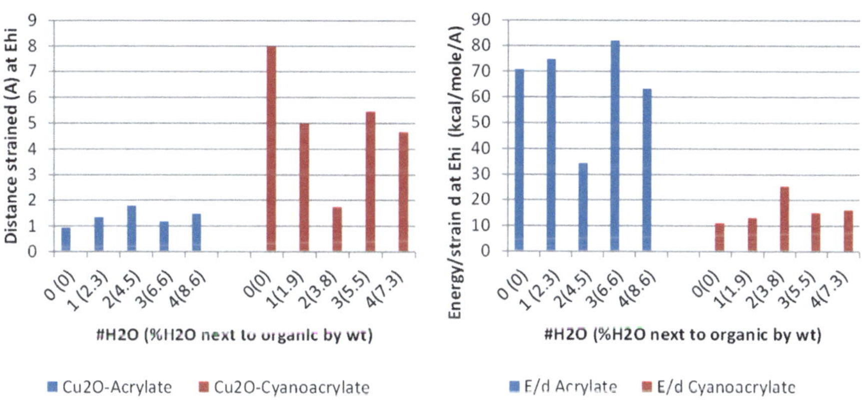

Fig. 5.11 Cu_2O adhesive analysis at the energy barrier: left, distance strained at the energy high; right, energy/strain change at the energy barrier in the trajectory curve

is at the trailing ends of the polymer, so that it is trying to stay together; only the competing polarity of the cyanoacrylate seems to pull the water apart. (Keep the water attraction to the cyanoacrylate in mind later, when water diffusion is discussed.) A third case of adhesion between the polymer strand types using one strand of the acrylate and one strand of the cyanoacrylate predictably showed the water would follow the cyanoacrylate and shed the acrylate side (Fig. 5.8). Energetically, the water level doesn't seem to change the energy change on the acrylate, and the Δenergy/Δdistance ratios are low (Fig. 5.9). The Δenergy/Δdistance ratios of the cyanoacrylate response is highly variable but starts off similar to the acrylate at low water levels but are high at the high water level. At the higher two levels of water, the strain (or deformation distance) plummets, suggesting that the interface is embrittling and explains the Δenergy/Δdistance result at the highest water level. That is, at certain moisture level, the cyanoacrylate should become embrittled.

For the copper oxide adhesive cases, Fig. 5.10 shows that for both the acrylate oligomer and the cyanoacrylate oligomer, the water competes the same for the polymer and polar surface. However, the energies are very different. Figure 5.11 indicates that the distance strained is very small for the acrylate but large for the cyanoacrylate, leading to very high adhesive energies for the acrylate but low adhesion for the cyanoacrylate. So, the acrylate-copper oxide is strong but brittle, and the cyanoacrylate-copper oxide interface is weaker put compliant, most likely due to the overdependence of the water. If the adhesive is a mixture of these polymers, then as the interface is moved, the acrylate should supply much of the energy resistance but fail first. The cyanoacrylate will supply additional resistance at the larger strain, but not as high as the acrylate and is more similar to the cohesion energy. Overall, the adhesive interface is stronger than the cohesive interface so cohesive-dominated failure is expected and was observed in test button-shear tests.

3.2 Structure and Architecture

Architecture issues are particularly important in defining the impact of specific interfacial molecular geometries, as the architecture will have a large impact on segment packing involved in interfacial interactions. Such interactions can in turn impact modulus, yield, strength, adhesion, glass transitions, diffusion resistance, and other properties that affect the reliability of the material by influencing the interaction densities. So as the model is built, attention must be paid to how well interfaces contact, the size of the contact and contact orientations, and any "defect" interactions which can lower the atomic contact density. Contact then can result from a combination of affinity and architecture. If the polymer is composed of structures with high affinity to the interface, then the optimized interface structure should hug the interface; but if there is an architectural defect that disrupts interfacial packing and contact (i.e., lowering the non-bond interactions), the interface will contain a gap (or molecular scale nano-void) at this molecular "defect."

Effect of architecture is especially confusing for amorphous polymers where structures will be random. This is in comparison to crystalline materials that already have defined structures and atomic architectures. Amorphous polymers on the other hand may have defined bonding, but do not have defined architectures. For example, we may know cross-link density, but we do not know exactly where, since cross-links are usually randomly generated. However, we typically know the chemical bond structure at the cross-link based upon the known components and the reaction chemistry. So the polymer considered for modeling should consider both the cross-link structure as well as the straight-chain part of the polymer, as each architecture is different.

A good example was found in the architecture of low k dielectric pore-wall structures, where both the architectural cross-linking as well as the pore size, had a significant effect on the modulus [34].

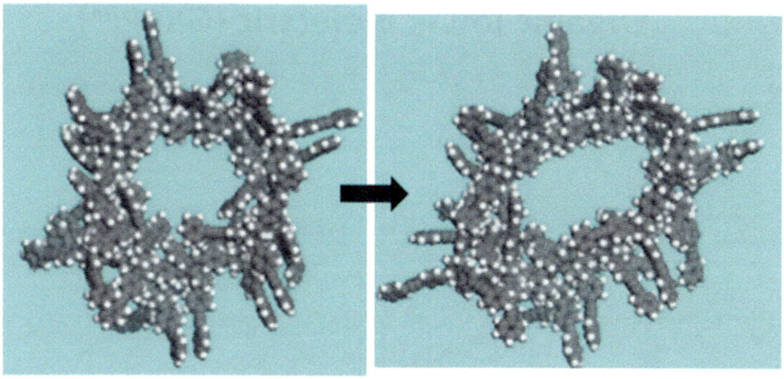

Fig. 5.12 Main "pore" for an organic low k dielectric GX3P before and after pore deformation. Wall structure cross-linking changes the modulus as seen in Fig. 5.13

Molecular models were constructed to represent an idealized open nano-pore (Fig. 5.12) but with different levels of cross-linking (or "defects") in the walls of the pore. The main pore was modeled based upon annular rings which were then deformed in order to simulate the energy response in order to extrapolate the modulus. Figure 5.13(top) shows an example of these results for the low k dielectric called GX3P™, and Fig. 5.13(bottom) shows the modulus results for both GX3P and NANOGLASS™ (an organosilicate-based low k dielectric). The results for both dielectric types show that there is a drop in modulus as the main pore increases in size ("pore volume" axis) but also as the cross-link density of the walls of the pore drops and bonding defects (deleting bonds) increase. NANOGLASS is interesting, because the choice of monomers can disrupt an ideal architecture. The triangles in the NANOGLASS chart show that modulus can be improved if the wall structure is ideal (highest cross-linking possible).

Spin-on-glass organosilicates are interesting examples because by nature, they cure to random and amorphous structures and encounter many of the modeling issues like organic polymers, but they contain specific architectural structures depending upon the formulation, so are excellent examples of how architecture can change performance film properties. Architecture may not seem like it should be so important in an amorphous system, but it was one of the variables found in the stability of cured ILD (interlayer dielectrics) silicates [37] during cleaning and in understanding the shrinkage of organosilicates used in flat panel displays [38]. The entire project will not be completely reviewed here, but molecular modeling was used to determine why we were obtaining up to 40% shrinkage in the film. FTIR analysis indicated that we were losing cage structures in the film composition, and quantum chemistry analysis suggested certain residual metals found by experimental analysis of the formulated polymer could catalyze depolymerization of these cage structures into more flexible ladder and ring structures. Three basic architectural structures were of interest: cages, ladders (multiple rings connected thru a common bond or several common bonds), and rings in order to show depolymerization, respectively (Fig. 5.14). Almost by definition of the 3D architecture, the cages

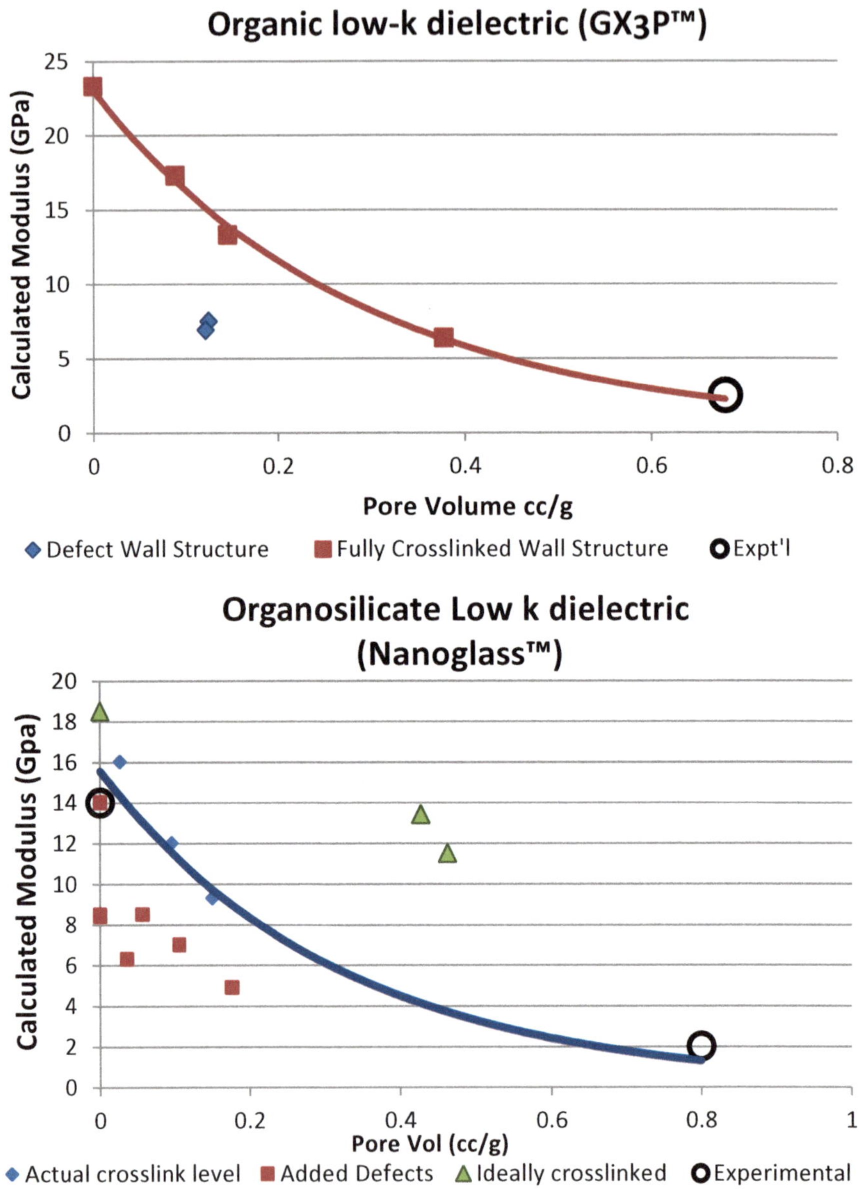

Fig. 5.13 Architectural effects on the modulus of an organic low k dielectric, GX3P™ (top) and an inorganic low k dielectric, NANOGLASS™ (bottom)

were found to have lower packing than the ladders and rings, with both disruption of packing by the corner cage substituents and lack of flexibility.

Initial cage structures were constructed in pairs and energy optimized. In order to represent depolymerized structures, different configurations of ladders and rings

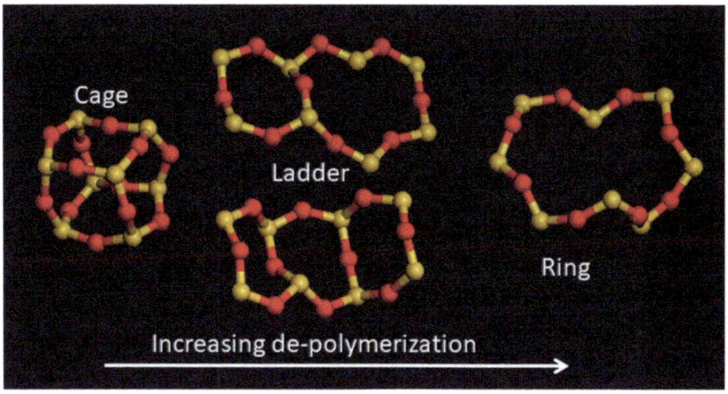

Fig. 5.14 Skeletal silicate architectural type examples [37, 38]

Table 5.1 Relative shrinkage in cluster models of depolymerized silicate cages [38]

Modeled fractional shrinkage from cage	Min	Max
Ladders (partial depolymerization)	0.29	0.35
Rings (full depolymerization)	0.21	0.41

were made from the original cage pairs, energy optimized, and the change in volume measured. What was found was that the shrinkage in these simple models mirrored the shrinkage found in experiment (Table 5.1), where "partial depolymerization" indicates different configurations of ladders and "full depolymerization" indicates rings. The relative modulus of interaction of each pair was also determined to understand how the depolymerization state impacts the mechanical resistance to shrinkage (Table 5.2 top) using the same methodologies discussed previously. As might be expected from a less-packed interface, the original "cage" structures had the lowest strength (defined from the highest energy barrier from the energy curves). So the volume may be expected to collapse due to rearrangement to the more highly packable depolymerized rings. A comparison was then made using the organosilicate compositions of the original formulation, against a new suggested composition comprising components with higher modeled cohesion. These new structures were indeed found to be more robust to compaction and film thickness loss (Table 5.2 bottom) by their higher relative modulus and strength and as a result had better mechanical properties than the original.

One area of architecture that is always investigated in organic polymers is that of the cross-link, since the cross-link helps to relatively confine sections of the polymer. So, from the aspect of small molecular models, the polymer must be thought of in terms of sections (the cross-link and the straight main chain), because of the potentially different interactions. Qualitatively it was found that cross-links composed of 3D tetrahedral centers usually have lower packing or lower contact at the

Table 5.2 Relative modulus and strength obtained from cluster dynamic models [38]

Relative property	Modulus	Strength
Original formulation	1	0.36
Original partial depolymerization	0.75	0.48
Original full depolymerization	1.45	0.68

Property improvement ratio	Modulus	Strength
Improved/original formulation		
Cage	1.3	2.1
Ladder partial depolymerization	1.4	1.3
Rings full depolymerization	1.2	1.2

cross-link itself, consistent with molecular steric hindrance of the tetrahedral architecture. However, if the cross-link is planar trigonal (such as that involving aromatic rings), there may be very good packing at the cross-link. The same may be said of the architecture of the linear part of the polymer as their structures also influence the intermolecular packing that should be considered for the best interactions, and the length of the straight main chain influences whether there is more random interaction, more packing interaction, or even highly crystalline orientation. This helps to explain the consistency of scale between the molecular and mesoscale models, as the molecular models consider the best interactions. This continuity of scale was especially captured in the mesoscale response to cross-linking and will be discussed in the next section. (The mesoscale is probably the most significant in that it will allow the full stress-strain curve to failure to be simulated.) This piecewise perspective also helps to explain the (often vexing to formulators) loss of strength of a polymer when over-cross-linked. From the small-scale component level, over-cross-linking can increase the number of 3D sites that decreases the interaction (contact) density and so decreases the interaction energies. Another architectural influence is crystallinity. However, because of their size, it is difficult to get significant crystalline spherulitic content represented in a molecular model to be realistically representative of their effect. In this case, the nature of the oriented chain interactions can be used in a relative manner to understand the effect of crystallinity.

3.3 Molecular-Mesoscale

The molecular to the "molecular-mesoscale" jump is merely a classic particle scale-up of the molecular scale, but instead of using atoms as the basic building blocks, groups of atoms are used that are parameterized into a new “bead.” Since mesoscale (or "molecular-mesoscale") and molecular models are both based upon classic Newtonian physics, the techniques to run them are similar in both cases. Bead parameterization has been discussed in previous papers for the epoxy being developed at Delft University [2, 3, 13–16], so just the use of the software to look at interface failure will be discussed in this section. Once the bead parameters have been defined in the force fields, the building can either be done bead by bead like molecular models or by redefining molecular models into the bead models. The important thing for these models is to recognize the bead definition and the relationship to the atom grouping that is represented in the force field. We have been using self-defined force fields for all of the work done for reliability, because of the desire to jump length scales further than the functional group level often used in typical mesoscale force fields.

The original study specially targeted to jump length scales to a 59 heavy atom repeat unit epoxy bead [13, 39]; and the interface was constructed using a periodic vacuum cell of the mesoscale polymer and copper oxide. The copper oxide was parameterized from molecular models using a 2×2×2 unit cell as the bead. As the copper oxide beads were relatively fixed during adhesion simulation, only the copper oxide-to-polymer interactions were parameterized. These bead sizes enabled larger mesoscale models representing millions of atoms of the polymer and interfaces ranging from 14 to 50 nm that could be handled on a low-end workstation. The beads were large enough that the most important parameters found were the geometry (bead-bond length, thru-space bead-bead distance), bead-bond strength and non-bond bead strength. This bead size jump was so large it was thought that specific atomistic parameters of the atomistic bonds (such as dihedral) were contained and averaged in the bead parameters determined from the molecular model parameterization process which employed pairwise stress analysis of small oligomers [13–15], and the bead dihedrals were left freely rotating. Although more work could have been done to isolate more parameters, these simple parameters were found to be key for this polymer. Most importantly to complete more realistic parameterization, a bead-bond failure criterion was imposed in these models that were determined from the molecular stress models, where the bead-bond length at failure was defined after the energy barrier in the molecular model at the point where the molecular bonds were unrealistic [14].

The model was then optimized and equilibrated at room temperature, and a short compaction test was run (where the copper oxide was stepped in small distance increments toward the polymer and re-equilibrated) to be sure the full interaction at the interface is achieved at the lowest energy of the model. The mesoscale strain models employing the vacuum periodic cells were run using a pseudo-NVT dynamic model and the copper oxide layer stepped into the vacuum at each step. Upon a short equilibration at room temperature, the epoxy was found to follow the copper oxide due to attraction to the interface. We are calling the procedure a pseudo-equilibrium NVT dynamic (also could be termed a nonequilibrium NVT), because the short temperature equilibration was not allowed to go long enough for complete equilibration but only long enough for the energies to start to bend toward equilibration. In this manner, it was thought that the system would always be in a state of pseudo-stability more representative of a system under dynamics, rather than in a stable state like that found in equilibrium processes (such as creep). The bead-bond failure criterion was imposed at each step of the mesoscale model, where all bead bonds were broken if they exceed the bond failure criterion. Using this process, like the molecular models, an energy vs deformation (i.e., stress-strain) curve could be constructed (Fig. 5.15) and the pseudo dynamic step process repeated until all bonds at the interface (either cohesive or adhesive) were broken. In this way, the simulation could be taken all the way to interfacial failure. The initial modulus (initial slope) yield and strain at failure could then be identified.

The first models concentrated on the adhesive separation of the flat epoxy-copper oxide interface (Fig. 5.16). Upon analysis, the same molecular-scale modulus and "molecular-mesoscale" modulus was found (as mentioned previously), which was also consistent with the modulus found in experiment. This continuity of scale between the molecular and mesoscale made sense, since molecular-scale models were used to parameterize the beads. This continuity helped to demonstrate that the source of modulus is probably from the small-strain interactions that could be captured in a molecular model. In addition, the yield strain in the mesoscale model was similar to the experimental, suggesting that the mesoscale was more quantitative than the small molecular interface models since the molecular models usually captured an

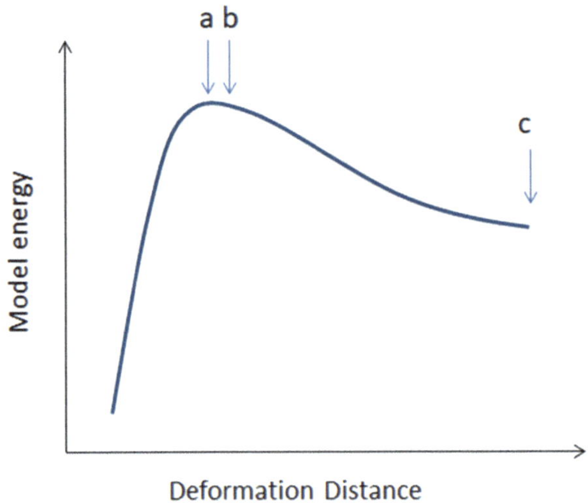

Fig. 5.15 Schematic of the energy deformation plot of a mesoscale model ((a) max energy; (b) max # bonds broken in a single step; (c) complete interface separation)

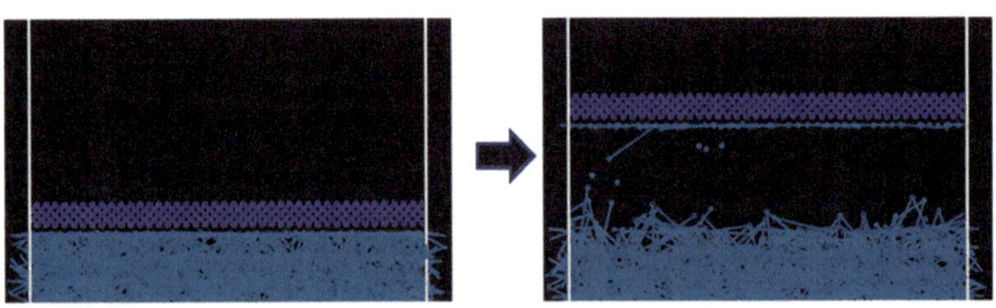

Fig. 5.16 Example of cohesive-side failure in a mesoscale model (before (left) and after (right) failure). Top layer is the copper oxide; bottom layer is the epoxy

energy barrier yield at higher strain levels than expected (even though the yield strength trends between molecular and mesoscale were similar). The difference suggested that the yield phenomenon was at a level too large to be captured by the small molecular model but could be captured by the "molecular-mesoscale" level which represented a larger bulk representation across a multitude of bead interactions and included bead-bond failure. Interestingly, the mesoscale models showed that the energy barrier deformation, or the global yield, roughly corresponded to the maximum number of bead bonds broken within a single step, suggesting that the evolution of failure for this epoxy is following atomistic failure. The results of the mesoscale epoxy-CuO interface models showed that the epoxy under investigation would have failure at the cohesive side of the interface, consistent with experiment. It also predicted that moisture would push the interface to lower yield and lower modulus, with more of the cohesive interface involved in the failure event (more epoxy would be left on the copper oxide when water was present), and push the absolute failure (complete separation of the interface) to higher strain (Table 5.3), which were all expected based upon experimental trends. In addition, the cross-link densities were compared. In the mesoscale models, as the cross-link densities increased, the modulus and yield strength increased, but the strain at break decreased. This was interesting, as the increase in modulus and strength is predicted based upon better architectural definition of the linear part of the polymer. The strain at break was not predictable from the architecture but probably expected, if there are more bonds being broken earlier in the deformation, as suggested by the bond-break trends mentioned earlier.

Further analysis was done using sawtooth interfaces [15] as exemplified in Fig. 5.17, to represent roughening of the copper oxide. The same general trends were found for the roughened interface as the flat interface, except the steeper the sawtooth, the more shear was involved and the lower the modulus and strength and less engagement of the cohesive side of the interface. It was interesting that the last

Table 5.3 Some properties found from the epoxy mesoscale model and the epoxy-CuO interface mesoscale model [2,3,13,14,15,16,39]

Modulus comparisons	Cohesive modulus (Gpa)	Cohesive yield %strain	%yield at void onset	CuO adhesive modulus (Gpa)
Mesoscale uncross-linked	0.97	2.9	3.8	~0.4
Mesoscale cross-linked	2.15	3.4	3.9	3.05

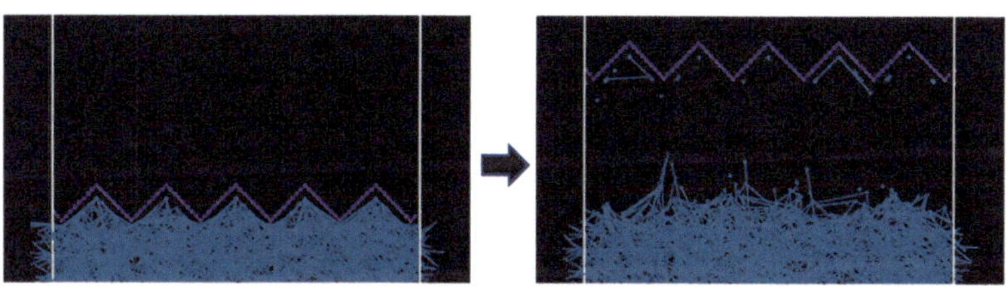

Fig. 5.17 Example of a sawtooth roughen interface before and after failure

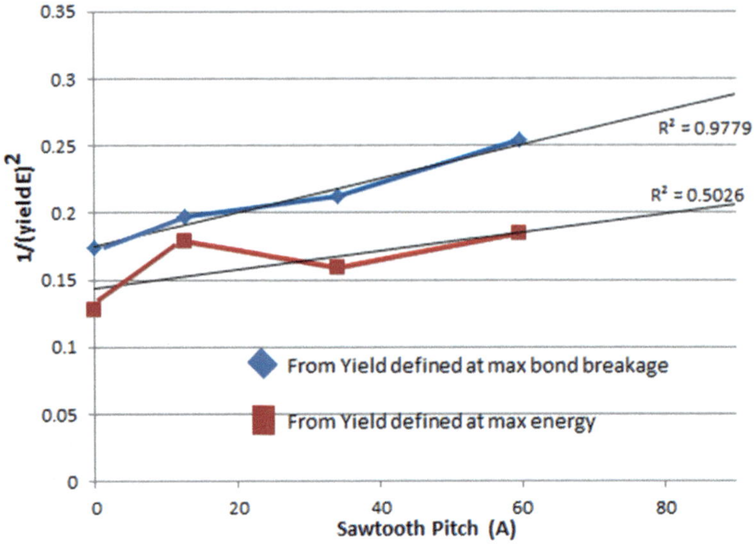

Fig. 5.18 Hall-Petch relationship of the mesoscale epoxy-copper oxide interface using models employing different sawtooth pitch

bead bonds failed tended to be located at sawtooth tips, suggesting that the highest resistance was from the most tensile interaction. Analyzed further, these mesoscale models also showed that the roughness followed a Hall-Petch relationship (Eq. 5.3) between the sawtooth pitch and the yield strength, as shown in Fig. 5.18 [2, 4], suggesting that molecular-mesoscale modeling could be used to investigate roughness morphology changes at the interface.

One interesting trend that was captured in this mesoscale study was that of strain orientational effects. When the original tensile models were repeated in full parallel interface shear, a much lower modulus resulted similar to the modulus found with the slotted "sawtooth" configuration. In fact, both the sawtooth interface models and the flat interface models repeated at different shear orientations yielded the same results. So the mesoscale models were capturing a more complicated accumulated interaction than the simple small molecular models would have predicted, even though the bead interactions were all derived from the molecular models. Even more compelling was a model that used a shear orientation across a sawtooth adhesive interface which would necessarily simulate shear, tension, and compaction at the same time depending upon the location on the sawtooth. This model resulted in higher energies than any of the previous cases, demonstrating that a complicated interface with a compilation of mixed modes could be responsible for better adhesion [4, 15].

$$\sigma = \sigma_0 + k \cdot \delta^{-0.5} \tag{5.3}$$

where σ = strength; δ = grain size

3.4 Using Static Quantum Mechanics to Understand Adhesion Effects in a Metal Diffusion Barrier

As mentioned previously, one good use of quantum mechanics calculations over classic Newtonian calculations is for looking at metal issues, where the force fields are usually not available. One very simple example employed the use of quantum mechanics to look at the static adhesion of diffusion barrier layers used in IC interconnect structures [40, 48]. It is briefly included in this chapter as a way of introducing how molecular modeling can comparatively look at layer effects and balances between cohesion and adhesion, as this perspective allowed prediction of failure in the metal barrier. (Another example of using comparative quantum mechanics the binder chemistry used in metal stereolithography will be also discussed in Sect. 5 that uses these underlying concerns of strength balance.)

It is well accepted in the industry that inadequate bonding increases the likelihood of copper diffusion into silicon even thru a diffusion barrier. We discovered that a TiZr or TiZrN diffusion barrier prohibited this diffusion (Fig. 5.19).

Static adhesive calculations were employed in order to find the general trends in both the barrier cohesive energies and the adhesive energies to the barriers (Si side and Cu side) Table 5.4. It was found that there is a large energy difference in the adhesive and cohesive interfaces when there is adhesive failure and significant copper diffusion (red cells in Table 5.5), whereas there is a low energy difference between the barrier cohesion and barrier adhesive interfaces for those barriers with good adhesion and no copper diffusion (green in Table 5.5).

From these static adhesion and cohesion calculations, it was clear that the stress distribution (or potential stress states, based upon differences in cohesion and adhesion) can be much lower for the TiZr barrier than even for the Ta barrier. As a result, TiZr barrier tests had no copper diffusion found (Fig. 5.19 right) [40].

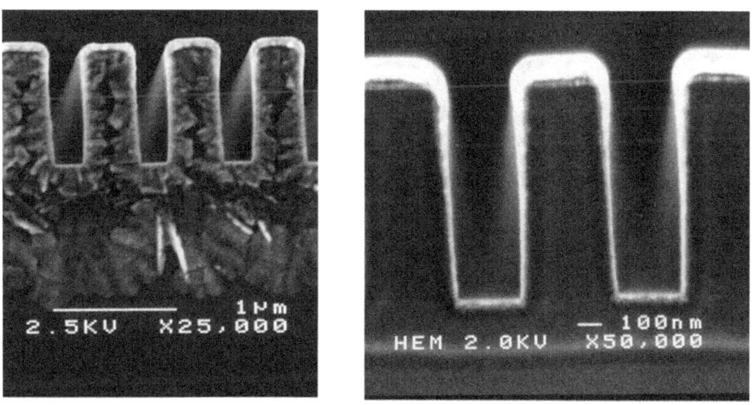

Fig. 5.19 Micrographs of interfaces showing copper diffusion down into silicon layer (left); and no diffusion (right) using a TiZr4 barrier with good adhesion

Table 5.4 Calculated static cohesive and adhesive energies (cal/m^2). Benchmark references are from cohesive energies [41]

Layer and interface	Interface adhesion CRC Ref. [41], cal/m^2	Model metal cohesion cal/m^2	Model barrier adhesion on Si cal/m^2	Model barrier adcohesion on Cu cal/m^2	Model barrier cohesion cal/m^2
Cu-Cu	0.182	0.15			
Ti-Ti	0.385	0.4			
Si-TiZr-Cu			0.334	0.274	0.573
Si-Ta-Cu			0.25	0.464	0.737

Table 5.5 Calculated stress energy change between cohesive and adhesive interfaces compared to the experimental layer stress

Multilayer description	Measured layer stress (MPa)	Modeled ΔE (Differential of Barrier cohesion to Si Adhesion) cal/m2	Modeled ΔE (Differential of Barrier Cohesion to Cu Adhesion) cal/m2
Si-TiZr-Cu	30	0.063	0.123
Si-Ta-Cu	50	0.487	0.273

4 Equilibrium Considerations

4.1 *Diffusion*

Diffusion considerations are important to reliability, when considering what compositions are present or what may be changing. For example, a change may be a wandering solvent or chemical reactant or water location and its impact. Diffusion simulations include internal and surface diffusion and may be handled using both quantum and classic methods.

4.1.1 Wetting and Surface Diffusion

Molecular models using classic dynamics have been discussed in other work [2–4, 29, 30, 44, 45, 46, 47, 48] for small molecule and atomic metal diffusion simulations, so examples will only be briefly described here. However, an example of quantum dynamics for metal diffusion will be more completely discussed in this section.

The surface diffusion simulations are interesting in that they have been instrumental in understanding wetting, as they mimic surface energy trends. Examples of wetting have been covered elsewhere [4, 29, 30], but they basically are similar to a contact angle experiment, in which a mass of the formulation (as represented by the molecular composition in the molecular model) is placed in contact with the substrate of interest, and this mass allowed to equilibrate for a set time. The extent of spreading has been found to mirror the wetting activity of the formulation, and the energy drop usually mirrors the formulation plus the formulation interfacial surface energy combination. From this perspective the molecularly derived trends generally follow simple surface theory (which says that good wetting is achieved when the substrate solid surface energy is greater than the combination of the liquid and the solid-liquid interfacial surface energy), as generalized by Eq. 5.4 [29, 30, 42-44]. So using a simple computational experiment, formulation components could be chosen based upon their wetting ability.

$$\gamma(s) \geq \gamma(sl) + \gamma(1) \tag{5.4}$$

This same modeling experiment can be altered somewhat to determine the wetting forces responsible for surface driven phenomena such as the capillary filling of filler particles in underfills, to be used to parameterize a more realistic particle dynamic model of the actual filling dynamics [43, 44]. For determining the forces active during underfilling, it was assumed that the major driver was the binder attraction (or repulsion) to the solid surfaces (filler particles or capillary wall). So, the force of two approaching surfaces, such as filler-filler or filler-wall, could be estimated by the distance dependence of the energy drop found of the binder formulation to the surface. That is, the energy drop over a specific equilibration period was determined using the same formulation mass, starting from different distances away from the solid surface (and then the energy drop at each distance plotted against the distance). This energy drop with distance was found to be steep if the surface energies were high and shallow when the surface energies were low; and the positive slopes could be used as the attraction force in parameterizing bulk particle dynamic simulations (discrete element models) of capillary flow.

When these energies were applied to a discrete element model of the underfill process, it was found that the underfill binder with the higher formulation surface energies tended to agglomerate, whereas those with low surface energies freely moved around the simulated solder bumps. This explained the heavy streaking from particle agglomeration in the acoustic images found in the formulation known to have binders with higher surface energies. Lowering the surface energy of the fillers with coatings appeared to help the fill process [43, 44].

Another example of surface diffusion dynamics has been well documented for the development of channel coatings in a micro-gas chromatography chip [45, 46, 47]. Although this is not a reliability case, it is mentioned here as another method to consider when looking at what interactions may influence the molecular definition of the model. In this case, the diffusion of analytes on the MEMS chip was modified by a surface coating on the walls of the channels. For the model, the analyte mixture was minimized as a mass and then the surface (representing channel walls) applied to the mass. In order to include directionality, a small forcing potential was applied at the center of mass for each molecule so that the molecules would flow in one direction and the whole system allowed to equilibrate at target temperatures. The most exciting result to come from this analysis was the same trend separation between mixtures of analytes was found in the simulations as that found in normal GC (gas chromatography) experiments. This model allowed prediction of a better material to coat MEMS channels and also predicted the contamination of the column due to analytes that had poor surface diffusion.

In the final example, quantum dynamics was employed to determine surface diffusion tendencies of metal-ceramic interfaces at elevated temperatures in order to understand problems associated with liquid metal wetting on ceramics. While this issue may seem like a macroscale issue, with new additive manufacturing tools being developed which employ depositing powder-resin formulations, nanoscale interfaces are formed, and these issues can quickly become molecular-scale issues for these emerging processes. In this example using a process employing additive manufacturing (3D printing) of a metal-deposited green part (before thermal treatment to obtain debinding and metal flow) in contact with a ceramic mold material, it was found that during the metal melting step, the nickel powder used did not have adequate wetting properties against the ceramic mold surface in order to reproduce the features of the part without defects. For the melt process under consideration, it appeared that dewetting created interfacial voids and deformities in the cooled part. So molecular modeling was engaged to understand the compositions that could be causing these dewet issues.

To answer this question, a method called quantum mechanics-molecular dynamics (QMMD) was used which uses a hybrid of quantum mechanics approach: quantum methods to define the atomic interactions, superimposed on classic equations of motion. As may be expected, the modeling method is slow but is more flexible as to the atom types that can be used in the model.

Extremely small molecular models were used to represent the metal particle being simulated, as shown in Fig. 5.20. This model was minimal in order to speed up calculation turnaround and starts with a 4-atom nickel metal cluster (starting from the nickel crystal structure for the metal of interest) on different surfaces that the metal may see. When the metal was highly attracted to the surface, the cluster spread out over the surface; when there was no attraction, the 3D cluster either moved further away or just rearranged the original cluster slightly to another cluster shape. One additional result from these models was the possibility of reaction since quantum mechanics allows prediction of bond rearrangement. So, if there was reactivity to the surface, the cluster and the surface would rearrange forming new bonds to the surface atoms. As shown in Fig. 5.20, nickel will wet to silica (SiO_2) through a reaction to form Ni-O-Si bonding, but it does not wet silicon carbide (SiC).

Hafnium, hafnium oxide, and hafnium carbide were also tried as a potential surface because chemical analysis suggested it might be present at the interface as a minor component of the nickel powder. (To explain hafnium at the interface, an additional QM/MD diffusion simulation of a nickel-hafnium mixture was done.

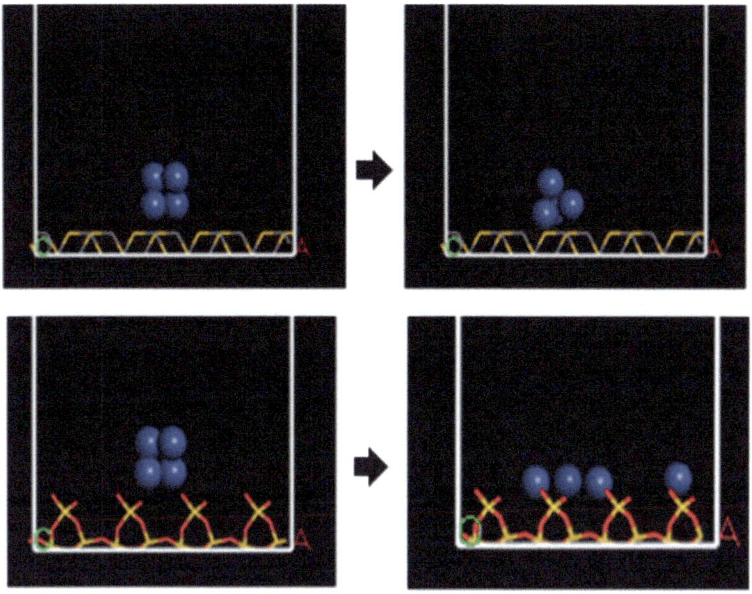

Fig. 5.20 Nickel diffusion simulation example on two different substrates. Top: Ni on SiC (no wetting). Bottom: Ni on SiO_2 (wetting with reaction)

Table 5.6 Qualitative wetting activity using QMMD

Surface (stationary)	Mobile atoms	Temperature (K)	Wetting activity on surface
SiO_2	Ni	1950	+spread (reacts)
SiC	Ni	1950	−clump
ZrO_2	Ni	1950	+spread
ZrC	Ni	1900	+spread
Hf	Ni	1950	+spread
HfC	Ni	1900	+spread
HfO_2	Ni	1900	−clump
W_2O_3	Ni	1900	−clump
WC	Ni	1900	+spread

This simulation showed that hafnium could diffuse to the surface of the nickel, explaining the chemical analysis.) In addition tungsten carbide and tungsten oxide were tried as possible surfaces present on the ceramic.

The general summary of the dynamic simulations are shown in Table 5.6. It shows that nickel has good wetting to possible ceramics present (silica, zirconia, zirconium carbide), but other oxides of metals that may be present such as hafnium oxide and tungsten oxide prevent nickel wetting. However, it was the activity of silicon carbide that was of major concern because of the organic binder used during additive manufacturing. The organic binder presented the possibility of being the carbon source for carbide formation from incomplete binder burnout, and from these simulations, indeed, it was found that silicon carbide would prevent nickel wetting and so was the major suspect in the process failure. This observation was partially confirmed using parts made using a different process where no binder was involved; when melted against the same ceramic as before, no dewetting defects were found. So whereas the trends found in the QMMD models were not surprising, they helped to explain observations from the compositional level.

4.1.2 Diffusion in and Across Layers

Diffusion inside a material is a standard tool in most molecular modeling software packages, in which a 3D periodic cell construction of the polymer (or solid) with its diffusant is built and equilibrated at the desired temperature. The root mean squared displacement of the diffusant is then calculated, and the diffusivity is extracted from the time-dependent plot.

The diffusivity gives you an average random 3D perspective on how the diffusant may move and so describes most of the story for cases when the volume is greater than the interfacial surface areas, so surface effects are minimized. However, in electronics, as the surface area effects become more important, the aspect of directional diffusion becomes important. For instance, it has been noticed that the diffusivities of penetration across the interface could be different than the internal diffusivities depending upon the composition. So by necessity we have extended the utility of the basic technique, for cases of interfacial diffusion to bias diffusion in one direction only for penetration. In this case, layers are built within a 3D periodic cell (infinite layers). The diffusant either is one of the layers or is enriched in one of the layers, and the whole system is allowed to equilibrate. One more added aspect of reality was added: that of directionality. That is, usually the diffusant flows in one direction. In order to ensure there is no back diffusion, either a barrier layer was added, or a forcing potential was placed on each diffusant molecule to define its one directional heading. The barrier layer could be any high density layer that could be used to act as a physical barrier and was always dynamically static in the simulation. Note that this model could have been constructed without the barrier layer which allowed a central diffusant layer diffusing in two directions, but because of mixing in the middle, a much larger unit cell would have had to be used. The directionality helped to simplify the calculation and the analysis afterward.

A mesoscale example that has been previously reported is that of amine catalyst diffusion across a photoresist layer [17] from the dielectric layer next to it. For this example the extent of diffusion was important because of the inhibiting effect on the photoresist cure. The diffusing amines and the polymer layer they were diffusing thru were parameterized much like the stress-dynamics mesoscale models. A big difference in the parameterization was consideration of the equilibrium process during diffusion. It was found that like the parameterization of the stress models, bead interactions had to be run at different rates, but the energy-strain curve was extrapolated back to zero strain, in order to represent this equilibrium process during

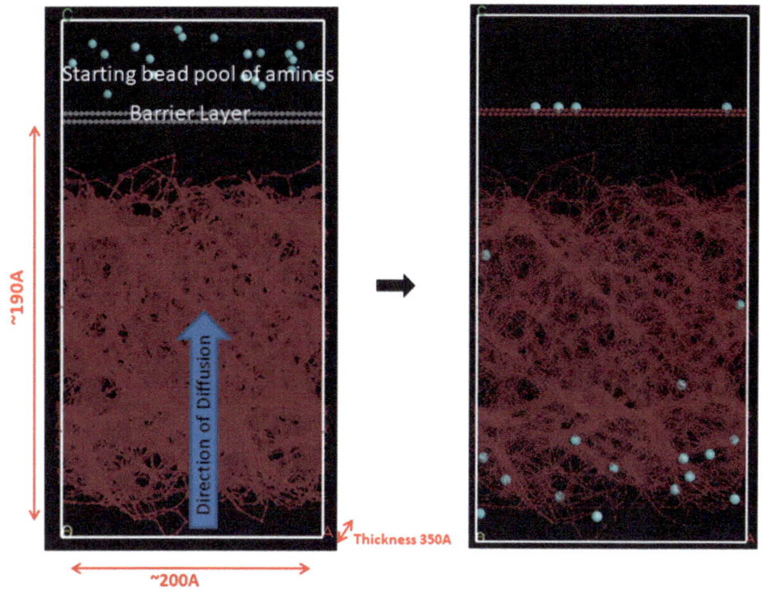

Fig. 5.21 Example of mesoscale diffusion of the smallest amine modeled (amine is small enough to be represented by one bead; larger amines are represented by multiple beads). Start configuration is at left. After 5000 ps NVT RT equilibration is at right [17]

diffusion (this parameterization was a little more difficult because with near zero strains and rates, the responses became more noisy).

The mesoscale model itself was a simple NVT equilibration, with the amine starting at one end of a PMMA (polymethyl methacrylate) polymer monolith representing an acrylate photoresist. The simulations showed that penetration into the photoresist could be slowed by the type of amine, water content, and also the size and architecture of the amine (i.e., branched and larger size tended to be slower), all which affected the level of agglomeration of the amine prior to entry into the photoresist. If the amine was small enough, it could enter and almost immediately diffuse thru the entire height of the photoresist as shown in Fig. 5.21. Figure 5.22 plots the vertical penetration distance and the population of amines in the polymer after equilibration indicating that the furthest penetration is almost all the way thru the polymer layer. Since the issue was poisoning of the catalyst in the photoresist, this penetration was important to capture.

One good example that should be mentioned is that of water diffusion from a polymer layer into the immediate interface between the polymer and copper as shown in Fig. 5.23 for two different acrylates [49]. This case does capture internal diffusion, as the water is started randomly distributed in the polymer layer, before the copper is introduced. What is interesting is the confining effect of the interface to accumulate water, depending upon the polymer. Figure 5.24 shows a bilayer of copper and an epoxy polymer containing water, before and after performing a 5000 ps equilibration at RT (NVT). As is shown in the figure, internal water

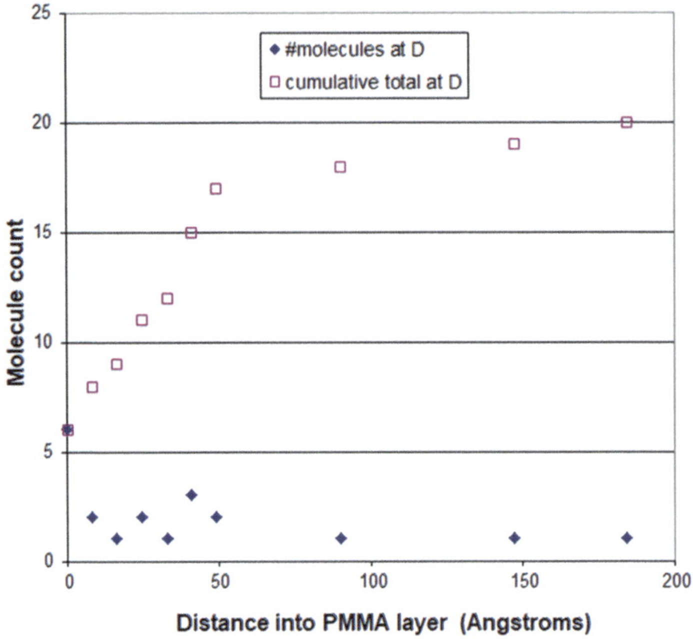

Fig. 5.22 Penetration distance into simulated photoresist (PMMA) layer showing the maximum penetration is almost all the way thru the thickness of PMMA

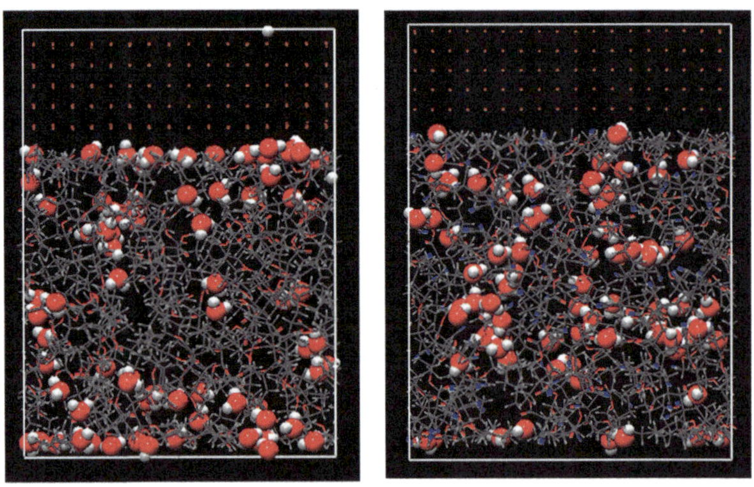

Fig. 5.23 Final distribution of water in an alkyl epoxy (left) and a cyanoacrylate (right). Water molecules have been increased in size for visualization [49]

diffusion occurs with accumulation at the copper-polymer interface. In this case, the water had been pre-optimized inside the polymer, before temperature dynamics was applied. In the manner of the surface diffusion models, all simulations used the same time and the same quantity of water. In all cases, the water diffused to the copper interface, enriching at the interface. It was found that when the cyanoacrylate (discussed in Figs. 5.8, 5.9, 5.10, and 5.11), was compared against the alkyl acrylate, much less water accumulated at the copper-polymer interface for the cyanoacrylate. This should be of no surprise because the simple molecular dynamics used for the cohesion and adhesion models showed a much higher preference of water for the cyanoacrylate, giving the water a stabilizing matrix. When bulk diffusivities were extracted from the simulation, it was found that both polymers showed similar diffusivities, even though the cyanoacrylate retained more water in the polymer. That is, water appeared to be acting more like an independent entity for the alkyl acrylate polymer allowing it to travel freely thru the polymer, to accumulate at the copper interface, whereas for the cyanoacrylate, water attraction kept most of the water diffusion to a local level. In addition, simulations at two different temperatures (260C vs RT) showed that the alkyl acrylate tended to attain twice the pressure rise as the cyanoacrylate when the temperature was raised from RT to 260C, suggesting that the rise in internal pressure was more contained in the cyanoacrylate, consistent to the containment of water movement. This further suggested that the interfacial strain due to the pressure rise could be controlled by the polymer. To further look at the impact of polymer, another case was run with the same epoxy used previously in the mesoscale models. It was found that a different pattern of accumulation appeared at the copper interface. Although the percentage of water accumulation was similar to the alkyl acrylate, the epoxy resin tended to have pockets of water accumulation in molecular-scale nano-void regions (Fig. 5.24).

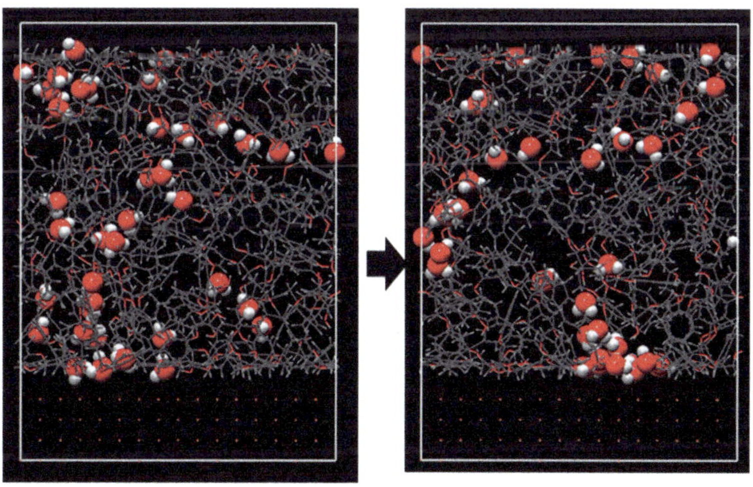

Fig. 5.24 Water diffusion in an epoxy from the randomly dispersed state to the copper interface. Water molecules have been increased in size for visualization [49]

5 Chemistry and Miscibility Involved Failure

This section was reserved for molecular models that are not tied directly to mechanical properties or how the molecules are moving in the material but how composition, along with understanding of the material, is part of the bigger picture in understanding and predicting whether the material will fail.

5.1 Chemical Stability

One final perspective of the molecular model in reliability that cannot be overlooked is the use of the quantum chemistry to understand the thermodynamic stability of the polymer. These are simple models in concept, in that the molecular structure of the components and the cure structure are calculated and the energy difference is determined. What is nice about the standard available quantum chemistry packages is that the thermodynamic free energies and kinetic transition states can be easily determined at different temperatures, showing the preference of the reaction to occur at the desired temperature. These types of models can be engaged up front to determine the most likely cure structures to be used in a stress model.

A good example that illustrates the use of quantum chemistry is from a recent study of the binder used in metal stereolithography (MSL), which uses highly filled metal powder formulations with a polymer binder to build (layer-by-layer) a three-dimensional green structure. The binder cures using UV exposure during each layer deposition. The green structure is then placed into a furnace under a thermal cycle to first burn out the binder (BBO or binder burnout) at temperatures low enough to avoid excessive metal oxidation (or other reactions like carbide formation) that might affect sintering and metal performance, and then the temperature is raised high enough for the metal particles to sinter.

Two issues being addressed at the time was incomplete binder burnout, leaving enough carbon to cause carbide defects (which was confirmed by elemental analysis) in the sintered metal, and incomplete cure which would make the structure vulnerable to thermal stresses during binder burnout. So the study involved looking at the possible cure compositions and their thermal decomposition temperatures. The binder used in this MSL study employed two basic types of monomer acrylate resins: one based upon polyol branched structures and another based upon triazine branches.

Binder burnout (BBO) calculations for the depolymerization chemistry of the main chain part of the acrylate clearly show that the more rigid and brittle triazine components would degrade at much higher temperatures than the 400 °C maximum required to minimize oxidation of the metal particles, whereas the more flexible polyol components decomposed at temperatures well below the binder burnout (BBO) target temperature of 400 °C (and accounted for much of the mass of this portion of the polymer). This analysis predicted why we were seeing carbon in the

Table 5.7 Temperatures needed for polymer degradation to start, as predicted from the calculated negative (favorable) free energy thermodynamics

Acrylate cross-link type	Temperature at negative G for cross-link cleavage	Temperature at negative G for acrylate main chain depolymerization reaction to
Triazine	>700 °C	>700 °C
Polyol	<<400 °C	~400 °C

Table 5.8 Relative transition state energies for acrylate reactions

Reaction of free radical from monomer acrylate 1 to acrylate monomer 2		ΔE at transition state (kcal/mole)
Acrylate 1 (initiated end)	Acrylate 2	
Triazine	Triazine	6.7
Triazine	Polyol	11
Polyol	Polyol	71
Polyol	Triazine	17

sintered part. This analysis also predicted a problem in stress distribution, as it was clear from the thermodynamics that the part could become too brittle during the higher-temperature sintering part of the process due to the brittle subcomponents left behind from BBO. That is, the brittle subcomponents predicted to form from the calculated negative (favorable) free energies of formation, could serve as stress-induced initiation centers (Table 5.7).

As mentioned earlier the MSL process under examination uses room temperature photoinitiated (UV) cured polymers. When looking at the thermodynamics of the free radical reactions during cure, it became apparent that the most flexible polymer (the polyol acrylate) would be needed to help the flexibility and mechanical stability of the part (i.e., help the metal binder interfaces survive CTE mismatches of the thermal treatment). However, (Table 5.8) the polyol acrylate would have more difficulty finishing its cure due to the higher transition state of the polyol acrylate, unlike the brittle triazine polymers. This suggested that the binder used for this MSL example probably cured with too brittle of a binder. Since temperatures needed for complete binder burnout (of the more brittle component) were above the oxidation temperatures of the metals involved, the simple thermodynamics explained the presence of carbon left in the metal after sintering, as well as explained why fabricated parts were warped due to the stronger and more brittle polymer dominating the dimensional stability (Fig. 5.25). That is, chemically, the binder system did not have high enough incorporation of the more flexible polymer to help reduce warpage at critical thermal steps during BBO. As a result significant warpage of the green part was found (Fig. 5.25), followed by significant cracking after binder burnout (Fig. 5.26).

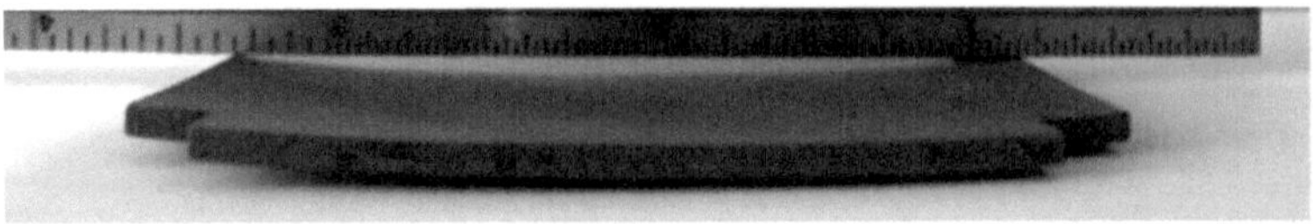

Fig. 5.25 Part before binder burnout, after low temperature anneal

Fig. 5.26 Part after binder burnout

5.2 Miscibility

Another equilibrium molecular model that is often used is the calculation of the solubility parameter. The solubility parameter calculation tools are usually available in most software packages and are very useful in understanding compatibility issues in formulations. Generally, the solubility parameter is the cohesive energy density (per volume) of a material so is derived from a temperature equilibrated randomly dispersed periodic unit cell of solvent or other material molecules under consideration. The energy difference between isolated to condensed state is determined and normalized with the volume. The classic way to use a solubility parameter is to match the cohesive energy densities; the closer the match, the more soluble.

One of the most interesting applications of the solubility parameter was the determination of a defect feature found on thin dielectric films [50] during IC manufacturing. As it was an optical defect and very difficult to experimentally find and reproduce, it affected the lithography process and reliability of the subsequent layers during the IC manufacturing layer buildup process. Additionally, it was observed that when the optical defect appeared, the films were thicker than normal. Solubility parameter analysis of the various components of the formulation did not yield an explanation. However, it was found the optical defect would go away if the bottle of formulation was allowed extra time to equilibrate from the freezer at room temperature.

From this small tidbit of information, it was deduced that the optical defect might be from an agglomerated material (like a gel) that could pass thru final filtration during the manufacturing process but depending upon the temperature conditions could form after filtration. It was further thought that anything that would cause this gel should also increase the film thickness, with an increase in effective MW. If this were so, we should no longer think of using the solubility parameter in the classic sense but use it in a relative sense to determine whether the oligomers within the formulation solution might be changing their association state with time and temperature depending upon whether they were getting closer or further away from the solvent mix.

With this in mind, a history-dependent solubility parameter of the most likely oligomer (estimated from previous thermodynamic cure calculations) was determined in a history-dependent cycle, starting at room temperature to freezer temperature and back. That is, the state of the unit cell of oligomer equilibrated at one temperature became the starting point for the next temperature equilibration step. The parameter at each step was determined before the next equilibration step. To represent a quick cool cycle, the total equilibration at each temperature step was kept small; for a long cool cycle, each temperature step had a long equilibration. The solubility parameter changes were subtle but distinctly different depending upon the length of equilibration at each temperature as indicated in Fig. 5.27 ("long" for a long equilibration step and "short" for a short equilibration step). Figure 5.27 shows a plot of these cycles.

The cycling clearly showed the temperature dependence of the solubility parameter: higher at the lower temperatures and lower at the higher temperatures. Keep in mind the actual solubility parameters of the solvent system in these formulations were above the curves in the figure. So with the slower the thermal cycles, the solu bility parameters stayed higher and closer to the solvent system parameters, than with quick thermal cycling. A quick freeze would also plummet the parameters to the low values, but if taken on a slow thaw back up in temperature, the beginning parameters were recovered. So, fast thermal cycles could relatively encourage agglomeration better than slow thermal cycles. This analysis predicted the better solubility of the long slow thaw used for formulation bottles that did not show the optical defect as their parameters were high, closer to the solvent. New tests designed around thermal freeze and thaw cycling were done to confirm these trends. It was found that indeed the faster the thermal history, the films were thicker than nominal, and with slower thermal history, the films were within specification. More importantly it was found that the lithography problems consistently came from bottles stored in a freezer with a faulty controller that maintained an unusually low temperature, necessarily causing the formulation bottles to experience an unusually quick freeze. This would then have caused them to have low solubility parameters

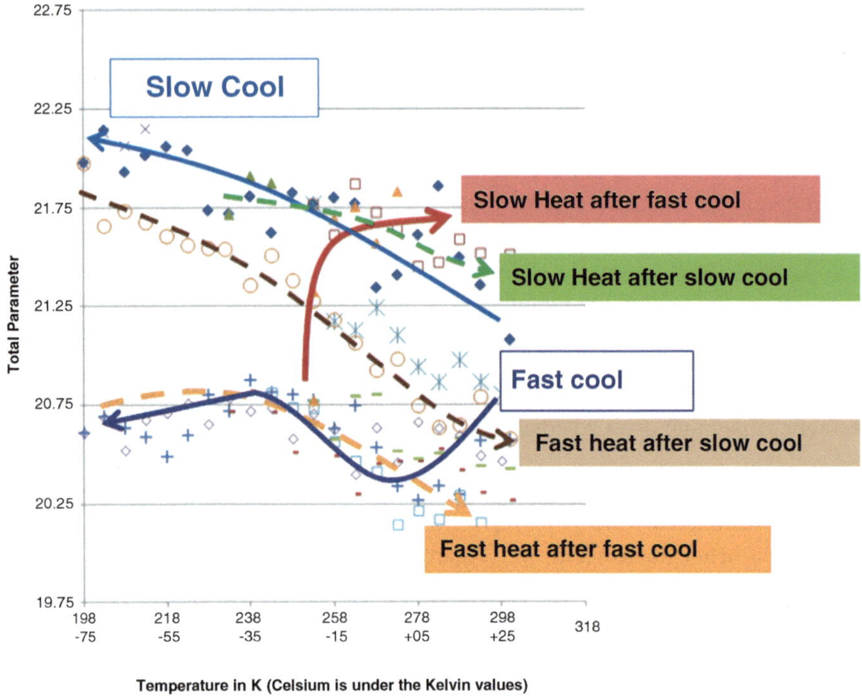

Fig. 5.27 Thermal history of a spin-on-glass (SOG) dielectric formulation using simulated history-dependent solubility parameters by molecular dynamics (direction of arrows show direction of heating or cooling simulation) [50]

which would require the observed extra time to thaw. The freezer controller issue was fixed, and the optical defect and film thickness problems went away. So, the solubility parameter told us the trends we needed to know to solve the problem, although not in the traditional manner of using a solubility parameter (Fig. 5.27).

6 Final Thoughts

This chapter was meant to give the modeler a broad molecular vision of reliability issues. As mentioned previously, many typical molecular modeling techniques engage periodic cell constructions that can be increased in size to the mesoscale. Periodic cell techniques are very useful when looking for properties and thus are well-embedded in most software packages. However, when trying to understand the direct molecular interaction contribution, periodic cells can be less useful, as mixed interfaces are usually represented at the same time. So, deconstructing individual molecular affects become more difficult, especially from amorphous materials like polymers where different interface compositions, stress orientations, and time domains may be present at the same time.

So this chapter took the molecular perspective one step deeper, to show methods that will allow the modeler to screen molecular components or understand the origin of larger scale observations, before engaging larger periodic cell techniques. These small models may not be as sophisticated as the larger periodic cell techniques but are useful in generating trends based upon molecular structure.

Using this perspective some key scale concepts have emerged. First, the small-strain mechanical properties, such as modulus, have a root basis in the molecular interactions, by the quality of their contact depending upon bonding, orientation, and molecular architecture. As such if carefully executed, there is a continuity of scale between the molecular and mesoscale models and the experimentally determined modulus. While the yield energy is not well represented in the small interface models, the trends are still present and can be used to qualify compositional effects. Secondly, the small-scale models appear to exhibit a molecular Coffin-Manson response that can be used to select molecular structural components based upon their ability to recover strain and which aids in the understanding of small-strain accumulation on reliability (even on the molecular scale). For failure energies at yield and for strain at failure, the best results were found by scaling to the mesoscale, where enough interactions can be involved to include the competing interactions and their balance and the impact of bond failure, cross-link architecture, and stress orientation. Additionally, our studies indicated that the yield strain reflects changes in bonding. Finally, there appears to be a mesoscale Hall-Petch aspect to the interface failure which starts to bring in aspects of the interface architecture. So with beads parameterized from molecular models, even interfacial roughness aspects may be probed, as well as morphology, and the combined multimode aspect of interfacial failure is found in the mesoscale (or "molecular-mesoscale").

In addition, this chapter was meant to show that the tried-in-true “normal” periodic cell molecular methods can be used to generate information that is not usually considered using these tools. Two good examples discussed were the stability of water in the cyanoacrylate. In this case the periodic cells showed that the attraction of water to the cyanoacrylate on the small model scale translated directly to stability in a molecular “bulk” periodic case. That is, compared to the alkyl acrylate, water formed its own pseudo-phase with the cyanoacrylate within which it was dynamically stable and could stay relatively confined within the layer (as compared to the movement of water in the alkyl acrylate). Another example is the time-temperature history-dependent solubility parameters. Without taking into consideration dwell-time and temperature effects, the subtle impact on the formulation would have never been discovered and the storage conditions corrected in order to correct a thin film layer manufacturing defect.

So, depending upon the investigation needed, different levels of molecular modeling may be engaged. If all levels are used, a fuller picture can be constructed of how the different structural and architectural levels of the molecular composition contribute to the total reliability issue under investigation. Such is the case of emerging manufacturing techniques employing additive manufacturing (also known as 3D printing and stereolithography), and it is hoped that many more emerging applications will be explored using molecular modeling.

References

1. L.J. Ernst, W. van Driel, O. van der Sluis, A. Corigliano, A.A.O. Tay, N. Iwamoto, H. Fan, M.M.F. Yuen, Fracture and delamination in microelectronic devices, in *Smart Systems Integration and Reliability*, ed. by B. Michel, K.-D. Lang, (© Goldenbogen, 2010)
2. N. Iwamoto, M. M. F. Yuen, H. Fan (eds.), *Molecular Modeling and Multiscaling Issues for Electronic Material Applications* (Springer, 2012)
3. A. Wymyslowski, N. Iwamoto, M. M. F. Yuen, H. Fan (eds.), *Molecular Modeling and Multiscaling Issues for Electronic Material Applications*, vol 2 (Springer, 2015)
4. O. van der Sluis, N. Iwamoto, J. Qu, S. Yang, C. Yuan, W.D. van Driel, G.Q. Zhang, Advances in delamination modeling of metal/polymer systems: atomistic aspects, in *Nanopackaging Nanotechnologies and Electronics Packaging*, ed. by J. Morris, (Springer, 2018)
5. N. Iwamoto, PBI hydration study in the development of methodology for the molecular modeling of mechanical property trends. Proceedings of the U.S. Army CRDEC Conference, Aberdeen Proving Ground, MD, 19–21 November 1991
6. N. Iwamoto, Molecular Modeling Study on the Mechanical Properties of Polybenzimidizole. NAWCPNS TP 8036, July 1992
7. N. Iwamoto, Chaos in the molecular dynamics study of polybenzimidazole. Proceedings of the U.S. Army CRDEC Conference, November 1992, Aberdeen Proving Ground
8. N. Iwamoto, Relative stabilities and self-alignment tendencies of peptide-based NLO materials. SPIE Proceedings conference 1853; January, 1993, Los Angeles, CA
9. N. Iwamoto, Simulations of melting point trends in liquid crystals using molecular dynamics. Proceedings of the U.S. Army ERDEC Conference, November 1993, Aberdeen Proving Ground, MD.
10. N. Iwamoto, A property trend study of polybenzimidazole using molecular modeling. Polym. Eng. Sci. **34**(434) (1994)
11. N. Iwamoto, Molecular dynamics of polybenzimidazole and contributions from Chaos. Polym. Eng. Sci. **34**, 1070 (1994)
12. N. Iwamoto, Molecular dynamics of binder-ADN interactions. Proceedings of the U.S. Army ERDEC Conference, November 1994, Aberdeen Proving Ground, MD
13. N. Iwamoto, Modeling mechanical properties of an epoxy using particle dynamics as parameterized through molecular modeling. Microelectron. Reliab. **51**, 1035–1045 (2011)
14. N. Iwamoto, Developing the stress-strain curve to failure using mesoscale models parameterized from molecular models. Microelectron. Reliab. **52**(7), 1291–1299 (2012)
15. N. Iwamoto, Molecularly derived mesoscale modeling of an epoxy/Cu interface: interface roughness. Microelectron. Reliab. **58**(8), 1101–1110
16. N. Iwamoto, O. Holck, S. Noijen, Multiscale perspectives of interface delamination in microelectronic packaging applications. Mol. Simul. **37**(8), 710–717 (2011)
17. N. Iwamoto, J.P. Kennedy, Diffusion simulation thru polymers using coarse grained models derived from molecular models. EuroSimE 2015 in Budapest, Hungary; April 20–22, 2014
18. W. Koch, M.C. Holthausen, *A Chemist's Guide to Density Functional Theory*, 2nd edn. (Wiley-VCH, 2001)
19. N.E. Iwamoto, S.J. Sanders, T.J. Callaghan, S. Yates, A. Taranta, J. Grooms, K.L. Warrensford, Stabilizing gas in an optical gyro based on an inorganic waveguide. US Pat 10,095,097B2, Oct 9, 2018
20. N.E. Iwamoto, S.J. Sanders, S. Yates, P. Fechner, A. Taranta, Stabilized gas environments in a proton-exchange lithium niobate optical chip. US Pat 9,817,254 B2 Nov 14, 2017
21. S.J. Marrink, H.J. Risselada, S. Yefimov, D.P. Tieleman, A.H. de Vries, The MARTINI forcefield: coarse grained model for biomolecular simulations. J. Phys. Chem. B **111**, 7812–7824 (2007)
22. S.J. Marrink, A.H. de Vries, A.E. Mark, Coarse Grained model for semiquantitative lipid simulations. J. Phys. Chem. B **108**, 750–760 (2004)

23. L. Monticelli, S.K. Kandasamy, X. Periole, R.G. Larson, D.P. Tieleman, S.J. Marrink, The MARTINI coarse grained forcefield: extension to proteins. J. Chem. Theo. Comp. **4**, 819–834 (2008)
24. W. Shinoda, R. DeVane, M.L. Klein, Multi-property fitting and parameterization of a coarse grained model for aqueous surfactants. Mol. Simul. **33**, 27–38 (2007)
25. H.-Y. Lee, J. Qu, Microstructure, adhesion strength and failure path at a polymer roughened metal interface. J Adhesion Sci. Technol. **17**(2), 195–2015 (2003)
26. O. Hoelck; E. Dermitazaki, B. Wunderle, J. Bauer, B. Michel, Molecular modeling of microelectronic packaging materials-basic thermo-mechanical property estimation of a 3D cross-linked epoxy/SiO2 interface. Eurosime 2010; Bordeaux, France April 26–28
27. B. Wunderle, E. Dermitzaki, O Hoelck, J. Bauer, H. Walter, Q. Shaik, K. Ratzke, F. Faupel, B. Michel, Molecular dynamics simulation and mechanical characterisation for the establishment of structure-property correlations for epoxy resins in microelectronics packaging applications. Eurosime 2009. April 26–29, 2009, Delft, Netherlands
28. E. Dermitzaki, B. Wunderle, J. Bauer, H Walter, B. Michel, Structure property correlation of epoxy resins under the influence of moisture and comparison of diffusion coefficient with MD simulations. Eurosime 2008; April 21–23, Freiburg-im-Breigau, Germany
29. N. Iwamoto, J. Pedigo, A. Grieve, M. Li, Molecular modeling as a tool for adhesive performance understanding. Proceedings of the MRS 98 Symposium J: Electronic Packaging Materials Science X, April 1998, San Francisco, CA, v. 515, pp. 23–30.
30. N. Iwamoto, J. Pedigo, Property trend analysis and simulations of adhesive formulation effects in the microelectronics packaging industry using molecular modeling. Proceedings of the 48th Electronic Components and Technology Conference, May 25–28, 1998; Seattle, WA, pp. 1241–1246
31. N. Iwamoto, Simulating stress reliability using molecular modeling methodologies. 32nd International Symposium on Microelectronics; Chicago, Il, Oct. 26–28, 1999; 1999 Proceedings pp. 415–420
32. N.E. Iwamoto, Advancing polymer process understanding in package and board applications through molecular modeling. Proceedings of the 50th ECTC May 21–24, 2000, Las Vegas, NV, pp. 1354–1359
33. N.E. Iwamoto, Advancing materials using interfacial process and reliability simulations on the molecular level. Proceedings of the 5th International Symposium and Exhibition on Advanced Packaging Materials, March 2000, Braselton, GA, pp. 14–17
34. N.E. Iwamoto, Understanding modulus trends in ultra low k dielectric materials through the use of molecular modeling. Proceedings of the 52nd Electronic Components and Technology Conference 2002; May 28–31, San Diego, CA 2002, pp. 1318–1322
35. N. Iwamoto, J. Kennedy, D. Varaprasad, S. Mukhophadhyay, S. Xie, Understanding process cycling in thin film dielectrics. Proceedings of Eurosime 2016, Montpellier, FR April 18–20, 2016
36. N. Iwamoto, C. Putman, G. Vernon, R.C. Horn, A. Bernreuther, Interface comparison involved in flexible electronics using molecular modeling. Proceedings of Eurosime 2017, Dresden, Germany, April 2–5, 2017
37. N. Iwamoto, D.Y.A. Larson, B. Palmer, T. Baldwin-Hendricks, Using molecular modeling to understand cleaner efficiency for BARC ("Bottom Anti-Reflective Coating") Layers after plasma etch in dual damascene structures. Proceedings of Eurosime 2007; London, April 16–18, 2007
38. N. Iwamoto, A. Krishnamoorthy, R. Spear, Performance properties in thick film silicate dielectric layers using molecular modeling. Microelectron. Reliab. **49**(8), 877–883 (2009)
39. N. Iwamoto, Mechanical properties of an epoxy modeled using particle dynamics, as parameterized through molecular modeling. Proceedings of Eurosime 2010 April 26–28, Boerdeaux, France
40. N. Iwamoto, E. Lee, N. Truong, New metal layers for integrated circuit manufacture: experimental and modeling studies. Thin Solid Films **469–470**, 431–437 (2004)

41. D.R. Lide, *77th ed CRC Handbook of Chemistry and Physics* (CRC Press, Boca Raton, 1996–1997)
42. W.F. Cooper, W.H. Nuttall, The theory of wetting, and the determination of the wetting power of dipping and spraying fluids containing a soap basis. J. Agric. Sci. **7**, 219–239 (1915)
43. N. Iwamoto, M. Li, S. McCaffrey, M. Nakagawa, G. Mustoe, Studies of underfill formulation effects using molecular dynamics and discrete element modeling. Proceedings of adhesives in electronics'98, Sept 28–30, 1998, Binghamton, NY, pp. 321–327
44. N. Iwamoto, M. Li, S.J. McCaffrey, M. Nakagawa, G. Mustoe, Molecular dynamics and discrete element modeling studies of underfill. 1998 Int. J. Microcircuits Electron. Packag. **21**(4), Fourth Quarter 1998, 322–328
45. N.E. Iwamoto, T.A. Ramos, R.R. Roth, U. Bonne, Stationary phase materials for micro gas analyzer. US Pat 7,744,818. June 29, 2010
46. N. Iwamoto, U. Bonne, Molecular modeling of analyte adsorption on MEMS GC stationary phases. Proceedings of Eurosime 2006, Como, Italy; April 24–26.
47. N. Iwamoto, U. Bonne, M. Rhodes, Molecular modeling of analyte adsorption on microelectromechanical systems gas chromatography stationary phases. Sensor Lett. **6**, 23–34 (2008)
48. N. Iwamoto, Molecular modeling studies of IC barrier concerns. Proceedings of Eurosime 2004; May 10–12, 2004, Brussels, Belgium
49. N. Iwamoto, Perspective of water effects in polymers through molecular dynamics. Proceedings of Eurosime 2018, Toulouse, France April 15–18, 2018
50. N. Iwamoto, T. Baldwin, Uncovering the origins of subtle solvation-based film defects. Proceedings of Eurosime 2014 Ghent Belgium April 7–9, 2014.

Chapter 6
Health Monitoring, Machine Learning, and Digital Twin for LED Degradation Analysis

Mesfin Seid Ibrahim, Zhou Jing, and Jiajie Fan

1 Introduction

Nowadays, light-emitting diode (LEDs) are widely used in different applications including general indoor and outdoor lighting lamps, automotive lighting [1], backlighting, robotics skin [2], medical and communication equipment, and so on. This is due to the many advantages, including longer lifetime (50,000–100,000 h), higher reliability, environmental friendliness, compactness in size, and quicker switching time when compared with traditional counter parts (incandescent and fluorescent) lighting sources [3–5].

Regardless of the many benefits and promising future applications that LED lighting sources provide, there are challenges facing LED manufacturers on the lack of a unified standard method to monitor in situ LED degradation and to gather reliability assessment information, thermal management, potential glare due to small size lamp, and color stability. In addition to this, there is also lack of accurate

M. S. Ibrahim
Department of Industrial and System Engineering, The Hong Kong Polytechnic University, Hung Hom, Hong Kong

College of Engineering, Kombolcha Institute of Technology, Wollo University, Kombolcha, Ethiopia

Z. Jing
College of Mechanical and Electrical Engineering, Hohai University, Changzhou, China

J. Fan (✉)
College of Mechanical and Electrical Engineering, Hohai University, Changzhou, China

Institute of Future Lighting, Academy for Engineering and Technology, Fudan University, Shanghai, China

EEMCS Faculty, Delft University of Technology, Delft, The Netherlands
e-mail: jiajie_fan@fudan.edu.cn

W. D. van Driel, M. Yazdan Mehr (eds.), *Reliability of Organic Compounds in Microelectronics and Optoelectronics*,
https://doi.org/10.1007/978-3-030-81576-9_6

remaining useful lifetime estimation and evaluation methods. This is due to the long lifetime and high reliability at normal operating conditions, various failure mechanisms, rapid technology advancement, and multicomponent features of LEDs compared to the traditional light sources [3, 6, 7]. However, this has brought another challenge for manufacturers in terms of obtaining sufficient failure data, determining reliability and estimating remaining useful lifetimes (RUL) in relatively short lifetime testing before the products are released to the market and with better prediction accuracy.

To address the challenges and shortcomings related to reliability assessment and lifetime prediction of LEDs, a number of research studies have been undertaken on the prognostics and lifetime estimation in academia and industry [6, 8–12]. In early 2001, a discussion was initiated by Narendran et al. [13] among the lighting industry experts concerning the standardization of definitions, procedures, and approaches in the process of useful lifetime estimation for LED products. Currently, LED manufacturers use IES-TM-28 [14], released by IESNA, to project lumen maintenance lifetime for LED lamps and luminaires where the required data is gathered according to industrial standard test report IES-LM-84 [15]. Previously, the IES-TM-21 standard [16] has been used to predict the lifetime of LED light sources based on the light output degradation data from the standard IES-LM-80 test report [17]. The approved IES-TM-21 procedure uses the nonlinear least-squares regression (LSR) approach to project lumen maintenance data to predict the lifetime (L50 or L70) of LED lighting sources. This lifetime testing method can be a good approach for comparing lifetime information of LEDs, but it does not provide detailed information regarding failure modes, mechanisms, and failure locations [3].

Recently, machine learning (ML) has emerged and is breaking new frontiers in reliability assessment and lifetime prediction studies due to systematic generation of large amount of data, newly introduced state-of-the-art algorithms, and an exponential increase in computing power. ML algorithms are a set of methods and procedures that can be used to capture, detect, and learn relevant information patterns from large amounts of data and then use the unhidden patterns for further decision-making in prognostics or predicting lifetime [18]. Thus, the ability of ML to learn from training data, generalize from historical data, and perform tasks without being explicitly programmed makes it tantalizing panacea for challenges in reliability analysis, anomaly detection, diagnostics, and prognostics.

There have been some reviews that studied the degradation mechanisms influencing the reliability of GaN-based white LEDs for different lighting purposes [3, 19–23]. An extensive review that mainly focused on failure causes, failure modes, and failure mechanisms of LEDs was presented by Chang et al. [3], while recently Sun et al. [23] have presented a literature review on recent trends in the prognostics of high-power white LEDs (HPWLEDs), including the failure modes, mechanisms, and some lifetime estimation approaches. Most of these reviews mainly focused on statistical-based data-driven approaches, failure modes, and mechanisms as well as physical degradation mechanisms of LEDs. While these topics are very important for the prognostics and health management (PHM) study of LEDs, it is not the focus of this study, which mainly focuses on the machine learning-based PHM approaches applicable for LEDs anomaly detection, diagnostics, and lifetime prediction. Thus, the main aim of this study is to review machine learning algorithms, methods and approaches and their pros and cons in the reliability assessment, failure or anomaly detection, and the remaining useful life prediction in general and focusing on LED light source products in particular.

2 PHM of LEDs

Nowadays, there is an increasing competition in the global market and the need to enhance customer satisfaction. In addition, huge advancements in technology, materials, and manufacturing processes are observed which facilitate the design and manufacturing of many consumer products that are highly reliable and have a longer lifetime before they fail. All of these factors lead to a shorter product development time, and that becomes challenging for manufacturers to evaluate the lifetime of high reliability items in a shorter period before being released to the market [24, 25]. This phenomenon is no different in the case of lighting products, especially for the high-power white light-emitting diodes (LEDs) that belong to highly reliable and long lifetime products that require a longer time to collect adequate degradation and/or failure data. That is why long-term lifetime estimation and reliability assessment of LEDs in a moderately shorter period of time before products are released to market have become challenging for LED manufacturers [26]. For this reason, PHM has evolved as an important method to solve the challenges in terms of increasing system reliability, availability, and maintainability, enhancing safety and decreasing life-cycle and operational costs of marketable products and systems in general and customer electronic systems in particular [27]. Thus, the reliability assessment and prediction of remaining useful life (RUL) studies have become an important aspect of PHM of many consumer electronic products, including high power white LEDs.

Basically, PHM is an engineering discipline that helps to prevent the failure of products, components, and subsystems which can lead to inadequate performance and safety concerns. It helps to anticipate problems in products and systems through signal and sensor data under actual application conditions [28]. PHM uses inputs such as information known about products/system, data collected from sensor measurements, and applies an algorithm or a set of algorithms to analyze and provide relevant outputs at various levels of prognostics, such as fault detection, diagnostics, and lifetime estimation, as depicted in Fig. 6.1.

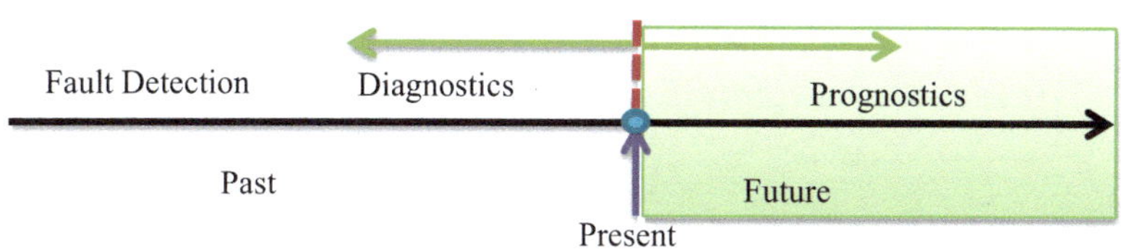

Fig. 6.1 PHM problem architectures (fault detection, diagnostics, and prognostics)

A well-organized prognostic health management framework should include data collection using sensors, data processing, security and integration, feature extraction, fault detection and recognition, damage models, physics of failure, reliability testing, physical models, prognostics, and so on [29], as illustrated in the PHM metro map shown in Fig. 6.2.

The main purpose of anomaly detection is to detect unusual or strange anomalous responses of systems and products through identification of deviations from normal healthy behavior, so that precautionary measures can be taken in advance to avoid potential failures. It is worth noting that anomalies may not necessarily indicate failure as changes in working or environmental conditions enable sensors to detect anomalous behavior. Diagnostics enable us to extract and gather failure magnitudes, failure modes, failure mechanisms, and other related data from anomalous behavior of a product/system through sensors. The term prognostics deals with the process of estimating the lifetime or predicting the future reliability of a product based on historic and current degradation data and assessing the degree of deviation from its normal operating conditions [27]. Prognostics can provide help in all product and/or system life cycles including design and development, production and ramp-up, product testing, operations and maintenance, as well as end-of-life phase (i.e., phase out and disposal) [30]. In this regard, the PHM of mechanical systems has been well studied, and as a result there is a considerable body of knowledge in the area. However, prognostics have only been applied to consumer electronic products/systems quite recently, and this is due to the fact that degradation is difficult to detect in electronic systems when compared with mechanical systems [31].

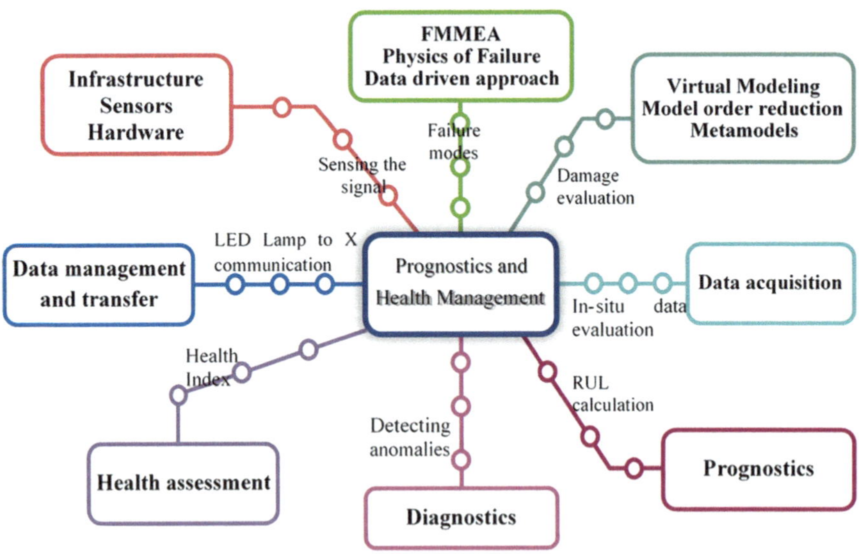

Fig. 6.2 A generic PHM metro map for products/system such as LED lighting, automotive parts, etc.

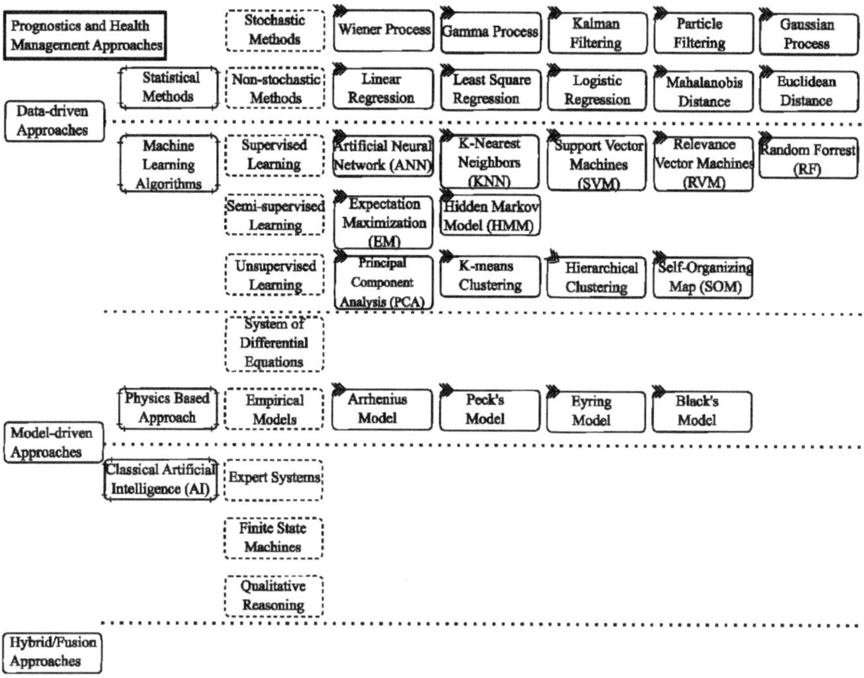

Fig. 6.3 Taxonomy of prognostics and health management approaches

Even though the expected lifetime for typical high-power LEDs can be rated up to 50,000 h, practical statistics indicate that about half the LED products failed to reach the rated lifetime [32, 33]. This has raised demands from experts in the LED sector, end-product manufacturers, and potential customers for dependable reliability information and remaining useful lifetime estimation approaches. Thus, through the application of PHM, the inadequate lifetime and reliability information provided by LED manufacturers should be addressed, and reliable approaches to monitor the status of LEDs and predict potential failures, especially for safety critical and emergency systems and products including the medical, aviation, automotive, and nuclear sectors, are needed. So far, many diagnostic and prognostic activities have been implemented and executed based on a variety of approaches and techniques. In general, the most commonly used approaches can be categorized as (i) model-based, also known as physics of failure (PoF) methods, (ii) data-driven methods, and (iii) hybrid (fusion) prognostic methods [34]. A more refined and detailed taxonomy of PHM approaches is presented in Fig. 6.3.

The data-driven (DD) methods are mainly dependent on large amounts of training data and/or degradation data collected through sensors in order to derive degeneration models for products and systems. The data collected in real time can be used to adjust and modify the model parameters. On the other hand, the model-based method requires prior mathematical models to describe the product's degeneration process based on physical laws. The DD methods are helpful for complex systems where component interaction is indeterminate and when large amounts of training data are available, while the model-based method demands knowledge of the physical laws governing the product degeneration expressed in mathematical models. Statistical-based and ML models and algorithms are used in DD approaches while physical models and classical AI methods implemented in model-based approach [35, 36]. Fusion/hybrid approaches that combine the benefits and eliminate the drawbacks of both DD and physics-based methods have also been implemented in prognostics studies [37]. The preferred choice of each algorithm depends on the different properties manifested for use in the intended analysis.

3 Model-Based Approaches

3.1 An Overview to Model-Based Approach

Model-based prognostics, also known as physics of failure (PoF) methods, makes use of information about a product's material characteristics, loading and stress settings, shape/geometry, and operational and working environmental conditions to assess reliability, identify failure modes and mechanisms, and so as to estimate the RUL. By using product life-cycle loading conditions (such as electrical, thermal, mechanical, chemical, electromechanical, etc.), the product geometry, and material properties, PoF is also used to design for reliability at the early stage of product design [38, 39]. The PoF-based approach has the benefit of identifying the root causes of system failure [23, 30]. However, sufficient knowledge about the product geometry, materials, properties, and operating conditions are required, and it may be difficult to obtain such information, especially for complex systems. For a certain product/system at a particular life-cycle loading condition, PoF focuses mainly on identification of potential failure locations, failure modes, as well as failure mechanisms. The stress at every failure location is obtained as a function of the life-cycle loading conditions, material properties, and product architecture/shape. Then fault generation and propagation are determined by damage models [28, 40]. Model-based approaches are also used to develop mathematical models in order to process and evaluate collected degradation data based on the prior knowledge of the product/system.

In the study of prognostics, PoF models implement the use and monitoring of performance parameters, physical characteristics, and operating and environmental conditions. These parameters are used to monitor the product during experiments and can be categorized according to their domains. For the prognostic analysis of LED products, the different impact (stress) factors such as electrical, thermal, humidity, mechanical, thermomechanical, and creep stress applied on the test sample can be monitored by sensors, and the PoF models with mathematical equations can be used for further analysis depending on the experimental plan. A brief summary of PoF models employed for LED products and systems is shown in Table 6.1. Pecht and Jie [38] studied the PoF-based prognostics for electronic and information-rich components. In their study, they criticized the use of old reliability handbooks due to prediction errors and uncertainties (in design, material, and operating conditions) caused and showed the growing trend of using PoF-based prognostics for electronic products so as to identify critical component failure modes and mechanisms.

The implementation approach framework for PoF-based PHM has been demonstrated in such a way that the first step is to undertake virtual life assessment. Virtual life assessment can be conducted using inputs from design data; failure mode, mechanisms, and effects analysis (FMMEA); expected lifetime conditions; and PoF models. During the product life cycle, high-priority failure mechanisms might be triggered by different severe and frequently occurring operational, environmental and loading conditions. The virtual life assessment which is the first phase in the physics of failure-based prognostics has been further investigated by Fan et al. [10]. Their study was focused on the investigation of failure sites, failure modes, and associated degradation mechanisms for high-power white LEDs (HPWLEDs). The sample selected for demonstration was a typical commercial HPWLED lamp according to "bottom-up" strategy at the chip, package, and system levels as shown in Fig. 6.4. Pictures in this figure are presented for the purpose of illustration.

Lu et al. [41] used the physics of failure-based approach to study down light color shift failure at the luminaire level conducted on the diffuser (PMMA), reflector and package parts of an LED lamp of 10 W, and CCT of 4000 k. The selected parts had undergone aging testing at room temp, 55 °C and 85 °C, irradiation testing at 85 °C, and humidity reliability test at 85 °C and 85% RH. The experimental results showed that LED packages have a greater contribution to color shift. Humidity and temperature also accelerate the color shift, where humidity has the stronger impact (Table 6.1).

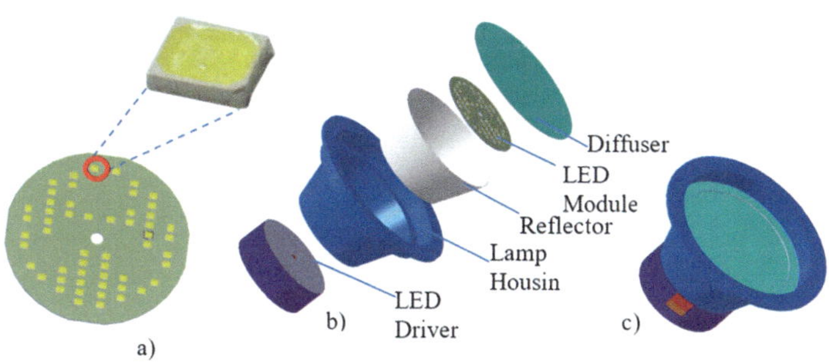

Fig. 6.4 LED lamp and components: (**a**) LED package and module, (**b**) LED lamps exploded, (c) LED lamp and lighting lamps

Table 6.1 PoF models for LED products and systems

Stress (impact) factors		PoF models	Performance indicators
Electrical (current) [42, 43]		Lumen degradation gradient [42]	Lumen flux depreciation [42],
		Inverse power law-Weibull [43]	
Thermal stress/shock [44, 45, 46, 43, 47]		Coffin-Mansion Eq. [44]	Lumen depreciation
		System reliability analysis [45]	Color shift over lifetime
		Hierarchical model (based on junction temperature) [45]	Junction temperature gradient
		Arrhenius Eq. [46]	
		Arrhenius-Weibull [43]	
		Finite element simulation using ANSYS and numerical analysis simulation	
Humidity/moisture [48, 49]		Luminous-efficiency gradient [48]	Lumen flux depreciation [48]
		Finite element simulation using ANSYS [49]	
Multi-physics	Thermal and humidity [41, 50, 51, 52, 49]	Chromaticity shift eq. [38]	Chromaticity shift, [38, 50] Lumen flux degradation [50, 51]
		Arrhenius Eq. [52]	
		Hallberg-Peck's model [5052]	
		Subsystem isolation method [51]	
		Finite element simulation using ANSYS [49]	
	Thermal and electrical (current) [53, 54, 55, 56, 43]	Junction temperature distribution	LED catastrophic failure for high thermo-electrical stress [54] Spectral power distribution (SPD) and Luminous intensity depreciation [53, 54, 56]
		Spectral power distribution (SPD) analysis [53]	
		Electrothermal simulation (junction temperature with Arrhenius equation)	
		Electrothermal simulation [56]	
		Generalized Eyring-Weibull [43]	
	Thermomechanical [57],	Thermal and mechanical stress on solder alloy	Solder joint fatigue
		Garafalo's hyperbolic creep model	Lumen depreciation
		Norris-Landzberg equation	
		Engelmaier equation for strain range	
	Thermomechanical [58] and hygro-mechanical stresses [58, 59]	Thermal and thermomechanical modeling (transient heat conduction equation)	Lumen flux depreciation
		Moisture diffusion and hygro-mechanical modeling (Fick's law of diffusion)	
		Finite element analysis (simulation)	
	Hygro-thermal-mechanical coupling modeling [58]	Heat conduction systems, Fick's law of diffusion and FEA simulation	Lumen flux depreciation

3.2 Failure Modes, Mechanisms, and Effects Analysis for LEDs

The failure modes, mechanisms, and effects analysis (FMMEA) could be considered as input to the PoF-based prognostic approach, as depicted in Fig. 6.5. The exposure of LED lighting products/systems to different loading and operational stresses such as electrical, thermal, mechanical, or chemical causes performance degradation and/or failure [60].

In LED systems, a failure mode is a recognizable way in which a failure of a package/lamp is noticed, and it can be classified as (i) loss of light output or open circuit, (ii) chromaticity shift (i.e., color shift), and (iii) degradation of luminous flux (decreasing in light output). Each failure mode could also be due to one or a combination of failure mechanisms which could be caused by thermal, mechanical, humidity, chemical, etc. Failure mechanisms can be described as thermal, mechanical, physical, chemical, or other processes that cause a failure. Failure mechanisms can be broadly classified as wear-out (gradual) and overstress (catastrophic) failures. The wear-out failures are caused by cumulative stresses (loads) for a prolonged period of time. On the other hand, overstress (catastrophic) failures occur as a result of a one type of stress /load condition that surpasses the optimal threshold of the product characteristic [38].

A comprehensive study was reported by Chang et al. [3] on the FMMEA at semiconductor, interconnect, and package levels for LED products. Subsequently, Fan et al. [11] conducted a study on the FMMEA of LED-based backlighting systems used for commercial displays and TVs. Since LED-based display systems are

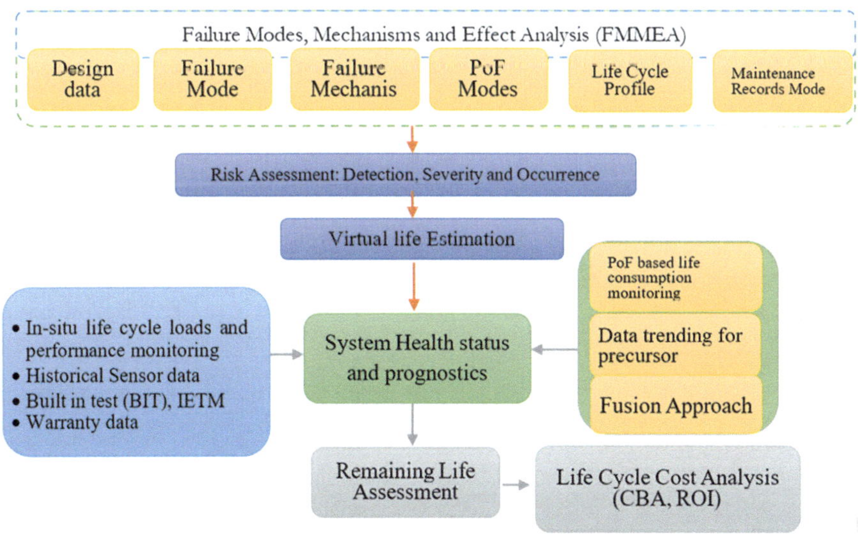

Fig. 6.5 PoF-based PHM Methodology [28, 61], with permission

formed by LED strips and electric driver systems, the study aimed to identify failure sites, failure modes and mechanisms at the chip (die/semiconductor), driver, package, and strip levels of LED backlight system. As an LED-based device, the failure modes for backlight units are lumen flux degradation, color change, and loss of light failure modes. In our review study, the FMMEA of LED products/systems are described by considering a more general architecture including chip (die/semiconductor) level, package (module along with interconnects), and system levels as presented in Table 6.2.

In general, the FMMEA of LEDs has been investigated at three levels: die/chip, interconnects, and LED package levels [3, 10]. At the chip level, an increased non-radiative recombination can cause a degradation of the active layer of LEDs which impacts in decreasing the luminous flux and power efficiency. Subsequently, the

Table 6.2 FMMEA of LEDs at different levels

LED failure site	Failure modes	Failure mechanisms
LED chip level (semiconductor/die) [11, 21, 22, 48, 49, 53, 62, 63]	Lumen flux depreciation [48, 49]	Propagation of defect and dislocation
	Light output off, short circuit [48]	Diffusion of impurities (dopants) in the quantum well
	Color shift	Cracking of chip/die
		Yellowing and cracking of the encapsulating lens [53]
LED module (package level including interconnects) – wire bond, bumps, attachments, encapsulate, lead frame, lens [22, 48, 50, 53, 58, 64, 65]	Lumen degradation and color shift [22, 50, 64, 65]	Propagation of defect and dislocation
	Delamination between chip and die, as well as lamp cup and outer shell [58]	Diffusion of impurities (dopants)
	Diffusion of moisture into the boundaries of packaging material [48]	Cracking of chip/die
		Yellowing and cracking of the encapsulating lens [53]
		Package epoxy browning [65]
System level (diffuser, reflector, electrical driver) [41, 54, 58, 64, 45, 51, 46, 66, 56, 47]	Lumen flux depreciation [58, 64, 46, 47]	Encapsulant yellowing
	Light output off, short circuit	Solder joint fatigue
	Plastic housing crack, glass bulb crack	
	Optical coating discoloration	
	Color shift [41, 54, 66]	

diffusion of dopants (impurities) in to the quantum well, defect propagation (due to defect/dark spot, propagation, and dislocation), and electromigration due to crystalline defects are the factors that play major roles to the non-radiative recombination.

At the package level, the commonly known failure mechanisms are delamination of the interface, encapsulant carbonization, encapsulant yellowing, thermal quenching of phosphor, solder joint fatigue, and lens cracking. These failures will eventually cause lumen flux depreciation and change the chromatic properties of the LEDs. The failure mechanisms at the interconnections can be fracture of the bond wire as well as fatigue on the wire ball bond due to thermal and electrical overstress, electrical contact degradation due metallurgical interdiffusion, and electrostatic discharge (resulting in rapid failure due to the open circuit). The failure mechanisms at different levels of LED devices will cause at least one of these failure modes to occur [48] [58].

4 Data-Driven Approaches

Data-driven (DD) approaches rely on the use of historical and observation data to learn intelligently without prior knowledge of the system, to obtain statistical and probabilistic lifetime estimates, and to provide help in making valuable decisions on system/product health and reliability. The DD approaches help to detect anomalies and predict RUL for a system based on the investigation of historical monitoring data collected from sensors [67]. It is assumed that the system statistical characteristics remain unchanged until an anomaly occurs in the LED product/system [34]. The DD approaches are usually considered as the black box approaches to PHM as they do not require prior knowledge on the system models. There are many ways to classify DD approaches; however, for simplicity, DD approaches can be categorized into two, statistical-based and machine learning-based DD, methods depending on the data analysis methods.

In the first case, statistical-based approaches rely on the use of empirical or analytical equations to build statistical models that helps to predict the degradation trend of LED performance parameters. These approaches are convenient to implement as they make use of primarily historical data and do not need to rely on expert knowledge. In fact, statistical-based data-driven methods depend not only on the availability of data but also on the nature of the data collected [35]. This approach has the capability of describing the uncertainties in performance degradation of LEDs by incorporating random and dynamic variances. On the other hand, machine learning (ML) algorithms refer to a set of methods and procedures that can be used to capture, detect, and learn relevant information patterns from large amount of data and use the unhidden patterns for further decision-making in prognostics or predicting the future lifetime [18].

The main advantage of the DD approach is that the methods and algorithms provide quick results and are computationally efficient. In addition, DD methods can also handle complex systems having multicomponent interaction, such as in the case of LED lighting systems, which are difficult to deal with using the physics-based method. On the other hand, one of the drawbacks of the DD approach is its dependency and demand for training (or historical) data to create correlations, understand patterns, and evaluate data trends and deliver accurate results [31]. In fact, statistical-based data-driven methods depend not only on the availability of data but also on the nature of the data collected [35]. In some cases where the products have a long lifetime, nonoperating, and standby systems, there will be insufficient training or operational data. In such conditions, data-driven approaches have to incorporate model-based approaches to bring a better prognostic solution. Commonly, data-driven methods are used in fault detection, diagnostics, and lifetime prediction. Even though the first two parts are mostly handled by using DD methods, the prediction part can also be handled with PoF approaches [29].

Assessing the reliability information of products (such as remaining useful lifetime, mean time to failure) plays a central role in the process of continuous quality and reliability improvement. This is especially true for highly reliable products such as LEDs, where it is time-consuming and expensive to assess their lifetime using traditional lifetime tests [68]. In such conditions, the quality characteristics of products whose degradation path (degradation data over time) is related to the reliability of the product can be collected and analyzed to infer important reliability information about the lifetime of the product. Lumen depreciation is the most common failure mode in LEDs [10]; thus the luminous flux maintenance lifetime, defined as the amount of time left until the initial light output falls below a failure threshold of 70%, is widely recognized as one of the critical characteristics for representing the LED's life and assessing its reliability (Fig 6.6).

LEDs belong to highly reliable electronic devices with long lifetimes (more than 50,000 h), provided that proper thermal management techniques are applied [32, 33]. Therefore, traditional reliability assessment methods based on failure data are not suitable for LEDs which have few failures even under accelerated conditions. Previously, the accelerated life test (ALT) was used to qualify the LED's reliability and was designed to cause the failure of LED packages/lamps at a faster pace compared to the usage under normal conditions [69]. However, there are two considerations when using ALT in the LED case; firstly, relating the real operation life and rated life under accelerated conditions is not easy for the LED case. Secondly, keeping the same failure modes and mechanisms under both normal operations and accelerated conditions is also difficult. In such situations, the use of degradation data to handle reliability assessment has been found to be a superior alternative compared with traditional censored failure data. It provides the benefits of identifying the degradation path as well as more reliability information (such as mean time to failure (MTTF), confidence intervals) that helps in maintenance decision-making before failures happen [70–75]. First introduced by Lu and Meeker [72], the general degradation path method was used to model degradation data in relation to time. Fan et al. [7] implemented the degradation data-driven-based PHM with statistical models into the high-power white LED to get additional reliability information

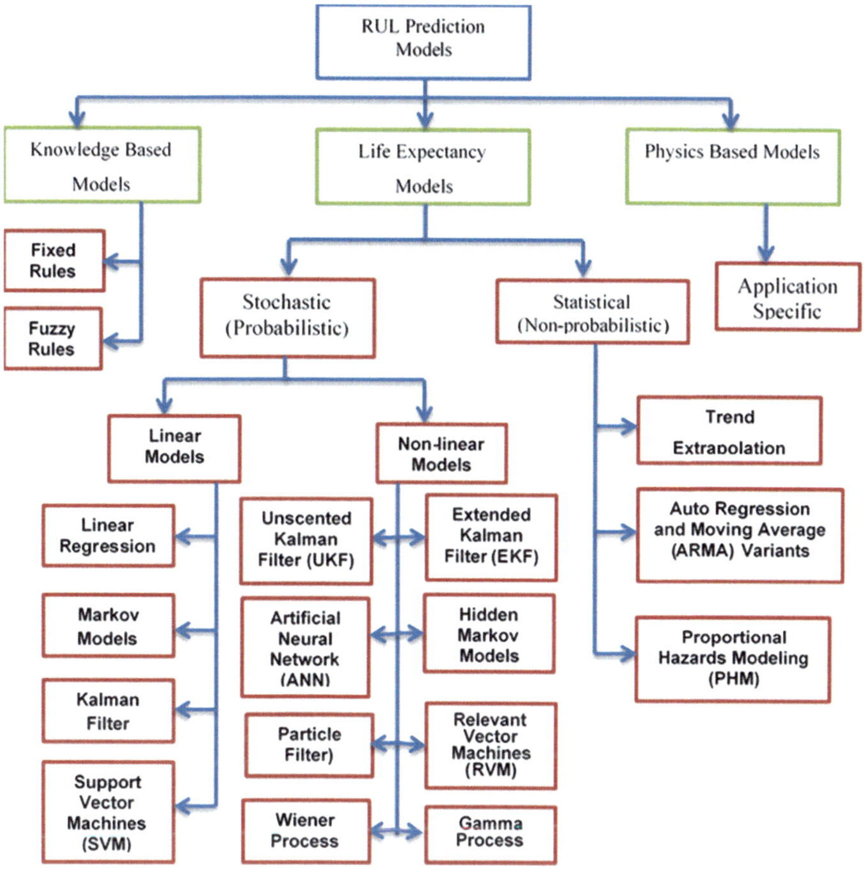

Fig. 6.6 Prognostic modeling techniques for remaining useful life

(such as reliability function, confidence interval, and MTTF) in addition to the luminous flux lifetime, the only information obtained from IES-TM-21-11.

Besides the deterministic statistical methods, stochastic modeling was also used to predict the lifetime of LEDs based on degradation data, where the degradation path was modeled as a stochastic diffusion process [25, 76]. Such stochastic degradation of products (e.g., lumen depreciation) is often modeled based on a failure rate function or a stochastic process such as random deterioration rate, Markov process, Brownian motion with drift (wiener process), or the gamma process [77]. Recently, Si et al. [78] and Wang et al. [79] proposed an improved remaining useful life estimation method in the diffusion degradation process, which can also be used to describe the LED's degradation path. Meanwhile, the Bayesian approach was also found to be an effective method to predict the residual life distributions from degradation data [73, 80]. In addition to dealing with degradation data, another data-driven-based PHM used in LED lighting is anomaly detection that uses distance measures to monitor the operating characteristics in LED (such as junction temperature, driving current). In this case, the health of an LED product/system can be described as the degree of depreciation or deviation from its anticipated typical performance. In order to evaluate the reliability of the product and predict the lifetime, the degree of deviation from the normal performance has to be evaluated precisely [81]. Therefore, distance measures were used to detect fault occurrence in a product's normal operation [82–85]. Based on this approach, Sutharssan et al. [86–88] applied distance measures (such as Euclidean and Mahalanobis distance) to do real-time health monitoring and determine remaining useful lifetime estimation for high-power LED.

In general, DD methods are based on statistical techniques, pattern recognition, deep learning and machine learning algorithms, and artificial intelligence approaches. These methods can be employed at the component, subsystem, or system levels [89]. Sikorska et al. [90] presented a comprehensive review on available prognostic modeling methods, strengths, and weaknesses that help to estimate remaining useful life and reliability of engineering assets. Some of these methods or approaches have been widely applied by researchers in the past few years. The appropriate application of these methods requires not only mathematical knowledge but also appropriate system understanding. The summary in Fig. 6.6, enhanced from Sikorska et al. [90] shows a general classification of most of the RUL prediction data-driven approaches that can be used for LED lighting system reliability assessment, failure analysis, and remaining useful life prediction. In the study of LED's reliability and lifetime prediction, many data-driven approaches can be found in the literature. The DD approaches can be categorized into different types depending on the nature of the degradation data (deterministic or stochastic), data training requirement (supervised, unsupervised, or semi-supervised), and so on. The data-driven approaches are widely used and the application spectrum is broader. A comprehensive summary of the machine learning algorithms is presented in Table 6.3, in the Appendix section. Many of the data-driven techniques that are found effective from other fields of study could be adapted and customized for the LED's lifetime estimation and reliability analysis with proper understanding as discussed in this section.

4.1 An Overview of Selected Statistical Data-Driven Methods

4.1.1 Wiener Process-Based Approach

A Wiener process is generally described as a drift component plus a diffusion component based on Brownian motion. A simple Wiener process with constant drift can be represented by Eq. (6.1):

$$X(t) = x(0) + \lambda t + \sigma \beta(t) \tag{6.1}$$

where $X(t)$ is degradation of performance characteristics (such as lumen maintenance, color shift, etc.), $x(0)$ is initial deterioration, $\lambda > 0$ is a drift parameter, $\sigma > 0$

is a diffusion coefficient, and $\{\beta(t), t > 0\}$ is a standard Brownian motion that represents the stochastic dynamics of the degradation process [91].

Degradation modeling with the Wiener process is mathematically important because the distribution of the first hitting time (FHT) at which the degradation process exceeds a threshold, i.e., lifetime (T) can be formulated analytically based on the inverse Gaussian distribution. That is why the Wiener process has been widely studied for lifetime prediction and reliability assessment [92–94], and the pdf of T can be given as:

$$f_T(t;\theta) = \frac{w}{\sqrt{2\pi\sigma^2 t^3}} \exp\left[-\frac{(w-\lambda t)^2}{2\sigma^2 t}\right] \tag{6.2}$$

where ω is a failure threshold and the mean and variance of T are $\theta = [\lambda, \sigma^2]$ and given as w/λ and $w\sigma^2/\lambda^3$, respectively [95].

A Wiener process is typically used to analyze degradation processes that vary bidirectionally over time with Gaussian noise, in other words, non-monotonic degradation processes, and it is one of the widely used degradation modeling approaches. The Wiener process was applied to predict the RUL of variable stress-accelerated degradation tests by pioneers Doksum and Hbyland [96]. Whitmore [97, 98] proposed a Wiener diffusion process to address measurement errors and a timescale transformation method to address the time-varying degradation drift. This method has been extensively applied in [99–104] to describe the degradation modeling of light-emitting diodes (LEDs), self-regulating heating cables [98], bridge beams [105], bearings [106], and so on. Peng and Tseng [99] employed the Wiener process to analyze the degradation path of LEDs and to estimate the equations for median life as well as MTTF. Liao and Elsayed [104] applied the Wiener process to model the degradation of electronic devices such as LEDs sources exposed to variable stresses under field conditions. Ibrahim et al. [107] investigated the lifetime estimation of high-power white LEDs based on lumen maintenance data using the Wiener process method. Jing et al. [108] used the constant drift Wiener process to model the radiation power degradation for ultraviolet LEDs. Recently, a modified Wiener process was proposed by Huang et al. [50] to handle dynamic and random variations of lumen degradation and color shift for mid-power white LEDs and predict their lifetime. The analysis of lumen maintenance and chromaticity shift of mid-power white LEDs with the modified Wiener process along with the cumulative distribution function (CDF) is shown in Fig. 6.7a–d.

Apparently, the Wiener process has been observed to entertain some variations as limiting cases. A common variation is a Wiener process with a linear drift which has been studied by Tseng et al. [100], Peng and Tseng [99], Tsai et al. [68], and Guo et al. [109]. On the other hand, a timescale-transformed Wiener process has been explored by Whitmore [97], Whitmore and Schenkelberg [98], and Wang [105]. Real-time reliability has been investigated by Xiaolin et al. [110] based on a

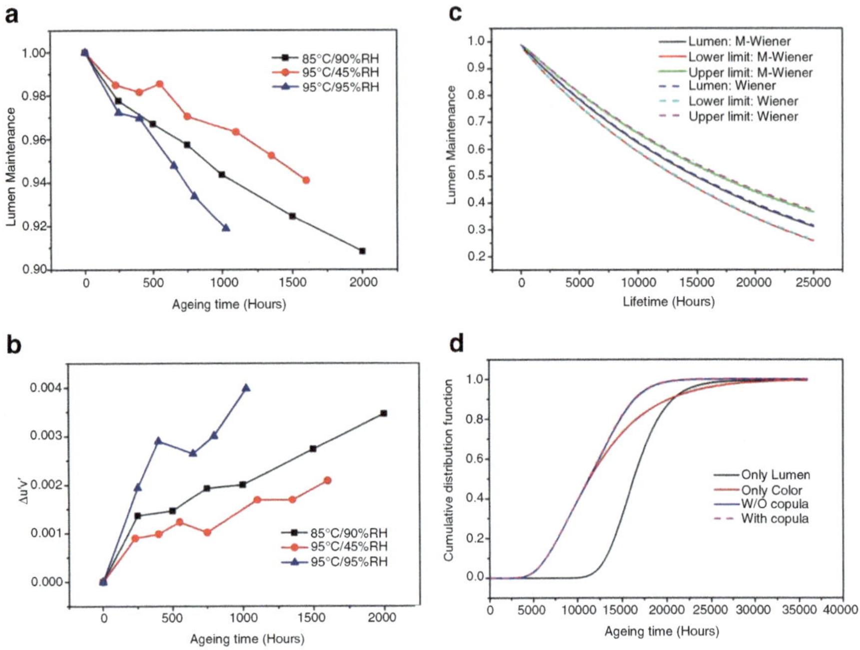

Fig. 6.7 (**a**) Luminous flux degradation, (**b**) chromaticity shift, (**c**) fitting lumen degradation data and lower and upper limits for Wiener process (WP) and modified WP, (**d**) cumulative distribution function for lumen and color with and without copula. [50] Copyright 2015 Optical Society of America

generalized Wiener process-based degradation model and validated based on a laser device and capacitor data. A comprehensive review on the Wiener process based methods and its implementation for degradation data analysis and lifetime estimation is given in Zhang et al. [111].

4.1.2 Gamma Process-Based Approach

The gamma process is one of the popular stochastic process models used for modeling nonnegative degradation increments taking place in a sequence of small step time increments. The gamma process is thus a suitable model for unidirectional degradation processes including crack growth, erosion, creep, fatigue, wear process, corrosion, swell, and related degrading health index or performance degradations [112]. The effectiveness of the gamma process for useful lifetime estimation and reliability assessment is due to relevant advantages. One of the main interesting features of gamma process in terms of lifetime prediction is that the mathematical calculations needed are fairly understandable and the underlying physical meaning is easy to comprehend [77]. The PDF for a degradation process *X*(*t*), which can be described in terms of the gamma process, is given according to the definition as:

$$f\left(x|\alpha,\beta\right)=\begin{cases}\dfrac{\beta^{\alpha}}{\Gamma\left(\alpha\right)}x^{\alpha-1}\exp\left(\beta x\right), x\geq 0\\ 0, x<0\end{cases} \tag{6.3}$$

where X(t) is a performance degradation parameter (such as luminous flux, color shift, etc.), α is a shape parameter, β is a scale parameter, and $\Gamma(\alpha)$ is the gamma distribution function.

The system/product's MTTF under this model M_G and failure threshold w has been approximated by Park and Padgett [113] as:

$$MTTF_G \cong \frac{w}{\alpha\beta}+\frac{1}{2\alpha} \tag{6.4}$$

Nevertheless, it is worth noting that the gamma process appears suitable for the monotonic degradation process, and this may restrict the application of the gamma process to some other dynamic degradation patterns. For this reason, incorporating the modified gamma process that uses the method of moments to estimate the model parameters can enhance degradation modeling and lifetime estimation process. Recently, the gamma process has been employed to model the lifetime of high-power white LEDs based on CCT shift [115]. Ibrahim et al. [114] also used gamma process to model reliability of phosphor-converted white LEDs by estimating the long-term lumen maintenance lifetime and validate by comparing with the NLS regression method. The results showed that the prediction accuracy of the gamma process was superior compared with the NLS regression-based approach. The plots demonstrating the luminous flux degradation, probability distribution with gamma, PDF at different time points, CDF, and reliability estimation are shown in Fig. 6.8.

4.1.3 Particle Filtering (PF) Approach

Particle filtering (PF) is a Monte Carlo simulation-based method which provides a convenient framework to handle Bayesian-framed prognostics. PF is a commonly used method to model and manipulate non-Gaussian processes and/or nonlinear performance degradations and measurement noise. PF uses a number of particles and set of weights associated with them to compute the prior distributions (probability densities) of the model parameters [116–118]. On the contrary, the IES TM-21 standard for projecting lumen maintenance lifetime uses the LSR to compute model parameters which depends on the minimization of the sum of errors or offsets between the estimated values by using proposed analytical equation and experimental or real measurements.

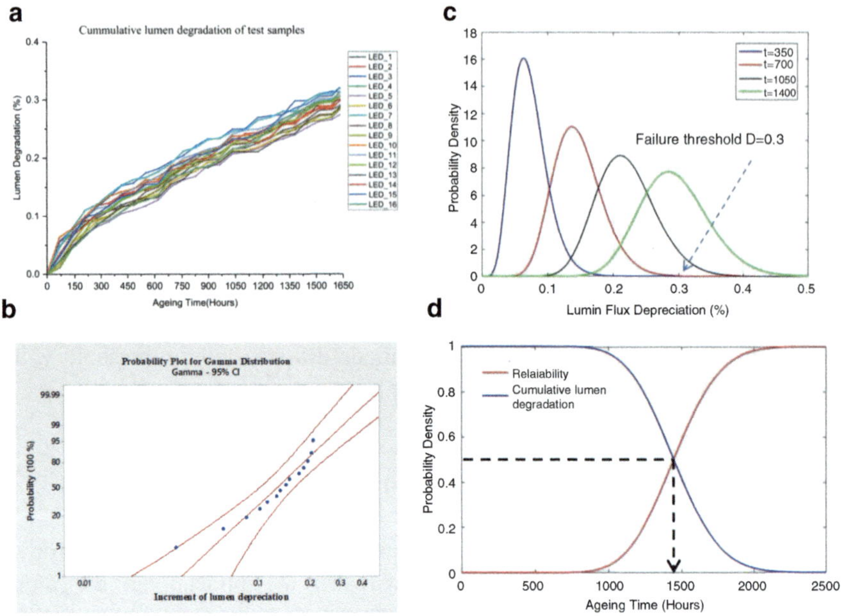

Fig. 6.8 Gamma process (**a**) average cumulative lumen degradation, (**b**) probability distribution plots with gamma distribution, (**c**) PDF at different times, (**d**) CDF and reliability plots [114]. Copyright, the authors

Due to its features, PF is found to be effective to model the lifetime of LED sources that are known to manifest dynamic and nonlinear performance deterioration, such as luminous flux and chromaticity shift. A typical procedure to apply PF method can be described according to Fan et al. [119], as follows: the first task is to choose a degradation model as suggested in the IES TM-21 standard (i.e., exponential-based decay model) to represent the performance degradation in the LED light source. Then the second step is to replace the LSR method used to estimate model parameters in TM-21 with Bayesian inference in PF approach. The Bayesian inference makes use of observations or experimental values to estimate the value of unknown model parameters and update their values in the form of distribution function. Within a proposed PF method, the procedure of the recursive state estimation and optimization with updated measurements can be performed in four steps: (i) initialize the model parameters, (ii) sample the model parameters and prediction, (iii) use the Bayesian inference algorithm to update values, and (iv) weight the particles and resample, as shown in Fig. 6.9a. At the end, the experimental measurements will be terminated at time t_p, and then the remaining useful life (RUL), with confidence interval limits, will be estimated by manipulating the updated model with measurement noise. Fan et al. [119] employed this PF method to project the lumen maintenance lifetime for high-power white LEDs. The feasibility of the PF method was validated, and its prediction accuracy was evaluated and

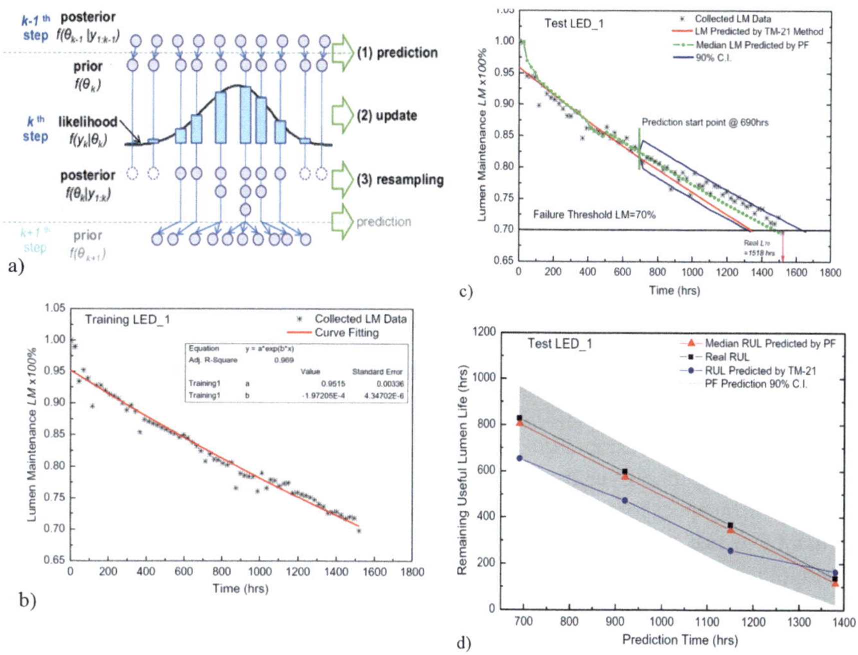

Fig. 6.9 Illustration of particle filter process to predict the lifetime of high-power white LEDs based on luminous flux degradation data (**a**) model parameter estimation process. [120] Copyright 2017 Springer International Publishing. Implementation of particle filtering algorithm (**b**) fitting all lumen degradation data to decay model as training samples, (**c**) prediction of lumen maintenance life, (**d**) PF method and IES-TM-21 LSR approach estimating RUL based on lumen maintenance data. [119] Copyright 2015, the authors

showed superiority over the current NLS regression-based TM-21 method. Illustration of the implementation of the PF approach to investigate the lumen maintenance lifetime for high-power LEDs is shown in Fig. 6.9.

As the main focus of this review is on the machine learning-based data driven approaches, the review on statistical approaches is limited to the updated and well-revised Wiener process, gamma process, and PF approaches. For other statistical-based data-driven approaches such as Mahalanobis distance, Euclidean distance, Kalman filter (KF), and unscented Kalman filtering (UKF), a brief review is given in Sun et al. [23]. The different types of ML algorithms employed to handle lifetime estimations of LED sources are presented in the next section.

4.2 *An Overview of Selected Machine Learning Methods for PHM*

Recently, an exponential increase in computing power, introduction of new state-of-the-art algorithms, and systematic generation of large data have been observed. Due to this, ML has emerged by breaking new frontiers in reliability assessment and lifetime prediction field of studies. ML algorithms are a set of procedures and methods that can be used to capture, detect, and learn relevant information patterns from large amounts of data and use the unhidden patterns for the process of decision-making in anomaly detection, diagnostics, and prognostics or predicting remaining useful lifetime [18]. ML can be defined as "the branch of artificial intelligence that deals with the development of algorithms and models that can automatically learn patterns from data and perform tasks without explicit instructions," according to Chen et al. [121]. A more engineering-oriented definition of machine learning was presented by Mitchell [122] as "a computer program is said to learn from experience E with respect to some class of tasks T and performance measure P if its performance at tasks in T, as measured by P, improves with experience E." In short, machine learning enables computers to learn through experience and improve performance without requiring explicit programming. For instance, if the task T is to identify the failure of LED systems, the training data such as lumen degradation and chromaticity shift can be considered as the experience E, and the failure prediction or estimation accuracy is the performance measure P. Depending on the amount and the type of human supervision required, it can be broadly categorized into supervised learning (predictive modeling), semi-supervised learning, and unsupervised learning (descriptive modeling).

4.2.1 Supervised Learning Approaches

In supervised learning, an output value or desired pattern can be estimated/predicted based on a classified or labeled set of input data. Depending on the output or response variable, the problem can be described as either classification (such as normal or abnormal) or regression (such as lumen degradation level, chromaticity shift, CCT degradation). As a result, the choice of the learning method is an important factor in achieving desired outputs or in discovering the group of input data. A typical supervised ML task is classification, and a diagnostic problem is a typical classification task. Due to this, the majority of supervised ML methods are used to address diagnostic problems (i.e., failure mode identification, normal, anomaly, etc.). However, supervised ML methods are also applicable in the estimation of remaining useful lifetime (RUL) which is a regression task [123]. Some authors recognize linear regression [124] [125] and logistic regression [126] as supervised machine learning methods. However, the well-known supervised machine learning approaches applied for the prognostics of systems include k-nearest neighbors (KNN), support vector machine (SVMs) [127], relevance vector machine (RVM)

[128], decision trees [129], artificial neural network (ANN), [18] [90] [130], and random forest. Some of the widely used machine learning methods are discussed as follows.

Artificial Neural Network

Artificial neural networks (ANN) form a set of mathematical algorithms conceived and modeled after the human brain's neurons structure and designed to recognize patterns [131]. A typical neural network and back-propagation learning [132, 133] is shown in Fig. 6.10. The working principle of the ANN algorithm mimics the human brain which connects many nodes in a complex structure. The nodes represent input, output, and hidden variables, while the links represent the weight parameters. The bias parameters are denoted by links coming from additional inputs and hidden variables x_0 and z_0, and more details about ANN are given in [132]. In an ANN, a network is modeled, and it learns an effective way to produce a desirable output by reacting to give inputs [35], as depicted in Fig. 6.10. In a back-propagation ANN, the learning process consists of forward propagation of the signals and backward propagation of the errors.

ANN is a popular ML approach used to perform many tasks such as prognostics (prediction/regression problems) and diagnostics (classification problems). ANN helps to compute a predicted output for the lifetime of a product explicitly or implicitly, from a mathematical representation of the product derived from measurement/experimental data rather than a physical understanding of the failure processes [90]. ANNs are known methods for modeling complex nonlinear systems effectively and

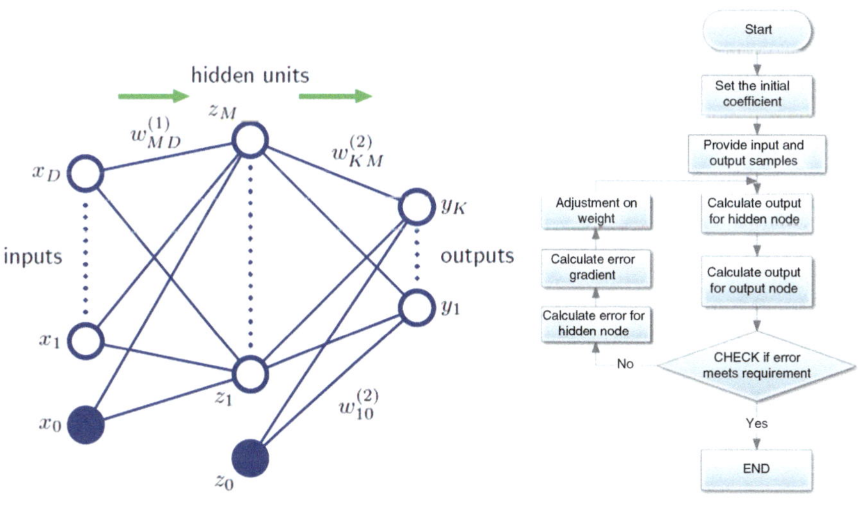

Fig. 6.10 A representative network diagrams for two-layer neural networks (left), flowchart for back-propagation learning algorithm

efficiently and can generalize and adapt solutions from a limited dataset [130]. Based on the mathematical operations and set of parameters required, ANN architecture can be of different types including feedforward neural network, back-propagation neural network, radial basis function neural network, recurrent neutral network (RNN), and self-organizing map. Although ANN has been widely applied in prognostics, it has two main limitations. The first is a lack of transparency or lack of documentation on how decisions are made in a trained network. The second one is related to optimization of results as there are no established methods to determine the optimal network structure.

As one of the popular approach in prognostics, ANN has been implemented to study transformers [135], aircraft actuator components [136], bearings [137], nuclear turbogenerators [138], electronic packages [139], etc. However, application of ANN methods for high-power white LEDs lifetime estimation was not very common until Sutharssan [140] demonstrated a basic neural network for lifetime prediction of LEDs. The model used consists of one hidden layer and two neuron nodes in the hidden layer. Recently Lu et al. [134] proposed and tested a model for lifetime prediction of high power as well as mid-power LED light sources. In their investigation, both the radial basis function network and back-propagation neural network were demonstrated. The AdaBoost algorithm is adopted to enhance backward propagation NN in training the weight points connecting input neurons with hidden layer neurons and predict the lifetime with a multidimensional input parameter such as lumen depreciation, color coordinates, driving current, and aging temperature. The BPNN data training, iterations, training errors, as well as predictions are shown in Fig. 6.11. In general, the performance of ANNs has good performance for lifetime estimation of systems due to the capability of learning complex relationships by training multilayer networks. However, it has few undeniable limitations, such as low transparency and the demand for high-quality data, which could be difficult for new products in industrial applications.

The recurrent neural network (RNN) is a type of ANN designed to recognized sequential data such as speech recognition, precise timing, and so on, due to its added feature of time dimension to NN model. However, RNN still suffers from gradient exploding or vanishing during the learning process [141]. With the capability of learning long-term dependencies, a special type of RNN called the long short-term memory (LSTM) architecture was found to be suitable to overcome the shortcomings of the traditional RNN architecture. Guo et al. [141] used LSTM architecture to predict the RUL of bearings, and, compared to SOM, the prediction performance of LSTM was found to be superior, as shown in Fig. 6.12. Similarly, Wu et al. [142] deployed the LSTM approach in prognostics and demonstrated a good prediction accuracy using aircraft turbofan engine's health performance data. While LSTM architecture RNN appears to be suitable for LED RUL estimation, application of this method has not been reported in the literature.

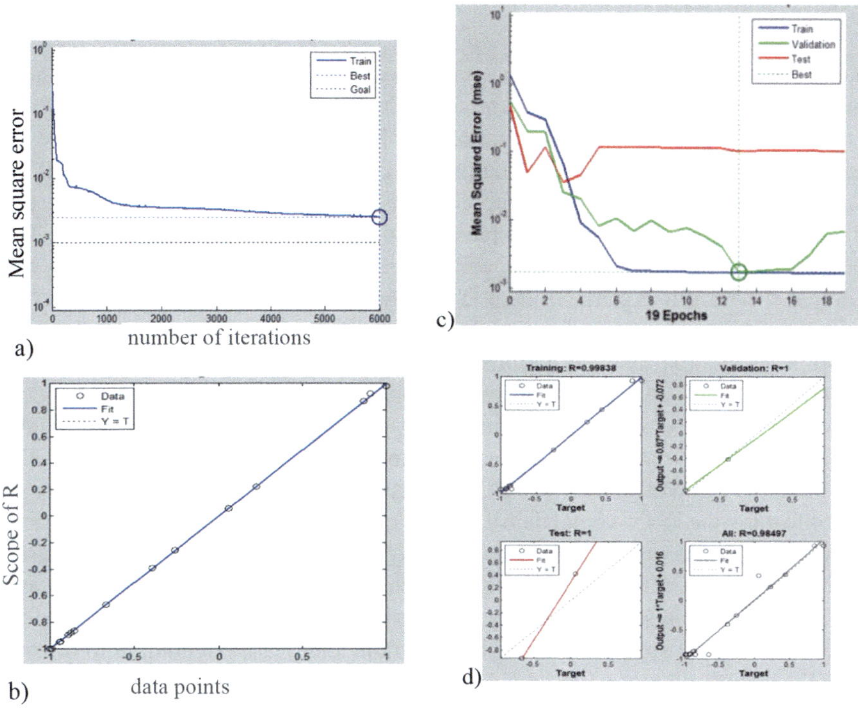

Fig. 6.11 A typical neural network (**a**) BP neural network training convergence curve, (**b**) effect of network training regression, (**c**) AdaBoost-improved BPNN curves of iterations and training error, (**d**) state of prediction regression [134]. Copyright 2018, the authors

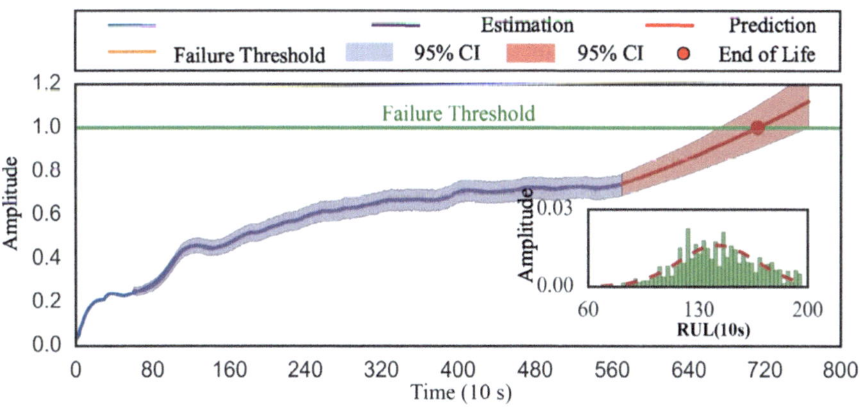

Fig. 6.12 RUL prediction result for a bearing [141]. Reproduced with permission from publisher/ Elsevier

K-Nearest Neighbors

K-nearest neighbors (KNN) is a supervised learning algorithm with a non-probabilistic property that belongs to similarity-based prognostics, and it has been employed in PHM for crack propagation [143], electromagnetic relay contact resistance [144], and printed circuit boards ball grid array solder joints [145]. As an emerging trend in the prognostics approach, KNN has been used as a lifetime estimation tool for reciprocating compressor valves based on regression [146]. The prognostic performance, precision, and accuracy of KNN regression (KNNR) was compared with self-organizing map (SOM) and multiple regressions using actual operating data of a valve from an industrial compressor. The result for all the approaches showed that the performance was relatively good and comparable to each other. A typical application for LED anomaly detection has been conducted based on the KNN-kernel density-based clustering algorithm [147]. In this study, peak analysis was used to extract features from spectral power distribution (SPD), the principal component analysis (PCA) was used for the reduction of dimensionality of feature, the KNN-kernel density-based clustering technique was used to partition the principal components datasets into clusters, and finally distance-based algorithm were used to detect anomalies. In this case study, the KNN algorithm was used to list *kth* nearest neighbor distances to each of the N single clusters formed by PCA. This typical application of KNN algorithm and related techniques to investigate the qualification of LEDs along with some results is illustrated in Fig. 6.13.

Support Vector Machine and Relevance Vector Machine

The support vector machine (SVM) is a modern and advanced technique used for classification problems (anomaly detection, diagnostics such as normal/anomaly) and regression (prediction) types of problems. It is a very successful approach in supervised learning using the flexible (i.e., multiparameter) linear kernel approach. Predictions are made in SVM based on a function of the form given as:

$$\mathrm{y}(\mathrm{x};\omega) = \sum_{n=1}^{N} \omega_n K(x,x_n) + \omega_0 \tag{6.5}$$

where w_n are the model weights and $K(x, x_n)$ is a kernel function. The target function of SVM has a key feature that attempts to reduce the number of errors on the training set while maximizing the margin between two classes in a classification study.

Due to this, it has the advantage of preventing over fitting that leads to good generalization and results in a sparse model dependent only on a subset of kernel functions [148]. The SVM classifier algorithm has been demonstrated in the problem of health evaluation and novelty detection. In [81], the Bayesian SVM was trained to model the posterior class probability in the absence of failure data (i.e., anomaly or negative class data), as in the case for a safety and mission critical

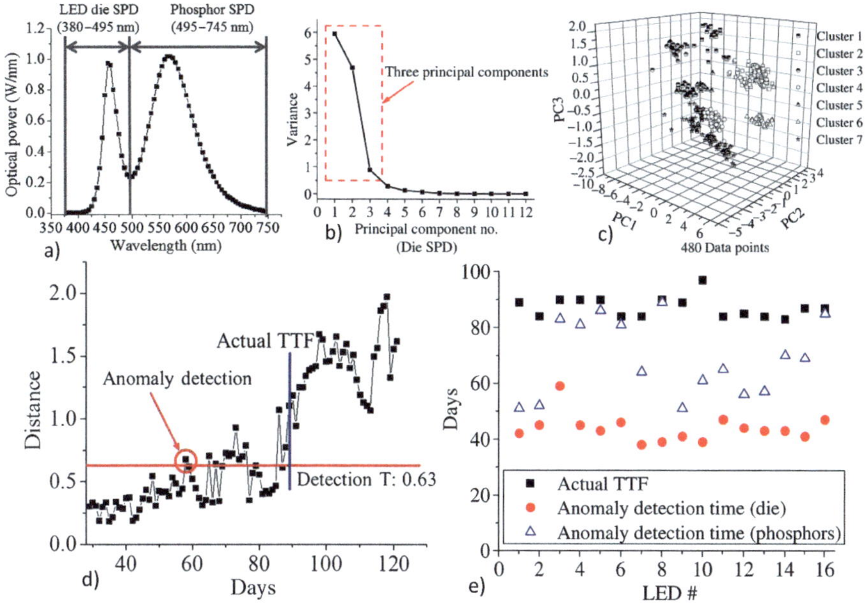

Fig. 6.13 (**a**) SPD feature extraction, (**b**) principal components from extracted features, (**c**) SPD training using KNN-kernel density-based clustering, (**d**) distance measure from cluster centroid to detect anomaly, (**e**) anomaly detection using die SPD [147]. Reproduced with permission from publisher/IEEE. Copyright 2014, the authors

system in Lockheed Martin equipment. In addition to this, a least-squares SVM combined with Bayesian inference was developed and used to investigate lifetime prediction of a microwave component [149]. In [149], the radial basis function NN (RBFNN) algorithm was also employed for RUL estimation and validation purposes, and the point and interval estimate of RUL based on least-squares SVM has been found to be more robust and stable compared with the RBFNN algorithm. Despite its success, SVM suffers from a disadvantage in terms of lack of probabilistic prediction outputs (for regression and classification problems) which is an important aspect in prognostics applications [148, 150].

The relevance vector machine (RVM) is an identical functional form to the SVM which has a probabilistic sparse kernel model as an additional feature. The RVM achieves this through the Bayesian approach and introduces a prior over the weights that are governed by a set of hyper-parameters. In addition to its generalization performance capability that is similar to SVM, the other feature of RVM is that it makes use of considerably fewer kernel functions compared to the SVM approach. In the PHM area, the RVM has been successfully explored to estimate the RUL of rotating equipment in an aerospace setting [128]. The RVM regression (i.e., a Bayesian machine learning technique) has also been implemented effectively to predict the RUL of LEDs, and the qualification result showed that the testing time for LEDs can be reduced from the IES standard (i.e., 6000 h) to hundreds of hours

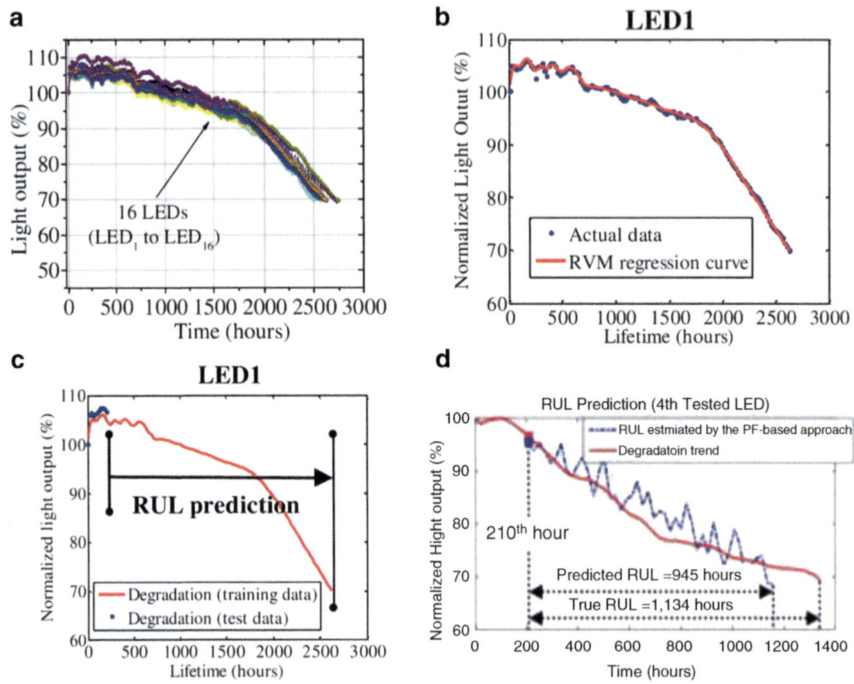

Fig. 6.14 (**a**) LED luminous flux degradation, (**b**) parameter measurement for RUL, (**c**) lumen degradation the RVM regression model, (**d**) PF-based lifetime prediction results [151]. Reproduced with permission from publisher / IEEE. Copyright 2017, the authors

(210 h). This approach was also reported to handle unit-to-unit variation and also has the capability of handling transient degradation dynamics. Due to this feature, the RUL prediction accuracy of the RVM approach has been reported to surpass the particle filtering approach [151]. The detailed results for the LED lifetime estimation based on RVM regression compared with the PF approach are depicted in Fig. 6.14.

In general, the SVM and RVM demonstrated superior performance compared to the ANN approaches for experiments with small sample sizes. Due to this, SVM and RVM may be suitable for lifetime prediction where limited measurements are available. On the other hand, challenges such as parameter estimation may slow down its wider application.

4.2.2 Unsupervised Learning Approaches

Unsupervised learning is a machine learning procedure where the input dataset is unlabeled, and also there is no classified or labeled target response value Y_i or response variable. In other words, there is no labeled output value to supervise the learning process of a learner, or there is no need of data to train algorithm. In

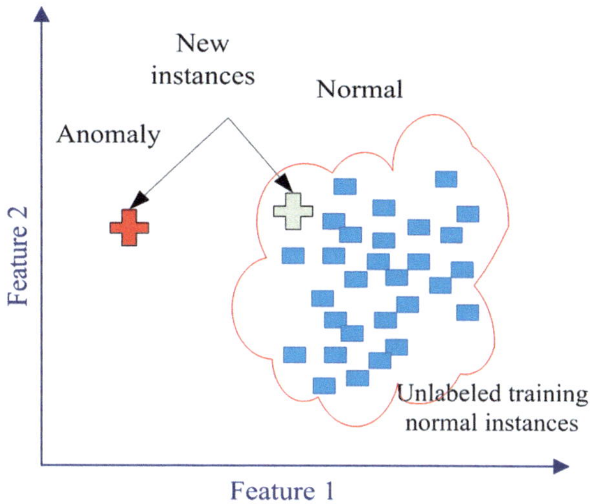

Fig. 6.15 An unlabeled training dataset for unsupervised learning

unsupervised learning methods, an unlabeled or unclassified set of data is used to find interesting patterns or outputs in the data. Due to this, the main tasks of unsupervised learning are clustering and dimensionality reduction, and the nature of these ML approaches enables the addressing of anomaly detection [152]. Some of the unsupervised algorithms are k-means clustering, principal component analysis (PCA), and hierarchical clustering. The unlabeled instances are used to train a model for representing normal behavior [123] as shown in Fig. 6.15. A few of these unsupervised learning approaches that have been investigated to conduct reliability assessment of LED products are described in this section.

Principal Component Analysis

Principal component analysis (PCA) is an exploratory data analysis technique used in dimensionality reduction to simplify the complexity of data while retaining patterns and trends. It performs this by transforming the original data into fewer comprehensive dimensions (indexes), which act as summaries of features [153]. Similar to clustering, PCA is an unsupervised learning method, and it finds patterns without reference to prior knowledge of the data. This approach was first introduced in 1933 by Hotelling [154] to transform the statistical dependency of groups of correlated variables in multivariate data to uncorrelated variables and to achieve optimal conditions.

The PCA method has been widely implemented in condition monitoring for mechanical systems. Wang and Zhang [155] used PCA to transform a set of variables for aircraft engine experimental observations to a new set of uncorrelated variables. The new set of data are known as principal components and then used in the aircraft engine lifetime recursive filtering-based prediction model. On the other hand, Ahmed et al. [156, 157] demonstrated PCA approaches for fault detection in reciprocating compressors by identifying 5 and 7 most important performance characteristics (PCs), respectively, from 9 and 14 original features.

The life of high-power LED is influenced by numerous parameters including series resistance, optical output saturation, junction temperature, and so on. Qiyan [133] adopted PCA to process the various parameters and select the principal components (parameters) for further processing using neural networks. Chang et al. [147] used PCA for dimensional reduction among 24 extracted features from LEDs die SPD (12 features) and phosphor SPD (12 features) to study anomaly detection of LEDs. The six principal components from 24 extracted features were further analyzed using a KNN-kernel density-based clustering technique. This study analyzed 480 and 640 training datasets and portioned into 7 and 8 clusters, respectively, and the results of feature extraction and principal component analysis are shown in Fig. 6.13 along with SVM/RVM plots.

K-Means Clustering

K-means clustering is an unsupervised learning fault detection approach which is widely used in industry because it can be applied without the need to be trained on data obtained from a faulty machine or system. In k-means a number of centroids are selected that define the number of clusters, and each data point is assigned to its closest centroid based on Euclidean distance. The k-means clustering helps to partition n number of objects into k clusters where each object will have the nearest mean distance from the cluster. The main objective of this method is to minimize the total distance between clusters or the square error function. This objective function can be formulated as follows:

$$J = \sum_{j=1}^{k}\sum_{i=1}^{n} x_i^{(j)} - c_j^{\,2} \tag{6.6}$$

where J is the objective function, n is number of objects, k is number of clusters, and $x_i^{(j)} - c_j^{\,2}$ is the chosen distance function among the data point $x_i^{(j)}$ and the cluster centroid c_j.

This method has been successfully applied for anomaly detection of mechanical components, such as rolling elements bearings [158], as well as for wind turbines [159]. In [159], data was collected from a normally operating turbine supervisory control and data acquisition system (SCADA) and fitted using the k-means clustering algorithm. This approach shows the suitability for employment in anomaly detection in LED systems as it does not require failure data or faulty system information. However, application of this approach for diagnostics and prognostics of LEDs was not found in the literature. Figure 6.16 shows a typical implementation

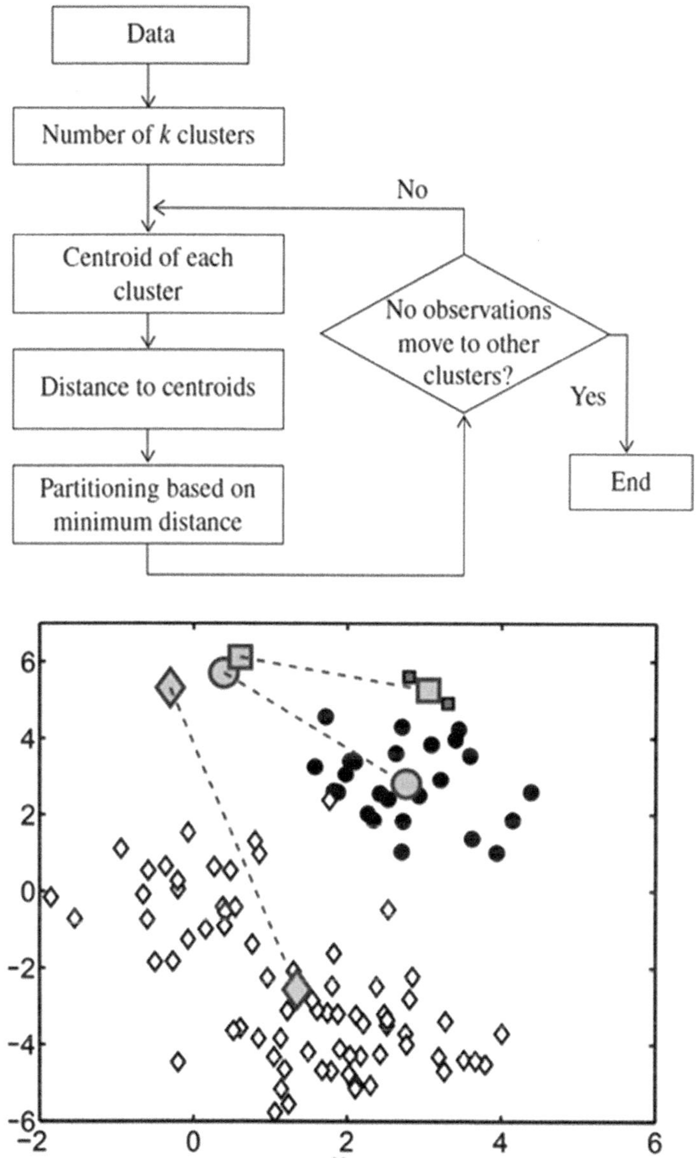

Fig. 6.16 K-means clustering procedure (top), clustering illustrated (bottom). [152]. Reproduced with permission from publisher/Wiley

procedure for this approach (left) and how trained data can find their clusters based on the distance from the centroid (right).

Self-Organizing Map (SOM)

First introduced by Kohonen [160], the self-organizing map (SOM) is one variety of ANN method mainly applied for unsupervised learning. The SOM has been employed to project high-dimensional data obtained from supervisory control and data acquisition system of a wind turbine into a two-dimensional space to capture the pattern of input training data. A Euclidean distance method was used to represent difference between new input data and target value as the indicator for system-level anomaly detection [161]. Tian et al. [162] demonstrated a SOM-based method for the purpose of anomaly detection with the k-nearest neighbor algorithm for the purpose of reducing sensitivity to noise in mechanical and electronic systems (cooling fan with ball bearing) data.

Recently, this approach has been applied as a lifetime estimation approach for compressor valve failure data, and the result was found to be relatively competitive with other approaches applied for purpose of comparison, such as KNNR and multiple regression [146]. The study claimed that the SOM was used for the first time as a standalone program for remaining useful lifetime estimation. Even though an implementation of this method was not found in the PHM of LEDs, the similarity of the nature of degradation data in the mechanical component observed in the study [146] suggests that this method appears to be promising for the RUL estimation of LED products [146]. The RUL prediction performance of SOP along with KNNR, multiple linear regression, and ensemble methods based on a historical failure data is depicted in Fig. 6.17.

4.2.3 Semi-supervised Learning Approaches

Semi-supervised learning paradigm is a ML approach that falls within supervised and unsupervised learning methods by introducing both labeled and unlabeled data for training. This approach has evolved recently and has been increasingly applied to automatically manipulate and exploit large amounts of unlabeled data and small amounts of labeled data for training without requiring human experts. The aim of semi-supervised learning is to classify a set of unlabeled data using the information set from the labeled data, and it is mainly applied for anomaly detection problems. For a typical semi-supervised learning, suppose a dataset $X = (x_i)_{i \in [n]}$ can be divided into two components: data points $X_j : (x_1, x_2, \ldots, x_j)$ for which labels $Y_j : (y_1, y_2, \ldots, y_j)$ are given and data points $X_k : (x_{j+1}, x_{j+2}, \ldots, x_{j+k})$, for which the labels are unknown [163]. The Semi-supervised learning methods are widely applied for speech analysis, web

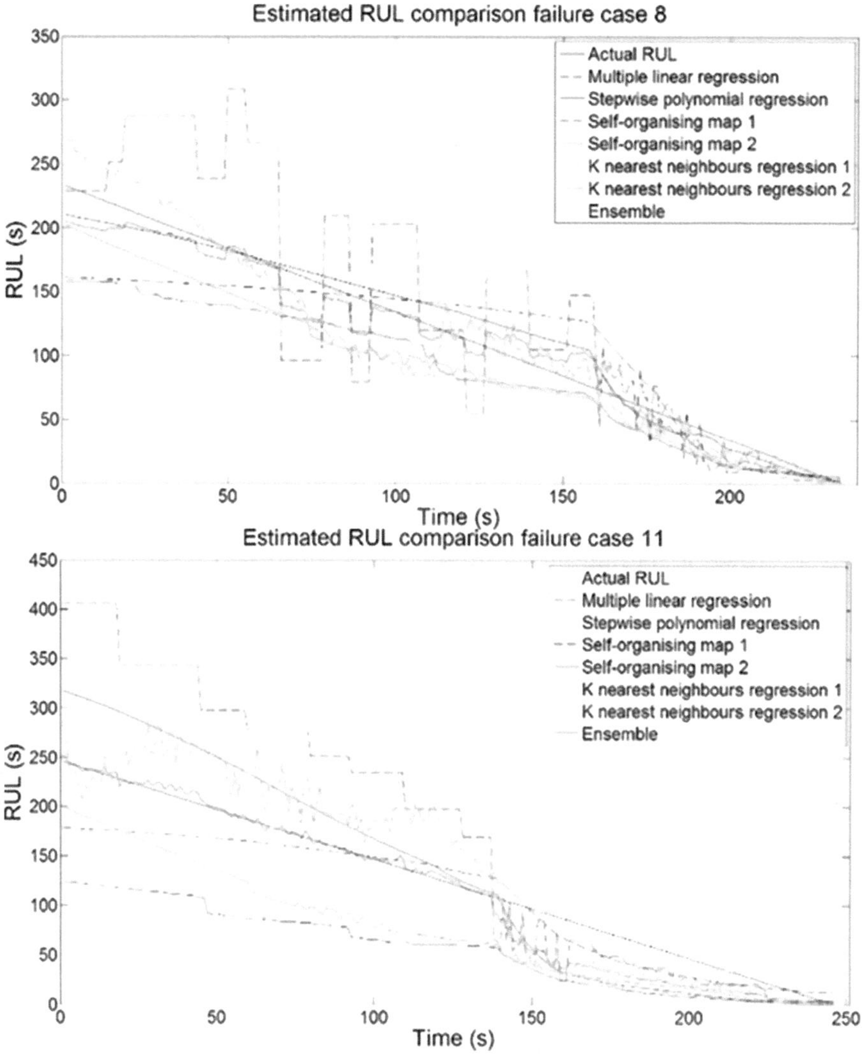

Fig. 6.17 RUL estimation based on SOP, KNNR, multiple linear regression, and ensemble methods [146]. Reproduced with permission from publisher/Elsevier

content classification, protein sequence classification, and recently in prognostics. Some of the examples that can be considered as semi-supervised learning algorithms include Hidden Markov Model, expectation maximization (EM) with generative mixture models, graph-based methods, and transductive SVM [164], and two of these methods that have been successfully applied in prognostics are discussed here.

Expectation Maximization

Expectation maximization (EM) is an iterative and general procedure employed to estimate model parameters in a parametric distribution. ML is often considered as a special case of maximum likelihood estimation where missing or incomplete data is examined and computed by alternating between (i) estimation of expectation (E-steps) and (ii) maximization during model re-estimation (M-steps) until it converges [165]. Although the EM algorithm is not widely seen in the PHM field, it is a very important algorithm, and a typical application of EM for use in a RUL prediction is presented by Si et al. [166]. In this study, linear and exponential-based closed-form degradation models were considered to demonstrate a degradation path approach for RUL prediction. The expectation maximization algorithm along with Bayesian updating was used to update the RUL distribution and model parameters when new degradation data was obtained [166]. In solid-state lighting, a recent work showed that expectation maximization (EM) has been applied to estimate the model parameters of the exponential decay model and to calculate the remaining useful lifetime of HPWLEDs [167] as shown in Fig. 6.18. In this study, the EM was applied to estimate the degradation model parameters for the state space model from unlabeled luminous flux degradation data. The RUL estimation results were claimed to be superior to TM-21 standard which is based on NLS regression method, and it showed a comparable accuracy to PF method (Fig. 6.18).

Hidden Markov Models

Hidden Markov Models (HMMs) are standard approaches for encoding, analyzing, and predicting patterns in multivariate and univariate observation data. Even though the HMM technique was developed in the late 1960s, it is still going through development and gaining popularity [168]. The HMMs are based on a stochastic model and Markovian hypothesis, where the current hidden (not observable) state of the model is influenced by its previous state. In HMM, each of the current model states (hidden) displays an outcome which is observable state. For instance, in case of LEDs, when estimating the lumen degradation or color shift state at time point t, the HMM considers not only the feature values $X(t)$ at time t but also the preceding value X_{t-1}.

The HMM is a semi-supervised approach, typically used for anomaly detection. However, HMMs can also address detection problems, decoding problems, as well as learning problems. This method was successfully applied for the first time in PHM study by Baruah and Chinnam [169], where the sensor signals from a machine were modeled using the HMM method to identify the health status as well as facilitate the remaining useful life estimation of cutting tools. The HMM has also been applied in PHM for mechanical parts, including hydraulic pumps [170, 171], helicopter gearboxes [172], as well for anomaly detection in an electronic component, insulated gate bipolar transistor (IGBT) [173]. A mixture of Gaussian Hidden Markov Models has also been employed to assess the current health status and

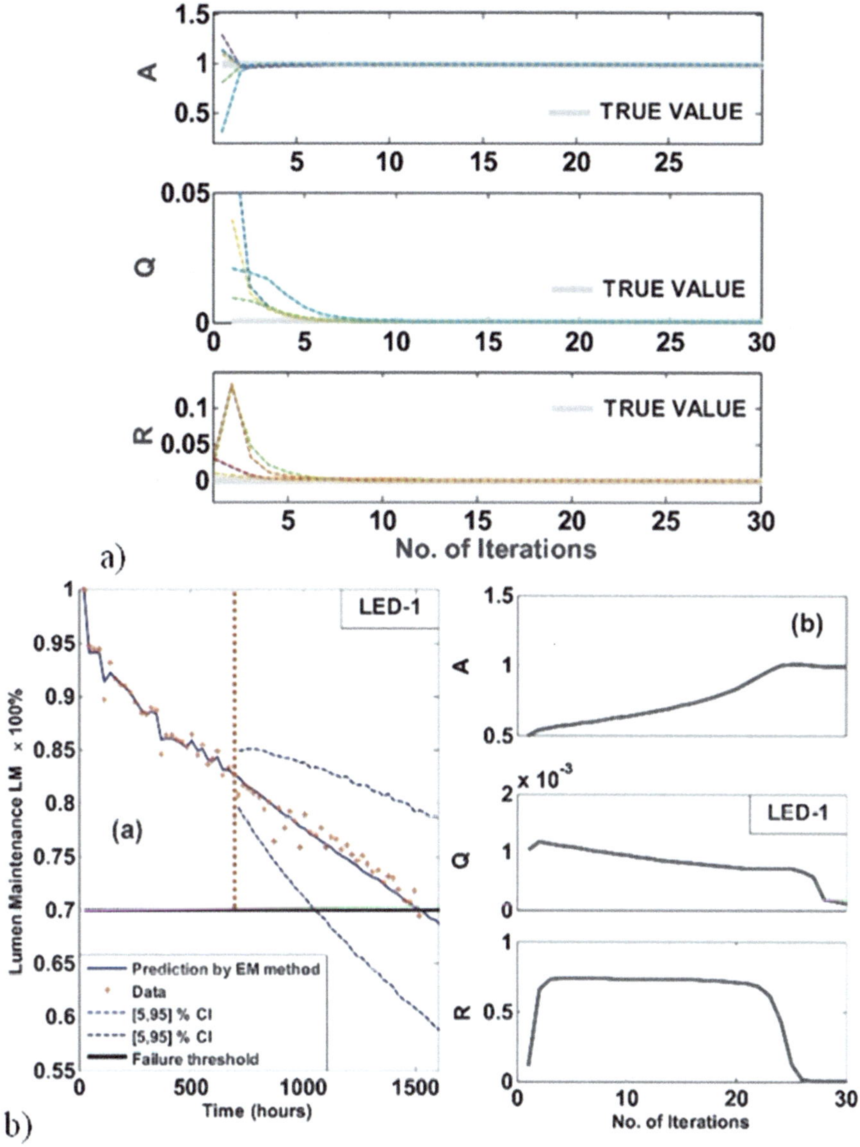

Fig. 6.18 (**a**) Parameter estimation using EM algorithm, (**b**) RUL prediction based on EM-estimated parameters and iteration trends values for parameters [167]. Reproduced with permission from publisher/Elsevier

estimate remaining useful lifetime of bearings [174]. Even though it has been applied for diagnostics and prognostics for mechanical parts and electronic components, its application has not been found in PHM for LED products and systems so far. A comprehensive theoretical explanation and step-by-step tutorials on the general HMM are given in Rabiner [175], while a review on the potential applications of HMM is demonstrated in [176].

The observation sequence $O = O_1\ O_2 \ldots O_T$ can be generated by HMM when appropriate values for *N*, *M*, *A*, *B*, and π are given. The compact notation for the discrete HMM model λ, when model parameters (*N and M*) and probability measures (*A*, *B and* π) specified are as:

$$\lambda = \{A, B, \pi\} \tag{6.7}$$

where *N*, *M*, *A*, and B are, respectively, number of hidden states in the model, number of distinct observations per state, state transition probability matrix, and the observation probability distribution of each state. The observed states are represented as *O* and *Q* which is hidden state at time *t*. The HMM can be represented graphically in different ways [176] as shown in Fig. 6.19. The first plot portrays a direct state transition graph, while the second illustrates the allowable transitions.

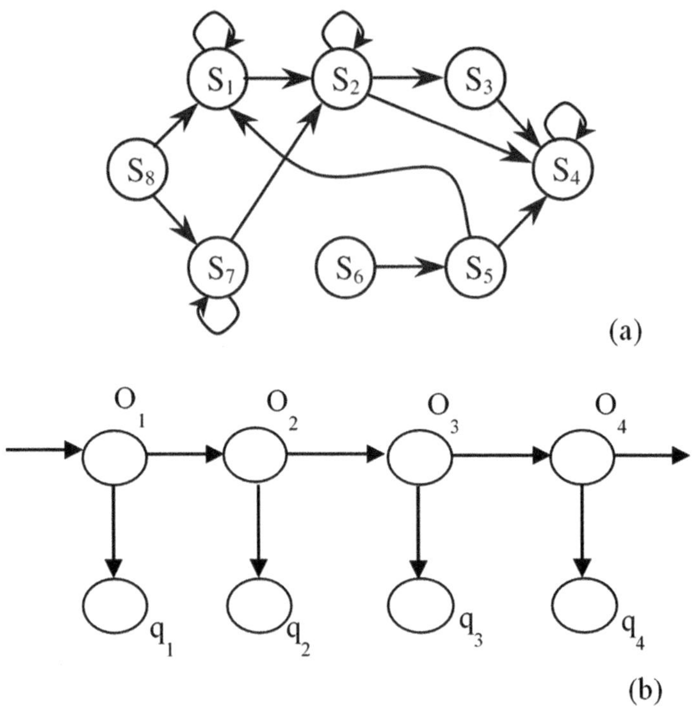

Fig. 6.19 HMM graphical description (**a**) stochastic finite-state automation view of a HMM, (**b**) a directed graphical model (DGM) [169]. Reproduced with permission from publisher/Taylor & Francis

In recent years, an increasing number of research studies can be found on prognostics using HMM. However, HMM still suffers from heavy computational workload problems, and consequently future research should focus on addressing the limitations and improve its applicability for complex and practical industrial systems and products including LEDs.

5 Fusion Prognostics Approach for Light-Emitting Diodes

Both the PoF- and DD-based PHM methods have been employed successfully in the prediction of failures in many devices and systems (e.g., machinery systems, LED lighting devices and systems, hybrid systems) [37, 177]. However, the PoF methods [178–183] require comprehensive knowledge of products in advance (e.g., materials and geometries, thermal, electrical, mechanical, life-cycle conditions, and other processes that lead to failures) that always increases the time and cost in actual applications. On the other hand, the data-driven approaches [119, 126, 184–194] need sufficient measurement or experimental data to estimate the health conditions and to predict trend thresholds from failure prognostics, but it is not easy to obtain these data in advance, especially for newly introduced LED lighting products. Thus, the fusion-based PHM is believed to solve these concerns by combining the advanced qualities and features of both the PoF and DD approaches. Fusion prognostics could apply PoF modeling, in situ monitoring procedures, and deployment of both statistics-based and ML-based DD methods to detect the performance deviation or degradation, predict the RUL, and assess the reliability for LED lighting products and systems. Because it uses in situ monitoring with the use of sensor technologies, fusion-based PHM can realize real-time failure diagnostics and RUL prediction in field applications.

The fusion (hybrid) prognostics approach combines the strengths of both PoF-based and data-driven methods, while eliminating their disadvantage to assess reliability, detect anomalies, and predict the lifetime of LED products and systems. The Fusion prognostics approach enables effective use of information from both methods for dynamic PHM and RUL prediction as well as to evaluate return on investment (ROI) [195–197] of LED product/systems [37]. Pecht and Jaai [34] assessed the state of applications in the PHM of electronic and information-rich products and presented a framework on the implementation of PHM for these products and systems by further illustrating a printed circuit board (PCB) case study. Cheng and Pecht [37] presented a fusion prognostic method to elaborate the useful lifetime of multilayer ceramic capacitors (MLCCs). They demonstrated this method with a special case study on the multivariate state estimation technique (MSET). Yao et al. [9] presented an implementation roadmap of PHM approaches for LED lighting systems. In their study, the LED lighting system was categorized into LED strings (including die, interconnect, and package) and the LED driving system (MOSFET, capacitor, etc.). However, for ease of understanding and the convenience of implementing prognostics approaches, the LED lighting product/system can be

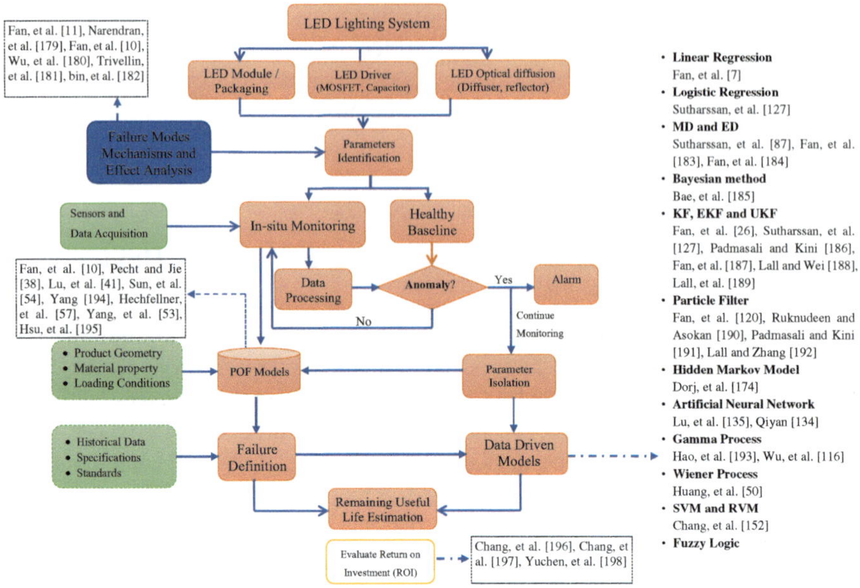

Fig. 6.20 Fusion prognostics for LED lighting system (adapted from Yao et al. [9], with permission)

categorized into three sub-parts as LED module/package (die, inter connect, encapsulates), LED driver (electrical part), and optical diffusion (diffuser and reflector parts). In general, the fusion prognostic approach based on PHM is an increasingly demanding method as it has not been well designed and developed for LED lighting. The detailed procedure for fusion prognostic approach implementation is shown in Fig. 6.20.

6 System-Level Reliability of Light-Emitting Diodes

As described in the previous sections, the diagnostics and prognostics of high-power white LEDs have been widely studied based on machine learning and statistical-based data-driven methods and algorithms, such as the Wiener process [50, 107], gamma process [114, 198], Kalman filter [186, 199], particle filtering [26], neural networks [134], expectation maximization [167], RVM regression [151], and so on. The reliability assessment and lifetime prediction of most of these studies are at the component level (such as package/module, LED driver, diffusers, and reflectors) using direct performance characteristics (i.e., lumen maintenance, color shift) and indirect characteristics (i.e., junction temperature and driving current) to examine the luminaire/system-level (such as LED luminaire, LED street lighting, LCD backlights, etc.) reliability. However, high-power white LEDs are complex products/systems composed of several subsystems/components, and it appears to be difficult to deduce the reliability of LED systems based on single component analysis as the product lifetime is affected by the health status of its components and their interaction. Due to this, LED manufacturers are facing challenges regarding system-level reliability assessment and remaining useful lifetime prediction of LED products/systems.

An LED system consists of several subsystems, including LED chip, electrical driver for power supply and control, thermal management module, optical part, and so on. One of the major challenges for a generic system-level approach for LED systems reliability is the large variety of products and applications [6]. In a high-power LED lamp system, the LED driver serves as the constant current source and optimizes the power to drive high-power LEDs [200]. Usually, the LED drivers are considered as the weakest part among all components in LED lighting products. Based on a family of outdoor luminaires failures, the US Department of Energy (DOE-US) [201] reported that the LED driver (power supply) is the weakest link in the LED lighting system, constituting 52% failure, LED package (10%), housing (31%), and control circuit driver (7%). On the other hand, Van Driel et al. [202] used the Monte Carlo approach to predict LED system-level reliability by taking both the failure mode of the sub-components and the operation conditions into account. The result showed that the LED emitters, solder interconnect, and driver accounted for 30%, 44%, and 26% failure rates, respectively, after 20,000 h of operation, showing that the solder interconnects are weakest parts in LED systems. Recently Ke et al. [64] introduced a subsystem isolation method to estimate the lumen degradation LED lamps, and the result showed lumen degradation of 70.5% due to the LED emitter, 21.5% the optical part, and 6.5% the driver, which contradicts the two studies (US DOE [201] and Van Driel et al. [202]) previously mentioned. Song et al. [45] also proposed a hierarchical life prediction model, which consists of component-level sub-physics-of-failure models, for the actively cooled LED luminaire system. In general, the results among studies based on subsystems and components for system-level lifetime analysis showed inconsistency.

In order to address the long-term reliability assessment concerns of complex and highly reliable products such as high-power LEDs and fulfill the guarantee of high prediction accuracy in less time and in a cost-effective manner, developing a system-level reliability assessment and lifetime prediction methods is necessary. Traditionally, graph model-based reliability block diagrams (RBD) and failure tree analysis (FTA) have been used to assess the system-level reliability of products and systems. However, these methods are based on deterministic relationships between components/subsystems. To address these concerns, the Bayesian network (BN) method, a probabilistic graphical machine learning method, appears to be a promising approach. The BN uses a directed acyclic graph (DAG) to represent the conditional and probabilistic relationship between component/subsystem relationships in a system [203]. As one of the popular modeling and reasoning tools, the BN model has been employed in the fields of machine learning, artificial intelligence, and uncertainty management [204]. The BN model has also been applied in the field of reliability engineering including software reliability [205], modeling maintenance [206], and fault diagnosis in systems [207, 208]. Recently, the BN model was found to be effective in estimating the system/product reliability of complex systems, such as high-speed trains [208], solar-powered unmanned aerial vehicles [209], and pitting degradation structural steel in marine systems [210].

In this section, a BN method that considers the intricacies of the high-power light-emitting diode (LED) lamp system and the functional interaction among components for reliability assessment and lifetime prediction is briefly introduced. This approach considers the parametric (degradation based) and catastrophic failure modes of each component in order to assess the system-level reliability, and it also requires the design of experiments to gather the required data. The functional and structural relationship analysis between components and the failure mode and effects analysis (FMEA) are considered [3] in order to construct a DAG for a BN model. In the BN model constructed in Fig. 6.21 (left), the variables which have no parents, such as LED_CAT, LED_DEP, Driver_CAT, Driver_DEP, Solder_CAT,

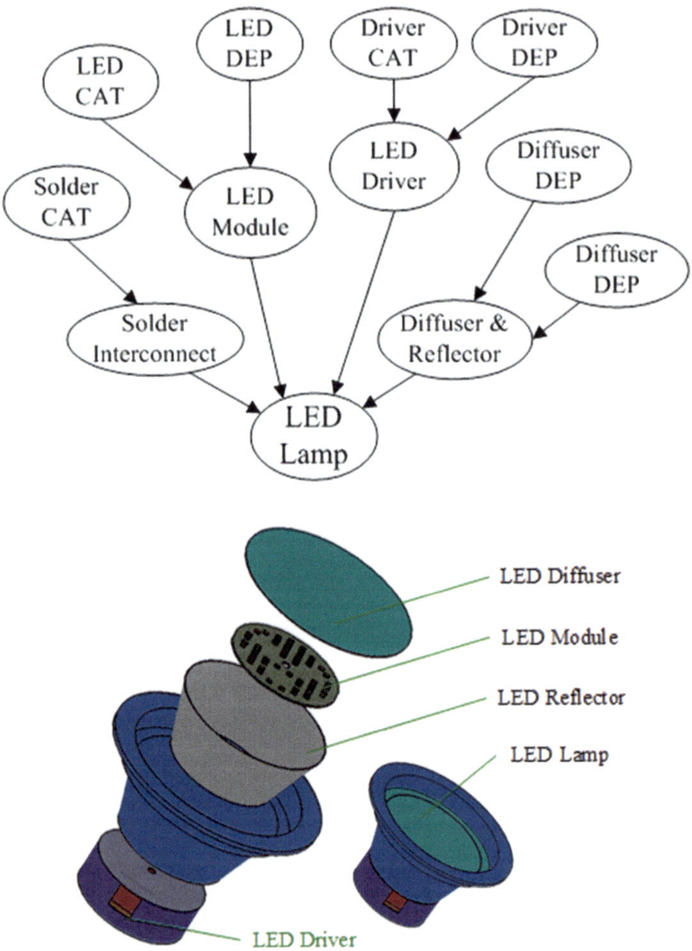

Fig. 6.21 DAG for product level LED light sources (top), 3D model exploded and assembly view (bottom)

DifRef_DEP, and DifRef_CAT, are referred as root nodes. On the other hand, the variables with no children are the leaf nodes (LED_Lamp), while the remaining variables are the intermediate nodes (LED_Module, LED_Diffuser, and LED_DifRef). The root nodes have unconditional probabilities, represented here as reliability state functions of the node X_i at time t $R_{Xi}(t)$, i = 1,...,p; the intermediate nodes as $R_{Mj}(t)$, j = 1,...,k; and the leaf node as $R_L(t)$. The BN model DAG analysis is based on the construction of a test sample, shown as a 3D model, with an exploded and assembled view Fig. 6.21 (right).

The reliability status of each root node or component is assessed based on the corresponding prediction model at a future time t_n, and the reliability state prediction matrix can be represented as follows:

$$R_{pn} = \begin{bmatrix} R_{11} & R_{12} & \cdots & R_{1n} \\ R_{21} & R_{22} & \cdots & R_{2n} \\ \vdots & \vdots & \ddots & \vdots \\ R_{p1} & R_{p2} & \cdots & R_{pn} \end{bmatrix} \tag{6.8}$$

The reliability state of the intermediate nodes can also be predicted based on the prediction models of the root nodes $U = \{R_1, R_2, \ldots, R_p\}$ and the assumption of conditional independence:

$$P\left(R_{Mj}(t)\right) = \sum_{U} P\left(R_{Mj}(t), R_{Xi}(t)\right) \tag{6.9}$$

Similarly, the reliability state of the leaf node can be predicted based on the probability of the intermediate and root nodes as follows, and the junction tree algorithm synchronizes the DAG of the BN model for product level lifetime prediction:

$$\begin{aligned} P\left(R_L(t)\right) &= \sum P\left(R_{X1}(t),\ldots,R_{Xp}(t),R_{M1}(t),\ldots,R_{Mk}(t),R_L(t)\right) \\ &= \sum_{Pa(L)} P(R_L(t) \mid Pa\left(R_L(t)\right). \sum_{Pa(Mj)} P(R_{M1}(t) \mid Pa\left(R_{M1}(t)\right)\ldots. \\ &\sum_{Pa(Mk)} P(R_{Mk}(t) \mid Pa\left(R_{Mk}(t)\right)\ldots..P\left(R_{X1}(t)\right).P\left(R_{Xp}(t)\right) \end{aligned} \tag{6.10}$$

Here $Pa(L)$, $Pa(M_j)$, *and* $Pa(M_k)$ are the parent nodes for leaf node L and intermediate nodes M_j and M_k respectively.

7 Challenges and Opportunities of Diagnostics and Prognostics Approaches

Recalling that PHM is a multifaceted engineering discipline that facilitates the safety, reliability, and maintenance aspect of components and systems, it helps to avoid unexpected product problems that can lead to products' performance deficiencies. Even though this approach has been widely accepted for product and system reliability assessment, lifetime prediction, and maintenance decision-making, it is still facing some challenges, especially for electronic systems, including LED lighting systems. The data-driven methods are based on the extraction of historical data collected from sensors, to exploit and learn the degradation behavior of the system through relevant feature identification using machine learning, AI, and statistical tools. On the other hand, model-based approaches implement a set of mathematical and analytical equations obtained from classical physics laws to represent the degradation behavior and predict the future behavior of physical components and systems.

The different approaches for the PHM in general need further improvement to be able to reduce the computational time, effort, and availability of historical data to accommodate the increasing demand in the reliability assessment and remaining useful life prediction in the LED light industry. There are always trade-offs in terms of accuracy, applicability, cost, and complexity while implementing DD approaches. While some approaches can handle complexity, it may be deficient in regard to computational time and accuracy and vice versa [211, 212]. Some algorithms, such as Hidden Markov Model and Gaussian process regression, consume longer computational time, while others such as artificial neural network, particle filtering, neuro-fuzzy systems, and Hidden Markov Model demand large amounts of historical data to perform prognostics. Accordingly, the advantages and disadvantages of the two main prognostics approaches are briefly summarized as follows:

Data-driven (statistical and machine learning methods)	Model-based (POF-based) approach
Assumptions or empirical estimations of physical parameters are not required	For a well-controlled system, predicting the future propagation of the degradation without prior knowledge about the mathematical model is possible
Less complex and more applicable than model-driven methods	Has higher accuracy if the systems'/products' physics of models remains consistent
Lower precision results compared with model-based approaches	Requires fewer data compared to data-driven approaches
Well-established theoretical basis and convenient to implement fast and accurate online pattern recognition	Usually complex and more stochastic to model system degradations
High-dimensional noisy data can be transformed in to lower dimensions convenient for prognostics	Might have difficulty to handle unit-unit variability in population and often provides overall estimate for entire sample
Relatively easy to calculate and predict future states	Might be difficult to get mathematical models for a particular kind of component or material
The more available information used, the better the accuracy	Computationally expensive

Data-driven (statistical and machine learning methods)	Model-based (POF-based) approach
Requires large amount of data to be more accurate in general	Requires simplifying assumptions
May lead to inaccurate time of change of predictions as it relies mainly on historical degradation	
Poor performance with high-dimensional data and longer learning time	

8 Digital Twin as Emerging LED Lifetime Analysis

In the past few years, dramatic advancement in information technology such as Internet of things, artificial intelligence, and big data has evolved which has led to an increasing interaction trend between virtual spaces and physical entities. This has led to the introduction of digital twins – a pragmatic method of cyber-physical fusion [213]. A digital twin is a dynamic and comprehensive virtual prototype of a physical product/system. The concept of digital twin was initially conceived and introduced by Vickers (NASA) and Grieves (University of Michigan) in 2003 [214].

In the past few years, many companies started using digital twin to increase their system operation efficiency, testing new products before deployment, and identifying problems [213]. According to prediction by Gartner, half of the large industrial corporations will be leveraging digital twin technology by 2021 to facilitate the assessment of system performance while gaining an improvement of 10% in system effectiveness [215]. The implementation and adoption of digital twin depend on the type of industry and products as there are no common standards, methods, or norms [216]. The National Aeronautics and Space Administration (NASA) built two identical spacecrafts for Apollo 13 mission with the idea of early "digital twin" where one was launched to space, while the other was kept on Earth to simulate and monitor the launched spacecraft. Later, with few technical improvements, NASA and the US Air Force introduced digital twins to the aerospace industry. Companies such as Chevron and General Electric also use digital twins to track operation of wind turbines [216]. Singapore is also creating a virtual copy of the entire city in partnership with Dassault Systemes, to assess, improve, and monitor utilities [217].

It can be recalled that PHM is very useful in the diagnostics and prognostic analysis of a product/system of a physical object. On the other hand, digital twins appear to have the capability to fill the gap in PHM by creating a link between the physical system and the virtual model. Recently, Tao et al. [218] introduced the application of digital twins in the PHM sector and demonstrated a case study on wind turbines. The implementation of PHM for products and systems in terms of fault detection, diagnostics, and prognostics is mainly based on the performance degradation and failure in the physical space which has a limited connection to the virtual model [218]. This gap could be filled with convergence of data from physical

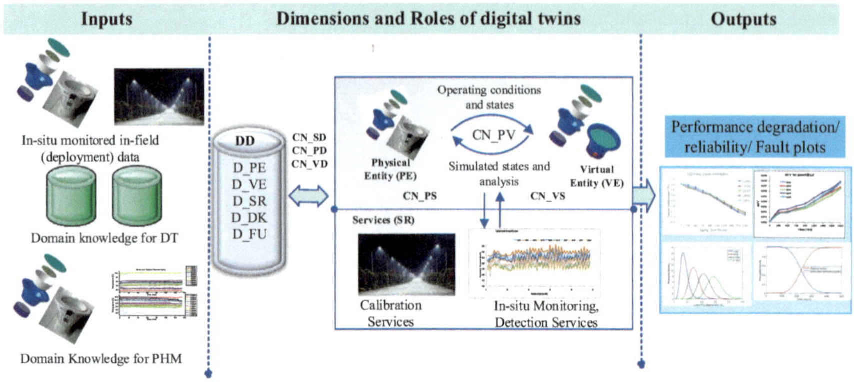

Fig. 6.22 Digital twin implementation framework with the five dimensions (PE, VE, DD, CN, SR) for LED products (adapted from Tao et al. [218]

and virtual space through digital twins to improve the PHM of systems/products seamlessly. Due to its comprehensive virtual representation of a physical object, digital twins can simulate the behavior and conditions of products and systems through mathematical models and data. Oftentimes, machine learning algorithms and artificial intelligence are employed to analyze system operation models and identify correlations among data generated in in situ and in-field (deployment) operation [216]. The machine learning algorithms used in digital twins include supervised learning (such as artificial neural network), unsupervised learning (such as clustering methods for virtual and real-world environment), and reinforcement learning approaches (during uncertain or partially observable operating environments) [219].

Leveraging the digital twin technology has the potential to enable real-time system performance assessment and improve PHMs of light-emitting diodes as well as other safety critical complex products and systems. Due to its potential to generate accurate data from physical and virtual space for lifetime assessment and real-time data and condition monitoring, digital twins represent the future technology for lifetime assessment of LED products/systems. Initially, Grieves proposed three dimensions of digital twins: physical entity (PE), virtual entity (VE), and the connection between physical and virtual systems (CN) [214]. Based on this, Tao et al. [218] extended the digital twins to a five-dimensional model of digital twins with the addition of services for the physical and virtual entity (SR) and digital twin data (DD). The extended five-dimensional (PE, VE, DD, CN, SR) digital twin concept along with a framework of implementation in PHM for light-emitting diode products and systems is highlighted in Fig. 6.22.

9 UV LED Degradation Modeling and Analysis

Ultraviolet (UV) light, with a wavelength between 250 and 350 nm, has numerous useful and attractive applications, such as virucide, air and water purification, photolithography, optical stimulus in drug activation, polymer curing, and laser surgery [220]. Due to its benefits of having a long life, compact size, and unimodal spectrum as well as being environmentally friendly, the III-nitride-based UV light-emitting diode (LED) is becoming a promising photoelectronic device to replace traditional UV light sources, such as mercury lamps [221, 222]. UV LEDs are important variations in the LED field. Especially after the global pandemic of coronavirus disease 2019 (COVID-19), ultraviolet rays from UV LEDs are being used for noncontact disinfection in an environment that is not high-temperature and cannot be wiped with alcohol [223]. Therefore, UV LEDs will continue to play a greater role in the future in phototherapy, sterilization, and related applications.

Radiation power is a critical physical index that reflects the optical radiation intensity of a photoelectronic device. The radiation power degradation of UV LEDs can cause by the UV LED chip degradation, the yellowing of packaging materials, and the cracking or delamination of interface [4]. The lifetime of UV LEDs is still challenged by the uncertainty of internal quantum efficiency, light extraction efficiency, and thermal management [224]. And due to the unclear failure physics and mechanisms of UV LEDs, the reliability and lifetime estimation methods are inconsistent [225]. This section focuses on modeling the dynamic nonlinear radiation power degradation process of UV LED packages in an accelerated degradation test. Firstly, we selected the exponential degradation model recommended in the TM-21 standard to describe the lumen radiation power degradation process of UV LEDs. Next, a LSTM neural network algorithm and two stochastic processing models, i.e., gamma process and Wiener process, are compared with the nonlinear least-squares (NLS) regression method recommended by the IESNA TM-21 standard. Finally, the prediction accuracy and robustness characteristics of the proposed methods are analyzed.

The TM-21 standard proposed by IESNA is a commonly used method in the industry to deal with the lumen maintenance of LED light sources and project long-term lifetimes [7, 16]. In this section, the radiation power of UV LEDs was adopted to evaluate the light output and predict long-term lifetime [108] (Fig. 6.23). And L_P is the lifetime when the UV LEDs mean radiation power maintenance decays to $P\%$ of the initial value ($P = 80$ is considered as an example).

The cumulative radiation power degradation of UV LEDs can be regarded as a time-dependent stochastic gamma process which can be used to model this degradation process. Assuming the L_{80} lifetimes of samples satisfied the two-parameter Weibull distribution, the mean of the distribution (MTTF-G1) can be predicted. Meanwhile, with the gamma process parameters of the whole group, the estimated L_{80} lifetimes (MTTF-G2) can be estimated by eq. (6.4).

In addition to the gamma process, the Wiener process is also a stochastic model method that is widely used to describe degradation processes. With the

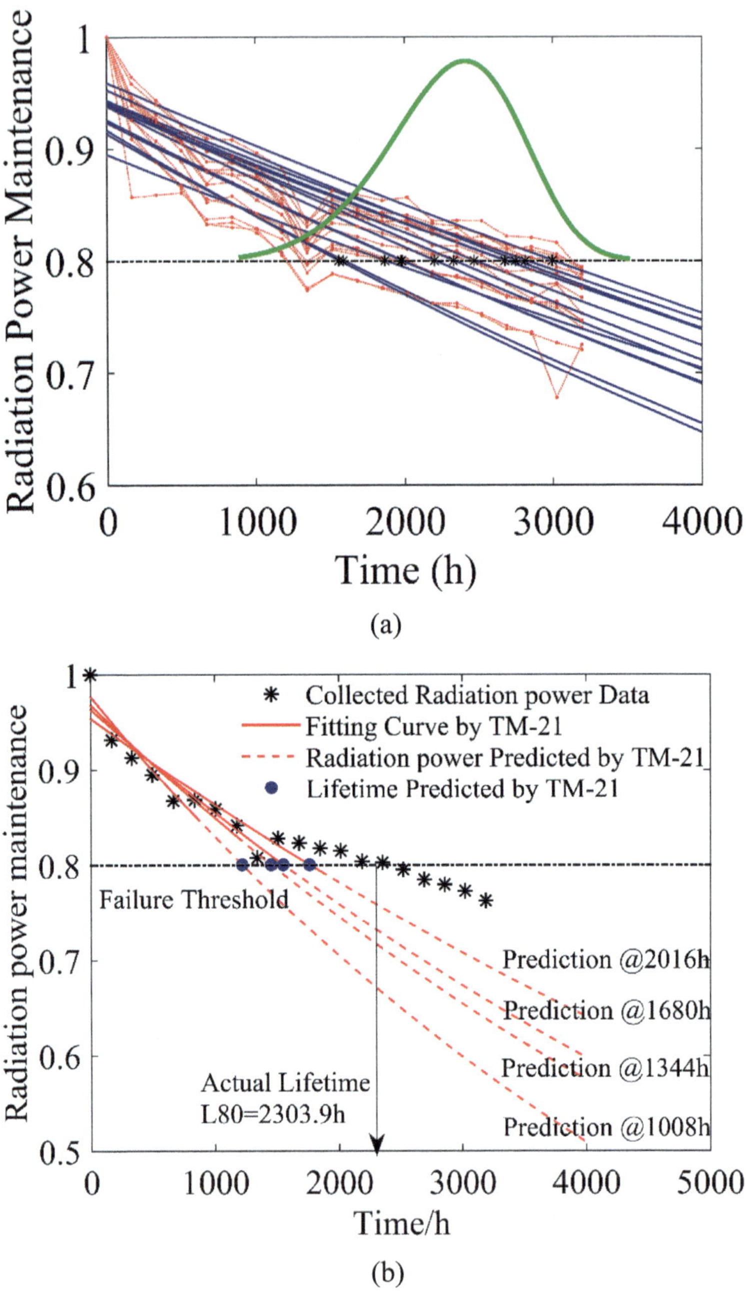

Fig. 6.23 (**a**) Lifetime calculation of 13 UV LEDs based on Weibull distribution; (**b**) NLS regression fitting based on IES-TM-21 standard

two-parameter Weibull distribution curve fitting of the predicted lifetimes of each sample, the estimated lifetimes, which were recorded as MTTF-W, can be estimated.

Also, ML approaches can be used to perform prediction. Neural network prediction does not need to determine the specific functional relationship between input and output. RNN can effectively and flexibly model the nonlinear relationship of dependent long time series data, that is, the current input is related to the previous input [226]. Hochreiter and Schmidhuber [227] proposed LSTM for solving the vanishing gradient problem in RNN.

h_t is a short-term state, which is equal to the output Y_t at time t. c_t stands for long-term memory, running horizontally above the cells in the hidden layer, with less interaction and better information maintenance. c_t determines what information is read, kept, and discarded in the long-term state of network learning. The output process can be expressed by the following eqs. [228]:

$$\begin{bmatrix} i_t \\ f_t \\ o_t \\ g_t \end{bmatrix} = \begin{bmatrix} \sigma \\ \sigma \\ \sigma \\ \tanh \end{bmatrix} \bullet W \bullet \begin{bmatrix} h_{t-1} \\ x_t \end{bmatrix} \tag{6.11}$$

$$c_t = f_t \otimes c_{t-1} + i_t \otimes g_t \tag{6.12}$$

$$h_t = o_t \otimes tanh\,(c_t) \tag{6.13}$$

where σ and *tanh* are the sigmod and *tanh* nonlinear activation functions in the neural network, respectively; W is the weight coefficient matrix.

Figure 6.24, shows a flowchart of the lifetime prediction of UV LEDs with the LSTM neural network. In this chapter, we have 9 input layers and 3 output layers, and the number of hidden layers is set to 18 after repeated trial calculations, as described in the Fig. 6.25, after multiple adjustments, the optimal weight matrix was obtained, and the lifetimes were predicted.

The comparison of prediction errors of each method is shown in Fig. 6.26. It can be seen that both the stochastic process method and the LSTM RNN method significantly improve the prediction accuracy compared to the TM-21 method. In general, the stochastic process method can achieve good prediction results as the prediction time increases. The LSTM neural network algorithm requires a small amount of test data to achieve better prediction accuracy compared with the other methods. It effectively reduces the collection and test time of UV LEDs and also has good robustness characteristics. It is found to be a very reliable and robust lifetime prediction algorithm for UV LEDs.

In a brief summary, by designing the experimental aging scheme and obtaining the actual lifetimes according to the two-parameter Weibull distribution, the NLS regression method, the stochastic process method, and the LSTM neural network algorithm were adopted to project the radiation power maintenance data to predict

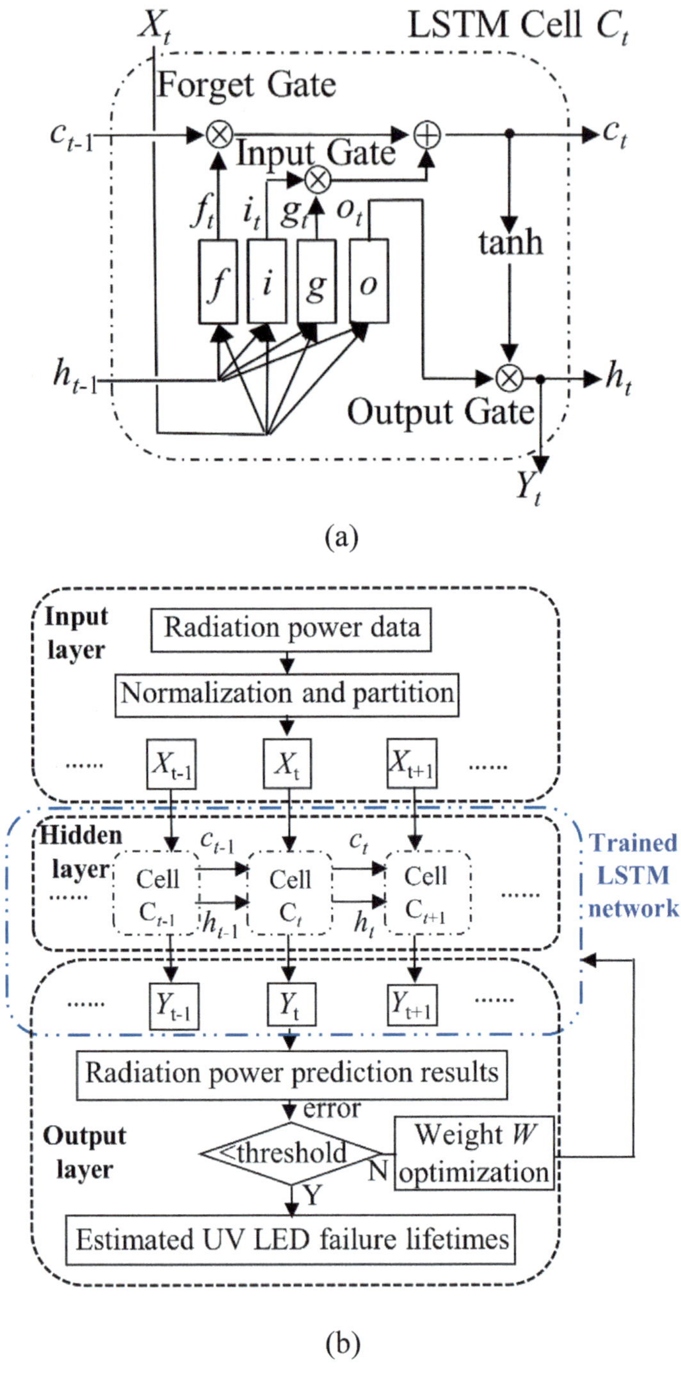

Fig. 6.24 (**a**) The basic structure of LSTM cell C_t in hidden layer; (**b**) the flowchart of the lifetime prediction of UV LEDs with the LSTM neural network

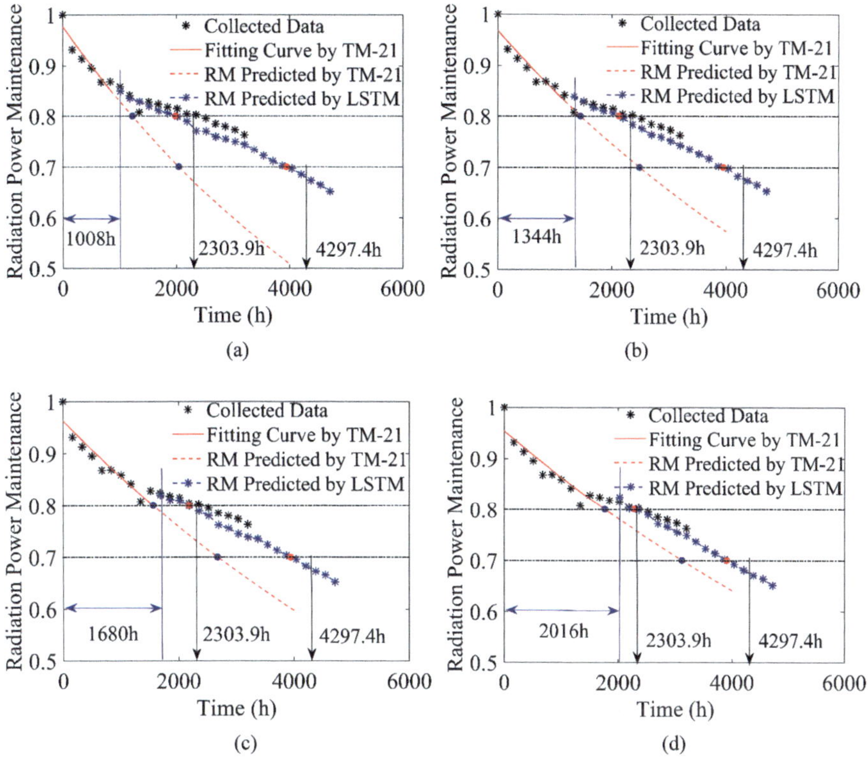

Fig. 6.25 The NLS regression method vs. the LSTM neural network method, lifetime prediction at different starting points: (**a**) 1008 h, (**b**) 1344 h, (**c**) 1680 h, (**d**) 2016 h

the lifetimes of UV LEDs. The results show that the prediction accuracy of the LSTM neural network is higher; the results also show that the stochastic process method and the LSTM neural network method have better robustness by varying the prediction starting points. Therefore, the LSTM neural network method can effectively project the UV LEDs' radiation power maintenance data with time series for lifetime prediction, which provides the feasibility for the rapid lifetime prediction of UV LEDs to accelerate the development of the UV LED industry and reduce the R&D costs.

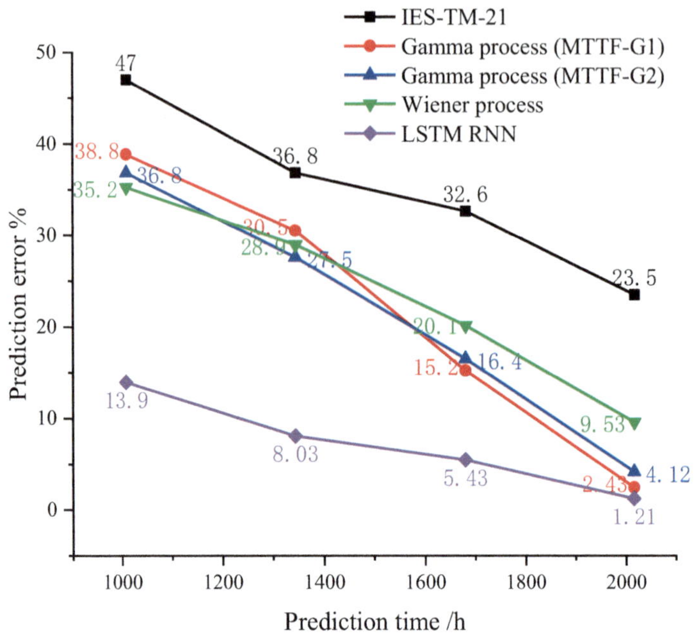

Fig. 6.26 Comparison of L_{80} lifetime prediction results: the TM-21 NLS regression method, the gamma process method, the Wiener process method, and the LSTM RNN method

10 Conclusions

In this study, the prognostic and diagnostic methods used in LED lighting have been reviewed, with due attention to machine learning-based data-driven approaches. Currently, there is an increasing number of studies on the reliability assessment and lifetime prediction of high-power white LEDs. However, the majority of conventional methods and approaches investigated have limitations in addressing the prognostics demand of the dynamic and unpredictable degradation behavior of LED systems. In addition to this, situations with sensor monitoring and data acquisition systems have shown an increasing trend in recent years. This has created opportunities as well as huge challenges to address the issues of diagnostics, RUL prediction, and extraction of useful information quickly from the abundantly generating operational and experimental big data. In the reliability study and lifetime prediction of LEDs, there are many machine learning algorithms that can help to provide lifetime prediction with improved accuracy. Some of the ML algorithms that have been employed in the study of mechanical components and systems can also be leveraged for LED lighting sources in future potential applications, including long short-term memory (LSTM) networks (a variety of recurrent neural networks), Hidden Markov Model (HMM), self-organizing maps (SOM), least-squares support vector machine, and fuzzy logic. An illustrative example is demonstrated on UV LED radiation degradation data based on NLS regression, Wiener, gamma, and LSTM neural network methods. The emerging trend in the application of digital twins for PHM with the focus on LEDs has also been briefly investigated.

Appendix

Table 6.3 A Brief Summary of machine learning algorithms for prognostics of LED products

Machine learning algorithm/method	Input data analysis and parameter estimations	Main study analysis results and findings
Artificial neural network [140, 134, 133] LSTM, recurrent neural network (RNN)[a] [141, 142]	Forward current (I_F, electrical) and temperature [140]	Probability of health status of LEDs (healthy 0.99 and not healthy 0.01)
	One hidden layer and two neuron nodes in the hidden layer [140]	Predict the lifetime of power LEDs with <5% error [133]
	MATLAB neural network toolbox, [134, 133]	
	Luminous flux, chromaticity coordinates u' and v', electric current and temperature [134]	Model can be used when the mean square error of datasets between estimated and expected life output narrow to the target
		R value of 0.985 and 0.974 for two dataset using AdaBoost BPNN.
K-nearest neighbors (KNN) kernel density-based algorithm [147, 146][a]	24 features from die and phosphor SPD clustered [147]	Anomaly detection conducted (two clusters for phosphor SPD and three clusters for die SPD) [147]
Relevance vector machine [151, 128][a]	LED light output (lumen maintenance and color shift) [151]	RUL lifetime prediction with error less than 5%, claimed to be better than PF
		Reduces qualification testing time (from 6000 h to 210 h) [151]
	Rotating component in aerospace setting (NASA)	Estimate the remaining useful life with acceptable accuracy [128] [a]
	Component feature damages (not specified) [128][a]	Not employed to anomaly detection
Support vector machine[a] [81][a], [149][a]	Dataset with 22 parameters for mission critical system from Lockheed Martin [81][a]	Identify system anomalies (with "healthy" and "unhealthy" class)
		Helps to manage false alarms
	Power gain degradation data of microwave [149]	Point and interval estimates of RUL obtained [149]
		Much more robust and stable as verified in comparison with RBF NN
Principal component analysis (PCA) [147, 133]	12 features from SPD (die and phosphor) considered for dimensional reduction [147]	PCA used to consider three features of SPD after reduction for further analysis (KNN) [147]
	Eight parameters considered for selection [133]	Four features with >85% contribution are reduced from 8 to use as an input BPNN [133]

(continued)

Table 6.3 (continued)

Machine learning algorithm/method	Input data analysis and parameter estimations	Main study analysis results and findings
Self-organizing map (SOM)[a] [161][a], [146][a]	Temp of gearbox, oil, nacelle, and rotor speed of wind of wind turbine [161] Failure data for an industrial reciprocating compressor [146]	System-level anomaly detection for WT [161], RUL estimation is obtained with good accuracy [146]
Hidden Markov Model (HMM)[a] [172][a], [173][a]	Experimental data for helicopter gearbox vibration from 68 operating conditions [172] Experimental data from IGBT from three operating conditions	Enabled defect level, defect type, and torque level classification for CBM [172] Anomalous behavior detection for IGBT based on Bayesian HMM classification [173]
Bayesian networks (BN)[a] [209]	Degradation data from unmanned aerial vehicles [209]	

[a] Shows that the machine learning algorithm has not been adopted yet for LEDs products

Acknowledgments The work described in this chapter was partially supported by a grant from the Research Committee of The Hong Kong Polytechnic University (under student account code RK21) and partially supported by the National Natural Science Foundation of China (Grant No. 51805147).

References

1. J.F. Van Derlofske, M.W. McColgan, White LED sources for vehicle forward lighting, in *Solid State Lighting II*, vol. 4776, (International Society for Optics and Photonics, 2002), pp. 195–206
2. F. Bergner, E. Dean-Leon, G. Cheng, Event-based signaling for large-scale artificial robotic skin-realization and performance evaluation, in *2016 IEEE/RSJ International Conference on Intelligent Robots and Systems (IROS)*, (IEEE, 2016), pp. 4918–4924
3. M.-H. Chang, D. Das, P. Varde, M. Pecht, Light emitting diodes reliability review. Microelectron. Reliab. **52**(5), 762–782 (2012)
4. W.D. van Driel, X.J. Fan, *Solid State Lighting Reliability Components to Systems* (Springer, New York, 2013)
5. E.F. Schubert, J.K. Kim, Solid-state light sources getting smart. Science **308**(5726), 1274–1278 (2005)
6. W. Van Driel, C. Yuan, S. Koh, G. Zhang, LED system reliability, in *Thermal, Mechanical and Multi-Physics Simulation and Experiments in Microelectronics and Microsystems (EuroSimE), 2011 12th International Conference on*, (IEEE, 2011), pp. 1/5–5/5
7. J. Fan, K.C. Yung, M. Pecht, Lifetime estimation of high-power white LED using degradation-data-driven method. IEEE Transactions on Device and Materials Reliability **12**(2), 470–477 (2012) Art no. 6166442
8. N. Narendran, N. Maliyagoda, A. Bierman, R.M. Pysar, M. Overington, Characterizing white LEDs for general illumination applications, in *Light-Emitting Diodes: Research,*

Manufacturing, and Applications IV, vol. 3938, (International Society for Optics and Photonics, 2000), pp. 240–249
9. B. Yao, C. Lai, Y. Huang, G. Lu, A prognostics and health management roadmap for LED lighting system, in *Reliability, Maintainability and Safety (ICRMS), 2014 International Conference on*, (IEEE, 2014), pp. 102–105
10. J. Fan, K.C. Yung, M. Pecht, Physics-of-failure-based prognostics and health management for high-power white light-emitting diode lighting. IEEE Transactions on Device and Materials Reliability **11**(3), 407–416 (2011) Art no. 5773483
11. J. Fan, K.C. Yung, M. Pecht, Failure modes, mechanisms, and effects analysis for LED backlight systems used in LCD TVs, in *2011 Prognostics and system health management conference, PHM-Shenzhen 2011*, (2011)
12. K.-C. Yung, B. Sun, X. Jiang, Prognostics-based qualification of high-power white LEDs using Lévy process approach. Mech. Syst. Signal Process. **82**, 206–216 (2017)
13. N. Narendran, J.D. Bullough, N. Maliyagoda, A. Bierman, What is useful life for white light LEDs? J. Illum. Eng. Soc. **30**(1), 57–67 (2001)
14. IES TM-28-14, *Predicting Long-Term Luminous Flux Maintenance of LED Lamps and Luminaires* (Illuminating Engineering Society, 2014)
15. IES LM-84-14, *Measuring Luminous Flux and Color Maintenance of LED Lamps, Light Engines, and Luminaires* (Illuminating Engineering Society, New York, 2014)
16. IES TM-21-11, *Projecting Long Term Lumen Maintenance of LED Light Sources* (Illuminating Engineering Society, New York, 2011)
17. IES LM-80-08, *Approved Method for Measuring Lumen Maintenance of LED Light Sources* (Illuminating Engineering Society, New York, 2008)
18. K.P. Murphy, *Machine Learning: A Probabilistic Perspective. 2012* (Cité en, 2014), p. 117
19. M. Meneghini, L.-R. Trevisanello, G. Meneghesso, E. Zanoni, A review on the reliability of GaN-based LEDs. IEEE Trans. Device Mater. Reliab. **8**(2), 323–331 (2008)
20. M. Meneghini, M. Dal Lago, N. Trivellin, G. Meneghesso, E. Zanoni, Degradation mechanisms of high-power LEDs for lighting applications: An overview. IEEE Trans. Ind. Appl. **50**(1), 78–85 (2014)
21. M. Meneghini, A. Tazzoli, G. Mura, G. Meneghesso, E. Zanoni, A review on the physical mechanisms that limit the reliability of GaN-based LEDs. IEEE Trans. Electron. Dev. **57**(1), 108 118 (2010)
22. G. Meneghesso, M. Meneghini, and E. Zanoni, "Recent results on the degradation of white LEDs for lighting," J. Phys. D Appl. Phys., vol. 43, no. /, p. 354007, 2010.
23. B. Sun, X. Jiang, K.C. Yung, J. Fan, M.G. Pecht, A Review of Prognostic Techniques for High-Power White LEDs. IEEE Trans. Power Electron. Rev. **32**(8), 6338–6362 (2017) Art no. 7593385
24. R. Kang, W. Gong, Y. Chen, Model-driven degradation modeling approaches: Investigation and review. Chinese J. Aeronaut. (2020)
25. S.-T. Tseng, C.-Y. Peng, Stochastic diffusion modeling of degradation data. J. Data Sci. **5**(3), 315–333 (2007)
26. J. Fan, K.C. Yung, M. Pecht, Prognostics of lumen maintenance for high power white light emitting diodes using a nonlinear filter-based approach. Reliab. Eng. Syst. Saf. **123**, 63–72 (2014)
27. M.K. Michael, G. Pecht, Introduction to PHM, in *Prognostics and Health Management of Electronics*, (2018), pp. 1–37
28. M.G. Pecht, *Prognostics and Health Management of Electronics* (Prognostics and Health Management of Electronics, 2008), pp. 1–315
29. A. Prisacaru, P.J. Gromala, M.B. Jeronimo, B. Han, G.Q. Zhang, Prognostics and health monitoring of electronic system: A review, in *Thermal, Mechanical and Multi-Physics Simulation and Experiments in Microelectronics and Microsystems (EuroSimE), 2017 18th International Conference on*, (IEEE, 2017), pp. 1–11

30. B. Sun, S. Zeng, R. Kang, M.G. Pecht, Benefits and challenges of system prognostics. IEEE Trans. Reliab. **61**(2), 323–335 (2012)
31. N.M. Vichare, M.G. Pecht, Prognostics and health management of electronics. IEEE Trans. Comp. Packag. Technol. **29**(1), 222–229 (2006)
32. N. Narendran, Y. Gu, L. Jayasinghe, J.P. Freyssinier, Y. Zhu, Long-term performance of white LEDs and systems, in *Proceeding of First International Conference on white LEDs and solid state lighting*, (2007), pp. 174–179
33. N. Narendran, L. Deng, R.M. Pysar, Y. Gu, H. Yu, Performance characteristics of high-power light-emitting diodes, in *Third International Conference on Solid State Lighting*, vol. 5187, (International Society for Optics and Photonics, 2004), pp. 267–276
34. M. Pecht, R. Jaai, A prognostics and health management roadmap for information and electronics-rich systems. Microelectron. Reliab. **50**(3), 317–323 (2010)
35. D. An, N.H. Kim, J.-H. Choi, Practical options for selecting data-driven or physics-based prognostics algorithms with reviews. Reliab. Eng. Syst. Saf. **133**, 223–236 (2015)
36. S. Yin, S.X. Ding, X. Xie, H. Luo, A review on basic data-driven approaches for industrial process monitoring. IEEE Trans. Ind. Electron. **61**(11), 6418–6428 (2014)
37. S. Cheng, M. Pecht, A fusion prognostics method for remaining useful life prediction of electronic products, in *2009 IEEE International Conference on Automation Science and Engineering, CASE 2009*, (2009), pp. 102–107
38. M. Pecht, G. Jie, Physics-of-failure-based prognostics for electronic products. Trans. Inst. Measure. Control **31**(3–4), 309–322 (2009)
39. J. Gu, M. Pecht, Predicting the reliability of electronic products, in *Proceedings of the Electronic Packaging Technology Conference, EPTC*, (2007)
40. S. Salemi, *Physics-of-Failure Based Handbook of Microelectronic Systems* (RIAC, 2008)
41. G. Lu et al., A POF based breakdown method for LED lighting color shift reliability, in *Solid State Lighting (SSLCHINA), 2015 12th China International Forum on*, (IEEE, 2015), pp. 59–62
42. T. Yanagisawa, T. Kojima, Long-term accelerated current operation of white light-emitting diodes. J. Lumin. **114**(1), 39–42 (2005)
43. K. Shailesh, C.P. Kurian, S.G. Kini, S. Tanuja, M.V. Kamath, Application of accelerated life testing principles to project long term lumen maintenance of LED luminaires, in *2012 International Conference on Emerging Trends in Electrical Engineering and Energy Management (ICETEEEM)*, (IEEE, 2012), pp. 483–488
44. X. Li, L. Chen, M. Chen, An approach of LED lamp system lifetime prediction, in *Quality and Reliability (ICQR), 2011 IEEE International Conference on*, (IEEE, 2011), pp. 110–114
45. B.M. Song, B. Han, A. Bar-Cohen, R. Sharma, M. Arik, Hierarchical life prediction model for actively cooled LED-based luminaire. IEEE Trans. Comp. Packag. Technol. **33**(4), 728–737 (2010) Art. no. 5549899
46. S. Koh, C. Yuan, B. Sun, B. Li, X. Fan, G. Zhang, Product level accelerated lifetime test for indoor LED luminaires, in *2013 14th International Conference on Thermal, Mechanical and Multi-Physics Simulation and Experiments in Microelectronics and Microsystems (EuroSimE)*, (IEEE, 2013), pp. 1–6
47. A.M. Colaco, C.P. Kurian, S.G. Kini, S. Colaco, C. Johny, Thermal characterization of multicolor LED luminaire. Microelectron. Reliab. **78**, 379–388 (2017)
48. X. Luo, B. Wu, S. Liu, Effects of moist environments on LED module reliability. IEEE Trans. Device Mater. Reliab. **10**(2), 182–186 (2010)
49. P. Singh, C.M. Tan, L.-B. Chang, Early degradation of high power packaged LEDs under humid conditions and its recovery—Myth of reliability rejuvenation. Microelectron. Reliab. **61**, 145–153 (2016)
50. J. Huang et al., Degradation modeling of mid-power white-light LEDs by using Wiener process. Optics Express **23**(15), A966–A978 (2015)
51. M. Cai et al., Step-stress accelerated testing of high-power LED lamps based on subsystem isolation method. Microelectron. Reliab. **55**(9–10), 1784–1789 (2015)

52. H.B. Fan, X.P. Li, J.X. Shen, M. Chen, An effective prediction method for LED lumen maintenance, in *2012 13th International Conference on Electronic Packaging Technology & High Density Packaging*, (2012), pp. 1560–1563
53. S.-C. Yang et al., Failure and degradation mechanisms of high-power white light emitting diodes. Microelectron. Reliab. **50**(7), 959–964 (2010)
54. B. Sun, X. Fan, J. Fan, C. Qian, G. Zhang, A PoF and statistics combined reliability prediction for LED arrays in lamps, in *2017 18th International Conference on Thermal, Mechanical and Multi-Physics Simulation and Experiments in Microelectronics and Microsystems (EuroSimE)*, (IEEE, 2017), pp. 1–5
55. J. Hu, L. Yang, M.W. Shin, Electrical, optical and thermal degradation of high power GaN/InGaN light-emitting diodes. J. Phys. D Appl. Phys. (3), 41, 035107 (2008)
56. B. Sun et al., A novel lifetime prediction for integrated LED lamps by electronic-thermal simulation. Reliab. Eng. Syst. Saf. **163**, 14–21 (2017)
57. R. Hechfellner, M. Kruger, T. Heemstra, G. Caswell, N. Blattau, V. More, A Solder Joint Reliability Model for the Philips Lumileds LUXEON Rebel LED Carrier Using Physics of Failure Methodology, in *International Symposium on Microelectronics*, vol. 2013, no. 1, (International Microelectronics Assembly and Packaging Society, 2013), pp. 000887–000892
58. H. Tang et al., Multi-physics modeling of LED-based luminaires under temperature and humidity environment, in *2012 13th International Conference on Electronic Packaging Technology & High Density Packaging*, (IEEE, 2012), pp. 803–807
59. J. Hu, L. Yang, M.W. Shin, Mechanism and thermal effect of delamination in light-emitting diode packages. Microelectron. J. **38**(2), 157–163 (2007)
60. M. Pecht, A. Dasgupta, Physics-of-failure: An approach to reliable product development. J. IES **38**(5), 30–34 (1995)
61. J. Gu, D. Lau, M. Pecht, Health assessment and prognostics of electronic products, in *Proceedings of 2009 8th International Conference on Reliability, Maintainability and Safety*, (ICRMS, 2009), pp. 912–919
62. T. Yanagisawa, Estimation of the degradation of InGaN/AlGaN blue light-emitting diodes. Microelectron. Reliab. **37**(8), 1239–1241 (1997)
63. M. Meneghini et al., Defect-related degradation of deep-UV-LEDs. Microelectron. Reliab. **50**(9–11), 1538–1542 (2010)
64. H. L. Ke et al., Lumen degradation analysis of LED lamps based on the subsystem isolation method. Appl. Opt. **57**(4), 849–854 (2018)
65. L. Trevisanello et al., Accelerated life test of high brightness light emitting diodes. IEEE Trans. Device Mater. Reliab. **8**(2), 304–311 (2008)
66. G. Lu et al., Degradation of microcellular PET reflective materials used in LED-based products. Opt. Mater. **49**, 79–84 (2015)
67. M. Pecht, S. Kumar, Data analysis approach for system reliability, diagnostics and prognostics, in *Pan Pacific Microelectronics Symposium*, (Kauai, Hawaii, USA, 2008), pp. 22–24
68. C.C. Tsai, S.T. Tseng, N. Balakrishnan, Mis-specification analyses of gamma and wiener degradation processes. J. Stat. Plan. Infer. **141**(12), 3725–3735 (2011)
69. W.B. Nelson, *Accelerated Testing: Statistical Models, Test Plans, and Data Analysis* (Wiley, 2009)
70. V. Crk, Reliability assessment from degradation data, in *Reliability and Maintainability Symposium, 2000. Proceedings. Annual*, (IEEE, 2000), pp. 155–161
71. M.B. Carey, R.H. Koenig, Reliability assessment based on accelerated degradation: A case study. IEEE Trans. Reliab. **40**(5), 499–506 (1991)
72. C.J. Lu, W.O. Meeker, Using degradation measures to estimate a time-to-failure distribution. Technometrics **35**(2), 161–174 (1993)
73. M.A. Freitas, E.A. Colosimo, T.R.d. Santos, M.C. Pires, Reliability assessment using degradation models: Bayesian and classical approaches. Pesquisa Operacional **30**(1), 194–219 (2010)

74. R. Pan, T. Crispin, A hierarchical modeling approach to accelerated degradation testing data analysis: A case study. Qual. Reliab. Eng. Int. **27**(2), 229–237 (2011)
75. V.R.B. De Oliveira, E.A. Colosimo, Comparison of methods to estimate the time-to-failure distribution in degradation tests. Qual. Reliab. Eng. Int. **20**(4), 363–373 (2004)
76. J.P. Kharoufeh, S.M. Cox, Stochastic models for degradation-based reliability. IIE Trans. **37**(6), 533–542 (2005)
77. J.M. van Noortwijk, A survey of the application of gamma processes in maintenance. Reliab. Eng. Syst. Saf. **94**(1), 2–21 (2009)
78. X.S. Si, W. Wang, C.H. Hu, D.H. Zhou, M.G. Pecht, Remaining useful life estimation based on a nonlinear diffusion degradation process. IEEE Trans. Reliab. **61**(1), 50–67 (2012)
79. W. Wang, M. Carr, W. Xu, K. Kobbacy, A model for residual life prediction based on Brownian motion with an adaptive drift. Microelectr. Reliab. **51**(2), 285–293 (2011)
80. N.Z. Gebraeel, M.A. Lawley, R. Li, J.K. Ryan, Residual-life distributions from component degradation signals: A Bayesian approach. IIE Trans. **37**(6), 543–557 (2005)
81. V.A. Sotiris, W.T. Peter, M.G. Pecht, Anomaly detection through a bayesian support vector machine. IEEE Trans. Reliab. **59**(2), 277–286 (2010)
82. G. Niu, S. Singh, S.W. Holland, M. Pecht, Health monitoring of electronic products based on Mahalanobis distance and Weibull decision metrics. Microelectron. Reliab. **51**(2), 279–284 (2011)
83. S. Kumar, V. Sotiris, M. Pecht, Health assessment of electronic products using Mahalanobis distance and projection pursuit analysis. Int. J. Comp. Inform. Syst. Sci. Eng. **2**(4), 242–250 (2008)
84. Y. Wang, Q. Miao, M. Pecht, Health monitoring of hard disk drive based on Mahalanobis distance, in *Prognostics and System Health Management Conference (PHM-Shenzhen)*, (IEEE, 2011), pp. 1–8
85. S. Kumar, T.W. Chow, M. Pecht, Approach to fault identification for electronic products using Mahalanobis distance. IEEE Trans. Instrum. Meas. **59**(8), 2055–2064 (2010)
86. T. Sutharssan, S. Stoyanov, C. Bailey, Y. Rosunally, Prognostics and health monitoring of high power LED. Micromachines **3**(1), 78–100 (2012)
87. T. Sutharssan, S. Stoyanov, C. Bailey, Y. Rosunally, Data analysis techniques for real-time prognostics and health management of semiconductor devices, in *EMPC-2011 - 18th European Microelectronics and Packaging Conference, Proceedings*, (2011)
88. T. Sutharssan, C. Bailey, S. Stoyanov, Y. Rosunally, Prognostics and reliability assessment of light emitting diode packaging, in *ICEPT-HDP 2011 Proceedings - 2011 International Conference on Electronic Packaging Technology and High Density Packaging*, (2011), pp. 937–943
89. J. Gu, N. Vichare, T. Tracy, M. Pocht, Prognostics implementation methods for electronics, in *2007 Annual Reliability and Maintainability Symposium*, (2007), pp. 101–106
90. J. Sikorska, M. Hodkiewicz, L. Ma, Prognostic modelling options for remaining useful life estimation by industry. Mech. Syst. Signal Process. **25**(5), 1803–1836 (2011)
91. W. Kahle, A. Lehmann, The Wiener process as a degradation model: Modeling and parameter estimation, in *Advances in Degradation Modeling*, (Springer, 2010), pp. 127–146
92. X.-S. Si, W. Wang, C.-H. Hu, D.-H. Zhou, Remaining useful life estimation–a review on the statistical data driven approaches. Eur. J. Oper. Res. **213**(1), 1–14 (2011)
93. R. Chhikara, *The Inverse Gaussian Distribution: Theory: Methodology, and Applications* (CRC Press, 1988)
94. D. R. Cox and H. D. Miller, "The Theory of Stochastic Processes," 1965.
95. X. Si, D. Zhou, A generalized result for degradation model-based reliability estimation. IEEE Trans. Autom. Sci. Eng. **11**(2), 632–637 (2014)
96. K.A. Doksum, A. Hbyland, Models for variable-stress accelerated life testing experiments based on wener processes and the inverse Gaussian distribution. Technometrics **34**(1), 74–82 (1992)

97. G. Whitmore, Estimating degradation by a Wiener diffusion process subject to measurement error. Lifetime Data Anal. **1**(3), 307–319 (1995)
98. G. Whitmore, F. Schenkelberg, Modelling accelerated degradation data using Wiener diffusion with a time scale transformation. Lifetime Data Anal. **3**(1), 27–45 (1997)
99. C.-Y. Peng, S.-T. Tseng, Mis-specification analysis of linear degradation models. IEEE Trans. Reliab. **58**(3), 444–455 (2009)
100. S.T. Tseng, J. Tang, I.H. Ku, Determination of burn-in parameters and residual life for highly reliable products. Naval Res. Logis. (NRL) **50**(1), 1–14 (2003)
101. T. Tsai, C. Lin, Y. Sung, P. Chou, C. Chen, Y. Lio, Inference from lumen degradation data under wiener diffusion process. IEEE Trans. Reliab. **61**(3), 710–718 (2012)
102. C.-M. Liao, S.-T. Tseng, Optimal design for step-stress accelerated degradation tests. IEEE Trans. Reliab. **55**(1), 59–66 (2006)
103. Z.-S. Ye, Y. Wang, K.-L. Tsui, M. Pecht, Degradation data analysis using wiener processes with measurement errors. IEEE Trans. Reliab. **62**(4), 772–780 (2013)
104. H. Liao, E.A. Elsayed, Reliability inference for field conditions from accelerated degradation testing. Naval Res. Logis. (NRL) **53**(6), 576–587 (2006)
105. X. Wang, Wiener processes with random effects for degradation data. J. Multivar. Anal. **101**(2), 340–351 (2010)
106. N. Gebraeel, A. Elwany, J. Pan, Residual life predictions in the absence of prior degradation knowledge. IEEE Trans. Reliab. **58**(1), 106–117 (2009)
107. M.S. Ibrahim, W.K.C. Yung, J. Fan, Lumen Maintenance Lifetime Prediction for Phosphor-converted White LEDs with a Wiener Process based Model, in *2018 20th International Conference on Electronic Materials and Packaging (EMAP)*, (2018), pp. 1–4
108. Z. Jing, M.S. Ibrahim, J. Fan, X. Fan, G. Zhang, Lifetime Prediction of Ultraviolet Light-emitting Diodes with Accelerated Wiener Degradation Process, in *2019 20th International Conference on Thermal, Mechanical and Multi-Physics Simulation and Experiments in Microelectronics and Microsystems (EuroSimE)*, (2019), pp. 1–8
109. C. Guo, W. Wang, B. Guo, X. Si, A maintenance optimization model for mission-oriented systems based on Wiener degradation. Reliab. Eng. Syst. Saf. **111**, 183–194 (2013)
110. W. Xiaolin, J. Ping, G. Bo, C. Zhijun, Real-time reliability evaluation with a general wiener process-based degradation model. Qual. Reliab. Eng. Int. **30**(2), 205–220 (2014)
111. Z. Zhang, X. Si, C. Hu, Y. Lei, Degradation data analysis and remaining useful life estimation: A review on Wiener-process-based methods. Eur. J. Operation. Res. (2018)
112. J. Lawless, M. Crowder, Covariates and random effects in a gamma process model with application to degradation and failure. Lifetime Data Anal. **10**(3), 213–227 (2004)
113. C. Park, W. Padgett, Accelerated degradation models for failure based on geometric Brownian motion and gamma processes. Lifetime Data Anal. **11**(4), 511–527 (2005)
114. M.S. Ibrahim, J. Fan, W.K.C. Yung, Z. Wu, B. Sun, Lumen degradation lifetime prediction for high-power white LEDs based on the gamma process model. IEEE Photon. J. **11**(6), 1–16 (2019)
115. Z. Wu et al., A gamma process-based prognostics method for CCT shift of high-power white LEDs. IEEE Trans. Electr. Dev. (2018)
116. M.E. Orchard, G.J. Vachtsevanos, A particle-filtering approach for on-line fault diagnosis and failure prognosis. Trans. Inst. Meas. Control. **31**(3–4), 221–246 (2009)
117. A. Doucet, N. De Freitas, N. Gordon, An introduction to sequential Monte Carlo methods, in *Sequential Monte Carlo Methods in Practice*, (Springer, 2001), pp. 3–14
118. A. Doucet, S. Godsill, C. Andrieu, On sequential Monte Carlo sampling methods for Bayesian filtering. Stat. Comput. **10**(3), 197–208 (2000)
119. J. Fan, K.C. Yung, M. Pecht, Predicting long-term lumen maintenance life of LED light sources using a particle filter-based prognostic approach. Expert Syst. Appl. **42**(5), 2411–2420 (2015)
120. N.-H. Kim, D. An, J.-H. Choi, Physics-Based Prognostics, in *Prognostics and Health Management of Engineering Systems: An Introduction*, (Springer, Cham, 2017), pp. 127–178

121. C. Chen, Y. Zuo, W. Ye, X. Li, Z. Deng, S.P. Ong, A critical review of machine learning of energy materials. Adv. Energy Mater. **10**(8), 1903242 (2020)
122. T.M. Mitchell, *Machine Learning (Engineering/Math)* (Mcgraw-Hill Science, 1997), p. 27
123. M. Kang, N.J. Jameson, Machine Learning: Fundamentals, in *Prognostics and Health Management of Electronics: Fundamentals, Machine Learning, and the Internet of Things*, (2018), pp. 85–109
124. J. Zhang et al., Life prediction for white OLED based on LSM under lognormal distribution. Solid State Electron. **75**, 102–106 (2012)
125. Y. Yoon, J. Hyung, U. Jeong, H. Lim, J. Jang, Life time comparison of LED package and the self-ballasted LED lamps by simple linear regression analysis. Microelectron. Reliab. **55**(9–10), 1779–1783 (2015)
126. T. Sutharssan, C. Bailey, S. Stoyanov, A comparison study of the prognostics approaches to Light Emitting Diodes under accelerated aging, in *2012 13th International Thermal, Mechanical and Multi-Physics Simulation and Experiments in Microelectronics and Microsystems*, (2012), pp. 1/8–8/8
127. A. Prisacaru, E.O. Guerrero, P.J. Gromala, B. Han, G.Q. Zhang, Degradation Prediction of Electronic Packages using Machine Learning, in *2019 20th International Conference on Thermal, Mechanical and Multi-Physics Simulation and Experiments in Microelectronics and Microsystems (EuroSimE)*, (IEEE, 2019), pp. 1–9
128. K. Goebel, B. Saha, A. Saxena, A comparison of three data-driven techniques for prognostics, in *62nd meeting of the society for machinery failure prevention technology (mfpt)*, (2008), pp. 119–131
129. M. Schwabacher, K. Goebel, A survey of artificial intelligence for prognostics, in *Aaai fall symposium*, (2007), pp. 107–114
130. M.S. Kan, A.C. Tan, J. Mathew, A review on prognostic techniques for non-stationary and non-linear rotating systems. Mech. Syst. Signal Process. **62**, 1–20 (2015)
131. B. Sadeghi, A BP-neural network predictor model for plastic injection molding process. J. Mater. Process. Technol. **103**(3), 411–416 (2000)
132. C.M. Bishop, *Pattern Recognition and Machine Learning (Information Science and Statistics)* (Springer, 2006)
133. Y. Qiyan, Research on thermal effect of power LED devices based on PCA and neural network, in *2017 29th Chinese Control And Decision Conference (CCDC)*, (IEEE, 2017), pp. 4324–4327
134. K. Lu, W. Zhang, B. Sun, Multidimensional data-driven life prediction method for white LEDs based on BP-NN and improved-Adaboost algorithm. IEEE Access **5**, 21660–21668 (2017)
135. Y. Zhang, X. Ding, Y. Liu, P. Griffin, An artificial neural network approach to transformer fault diagnosis. IEEE Power Eng. Rev. **16**(10), 55–55 (1996)
136. C.S. Byington, M. Watson, D. Edwards, Data-driven neural network methodology to remaining life predictions for aircraft actuator components, in *Aerospace Conference, 2004. Proceedings. 2004 IEEE*, vol. 6, (IEEE, 2004), pp. 3581–3589
137. A.K. Mahamad, S. Saon, T. Hiyama, Predicting remaining useful life of rotating machinery based artificial neural network. Comp. Math. Appl. **60**(4), 1078–1087 (2010)
138. B.-j. Ge, W. Guo, D.-h. Zhang, Application of BP neural network in turbo-generator harmonic analysis under negative-sequence loss conditions, in *Measurement, Information and Control (ICMIC), 2013 International Conference on*, vol. 2, (IEEE, 2013), pp. 959–963
139. A. Prisacaru, A. Palczynska, A. Theissler, P. Gromala, B. Han, G.Q. Zhang, In situ failure detection of electronic control units using Piezoresistive stress sensor. IEEE Trans. Compon. Packag. Manuf. Technol. **8**(5), 750–763 (2018)
140. T. Sutharssan, *Prognostics and Health Management of Light Emitting Diodes* (University of Greenwich, 2012)
141. L. Guo, N. Li, F. Jia, Y. Lei, J. Lin, A recurrent neural network based health indicator for remaining useful life prediction of bearings. Neurocomputing **240**, 98–109 (2017)

142. Q. Wu, K. Ding, B. Huang, Approach for fault prognosis using recurrent neural network. J. Intell. Manuf., 1–13 (2018)
143. F. Di Maio, E. Zio, Failure prognostics by a data-driven similarity-based approach. Int. J. Reliab. Qual. Saf. Eng. **20**(01), 1350001 (2013)
144. M. You, G. Meng, A generalized similarity measure for similarity-based residual life prediction. Proceedings of the Institution of Mechanical Engineers, Part E: Journal of Process Mechanical Engineering **225**(3), 151–160 (2011)
145. L.L. Li, D.J. Ma, Z.G. Li, Residual useful life estimation by a data-driven similarity-based approach. Qual. Reliab. Eng. Int. **33**(2), 231–239 (2017)
146. P. Loukopoulos et al., Abrupt fault remaining useful life estimation using measurements from a reciprocating compressor valve failure. Mech. Syst. Signal Process. **121**, 359–372 (2019)
147. M.H. Chang, C. Chen, D. Das, M. Pecht, Anomaly detection of light-emitting diodes using the similarity-based metric test. IEEE Trans. Indus. Inform. **10**(3), 1852–1863 (2014) Art. no. 6846338
148. M.E. Tipping, The relevance vector machine, in *Advances in Neural Information Processing Systems*, (2000), pp. 652–658
149. F. Sun, X. Li, H. Liao, X. Zhang, A Bayesian least-squares support vector machine method for predicting the remaining useful life of a microwave component. Adv. Mech. Eng. **9**(1) (2017)
150. M.E. Tipping, Sparse Bayesian learning and the relevance vector machine. J. Machine Learn. Res. **1**(Jun), 211–244 (2001)
151. M.H. Chang, M. Kang, M. Pecht, Prognostics-Based LED Qualification Using Similarity-Based Statistical Measure with RVM Regression Model. IEEE Trans. Indus. Electron. **64**(7), 5667–5677 (2017) Art. no. 7869350
152. M. Kang, Machine Learning: Anomaly Detection, in *Prognostics and Health Management of Electronics*, (2018), pp. 131–162
153. J. Lever, M. Krzywinski, N. Altman, Principal component analysis. Nat. Methods **14**, 641 (2017)
154. J.E. Jackson, *A user's guide to principal components* (Wiley, 2005)
155. W. Wang, W. Zhang, A model to predict the residual life of aircraft engines based upon oil analysis data. Naval Res. Logis. (NRL) **52**(3), 276–284 (2005)
156. M. Ahmed, M. Baqqar, F. Gu, A.D. Ball, Fault detection and diagnosis using Principal Component Analysis of vibration data from a reciprocating compressor, in *Proceedings of 2012 UKACC International Conference on Control*, (2012), pp. 461–466
157. M. Ahmed, F. Gu, A. Ball, Fault detection of reciprocating compressors using a model from principles component analysis of vibrations. J. Phys. Conf. Ser. **364**(1), 012133 (2012) IOP Publishing
158. G. Wang, C. Liu, Y. Cui, Clustering diagnosis of rolling element bearing fault based on integrated autoregressive/autoregressive conditional heteroscedasticity model. J. Sound Vib. **331**(19), 4379–4387 (2012)
159. Z. Zhang, A. Kusiak, Monitoring wind turbine vibration based on SCADA data. J. Sol. Energy Eng. **134**(2), 021004 (2012)
160. T. Kohonen, The self-organizing map. Proc. IEEE **78**(9), 1464–1480 (1990)
161. M. Du, L.B. Tjernberg, S. Ma, Q. He, L. Cheng, J. Guo, A SOM based anomaly detection method for wind turbines health management through SCADA data. Int. J. Progn. Health Manag. **7**, 1–13 (2016)
162. J. Tian, M.H. Azarian, M. Pecht, Anomaly detection using self-organizing maps-based k-nearest neighbor algorithm, in *Proceedings of the European Conference of the Prognostics and Health Management Society*, (Citeseer, 2014)
163. O. Chapelle, B. Scholkopf, A. Zien, *Semi-supervised Learning*, IEEE Transactions on Neural Networks, no. 3 (2009), p. 524
164. X.J. Zhu, *Semi-Supervised Learning Literature Survey* (University of Wisconsin-Madison Department of Computer Sciences, 2005)

165. C.B. Do, S. Batzoglou, What is the expectation maximization algorithm? Nat. Biotechnol. **26**(8), 897–899 (2008)
166. X.-S. Si, W. Wang, M.-Y. Chen, C.-H. Hu, D.-H. Zhou, A degradation path-dependent approach for remaining useful life estimation with an exact and closed-form solution. Eur. J. Operation. Res. **226**(1), 53–66 (2013)
167. P.L.T. Duong, H. Park, N. Raghavan, Application of expectation maximization and Kalman smoothing for prognosis of lumen maintenance life for light emitting diodes. Microelectron. Reliab. **87**, 206–212 (2018)
168. L.E. Baum, T. Petrie, Statistical inference for probabilistic functions of finite state Markov chains. Ann. Math. Stat. **37**(6), 1554–1563 (1966)
169. P. Baruah, R.B. Chinnam, HMMs for diagnostics and prognostics in machining processes. Int. J. Prod. Res. **43**(6), 1275–1293 (2005)
170. M. Dong, D. He, Hidden semi-Markov model-based methodology for multi-sensor equipment health diagnosis and prognosis. Eur. J. Oper. Res. **178**(3), 858–878 (2007)
171. M. Dong, Y. Peng, Equipment PHM using non-stationary segmental hidden semi-Markov model. Robot. Comput. Integr. Manuf. **27**(3), 581–590 (2011)
172. C. Bunks, D. McCarthy, T. Al-Ani, Condition-based maintenance of machines using hidden Markov models. Mech. Syst. Signal Process. **14**(4), 597–612 (2000)
173. E. Dorj, C. Chen, M. Pecht, A Bayesian Hidden Markov Model-based approach for anomaly detection in electronic systems, in *2013 IEEE Aerospace Conference*, (IEEE, 2013), pp. 1–10
174. D.A. Tobon-Mejia, K. Medjaher, N. Zerhouni, G. Tripot, Hidden Markov models for failure diagnostic and prognostic, in *2011 Prognostics and System Health Management Conference*, (IEEE, 2011), pp. 1–8
175. L.R. Rabiner, A tutorial on hidden Markov models and selected applications in speech recognition. Proc. IEEE **77**(2), 257–286 (1989)
176. J.A. Bilmes, What HMMs can do. IEICE Trans. Inf. Syst. **89**(3), 869–891 (2006)
177. A.S.S. Vasan, B. Long, M. Pecht, Diagnostics and prognostics method for analog electronic circuits. IEEE Trans. Ind. Electron. **60**(11), 5277–5291 (2013)
178. N. Narendran, Y. Gu, J. Freyssinier, H. Yu, L. Deng, Solid-state lighting: Failure analysis of white LEDs. J. Cryst. Growth **268**(3), 449–456 (2004)
179. F. Wu, W. Zhao, S. Yang, C. Zhang, Failure modes and failure analysis of white LEDs, in *Electronic Measurement & Instruments, 2009. ICEMI'09. 9th International Conference on*, (IEEE, 2009), pp. 4-978–4-981
180. N. Trivellin, M. Meneghini, M. Buffolo, G. Meneghesso, E. Zanoni, Failures of LEDs in real-world applications: A review. IEEE Trans. Device Mater. Reliab. **18**(3), 391–396 (2018)
181. Y. Bin, Z. Zhenwei, L. Guoguang, S. Zhenyu, L. Canxiong, H. Yun, Prognostics and health monitoring of LED lighting system, in *2017 Prognostics and System Health Management Conference (PHM-Harbin)*, (2017), pp. 1–4
182. D. Yang, Physics-of-failure-based prognostics and health management for electronic products, in *Proceedings of the Electronic Packaging Technology Conference*, (EPTC, 2014), pp. 1215–1218
183. Y.-C. Hsu et al., Failure mechanisms associated with lens shape of high-power LED modules in aging test. IEEE Trans. Electron Dev. **55**(2), 689–694 (2008)
184. J. Fan, K.C. Yung, M. Pecht, Anomaly detection for chromaticity shift of high power white LED with mahalanobis distance approach, in *14th International Conference on Electronic Materials and Packaging*, (EMAP, 2012)
185. J. Fan, C. Qian, X. Fan, G. Zhang, M. Pecht, In-situ monitoring and anomaly detection for LED packages using a Mahalanobis distance approach, in *Proceedings of 2015 the 1st International Conference on Reliability Systems Engineering*, (ICRSE, 2015)
186. S.J. Bae, T. Yuan, S.J. Kim, Bayesian degradation modeling for reliability prediction of organic light-emitting diodes. J. Comput. Sci. **17**, 117–125 (2016)
187. A. Padmasali, S.G. Kini, LED life prediction based on lumen depreciation and colour shift. Light. Res. Technol. **49**(1), 84–99 (2017)

188. J. Fan, K.C. Yung, M. Pecht, Prognostics of chromaticity state for phosphor-converted white light emitting diodes using an unscented kalman filter approach. IEEE Trans. Dev. Mater. Reliab. **14**(1), 564–573 (2014) Art. no. 6609048
189. P. Lall, J. Wei, Prediction of L70 life and assessment of color shift for solid-state lighting using Kalman filter and extended Kalman filter-based models. IEEE Trans. Device Mater. Reliab. **15**(1), 54–68 (2015)
190. P. Lall, H. Zhang, L. Davis, Prognostication of LED remaining useful life and color stability in the presence of contamination, in *2015 IEEE Conference on Prognostics and Health Management: Enhancing Safety, Efficiency, Availability, and Effectiveness of Systems Through PHAf Technology and Application*, (PHM, 2015)
191. F. Ruknudeen, S. Asokan, Application of Particle Filter to On-Board Life Estimation of LED Lights. IEEE Photon. J. **9**(3) (2017) Art. no. 7912371
192. A.N. Padmasali, S.G. Kini, Prognostic algorithms for L70 life prediction of solid state lighting. Light. Res. Technol. **48**(5), 608–623 (2016)
193. P. Lall, H. Zhang, Assessment of lumen degradation and remaining life of light-emitting diodes using physics-based indicators and particle filter. J. Electron. Packag. Trans. ASME **137**(2) (2015) Art. no. 021002
194. H. Hao, C. Su, C. Li, LED lighting system reliability modeling and inference via random effects gamma process and copula function. Int. J. Photoenergy (2015)
195. M.H. Chang, P. Sandborn, M. Pecht, W.K.C. Yung, W. Wang, A return on investment analysis of applying health monitoring to LED lighting systems. Microelectr. Reliab. **55**(3–4), 527–537 (2015)
196. M.H. Chang, M. Pecht, W.K.C. Yung, Return on investment associated with PHM applied to an LED lighting system, in *PHM 2013–2013 IEEE International Conference on Prognostics and Health Management, Conference Proceedings*, (2013)
197. L. Yuchen, Z. Shunong, K. Rui, M. Pecht, Return on Investment of a LED lighting system, in *2011 Prognostics and System Health Management Conference*, (2011), pp. 1–5
198. Z. Pan, N. Balakrishnan, Reliability modeling of degradation of products with multiple performance characteristics based on gamma processes. Reliab. Eng. Syst. Saf. **96**(8), 949–957 (2011)
199. P. Lall, J. Wei, L. Davis, Prediction of lumen output and chromaticity shift in LEDs using Kalman Filter and Extended Kalman Filter based models, in *2013 IEEE Conference on Prognostics and Health Management (PHM)*, (2013), pp. 1–14
200. Z. Yuege, L. Xiang, Y. Xuerong, Z. Guofu, A remaining useful life prediction method based on condition monitoring for LED driver, in *Proceedings of the IEEE 2012 Prognostics and System Health Management Conference (PHM-2012 Beijing)*, (2012), pp. 1–5
201. US DOE, Lifetime and reliability. Fact Sheet, 4 (2013)
202. W. Van Driel, F. Evertz, J. Zaal, O.M. Nápoles, C. Yuan, An introduction to system reliability for solid-state lighting, in *Solid State Lighting Reliability*, (Springer, 2013), pp. 329–346
203. H. Langseth, L. Portinale, Bayesian networks in reliability. Reliab. Eng. Syst. Saf. **92**(1), 92–108 (2007)
204. J.P. González-Brenes, J.T. Behrens, R.J. Mislevy, R. Levy, K.E. DiCerbo, Bayesian networks. The Wiley Handbook of Cognition and Assessment, 328–353 (2016)
205. H. Boudali, J. Duga, A new Bayesian network approach to solve dynamic fault trees, in *Annual Reliability and Maintainability Symposium, 2005. Proceedings*, (IEEE, 2005), pp. 451–456
206. H. Langseth, B.H. Lindqvist, A maintenance model for components exposed to several failure mechanisms and imperfect repair, in *Mathematical and Statistical Methods in Reliability*, (World Scientific, 2003), pp. 415–430
207. O. Kipersztok and G. Provan, A Framework for Diagnostic Inference of Commercial Aircraft Systems. Presented at the First Bayesian Applications Modeling Workshop, 2003.

208. S. Chen, Z. Qi, D. Chen, L. Guo, W. Peng, Investigation of Bayesian network for reliability analysis and fault diagnosis of complex systems with real case applications. Adv. Mech. Eng. **9**(10), 1687814017728853 (2017)
209. J. Wang, X. Wang, L. Wang, Modeling of BN lifetime prediction of a system based on integrated multi-level information. Sensors **17**(9), 2123 (2017)
210. J. Bhandari, F. Khan, R. Abbassi, V. Garaniya, R. Ojeda, Pitting degradation modeling of ocean steel structures using Bayesian network. J. Offshore Mech. Arctic Eng. **139**(5) (2017)
211. F. Ahmadzadeh, J. Lundberg, Remaining useful life estimation: Review. Int. J. Syst. Assur. Eng. Manage. **5**(4), 461–474 (2014)
212. D.A. Tobon-Mejia, K. Medjaher, N. Zerhouni, G. Tripot, A data-driven failure prognostics method based on mixture of Gaussians hidden Markov models. IEEE Trans. Reliab. **61**(2), 491–503 (2012)
213. F. Tao, M. Zhang, A.Y.C. Nee, *Digital Twin Driven Smart Manufacturing* (Academic, 2019)
214. M. Grieves, Digital twin: Manufacturing excellence through virtual factory replication. White paper **1**, 1–7 (2014)
215. C. Pettey. Prepare for the Impact of Digital Twins. Available: https://www.gartner.com/smarterwithgartner/prepare-for-the-impact-of-digital-twins/ (2017, June 4)
216. B. Marr. 7 Amazing Examples of Digital Twin Technology in Practice. Available: https://bit.ly/2Y2rPvF (2019, June 5)
217. Dassault Systems. Meet Virtual Singapore, the city's 3D digital twin. Available: https://govinsider.asia/digital-gov/meet-virtual-singapore-citys-3d-digital-twin/ (2018, June 5)
218. F. Tao, M. Zhang, Y. Liu, A. Nee, Digital twin driven prognostics and health management for complex equipment. CIRP Ann. **67**(1), 169–172 (2018)
219. A.M. Madni, C.C. Madni, S.D. Lucero, Leveraging digital twin technology in model-based systems engineering. Systems **7**(1), 7 (2019)
220. M. Kneissl, A Brief Review of III-Nitride UV Emitter Technologies and Their Applications. 2016.
221. K. Song, M. Mohseni, F. Taghipour, Application of ultraviolet light-emitting diodes (UV-LEDs) for water disinfection: A review. Water Res. **94**, 341–349 (2016)
222. A. Kheyrandish, M. Mohseni, F. Taghipour, Development of a method for the characterization and operation of UV-LED for water treatment. Water Res. **122**, 570–579 (2017)
223. Diagnosis and treatment programme of coronavirus pneumonia (Trial version 7). J. Tianjin Univ. Tradition. Chinese Med. **39**(2), 121–127 (2020)
224. Y. Muramoto, M. Kimura, S. Nouda, Development and future of ultraviolet light-emitting diodes ~UV-LED will replace UV lamp~, in *Summer Topicals Meeting Series*, (2014)
225. M. Liu, S. Zhou, X. Liu, Y. Gao, X. Ding, Comparative experimental and simulation studies of high-power AlGaN-based 353 nm ultraviolet flip-chip and top-emitting LEDs. Jpn. J. Appl. Phys. **57**, 031001 (2018)
226. J. Liu, A. Saxena, K. Goebel, B. Saha, W. Wang, An Adaptive Recurrent Neural Network for Remaining Useful Life Prediction of Lithium-ion Batteries. Annual Conference of the Prognostics and Health Management Society, PHM **2010**, 01/01 (2010)
227. S. Hochreiter, J. Schmidhuber, Long Short-term Memory. Neural Comput. **9**, 1735–1780 (1997)
228. H. Yan, Y. Qin, S. Xiang, Y. Wang, H. Chen, Long-term gear life prediction based on ordered neurons LSTM neural networks. Measurement **165**, 108205 (2020)

Chapter 7
Reliability and Failures in Solid State Lighting Systems

W. D. van Driel, B. J. C. Jacobs, G. Onushkin, P. Watte, X. Zhao, and J. Lynn Davis

1 Introduction

The penetration of LED-based products has significantly increased in the past years [1–6]. Here, an LED-based product is an apparatus that distributes, filters, or transforms light transmitted from one or more LED light sources. It can be seen as a system that includes all the parts necessary to support, fix, and protect light sources and (where necessary) circuit auxiliaries, along with the means to connect them to the supply but not the light sources themselves. The global LED lighting market grew by 3.2% from 2018 to almost 60BEuro in 2019 [1]. It is expected that the market will grow at a compound annual growth rate (CAGR) of 2.8% largely based on the expected growth in healthcare and industrial applications [1]. Several reports in the past years [4–6] predict that, compared to conventional incandescent, halogen, fluorescent, and high-intensity-discharge white-light sources, the rate of LED market penetration will increase steadily, rising to 75–85% percent by 2030 (see Fig. 7.1). This prediction year-on-year is adapted with the latest prediction values rather 75% [6] than 85% [4], see Fig. 7.1.

Figure 7.2 depicts the total LED lighting market by the percentage revenue for these applications in 2019. Application-wise, lighting for health and wellbeing will drive the market for LED lighting in the coming period. Healthcare, industrial, and

W. D. van Driel (✉)
Signify, HTC7, Eindhoven, The Netherlands

Delft University of Technology, Delft, The Netherlands
e-mail: willem.van.driel@signify.com

B. J. C. Jacobs · G. Onushkin · P. Watte · X. Zhao
Signify, HTC7, Eindhoven, The Netherlands

J. L. Davis
RTI International, Research Triangle Park, Durham, NC, USA

W. D. van Driel, M. Yazdan Mehr (eds.), *Reliability of Organic Compounds in Microelectronics and Optoelectronics*,
https://doi.org/10.1007/978-3-030-81576-9_7

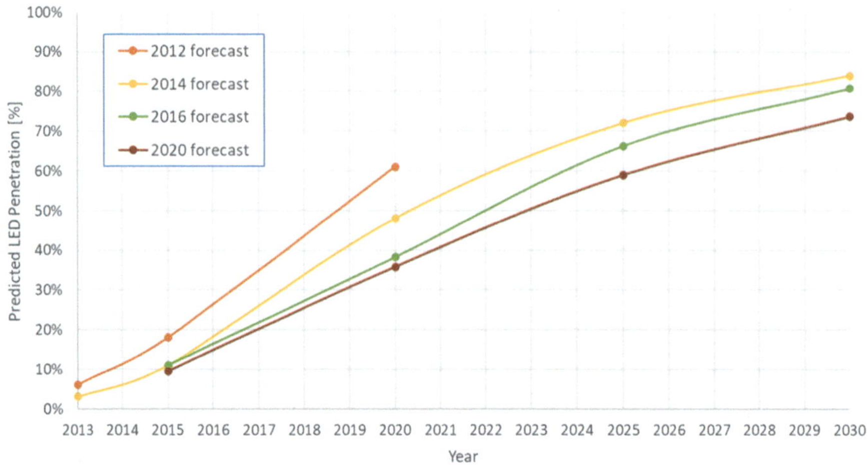

Fig. 7.1 Predicted LED penetration levels as a function of year (taken from [4–6]; with permission)

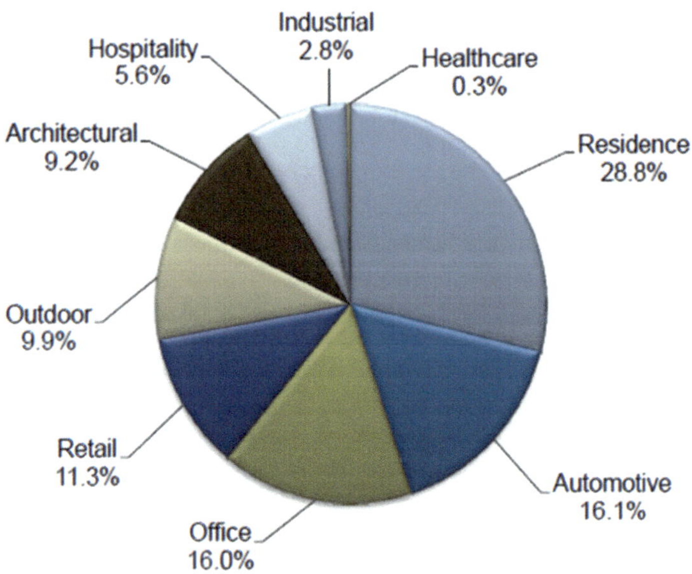

Fig. 7.2 Total LED Lighting Market: Percent Revenue by Applications, Global, 2019 (taken from [1]; with permission)

office segments will be the major adopters of such lighting in the future. Architectural lighting solutions or products are used in all applications, including outdoor lighting for city beautification and urban lighting displays. Smart cities have the capability to transform daily life and IoT plays a key role in the development of smart cities.

Major global economies and developing economies have invested heavily in smart city development. Accompanied with the LED penetration, the lighting industry also experiences an exponential increasing impact of digitization and connectivity of its lighting systems [1, 7]. The impact of this digitization is far beyond the impact on single products and extends to an ever-larger amount of connected systems. Continuously, more intelligent interfacing with the technical environment and with different kinds of users is being built-in by using more and different kinds of sensors, (wireless) communication, and different kinds of interacting or interfacing devices.

New business models, like Light as a Service (LaaS) and/or Pay-per-Lux (PPL), are widely adopted that repurpose LED lights and make them re-usable. LaaS and PPL provide a cost-effective lighting solution for LED applications by shifting from an "asset ownership" model to an "as a service" model. These systems save upfront costs associated with installing energy-saving lighting. By planning for longevity rather than a "fit and forget" approach, it provides the most efficient and cheapest lighting possible – which encourages the uptake of energy-saving lighting. At the end of the contract, products can be returned to the production process again, reusing the raw materials, optimizing recycling, and reducing waste. As such, it maximizes the asset value for service providers and reduces the initial investment and risk of ownership for customers [8, 9, 10]. Smart LED-based products have rapidly entered the building market and widespread commercial and residential LED adoption will be centered on smart LEDs. Ease of use, aesthetic value, and affordable cost are enablers for the quick adoption of smart LEDs. IoT-enabled LED lighting will continue to drive the growth of connected buildings and related services.

The use of information from above-mentioned connected sources can be described as a revolution named big data. With big data, data analytics from live connections of "intelligent" systems can be used to determine the system prognostics. Due to these changes in technology, the next generation of product data will be much richer in information [11]. From a reliability perspective, it all means that the service life of LED-based products will simply increase and 10–15 years will be normal. In that extensive time frame, one should be able to understand which possible failure modes can occur or can be triggered. With the continuous introduction of new processes and new materials this is not without challenges as these will introduce a new series of new and unknown failure modes in connected LED-based products. In this chapter, we will describe our current understanding of the reliability and known failure modes in these products.

2 Customer View: Catastrophic vs Degradation Failures

Failure modes in LEDs are well described in the literature. Pecht and Chang discuss thirteen different types of failure mechanisms of LEDs based on previously published papers and opinions of experts in the LED industry [12, 13]. These failure modes are dislocations, die cracking, dopant diffusion, electromigration, overstress, electro-static discharge (ESD), carbonization, delamination, yellowing, cracking, thermal quenching, and solder joint failure. Extensive work on the LED epitaxial degradation level was done by the group of Prof Zanoni, from the Electronics University of Padova [14, 15]. Their work concentrates on light output degradation due to nonradiative recombination at epitaxial defects and shifted electrical parameters due to increased reverse leakage currents. According to their findings, the lifetime and performance of LEDs are limited by crystal defect formations in the epitaxial layer structure. Crystal defects are mainly generated in contacts and in the active regions and result in a reduction in the lifetime of non-equilibrium electron-hole pairs and an increase in multi-phonon emissions under high drive currents. Multi-phonon emissions result in strong vibration of defect atoms and reduce the energy barrier for defect motions such as migration, creation, or clustering. Another great overview of LED failure modes was given by Caers and Zhao [16]. They distinguish catastrophic and degradation failure modes on all LED product levels ranging from LED package to LED products to LED systems. Unfortunately, a document like JEP122F, Failure mechanisms and models for semiconductor devices [17], does not exist in the Lighting industry.

It is correct to distinguish failure modes having a catastrophic or degradation behavior in LED-based systems. Here catastrophic refers to:

a catastrophic failure is a sudden and total failure from which recovery is impossible.

and degradation refers to:

A failure which is gradual or partial; it does not cease all function but compromises that function. It may lower output below a designated point, raise output above a designated point or result in erratic output. A degraded mode might allow only one mode of operation. If left unattended, the degraded mode may result in a catastrophic failure.

Figure 7.3 illustrates the different failures in a LED-based system. Degradation may not only be reflected by a reduction in light output, it may also change the color output. The general approach in the industry is that degradation due to lumen maintenance is the basis for commercial claims of LED-based products. LED lifetime specifications are based on the time to, e.g., 70% (L70) or 80% (L80) of light output degradation at application temperatures. Here, IES LM-80-08 [18] serves an approved method for measuring lumen maintenance of LED lighting sources. And the IES standard TM-21-11 [19] provides a guideline for lifetime prediction of LED devices. On system level, however, Fig. 7.3 illustrates that the further reason for degradation effects can be either due to differential light output reduction, by catastrophic failures of single LED package, by degradation of the optical materials in the system, and even by many more causes.

IEC62681 [20] clearly state that LED-based systems depend generally on how balanced its principal components are in terms of their reliability. It is not only the LED components that determine product performance, but also other parts (or components) of the LED product play an equally important role. The technical document specifies the accelerated tests for each other parts, e.g., electronics, mechanical parts, that need to be conducted to determine the reliability of the system. Each other part or component may "equally" contribute to the failure as such.

For the case where catastrophic and degradation failure modes are considered simultaneously, system simulation techniques do exist [21]. Pettit and Young [22]

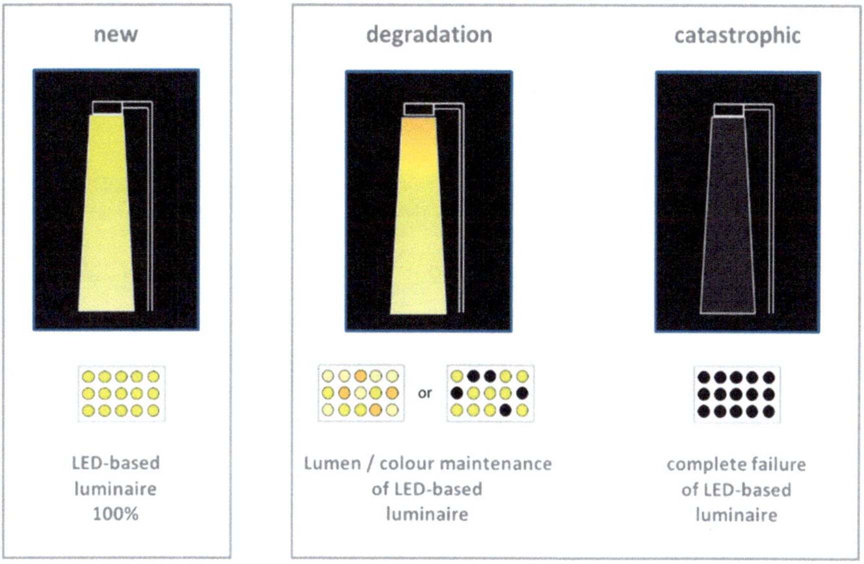

Fig. 7.3 Degradation and catastrophic failures can have unexpected effects on the lighting function

developed Bayesian procedures for simultaneously analyzing the lifetimes of failed units during the test and the degradation measurements of un-failed units observed at the end of the test. This work can be extended to the case of accelerated testing and repeated measures. Schuld et al. presented different system-level techniques to calculate the joint reliability by, e.g., Monte Carlo techniques [23]. Certainly, more work is available on this topic, but this is not our interest in the chapter.

To summarize the above: from an observation point of view, two effects can be noted by the user, being catastrophic of degradation failure, but the further substantiation of this observation can be more complex. Again, Fig. 7.3 illustrates this point.

3 From Observation to Malfunctioning

As mentioned above, the user will either notice a total or a degraded loss of the lighting function. The true root cause of that observation is not so straightforward and is determined by the components in the system. LED-based systems are composed of [20]:

- LED package and interconnects (the lightsource), e.g.
  - High-, mid-, or low-power LEDs
  - Solders, SAC or a-like
  - PCBs like FR4, CEM3, MCPCB

- Optical materials, e.g.
  - Polycarbonate (PC)
  - Polymethylmethacrylaat (PMMA)
  - Silicones
- Electronic subassemblies, e.g.
  - Drivers
  - Controllers
- Cooling systems, e.g.
  - active (e.g., fans)
  - passive (e.g., thermal interface material);
- Construction materials, e.g.:
  - Mechanical parts: connectors (e.g., screws, clips), wires
  - Housing: metal casing, plastic boxes
- Digital solutions, e.g.
  - Software
  - Data storage
  - Sensors
  - Wireless connectivity

Field quality control is a process by which entities review the quality of all factors involved in product installation. Field quality engineers regularly inspect, e.g., on yearly basis, installations toward possible failures and/or malfunction of the product. Products that do not fulfill the specified function anymore are send back to the supplier, in our case this will be the manufacturing site. At the manufacturing site the product in inspected and those components that have failed are registered in a cloud-based database. Figure 7.4 shows the pareto made from this database from an installation of >10 M pieces sold, in a time frame of 5 years, and a failed quantity of 19 k pieces (reflecting 0.2%). Percentage-wise the failures per component are listed as:

- Lightsource 23.4%
- Optical materials 4.3%
- Electronics 34.0%
- Cooling system 0.3%
- Construction materials 34.7%
- Digital solution 3.3%

Most of the failures are either due to the electronics or due to the construction materials. Lightsource malfunction is third in the row, the other 3 components occur significantly less. Field quality engineers also report out the reason for the

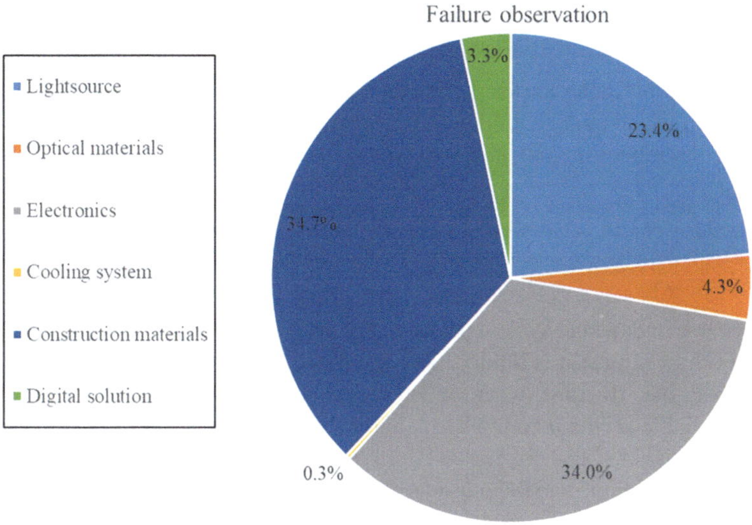

Fig. 7.4 Pareto's of Quality control (QC) is a process by which entities review the quality of all factors involved in production

Table 7.1 Further substantiation of the component malfunctioning

Component	Substantiation of field issue
Lightsource	Bad contact; broken; burnt / overheated; damaged; flickering; inoperative; missing; not controllable / programmable; out of expectation; randomly off; visual defect (aspects); wrong color; wrong light level
Optical materials	Visual defect (aspects); wrong; broke; damaged; loose; missing
Electronics	Bad contact; burnt / overheated; damaged; fallen; flickering; ingress of water/ dust; inoperative; missing; not controllable / programmable; out of expectation; randomly off; wrong light level; wrong programming
Cooling system	Damaged; deformed
Construction materials	Broken; damaged; deformed; ingress of water/dust; missing; bad contact; burnt / overheated; loose; bad mounting; ingress of water/dust; out of expectation; oxidation/corrosion; visual defect (aspects); wrong dimensions; bad machining; fallen
Digital solution	Broken; inoperative; out of specification; missing data; out of expectation

product malfunctioning. Table 7.1 lists further substantiation of the reason for the product malfunctioning. It ranges from missing, to broken, to lose, to fallen, to wrong color. For construction materials, the list is extensive and an item like "wrong dimensions" may also appear as reason. Clearly, the below substantiation does not.

4 From Malfunctioning to Root Cause

Physics-of-failure, also known as reliability physics, is a technique that leverages the knowledge and understanding of the processes and mechanisms that induce failure to predict reliability and improve product performance [24]. The most used definition is:

> *The process of using modeling and simulation based on the fundamentals of physical science (physics, chemistry, material science, mechanics, optical) to predict reliability and prevent failures.*

It helps to understand system performance and reduce decision risk during design and after the equipment is fielded. This approach models the root causes of failure such as fatigue, fracture, wear, and corrosion. An approach able to design and develop reliable products to prevent failures, based on the knowledge of root cause failure mechanisms. The concept is based on the understanding of the relationships between requirements and the physical characteristics of the product and their variation in the manufacturing processes, and the reaction of product elements and materials to loads (stressors) and interaction under loads and their influence on the fitness for use with respect to the use conditions and time.

The application of this concept to solid-state lighting products is founded on the conviction that the failure of LED-based products is governed by optical, mechanical, electrical, thermal, and chemical processes. As such, potential problems in new and existing technologies can be identified and solved even before they occur, by understanding the possible failure mechanisms [24]. For LED-based systems, the concept is used, and results are carved in the so-called Failure Mode Handbook. This handbook consists of summary sheets for each newly discovered failure mode, see Fig. 7.5, detailing out:

- Failure mode description

Short description of the failure mode, what is the observation? Accompanied, if possible, with a picture.

- Root cause/failure mechanism

What is the true cause of the failure mode, which physical mechanism is behind it?

- Solutions

What are possible solutions, how can one prevent the failure modes, what are the design rules to be obeyed?

- Lifetime model/acceleration

Under give testing conditions, what acceleration factors can be reached and what (lifetime) model is applicable.

- Testing method

Date:	Index	
Failure mode description	**Root cause / failure mechanism**	**Solutions**
Lifetime model / accelerators	**Testing methods**	**Reference to technical documents and experts**

Fig. 7.5 Summary sheet for failure modes, as part of the failure mode handbook

Date: June 2016	Organic material degradation	
Failure mode description	**Root cause / failure mechanism**	**Solutions**
Optical material discolors to get yellowish color.	• Thermal degradation of encapsulants induced by high junction temperature between LED die and lead frame • Thermal degradation of exit window induced by high temperatures • Photo degradation of encapsulants induced by UV radiation from LED dies and outdoor radiation • UV degradation of plastic materials • Thermal oxidation	• Higher grade lens material • Lower temperature • Apply UV coating • Reduce exposure to UV radiation
Lifetime model / accelerators	**Testing methods**	**Reference to technical documents and experts**
Temperature and UV radiation exposure (with Intensity I) are accelerators.	• HTSL • UV exposure • Refer to QS-000221 Optical Material Reliability Release Procedure	• Web search will reveal significant amount of references. • Work from PhD Maryam Yazdan Mehr • PR-TN 2017/00088 Color maintenance of LED-based products - Towards a system level prediction method • QS-000221 Optical Material Reliability Release Procedure *Experts: Willem van Driel, Boudewijn Jacobs, Guido Crick, William Yao* *External: Eurofins EHV*

Fig. 7.6 Summary sheet for the failure mode organic material degradation

Following IEC62861 [20], *or alike, which (accelerated) test provokes the failure mode?*

- Reference to technical documents and experts

Internal or external document and/or experts are mentioned as touchpoints for further details.

A pre-filled example is shown in Fig. 7.6: organic material degradation. This failure is well described in an open access review paper [25].

Since 2011 the physics-of-failure concept is applied to LED-based products and systems. Both accelerated testing results prior to commercial release and actively monitoring field response (see the former paragraph) have yielded a total number of 88 unique failure modes since then. Figure 7.7 depicts the detection of new failure modes in a 10-year period. On average 10 new failure modes are discovered per year. As it is expected, due to the growing maturity level of solid-state lighting products, this trend will once flatten out, the data is fitted with the Goel-Okumoto maturity growth model. This model is well-known for predicting the reliability of software [26]. Eventually, a total number of approximately 130 unique failure modes is to be discovered.

Further data analysis is feasible as each unique failure modes is well described. Figure 7.8 depicts a pareto of the number of failure modes per product type. Examples are:

- LED package

Browning of LED silicone, Chip moisture corrosion, Dome melting/deforming, LED Vf shift, Silver mirror corrosion, Sticky silicone dome

- LED product

BOM outgassing, Color shift, Driver induced LED failures, Zener burn-out,

- LED system

Battery failure, Software reliability, Surge issues, Water ingress.

Table 7.2 lists the classification toward the component that failed and how it failed, either in a catastrophic manner or if any signs of degradation yielded to its failure. The numbers clarify the following:

- Degradation is a dominant failure mechanism within solid-state lighting products. This by itself is not a surprise as these products are intended for long-term usage.

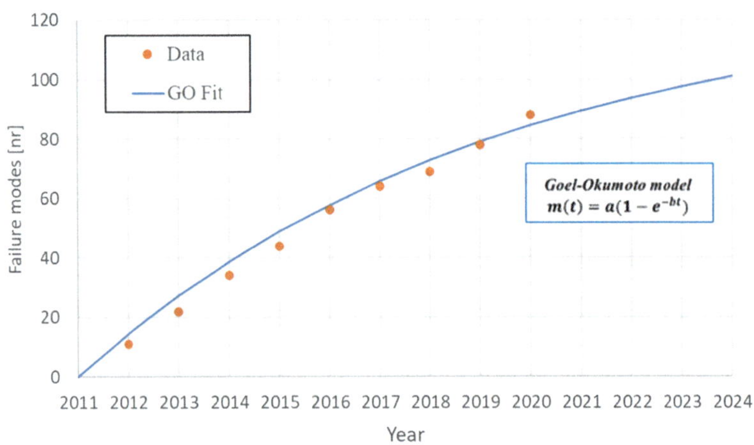

Fig. 7.7 Number of unique failure modes as function of years

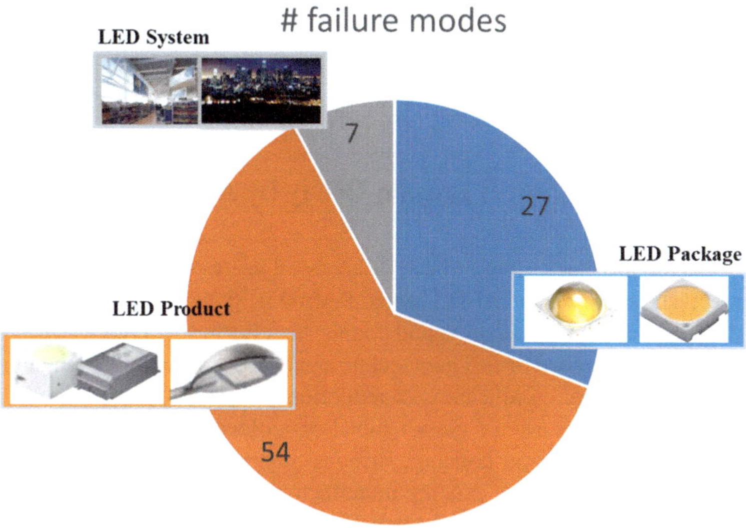

Fig. 7.8 Pareto for the number of failure mode vs product type

Table 7.2 Failure mode classification towards component, catastrophic, and degradation

Component	Catastrophic	Degradation	Total
Lightsource	16	16	32
Optical materials	3	10	13
Electronics	12	11	23
Cooling system	0	2	2
Construction materials	5	10	15
Digital solution	2	1	3
Grand Total	39	49	88

- As seen in the field response, the components that contribute the most to product failure are the lightsource, the electronics, and the mechanical construction.
- Failure modes in digital solutions (sensors, software) remain low and it is expected that this number will grow in the coming years due to the extension of the connected portfolio.
- Failure modes in the cooling system seem rare, which was also concluded from the 5-year field response (see Fig. 7.4).
- In comparison with the 5-year field response data, the optical material degradation surprisingly pops up in our physics-of-failure concept. Degradation is leading for this component with discoloration, yellowing, browning, and corrosion as long-term events to occur.

In summary, whatever is observed by the user, after further substantiation to component level and physics-of-failure root cause may lead to a surprisingly complex failure mechanism. The number of failure modes increases over time, not only due to newly introduced technologies but also because degradation mechanisms start to appear after 5–10 years. Figure 7.9 illustrates the before-mentioned approach. In next paragraphs, several failure modes will be further described.

5 Degradation Failure: Lumen Decay by Absence of Oxygen

In homes, available sockets like the E19 and E27 are to be replaced by an LED alternative. These are named LED Lamps. Per today, the penetration is in the area of 50–80% depending on the region and country and further cost reduction is a must to increase the acceptance of this new technology. LED Lamps development is currently in the process of significant cost reduction in materials, component design, and manufacturing. The use of glass instead of molded plastic is seen as a cost-efficient alternative for lamp housing. Besides that, glass is used in conventional bulbs meaning leverage of the existing manufacturing lines. As such, several major lighting companies did investigate the concepts for glass LED bulbs [27]. The products that are currently available on the market are depicted in Fig. 7.10.

The ongoing cost reduction of LED bulbs led to the re-use of the installed manufacturing base used in conventional lighting. As mentioned above, glass is re-introduced. These bulbs, in general, have a general design which is depicted in Fig. 7.11. They resemble typical convention E27 sockets, but in this case the inner part of the glass bulb consists either of a printed circuit board (PCB) with LEDs

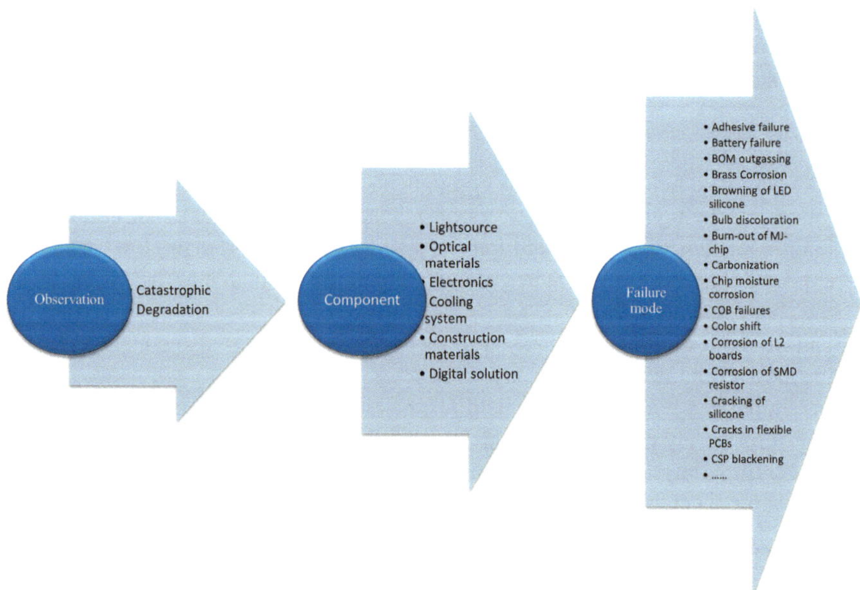

Fig. 7.9 From observation to failure mode

Fig. 7.10 Examples of glass LED bulbs, (**a**) transparent with filaments, (**b**) not transparent with an LED light source on PCB

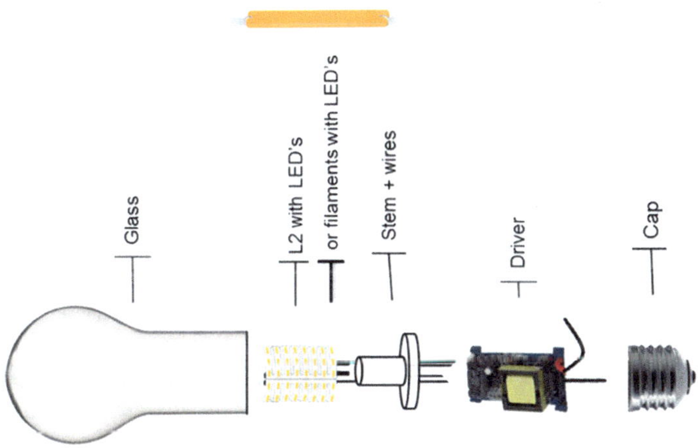

Fig. 7.11 Architecture of glass LED bulbs. Either with L2's or with filaments

mounted on top (l2) or of a filament with LEDs mounted on a substrate and covered with phosphor-embedded silicone. Several concepts are possible for the LED engine consisting of different PCB or substrate types (CEM, metal, glass), different LED technologies (mid-power or low power). The LED engine is mounted on top of a stem, placed into the glass which is sealed at the bottom. As such, the LEDs are placed in an enclosed environment with no direct (chemical) contact with the outer world. Furthermore, glass is highly transparent, cheap, and fire safe.

Glass LED bulbs' performance strongly relies on its lumen depreciation in which the light source gradually but slowly degrades over time. For thermal reasons, the LEDs are placed in an enclosed environment that is filled with helium. This enclosed environment is sealed from outside meaning no natural flow of gasses may occur. A drawback of the design is severe discoloring of the LEDs over life in absence of oxygen. The root cause for this discoloring can be two folded:

- Yellowing/browning: degradation of silicone lenses due to photo-oxidation [28–30]
- Carbonization: discoloration at the epitaxial–silicone interface by (volatile) organic compounds [12, 25]

This problem can be overcome by dosing a mixture of oxygen and helium, typically 10–20% O_2/He. The drawback of this measure is less cooling. As the dosed oxygen is consumed over lamp life, an optimum mixture should be found coping with the trade-off:

Sufficient cooling ↔ sufficient lifetime

Here, the lifetime is reached below lumen maintenance values <70% which may occur once the oxygen is consumed. For determination of the proper trade-off setting, knowledge on both thermal management in O2/He mixtures and oxygen depletion is needed.

The heat transport inside the glass LED bub is a combination of convection in the gas mixture and thermal radiation between the surfaces. The convective transport is depending on the gas mixture composition, while the radiation transport depends on the surface emission coefficient. The gas mixture is not taking part in the radiation transport. The heat transport is analyzed with a CFD simulation model (ANSYS CFX) of the bulb geometry. In this model, the gas mixture is described as an ideal mixture with temperature-dependent properties. The thermal resistance from "heat spreader" (substrate or PCB) to ambient, $Rth_{hs\text{-}amb}$, is calculated with the model. It is based on the average surface temperature of the engine. The goal of the simulations is to investigate the impact of various gas mixture compositions on the $Rth_{hs\text{-}amb}$ value within the same geometry.

Lifetime tests involve lamps put on traditional burning racks, see Fig. 7.12. In these tests, bulbs are operated in a well-controlled environment with constant ambient temperature at 25C, 35C, and 45C, respectively. Accelerated tests included dedicated degradation experiments under flushed N_2 or He environment, without any presence of oxygen. During the O_2 lifetime test, the gas mixture composition is measured at several moments. The results of these tests showed that during lifetime:

- the pressure in the glass LED bulb is decreasing
- the amount of O_2 is decreasing, while CO_2 is increasing
- the amount of He (mol/m^3) is almost constant.

The consumption of O_2 appeared to be a rather complicated process as there are many different hydrocarbons, from several sources inside the lamp located at positions with different temperatures. Here, a scientific description of the physical-chemical processes will not be discussed rather we present degradation models that describe the O_2 depletion in terms of design and application parameters. For the gas analysis, at least one glass LED bulb is measured to obtain the initial oxygen

Fig. 7.12 Lifetime tests with glass LED bulbs put into traditional burning racks

concentration level before the lifetime test is started. The tested bulbs are taken out for gas analysis at different test times. Since the gas analysis is a destructive analysis, all gas analysis results are from different bulbs. The gas measurements were performed on a mass spectrometer. Before the measurement, the lamp is brought in a stainless (pre-cleaned) cracker and purged with ultra-pure argon for at least 1 hour with a purge flow of several litters/minute. After the initial purge time, the cracker has been connected to the mass spectrometer and the composition of the argon filling gas has been analyzed. After another 30 min purge time, the lamp has been broken in the Argon environment. The composition of the gas is calculated by using the calibration factors of the Mass Spectrometer. Results are calculated without the Argon content and normalized to 100%. During the lifetime test, as mentioned before, with the decreasing O_2 and the increasing CO_2, the gas pressure inside the lamps decreases while the volume of He increases, see Fig. 7.13. To make the calculation of the $_{O2}$ concentration after burning consistent with the initial situation before the test, the "absolute" O_2 concentration is calculated by correcting for He following this relationship (in volume %):

$$O_{2\_abs} / O_{2\_measured} = He_{initial} / He_{measured} \tag{7.1}$$

The first step in our analysis is to correlate the failure criterium (L70) to the value of O_2 concentration. For that purpose, survivals and none-survivals are analyzed toward remaining O_2 levels, results are listed in Table 7.3. In the table, bulbs 1 to 4

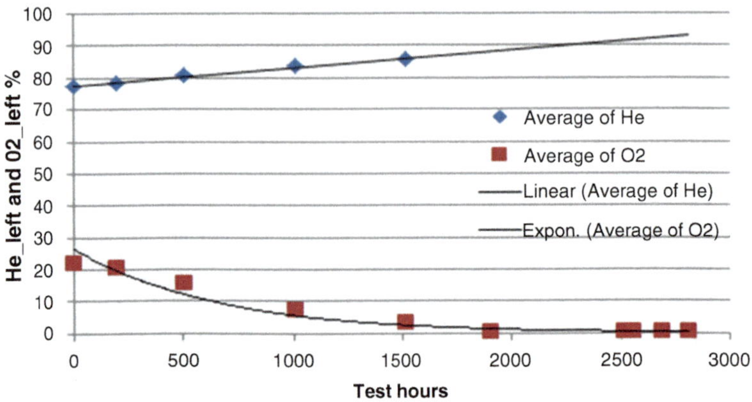

Fig. 7.13 Trends of measured O_2 concentration and He volume during operational life test

Table 7.3 Estimated minimum oxygen levels to keep the bulb in the L70 target

Nr	P [W]	O_2 initial [%]	O_2 after test [%]	L70 criterium
1	60	21	0.40	Pass
2	60	10	0.38	Pass
3	60	11	0.64	Pass
4	40	2	0.66	Pass
5	60	20	0.07	Fail
	60	20	0.08	Fail

showing more than 0.4% O_2 concentration give still enough lumen output while numbers 5 and 6 with the amount of O_2 concentration below 0.1% give weak lumen output below the target. Therefore, the needed O_2 concentration to prevent severe degradation is between 0.1% and 0.4%.

The O_2 consumption is determined by reacting with organic volatiles or VOC by: $C_xH_y + O_2 \rightarrow CO/CO_2$, a well-known reaction. Here, the O_2 consumption can be expressed as:

$$-d\frac{[O2]}{dt} = k.[O2].[VOC] \tag{7.2}$$

Assuming the level of VOCs in the glass bulb is fixed for any fixed design, the above equation is equivalent as

$$\int^{[O2]_i} \frac{d[O2]}{[O2]} = \int -k.dt \tag{7.3}$$

Assuming k is constant during whole lifetime, the above equation is integrated to

$$[O2] = [O2]_i \cdot e^{-kt} \tag{7.4}$$

or:

$$t = -\ln\left(\frac{[O_2]}{[O2]_i}\right) / k \tag{7.5}$$

where

k: reaction rate [1/hr]
t: lamp burning time [hrs]
$[O_2]$: oxygen content [%]
$[O_2]_i$: Initial oxygen content [%].

Using the lifetime tests listed above we have predicted the lifetime of glass bulbs as a function of temperatures and oxygen concentration. Here, we assume the reaction rate follows the well-known Arrhenius model:

$$k = k_0 e^{-E_a/RT} \tag{7.6}$$

with:

k0: constant [1/hr]
Ea: activation energy [eV]
T: absolute temperature [K].

In order to predict the lifetime of glass LED bulbs, we have determined E_a and K_0 by fitting the below equation to the experimental data:

$$t = -\ln\left(\frac{[O_2]}{[O2]_i}\right) / \left(k_0 e^{-\frac{E_a}{RT}}\right) \tag{7.7}$$

Figure 7.14 depicts such a fitted curve for the case of an initial O_2 level of 20%, tested at an ambient temperature of 35C. The data is well described within a confidence interval of 95%.

For all variations we tested, the activation energy was found to be between 0.37 eV and 0.58 eV. The influence of temperature and initial O_2 level is depicted in Fig. 7.15. When the target lifetime of the bulb is 10khrs, these plots will give the design guidelines in order to achieve those values.

In summary, the O_2 consumption in glass LED bulbs can be modeled and understood from a physical point of view and is directly related to the lifetime of the product.

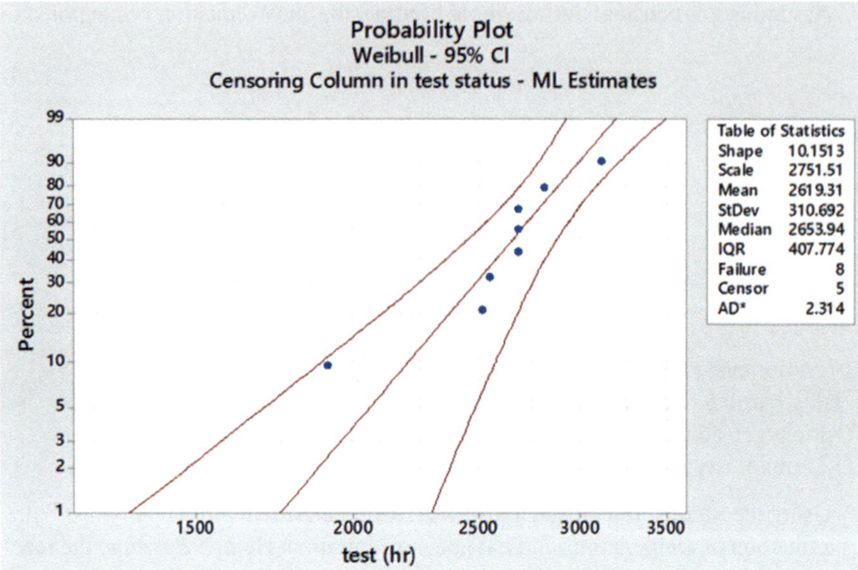

Fig. 7.14 Weibull plot of tested O2 consumption lifetime for 60 W, 20% [O2]-96 mA-35C ambient temperature

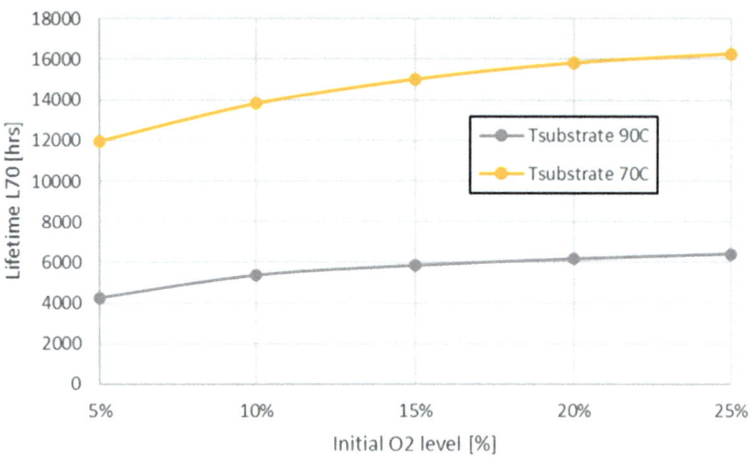

Fig. 7.15 Lifetime as a function of temperature and initial O_2 level

6 Degradation Failure: Color Maintenance

A recently appearing system failure mode in LED-based products is color maintenance. Color maintenance problems are insidious, because they are poorly understood, and only appear after many hours of operation [31, 32]. Lumen maintenance failure of SSL products is generally characterized by L70 life or 70% degradation of the lumen output. However, no failure criterion for color maintenance has been defined specifically in the application field except that ENERGY STAR® program mandates that Δu'v' at 6000 hours of operation should not exceed 0.007 [18], which is perhaps the only industry-wide criterion. It is a reasonable starting point but may not be strict enough to ensure very high-quality lighting, especially since the lifetimes of LED products routinely far exceed 6000 hours [2, 3]. Poor color maintenance can be a substantial problem in applications where color quality is important, including museum and gallery lighting, architectural facade lighting, retail display lighting, healthcare lighting, hospitality applications, cove and wall wash lighting, down lighting in commercial and residential applications [33–35].

Color maintenance is analogous to lumen maintenance and is the change in chromaticity of a light source with respect to the chromaticity at the beginning of the lamp's life. It is typically measured as Δxy or Δu'v', in the CIE color coordinate systems [36, 37]. Energy Star specifies that color maintenance must not exceed Δu'v' = 0.007 on the CIE u'v' diagram, after 6000 hours of operation. This is a liberal allowance for color shift, and we expect that LED-based products should do better than this to maintain a positive perception in the marketplace.

Color consistency is the variation in chromaticity, at the start of product life, among a population of products. For example, a product may be made from LEDs that are binned to fall within 3 MacAdam [36] steps of a target chromaticity. The LEDs have a color consistency of 3 steps. Color consistency can also be defined in terms of xy or u'v'. The color consistency of lamps built from these LEDs may be worse than three steps, because of temperature variations, current variations, or other factors. Color stability describes how the entire spectrum changes over time, and is closely related to color maintenance, but encompasses more detail. Color maintenance will be used in this paragraph, an example is shown in Fig. 7.16. Origins of color maintenance are listed in Table 7.4 [31, 38].

What we need to develop color maintenance predictions on system level are:

- System-level approach that can combine (known) component-level color shift functions.
- An approach that can deal with the product (optical) design.
- Test data on component, sub-component, and system level to validate and verify the system level predictions.

So far there has been one attempt published that is able to do product-level prediction in this area [32]. Davis' method also considers changes in the LEDs and the optical components (i.e., lenses and reflectors) of the luminaire. In addition, the relative impacts of lens and reflector degradation will depend on the design of the

Fig. 7.16 Example of color maintenance drift for an indoor installation

Table 7.4 Origins of color maintenance

Color maintenance Root cause	Examples
Material degradation	Degradation of direct optical path from LED die to air Degradation of reflective surfaces within the LED component Degradation of the optical materials with the system, be it MCPET, white solder resist, poly carbonate or PMMA.
External contaminants	Contaminations in the direct optical path such as browning of the optical path due to VOCs or residual flux after reflow on the exterior of the LED package. Change in the reflective surface properties of materials within the LED component, including, for example, tarnishing of silver. Carbonization due to lack of oxygen. Sedation of particles onto any optical surface, for example, onto the silicones.
Interface delamination	Separation between different material interfaces such as substrate and optical path materials. Material cracking, for example, in the MCPET reflector due to brittleness.

luminaire. This phenomenon can be studied by using optical simulation tools for nearly any luminaire design. This simulated or "virtual" luminaire approach, which is illustrated here, involves designing a luminaire in the simulation tool that includes the selection of the initial optical properties of the lenses and reflectors. Then, the properties of these materials can be degraded in a systematic manner, and the impacts on luminaire performance can be determined. For this analysis, any luminaire design can be modeled for any expected physical parameters, including size and optical cavity depth. Aging of materials used in luminaire could be easily accommodated in the simulation by attenuating the optical properties of the materials. For example, the impact of a 10% drop in lens transmittance can be simulated by reducing normalized lens transmittance by 10%. Transmittance is normalized by the initial transmittance of the pristine material. By systematically changing the design parameters and simulating aging of the optical surfaces by introducing new values for normalized transmittance and reflectance, a model can be created to determine the change in luminous flux produced by the luminaire during aging. A simple power-law model of the form shown in the equation below captures the impact of optical materials degradation on lumen maintenance for the LED device:

$$\Phi_{tot}(t) = \Phi_{init} F_{LEDs} \left[L(t)\right]^{m} \left[R(t)\right]^{n} \tag{7.9}$$

where

- Φ_{tot} = Total luminous flux from the luminaire at time t
- Φ_{init} = Initial luminous flux from the luminaire at time zero. This value is also the product of the luminous flux from the light engine and the luminaire efficiency.
- F_{LEDs} = Lumen maintenance factor of the LED at time t
- $L(t)$ = Change in the normalized lens transmittance $[\%T(t)/\%T(t = 0)]$ at time t
- m = Design-dependent factor for the lens as determined from the simulation
- $R(t)$ = Change in the normalized reflector reflectance $[\%R(t)/\%R(t = 0)]$ at time t
- n = Design-dependent factor for the reflector as determined from the simulation.

Regarding color maintenance of LEDs, TM35 describes the projection of long-term chromaticity coordinate shift of LEDs [39]. Ignoring these effects of taking them as a given, Eq. 7.9 can be recast into a model for luminaire optical efficiency degradation as shown in the following eq.

$$LE(t) = LE(t = 0)\left[L(t)\right]^{m} \left[R(t)\right]^{n} \tag{7.10}$$

where

- $LE(t)$ = Luminaire efficiency at time t
- $LE(t = 0)$ = Initial luminaire efficiency

The parameters m and n consider the impact of the luminaire design on the relative contributions of lens and reflector aging upon degradation in luminaire efficiency. For example, for a 2 × 2 m^2 troffer design, m is approximately 2 and n is approximately 0.5 [32], indicating that the degradation of the lens has a much greater impact on changes in luminaire efficiency and be extension chromaticity stability than the degradation of the reflectors. However, for luminaires with smaller apertures and proportionally deeper optical mixing cavities, the value of n can equal or surpass the value of m, indicating that reflector degradation has a significant impact on overall luminaire efficiency changes in some designs. Conceptually, the value of n can be thought of as being proportional to the likelihood that light emitted by an LED at the base of an optical cavity will strike the reflector in that optical cavity and be impacted by reflector degradation. Consequently, the impact of reflector degradation on lumen maintenance and chromaticity stability can be reduced by shrinking the optical mixing cavity, but this may also change the distribution and homogeneity of the light emitted by the luminaire.

A more generic approach for deducting the product design parameters m and n is based on the so-called view factor method [38]. To help illustrate this approach, a schematic diagram of light paths for a LED-based product is shown in Fig. 7.17. Each exchange of light in the light paths can be considered as a contribution to color maintenance.

As the color maintenance of the light out of the product is the result of those coming from the LEDs, reflector, PCB, and optics, the interaction between these optical elements needs to be included. A complete ray tracing model will result in a too complex model to be used by designers and reliability engineers. The view factor approach assumes that a certain fraction of the light emitted or reflected by an optical part is radiated to another part (or even itself again). Each part that reflects light can also give this light a small color shift or spectral change.

The set of equations describing the light interaction between the 4 optical elements is as follows:

$$\begin{pmatrix} \Phi_{\text{i to LEDs}} \\ \Phi_{\text{i to pcb}} \\ \Phi_{\text{i to housing}} \\ \Phi_{\text{i to exit}} \end{pmatrix} = \begin{pmatrix} F_{L-L} & F_{p-L} & F_{h-L} & F_{e-L} \\ F_{L-p} & F_{p-p} & F_{h-p} & F_{e-p} \\ F_{L-h} & F_{p-h} & F_{h-h} & F_{e-h} \\ F_{L-e} & F_{p-e} & F_{h-e} & F_{e-e} \end{pmatrix} \begin{pmatrix} \Phi_{\text{i from LEDs}} \\ \Phi_{\text{i from pcb}} \\ \Phi_{\text{i from housing}} \\ \Phi_{\text{i from exit}} \end{pmatrix} \quad 7.11$$

Here the letter *i* denotes how many interactions (= reflections) the light has already undergone. The matrix with the numbers *F* is the view factor matrix, and as we assume that all light emissions and reflections have Lambertian distribution, the view factor matrix is independent of *i*. The sum of the elements in each column is equal to 1. The part "housing" could also be described as the "reflector". Part of the

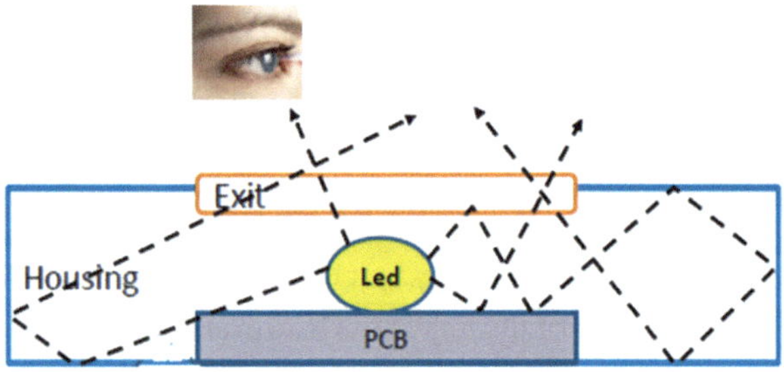

Fig. 7.17 Schematic diagram of light paths representing the view factor approach

light hits upon each optical element is also reflected again. This can be expressed in the next equation:

$$\begin{pmatrix} \Phi_{\text{i+1 from LEDs}} \\ \Phi_{\text{i+1 from pcb}} \\ \Phi_{\text{i+1 from housing}} \\ \Phi_{\text{i+1 from exit}} \end{pmatrix} = \begin{pmatrix} R_L & & & \\ & R_p & & \\ & & R_h & \\ & & & R_e \end{pmatrix} \begin{pmatrix} \Phi_{\text{i to LEDs}} \\ \Phi_{\text{i to pcb}} \\ \Phi_{\text{i to housing}} \\ \Phi_{\text{i to exit}} \end{pmatrix} \quad 7.12$$

At each reflection R the light can undergo a color change. This color change can be described in two ways:

- color shift Δx and Δy, independent of the color point (spectrum) of the incoming light
- spectral change, the reflectivity depends on the wavelength of the light

Method 2 is more realistic, but in that case the spectra of the light need to be included in the calculations. In case of small color changes (small changes of the spectrum upon reflection) and low reflectivity's *R* (fast damping), method 1 can be good enough. Therefore, this method is preferred. We note the initial condition:

$$\begin{pmatrix} \Phi_{\text{0 from LEDs}} \\ \Phi_{\text{0 from pcb}} \\ \Phi_{\text{0 from housing}} \\ \Phi_{\text{0 from exit}} \end{pmatrix} = \begin{pmatrix} \Phi_{\text{LEDs}} \\ 0 \\ 0 \\ 0 \end{pmatrix} \quad 7.13$$

And then the light coming out of the lamp/luminaire by means of transmission through the exit window:

$$\Phi_{\text{from exit}} = \left(1 - R_e - A\right)\Phi_{\text{to exit}} \quad (7.14)$$

where *A* stands for absorption and the light is summed over enough *i*. The transmission of the light can also be accompanied with a color change of the light. As the light out of the module is also calculated, the approach also gives the optical efficiency as output so it can be used to estimate the optical efficiency.

For 2 common geometries, a rectangular and an axis symmetric product, the view factors for the users are known [40], following the conservation law and reciprocity, see Fig. 7.18:

$$\sum_j F_{\text{from } i \text{ to } j} = \sum_j F_{ij} = 1$$

$$A_i F_{ij} = A_j F_{ji} \quad (7.15)$$

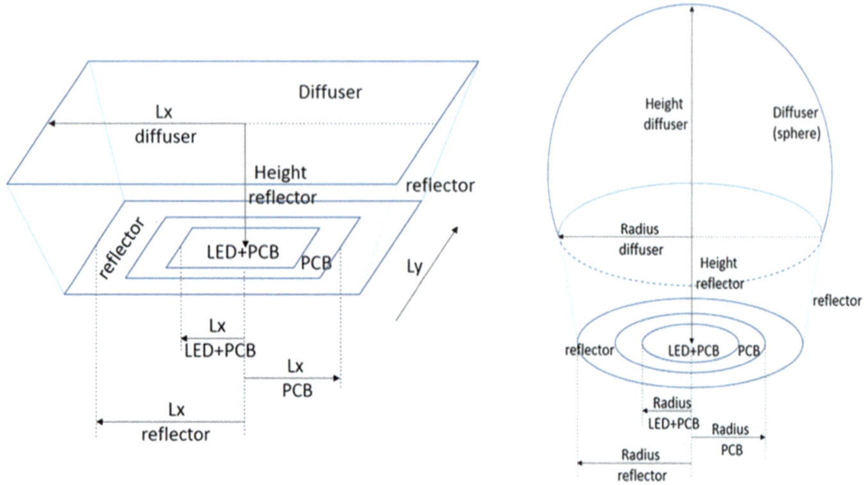

Fig. 7.18 Rectangular product geometry (left) vs axis symmetric product geometry (right)

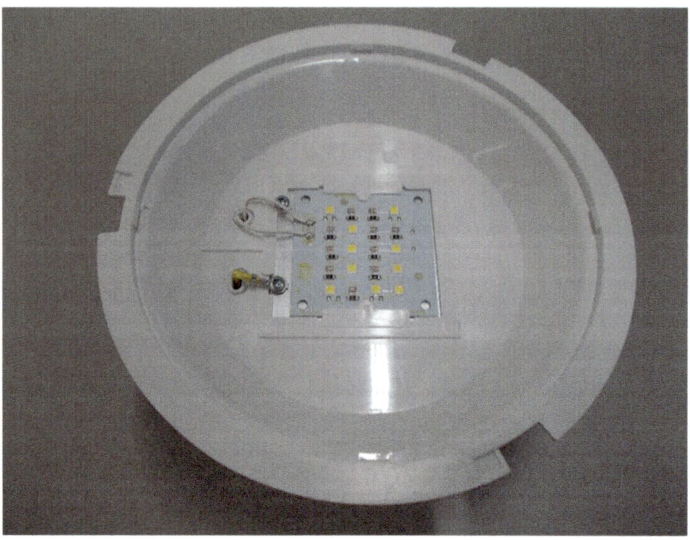

Fig. 7.19 Photo of carrier product with the diffusive flat exit window taken out

The carrier product used for validation of the approach is depicted in Fig. 7.19. It perfectly fits in the "axis symmetric" shape of which the view factors can be easily calculated, see Table 7.5. Most of the light coming from the LEDs will travel through the diffuser (91.7%). Reflection from the PCB is not negligible (13.2%). For purpose of validation, the carrier product is tested at elevated temperatures of 110C and 120C. All optical materials will degrade over time. Regarding the aging of these plastics, the correlation between aging time and aging temperature is

Table 7.5 Calculated view factors for the carrier product

View factor [%]		From			
TO		LEDs	PCB	Reflector	Diffuser
	LEDs	0.0	0.0	0.1	0.8
	PCB	0.0	0.0	1.2	13.2
	Reflector	8.3	8.9	20.9	86.0
	Diffuser	91.7	91.1	77.8	0.0
	SUM	100	100	100	100

described by the principle of time-temperature superposition or equivalently introducing the effective time variable:

$$\Delta t_{eff} = \Delta t / a_T$$

$$a_T = \exp\left[\frac{\Delta H}{R}\left(\frac{1}{T} - \frac{1}{T_0}\right)\right] \tag{7.16}$$

where T is the temperature in Kelvin (T_0 the reference temperature), R the gas constant 8.314 J/mol.K, and ΔH the activation energy, used to fit the data at several temperatures (i.e., sensitivity to temperature).

Each part in the carrier product is aged during a period of approximately 4000 hrs. Color changes in terms of coordinates Δx and Δy are found to follow:

$$\Delta x = \Delta x_t \left(\frac{t}{t_0}\right)^{n_x}$$

$$\Delta y = \Delta y_t \left(\frac{t}{t_0}\right)^{n_y}$$

where t denotes time, t_0 a reference time, an example is given in Fig. 7.20.

Each part in the carrier product exhibits its own degradation. Figure 7.21 depicts the testing results after approximately 4000 hrs at elevated temperatures in terms of changes in color coordinates (the arrow gives the direction). Notice that the LED color changes are opposite in direction from the optical parts.

With the given color maintenance view factor approach, predictions of both the lumen decay as the color shift were made. Now the temperature-dependent (effective time) models for color shift, lumen decay, reflection, and transmission change are used. As a function of time and temperature the parameters of the LEDs, PCB, reflector, and exit window can be calculated. Results are depicted in Figs. 7.22 and 7.23. From these figures we read that the LED color shift starts faster than the color shift due to yellowing of the PCB, reflector, and exit window. In the end the

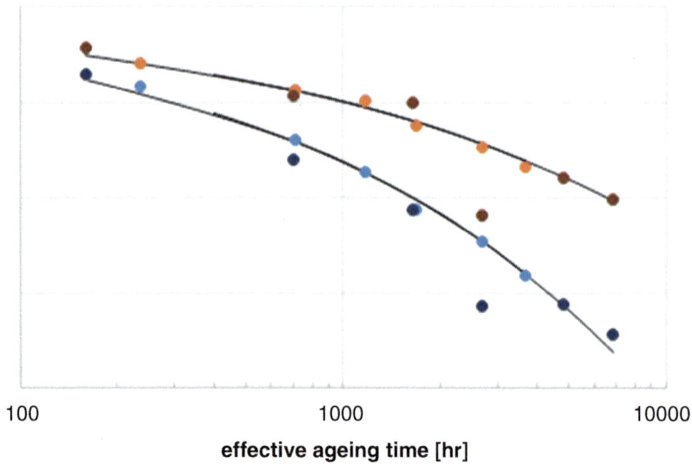

Fig. 7.20 Measured x (blue) and y (orange) color shifts for 2 carrier products with a reflector but without the diffuser

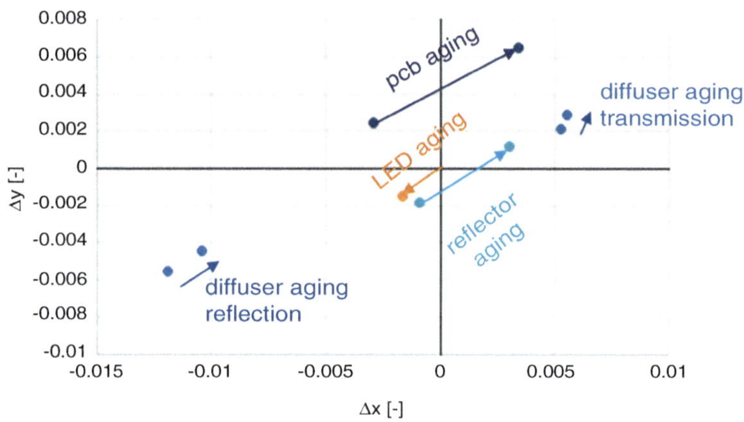

Fig. 7.21 Colors changes after ~4000 hrs testing in the product parts depicted in x, y color coordinates, indicated with an arrow

yellowing "wins" even with 90 °C LED temperature and 70 °C reflector and exit window temperature. As the LED color shift is toward blue, the product color point first moves a bit toward blue and later moves in the opposite direction toward yellow. The result is that if the LEDs would give a fixed color (no aging) the color shift of the product will be larger. One can also see that lumen decay due to deteriorating of the optics is small compared to the lumen decay caused by the LED degradation.

When one asks which criterion is limiting, lumen decay (80%) or color shift (7 SDCM), then Fig. 7.23 shows that lumen decay is most critical. But if one thinks of

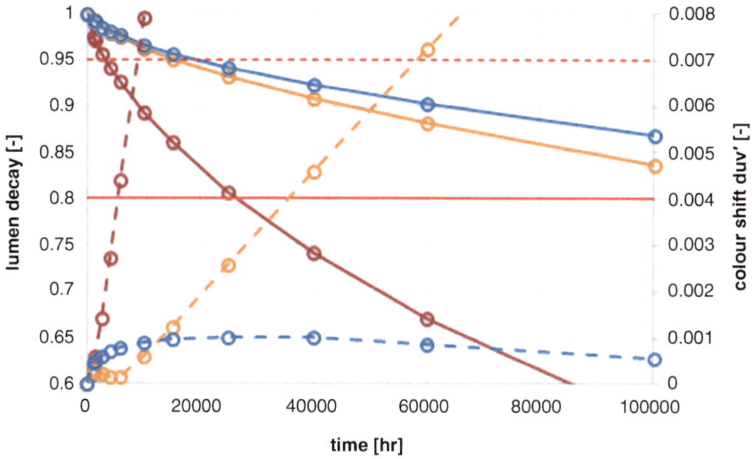

Fig. 7.22 Calculated uv' color shift (dotted line) and lumen decay (solid line) as a function of aging time. Red: all parts 110 °C; Orange: all parts 90 °C; Blue: LEDs 90 °C, PCB 80 °C, reflector & exit window 70 °C

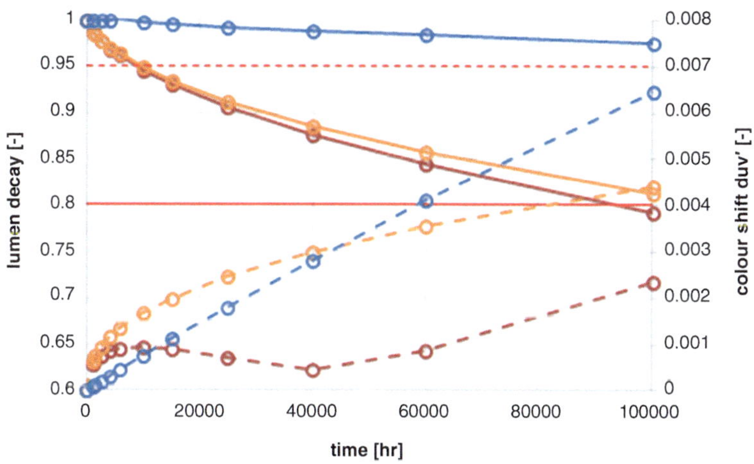

Fig. 7.23 Calculated uv' color shift (dotted line) and lumen decay (solid line)as function of (aging) time; LEDs 100 °C, PCB 90 °C, reflector & exit window 80 °C. Red: all parts aging; Orange: only LED aging; Blue all but LED aging

stricter specified products with only 2 SDCM color shift allowed then the 80% lumen decay and the 0.002 uv' color shift occur both at ~90 khr.

In summary, we presented the results of this newly developed method, including the validation experiments that were executed in other to compare predictions with reality. The results indicate that:

- Color maintenance in LED-based products is a combination of color maintenance in LEDs, optical materials, and the construction itself.
- Temperature is the main driver for the color maintenance.
- The optical materials play an equally important role in controlling the amount of color maintenance.

For fixed constructions, color maintenance on system level is just a balance between the LED and the optical materials.

7 Final Remarks

In the past ten years we have witnessed a substantial change in the lighting industry. Traditional companies have changed their strategy and upcoming competition has pushed down prices for LED-based products considerably. LED penetration levels increased so as the diversity of commercially available replacement products. New processes and materials were introduced, and consequently new failure modes appeared. This trend has continued in the past four years as the lighting industry is getting connected and large amounts of user data are being analyzed. New components are needed to deliver this functionality (sensors, actuator IoT modules) and, as such, the diversity from an architectural point of view will also increase. In this chapter, we have presented the currently known reliability and failures found in these solid-state lighting systems. It includes both degradation and catastrophic failure modes from observation toward a full description of its mechanism obtained by extensive use of acceleration tests using knowledge-based qualification methods. A total number of 88 failures modes are found, from which 60% are related to degradation. This indicates the importance of monitoring the degradation process in these products, as longer lifetimes and warrantees are industry targets. As such, gradually but slowly the term reliability in the lighting industry will be replaced by availability and "smart" maintenance will distinguish good from bad products.

Two degradation cases are presented, O_2 consumption in glass LED bulb and color maintenance in LED-based products. The developed algorithms describe the degradation processes in detail and, thus, can be used for monitoring the product lifetime [41].

Acknowledgments This project has received funding from the ECSEL Joint Undertaking (JU) under grant agreement No 876659. The JU receives support from the European Union's Horizon 2020 research and innovation programme and Germany, Austria, Slovakia, Sweden, Finland, Belgium, Italy, Spain, Netherlands, Slovenia, Greece, France, and Turkey.

References

1. Frost & Sullivan, 2020 Annual Update of Global LED Lighting Market, September 2020
2. W.D. Van Driel, X.J. Fan (editors), Solid State Lighting Reliability: Components to System. 01/2013; ISBN 978-1-4614-3067-4 Springer New York
3. W.D. Van Driel, X.J. Fan and G.Q. Zhang (editors), Solid State Lighting Reliability: Components to System Part II. 06/2017; ISBN 978-3-319-58174-3 Springer New York
4. Navigant Consulting, Inc., Energy Savings Forecast of Solid-State Lighting in General Illumination Applications, report prepared for the U.S. Department of Energy, September 2016
5. Estimated LED penetration of the global lighting market from 2010 to 2020, available at http://www.statista.com/statistics/246030/estimated-led-penetration-of-the-global-lighting-market/ (last visited on 8/25/2016)
6. Market penetration predicted for white light, freely available at http://edisonreport.com/files/7613/7631/7460/SSL_Energy-Savings_Predictions.pdf (last visited on 10/16/2020)
7. D. Schenkelaars, W.D. van Driel, M. Klompenhouwer, I. Flinsenberg, R. Duijve, Towards Prognostics & Health Management in Lighting Applications, European Conference of the Prognostic and Health Management Society 2016, open access journal, available at: http://www.phmsociety.org/node/2090/, Volume 7, Page count: 7, 2016
8. T. Tuunanen, H. Cassab, Service process modularization: reuse versus variation in service extensions. Journal of Service Research (2011) 1094670511413912
9. L.L. Berry, V. Shankar, J. Turner Parish, S. Cadwallader, T. Dotzel, Creating new markets through service innovation. Sloan Management Review **47**(2) (2006)
10. J. Björkdahl, M. Holmén, Editorial: Business model innovation–the challenges ahead. International Journal of Product Development **18**(3/4), 213–225 (2013)
11. Meeker, William Q. and Hong, Yili, Reliability Meets Big Data: Opportunities and Challenges (2013). Statistics Preprints. Paper 82. http://lib.dr.iastate.edu/stat_las_preprints/82
12. Michael G. Pecht, Moon-Hwan Chang, Failure Mechanisms and Reliability Issues in LEDs, In: W.D. Van Driel, X.J. Fan (editors), Solid State Lighting Reliability: Components to System.. 01/2013; ISBN 978-1-4614-3067-4 Springer New York
13. M-H. Chang, D. Das, P. V. Varde, and M. Pecht, Light Emitting Diodes Reliability Review, Journal of Microelectronics Reliability, Article in Press, 2011, https://doi.org/10.1016/j.microrel.2011.07.063
14. G. Meneghesso, S. Leveda, E. Zanoni, G. Scamarcio, G. Mura, S. Podda, M. Vanzi, S. Du, I. Eliashevich, Reliability of Visible GaN LEDs in Plastic Package. Microelectronics Reliability **43**, 1737–1742 (2003)
15. C. De Santi, D. Monti, P. Dalapati, M. Meneghini, G. Meneghesso, E. Zanoni, Reliability of Ultraviolet Light-Emitting Diodes, in *Light-Emitting Diodes, Solid State Lighting Technology and Application Series 4*, ed. by J. Li, G. Q. Zhang. https://doi.org/10.1007/978-3-319-99211-2_11
16. J.F.J.M. Caers, X.J. Zhao, Failure Modes and Failure Analysis, in *Solid State Lighting Reliability: Components to System*. 01/2013; ISBN 978-1-4614-3067-4, ed. by W. D. Van Driel, X. J. Fan, (Springer, New York)
17. JEP122F, Failure mechanisms and models for semiconductor devices, JEDEC publication, March 2009
18. IES LM-80-08: Approved method for measuring maintenance of Led light sources, 2019
19. IES TM-21-11: Projecting Long Term Lumen Maintenance of LED Light Sources, 2019
20. IEC/TS 62861 Ed. 1: Guide to principal component reliability testing for LED light sources and LED luminaires, technical specification (under creation)
21. S.-H. Kim, S.-I. Sung, Modeling and analysis of the catastrophic failure and degradation data. Microelectronics Reliability (2020) in press
22. L.I. Pettit, K.D.S. Young, Bayesian analysis for inverse Gaussian lifetime data with measures of degradation. Journal of Statistical Computation and Simulation **63**(3), 217–234 (1999). https://doi.org/10.1080/00949659908811954

23. M.H. Schuld, B.F. Schriever, J.W. Bikker, Solid State Lighting System Reliability, in *Solid State Lighting Reliability: Components to System*. 01/2013; ISBN 978-1-4614-3067-4, ed. by W. D. Van Driel, X. J. Fan, (Springer, New York)
24. M. Pecht, A. Dasgupta, *Physics-of-failure: An approach to reliable product development, Proceedings Integrated Reliability Workshop* (1995). https://doi.org/10.1109/IRWS.1995.493566
25. M. Yazdan Mehr, A. Bahrami, W.D. van Driel, X.J. Fan, J.L. Davis, G.Q. Zhang, Degradation of optical materials in solid-state lighting systems. International Materials Reviews **65**(2), 102–128 (2020). https://doi.org/10.1080/09506608.2019.1565716
26. W.D. van Driel, J.W. Bikker, M Tijink, A Di Bucchianico, Software Reliability for Agile Testing, Mathematics 2020, 8(5), 791; https://doi.org/10.3390/math8050791
27. J.D. Hooker, W. Schaaf, J. Achten, W. Vinckx, L. Derhaeg, Development of Advanced Gas-Cooled 'LED Filament' Lamps, Proceedings LS14 conference, 2014
28. H.M. Le Huy, V. Bellenger, J. Verdu, Thermal oxidation of anhydride cured epoxies. 1--mechanistic aspects. Polymer Degradation and Stability **35**, 77–86 (1992)
29. H.M. Le Huy, V. Bellenger, M. Paris, J. Verdu, Thermal oxidation of anhydride cured epoxies II Depth distribution of oxidation products. Polymer Degradation and Stability **35**, 171–179 (1992)
30. X. Buch, M.E.R. Sha, Thermal and thermo-oxidative ageing of an epoxy adhesive, Polymer Degradation and Stability. **68**, 403–411 (2000)
31. M. Yazdan Mehr, W.D. van Driel, G. Q. (Kouchi) Zhang, Progress in Understanding Color Maintenance in Solid-State Lighting Systems. Engineering **1**(2), 170–178, ISSN 2095-8099 (2015). https://doi.org/10.15302/J-ENG-2015035
32. J.L. Davis, Color shift in LEDs and SSL luminaires, presentation at the 2014 DOE Solid-State Lighting Manufacturing R&D Workshop, San Diego, CA, May 8, 2014
33. J.L. Davis, Solid-state lighting luminaire reliability, presentation at Delft University, Delft, the Netherlands, April 10, 2014
34. J.L. Davis, K. Mills, M. Lamvik, R. Yaga, S.D. Shepherd, J. Bittle, N. Baldasaro, E. Solano, G. Bobashev, C. Johnson, A. Evans, *System reliability for LED-based products, in Proceedings of the 2014 15th International Conference on Thermal, Mechanical, and Multi-physics Simulation and Experiments in Microelectronics and Microsystems (IEEE EuroSimE)* (Ghent, Belgium, 2014)
35. J.L. Davis, M. Lamvik, J. Bittle, S. Shepherd, R. Yaga, N. Baldasaro, E. Solano, and G. Bobashev, Insights into accelerated aging of SSL luminaires, Proceedings of SPIE: LED-based Illumination Systems. 8835 (2013) 88350L-1–88350L-10
36. D.L. MacAdam, *Color Measurement, Theme and Variations*, 2nd edn. (Springer-Verlag, New York, 1985)
37. Y. Ohno, Color quality, In: W.D. Van Driel, X.J. Fan and G.Q. Zhang (editors), Solid State Lighting Reliability: Components to System Part II. 06/2017; ISBN 978-3-319-58174-3 Springer New York
38. L.U. Guangjun, W.D. van Driel, X. Fan, J. Fan, G.Q. Zhang, LED based Luminaire Color Shift Acceleration and Prediction, in *Solid State Lighting Reliability: Components to System Part II*. 06/2017; ISBN 978-3-319-58174-3, ed. by W. D. Van Driel, X. J. Fan, G. Q. Zhang, (Springer, New York)
39. IES – TM-35, Projecting Long-Term Chromaticity Coordinate Shift of LED Packages, Arrays, and Modules, 2019
40. JOHN R. HOWELL, A catalog of radiation heat transfer configuration factors, available at www.thermalradiation.net, (last visited on 25/11/2020)
41. M.S. Ibrahim, J. Fan, W.K.C. Yung, A. Prisacaru, W. Van, X. Fan, G. Zhang, Machine Learning and Digital Twin Driven Diagnostics and Prognostics of Light-Emitting Diodes. Laser & Photonics Reviews, 2000254 (2020). https://doi.org/10.1002/lpor.202000254

Chapter 8
Degradation and Failures of Polymers Used in Light-Emitting Diodes

M. Yazdan Mehr, W. D. van Driel, and G. Q. Zhang

1 Introduction

Solid-state lighting (SSL) technology is one of the novel technological accomplishments in the last decades. Light-emitting diode (LED), based on SSL technology, is rapidly increasing its share in the general lighting market [1]. LEDs are used widely in indoor lighting and also are more widely used in outdoor and industrial applications [1, 2]. Higher light output, smaller size, longer lifetime, eco-friendly, and more color controllability are the LED advantages compared to more traditional light sources (fluorescent lamps and tubes). In addition, LEDs have the lowest energy consumption per lumen produced among all available lighting sources [3, 4].

An LED lighting system is a complex system that consists of optical, electrical, or thermal functionality components at different levels. These components have dependence on the roles they have in the system. Around 30 failure modes have been reported in SSL devices, and these failures can occur at any system level [5]. These failure modes are unique to LED-based lighting products and are not comparable with the failure modes of other lighting devices.

Most commonly failures and most dramatic ones are limited to the LED chip, the LED package, and its solder interconnects [6, 7]. It is required for microelectronic

M. Y. Mehr (✉)
EEMCS Faculty, Delft University of Technology, Delft, The Netherlands
e-mail: m.yazdanmehr@tudelft.nl

W. D. van Driel
EEMCS Faculty, Delft University of Technology, Delft, The Netherlands

Signify, HTC48-1A, Eindhoven, The Netherlands

G. Q. Zhang
EEMCS Faculty, Delft University of Technology, Delft, The Netherlands

State Key Laboratory of Solid State Lighting, Changzhou, China

W. D. van Driel, M. Yazdan Mehr (eds.), *Reliability of Organic Compounds in Microelectronics and Optoelectronics*,
https://doi.org/10.1007/978-3-030-81576-9_8

components to protect against the aggressive system and environmental stresses. Protection is usually done by covering chips and other microelectronic components with a package that usually includes polymeric encapsulants to prevent the LED chip from being in direct contact with said environmental stresses. So, it is not unexpected that many of LED failures are related to the package and encapsulation. With that said, the best selection for packaging material is of crucial importance. The organic material used in LED applications should be stable enough to resist the heat and high photon fluxes that occur in LEDs.

The main purpose of this chapter is studying the degradation and aging mechanisms of different optical materials in solid-state lighting systems. In this chapter, first currently used optical polymers [8–15] are introduced. Then common optical degradation mechanisms that happened in the LED polymer lens and encapsulants are explained.

2 Optical Materials Used in LED-Based Systems

Basic knowledge of the structure of optical materials used in LED packages and lighting device is necessary to exploring different failure mechanisms in LEDs. Combination of thermal, optical, electrical, and mechanical properties are expected for organic materials used in LED devices. In order to have a reliable light source, lots of complexities and the technical challenge should be solved, and to solve material-related challenges optimization of materials are required. Silicone, epoxy, Bisphenol A-Poly carbonate (BPA-PC), and Poly Methyl Methacrylate (PMMA) are more common optical materials which are used widely in LEDs and LED-based products which will be briefly introduced. In next section (Sect. 3) the main failure mechanism of each organic materials will be explained thoroughly. Phosphor as a non-organic material has an essential role in the LED and LED-Based product to produce white light, and it also degrade during service life, but in this chapter, we do not explain the degradation of the phosphor.

2.1 Epoxy

Widely used polymer in electronic is epoxy, different factors make epoxy resins unique between all the thermoset resins [16–18], These factors can be listed as, Low fabrication pressure, low cure shrinkage and hence low residual stress in the cured product, use of a wide range of temperatures which leads to good control over the degree of crosslinking, low price, mechanical stabilities, and moisture resistance [19–22]. Because of these unique characteristics and useful properties of the network polymers, epoxy resins are widely used [23, 24] in different applications such as structural adhesives, surface coatings, engineering composites, and electrical laminates. Epoxy materials have the most significant share in the LED packaging market [24]. Epoxies are commonly used as die-attach and underfill adhesives and are used in many printed circuit boards (PCBs).

Epoxy resins are known to have two significant weaknesses as LED encapsulants. These issues are brittleness of this resin and photo-degradation of epoxy under radiation at high temperatures. Huge chain scission takes place at said condition, causing in a discoloration. The main consequence of discoloration is reduction of transparency and a light output degradation.

2.2 Polycarbonate (BPA-PC)

Another important class of polymers which are extensively used in microelectronic is polycarbonates. The first polycarbonate, a cross-linked material, was discovered in 1898, but the first linear thermoplastic polycarbonate was not made until 1953 and brought into commercial production in 1960 [25]. The polycarbonates are sturdy, engineering materials that will withstand a wide range of temperatures.

Bisphenol-A polycarbonate (BPA-PC) is one of the most common polymers which is used in LED and LED-based products as a secondary optics, due to its high elastic modulus, high fracture resistance, and excellent thermal stability. Much information about the structure, properties, and thermal/optical degradation mechanisms of polycarbonates is reported [26–36]. The main disadvantages of polycarbonate are yellowing and discoloration under exposure to radiation at high temperatures. Discoloration leads to decreasing light output. Polycarbonate degrades by an oxidation process that is strongly dependent depends on the composition of the polycarbonates.

2.3 Silicone

Silicone is a unique type of polymer, because of a combination of organic groups (methyl, vinyl, …) with an inorganic backbone (Si—O) which leads to the semi-organic structure. Fig. 8.1 illustrates the schematic of silicone , where R can be methyl (CH_3), phenyl (C_6H_5), vinyl ($CH{=}CH_2$), or trifl uoropropyl ($C_3H_4F_3$).

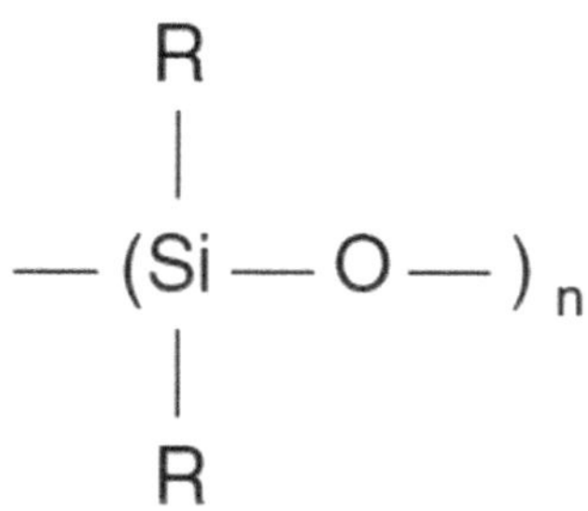

Fig. 8.1 Chemical structure of a general silicone

Silicone has unique properties such as higher purity, stronger stress absorbing, and more biocompatibility than other polymers. Also, silicone maintains its excellent electrical properties at high temperatures and under humid environments [37].

On the other hand, lower glass-transition temperature (T_g), a larger coefficient of thermal expansion (CTE), and poor adhesion to the housing listed as disadvantages of silicone. Siloxane-modified LED transparent encapsulant is usually used in order to improve the thermal and mechanical properties of silicone

2.4 Poly Methyl Methacrylate (PMMA)

Another commonly used thermoplastic polymer in LED lens, which is the cheapest options in the market and packaging is polymethylmethacrylate (PMMA) [38–40]. Processing and manufacturing of PMMA are straightforward than other polymers because of the glass transition temperature of PMMA, Tg-between 105 to 110 °C, which leads to the extensive use of PMMA.

Main advantage of PMMA is an excellent light/UV transmittance [41–45]. On the other hand, PMMA discolor under high water absorption, low heat resistance, and low abrasion resistance which is the disadvantage for PMMA [40–45].

3 Degradation of Optical Materials

Among different failures in LED system, package-related failure mechanisms result in an optical degradation, color change, and severe discoloration of the encapsulant/lens Materials.

Chromaticity shift in the output light is one of the most common failures in LEDs [12–15]. It is considered as one of the most critical measures of quality and is often studied and reported [13–15]. LEDs have to work in different conditions such as high temperatures and humidity environments, moisture, ionic contaminants, heat, radiation, and mechanical stresses. Working in such harsh conditions can be highly detrimental to LEDs and LED-based products and may lead to device failures. Optical/packaging materials are expected to have some characteristics to withstand these harsh environments, to increase the light extraction efficiency, minimize the heat generated, conduct more heat out of the package, and resist heat and UV light. In this chapter, attention will be paid to essential failure mechanisms, with all having significant influences on the packaging failures. The focus will be on yellowing, lens cracking, delamination, and carbonization of polymeric materials used as encapsulants in LED packaging. Most observed and reported chromaticity shifts are due to either of these mechanisms or a combination of them, Table 8.1 summarizes mentioned failure mechanisms. Each mechanism will be explained in detail for some of the packaging materials in the next section.

Table 8.1 Summary of main failure mechanisms, which will have contribution to the color shift [12]

Failure mechanisms	Description
Discoloration	Oxidative degradation of the encapsulant/lens materials (BPA-PC, PMMA, silicone, and epoxy), resulting in yellowing • Oxidation degradation of reflective surfaces within the LED device
Cracking	Mechanical defects in the LED which change the average photon path through the phosphor
Delamination	Separation between different material interfaces, such as between the LED substrate and optical path materials
Carbonization	Over-absorption of blue light by the phosphor • Joule heating • Self-heating of phosphors • Radiation of lights with shorter wavelength

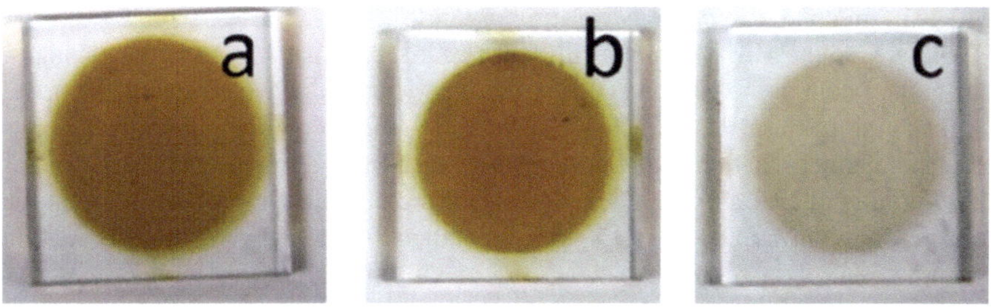

Fig. 8.2 Picture of the sandwiched samples of the silicone glues after UV-exposure under vacuum using the ESTEC SUV facility [40, Reproduced with permission from authors]

3.1 *Yellowing of Encapsulant/Lens*

Chemical reactions of the polymeric materials under continued exposure to wavelength emission (blue/UV radiation), and working at high-temperature environment leads to discoloration and yellowing of optoelectronic materials. Enhanced molecular mobility, using additives such as flame retardant, scattering agent, and stabilizers are the main chemical reasons for discolorations in the opto-electronic materials [46]. In this section, different illustrations of the discoloration of different optoelectronic materials used in LED and LED-based products will be studied.

Photo degradation is influenced by the exposure time and the intensity of radiation. Besides, thermal effects, related to the junction temperature of the LED chip self-heating of the package [12], also play a significant role in polymer discoloration in LED-based products [12].

One of the examples of silicone yellowing is shown in Fig. 8.2, it is evident that the discoloration upon UV irradiation seems to be independent from the molecular structure of the silicones.

Another reason of discoloration is a failure because of the chemicals and contaminations which leads to the browning of polymeric materials. Figure 8.3 shows the degradation effect from a volatile organic compound (VOC) that permeated the silicone encapsulant and covered the LED chip [41].

Another reason for discoloration of the optoelectronics materials is humidity. Moisture and humidity have an accelerating effect on the rate of yellowing in epoxy molding compound (EMC) used in some LED packages [12]. An example of the effect of the humidity on the yellowing of EMC is shown in Fig. 8.4. The sample in the wet high temperature (95°C) operation lifetime (WHTOL 95°C/95%RH) showed slightly more yellowing the one which is only thermally aged at 105°C (HTOL 105 °C).

Hydrothermal degradation of silicone is together with the absorption of water, leads to the increasing the weight of the EMC epoxies.

Diffusion of the water in opto-electronics materials follows a Fickian law and the degradation reaction is controlled by hydrolysis, a model with a perfect agreement with experimental observations was proposed [48].

BPA-PC and PMMA are typically used as a secondary lens in LED packages, due to their comparatively lower costs. Yazdan Mehr et al. have extensively worked on the aging of the BPA-PC [46–48]. Thermal aging at low-temperature regime is associated with oxidation and rearrangement reactions, both having implications for discoloration of LEDs. A significant consequence of yellowing is a decrease in the transmissivity in LED lenses [50–55]. Whether the rearrangement or oxidation reactions control the discoloration depends on the temperature and oxygen contents.

Lu et al. [51, 52] examined the influence of temperature- and blue light radiation-induced degradation of BPA-PC and PMMA on their chromaticity shifts of LED devices. Figure 8.5 illustrates the differences between the yellowing of PC and PMMA samples [52].

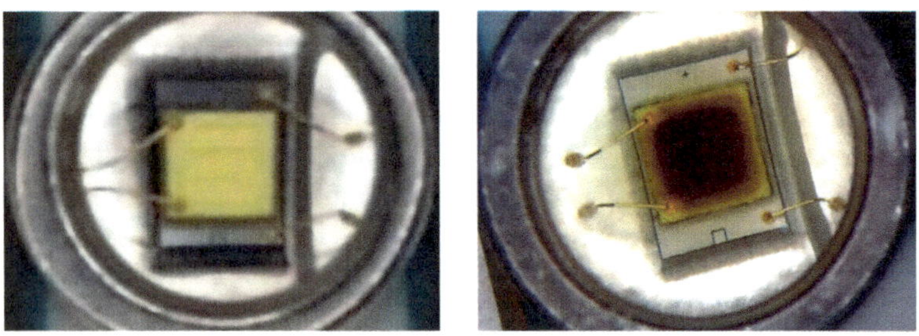

Fig. 8.3 Influence of thermal aging on the discoloration of LED optical package [47, Reproduced with permission from authors]

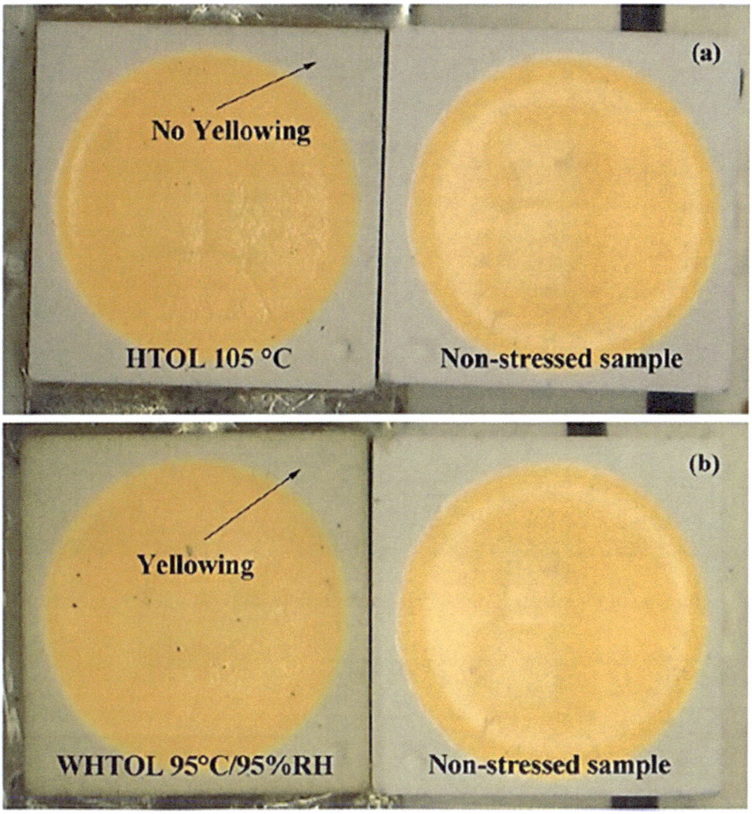

Fig. 8.4 Influence of humidity of the yellowing of EMC [10, reproduced with permission from authors]

3.2 *Crack and Delamination*

Delamination at the chip/phosphor interface is another crucial failure mechanism in high-powered LEDs, which leads to discoloration and chromaticity shifts in the yellow direction. Likewise, differences between the coefficients of thermal expansion (CTE) as well as temperature cycles during production and service are another commonly observed failure modes in LEDs debonding at the soldered areas. Figure 8.6 reported the severe delamination at the interface of lead frames, and epoxy molding compound (EMC) used to encapsulate the lead frame [23].

Delamination at the polymer/lead frame interface is shown in Fig. 8.7. Absorbed moisture accelerated the influence of oxygen and hydrogen ions [12].

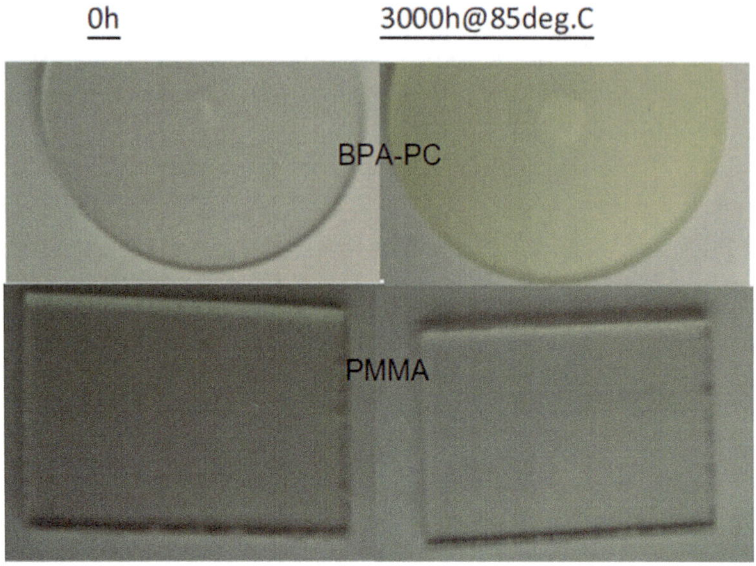

Fig. 8.5 Yellowing of PC versus PMMA [52, Reproduced with permission from authors]

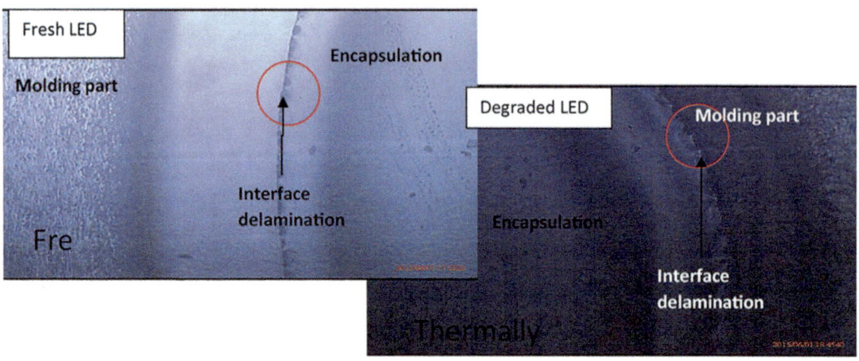

Fig. 8.6 Optical microscope images of interface delamination in white high-power LEDs [23, Reproduced with permission from Elsevier]

3.3 Carbonization

Another important reason for the discoloration and failure of the optical materials in LED and Led-based products is carbonization [12]. Silicone can undergo severe carbonization by applying adequate electrical current to the LED package [12].

Electrical overstress which leads to the formation of a conductive path across the LED is because of carbonization of the polymer material on the chip surface. Reduction of the insulation resistivity is also due to the carbonization of the polymeric materials used as encapsulants in LED-based products [12]. The

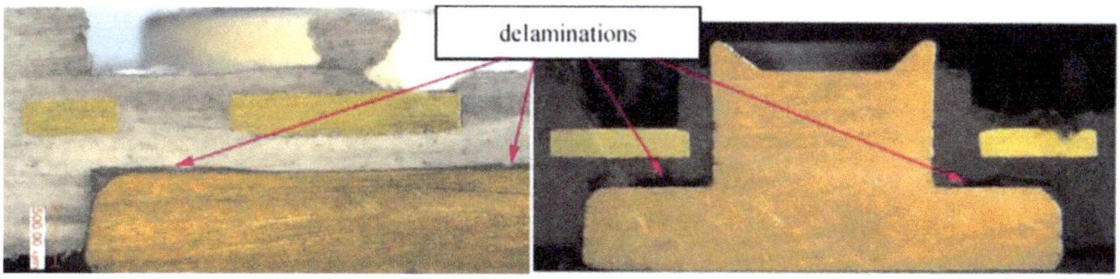

Fig. 8.7 Delaminations at the polymer packages of Luxeon [12, Reproduced with permission from authors]

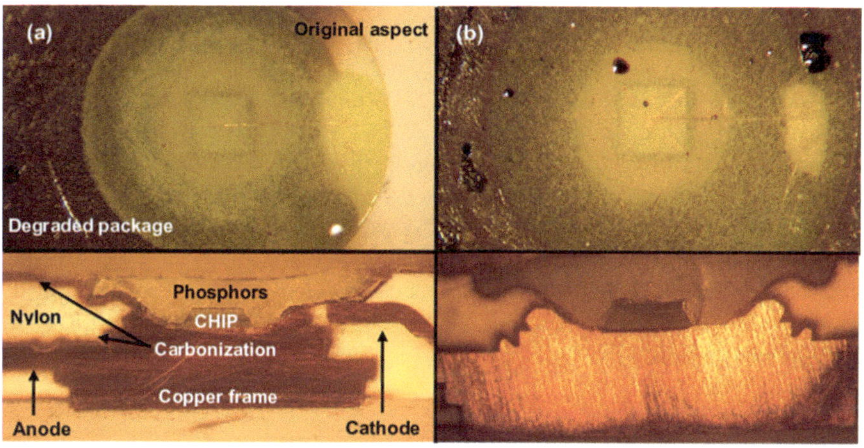

Fig. 8.8 Carbonization in LED package, (**a**) aged at 400mA and (**b**) aged at 220°C [55, Reproduced with permission from authors]

carbonization of the encapsulant leads to light output degradation and color shifting. Figure 8.8 shows the carbonization of the silicone materials in the LED package. Silicone encapsulant in the device is aged under temperature and current.

4 Color Shifting

The main result of above-explained failure mechanisms is the reduction of light output and color shifting of the light.

One of the most critical measures of quality of light is chromaticity shift [56–58]. Chromaticity shift is explained as the variation of chromaticity coordinate from its original values.

Color shift is defined as the change in the chromaticity coordinates from its original values. It is named chromaticity shift, in case CIE1976 (u´,v´) chromaticity diagram is used. The (u´,v´) coordinates are related to the CIE 1931 (x, y) coordinates. The chromaticity coordinates of a source provide a numerical representation of the color of the light.

One of the examples of color shifting is shown for the Philips Lumileds Luxeon Rebel LED package in Fig. 8.9 [58]. Color shifting in these products happens in 3 stages which is due to the elevated temperature. It is illustrated that there is initially a fast shift in the chromaticity at all temperatures, followed by a plateau in which the shift is almost negligible. The third stage is accompanied by the decrease in $\Delta u'v'$. It is shown that by increasing the temperature plateau stage is shorter.

Degradation of optical components in LED packages is one of the most significant reasons for the chromaticity shift of LED devices, which means that optical components in LED packages and LED-based products have an essential role in the optical performance of LED devices by controlling the emission of light from the device.

The chromaticity shift can be determined by dominant degradation mechanisms occurring in the package, which depends on the package materials and methods of construction. Directions of chromaticity shift can be different from case to case, as shown in Fig. 8.10, and can be toward the yellow, blue, green, or red colors. Different package systems have shown distinct differences in the chromaticity shift signatures. Table 8.2 illustrates different commonly seen chromaticity shift mechanisms that can happen in LED lighting systems which are related to both organic materials (polymers) and inorganic materials(phosphors) which are widely used in LED and LED-based products.

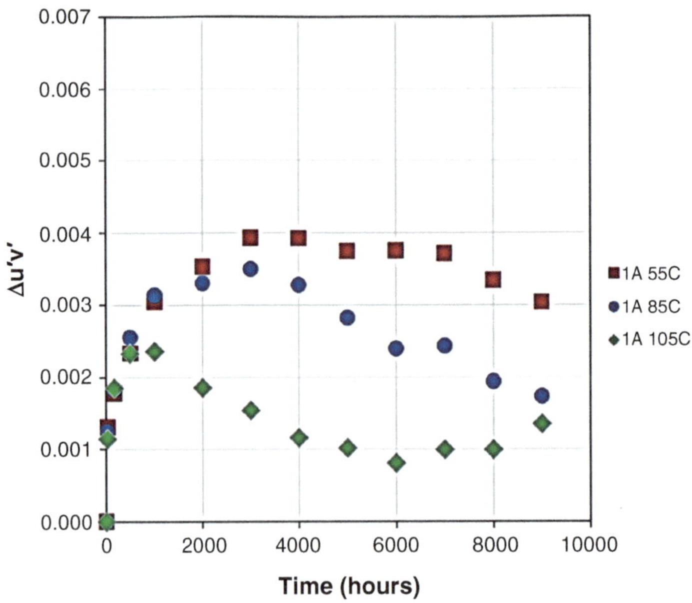

Fig. 8.9 Effects of temperature on the chromaticity shift ($\Delta u'v'$) of Luxeon LEDs operated at 1 amp [59, Reproduced with permission]

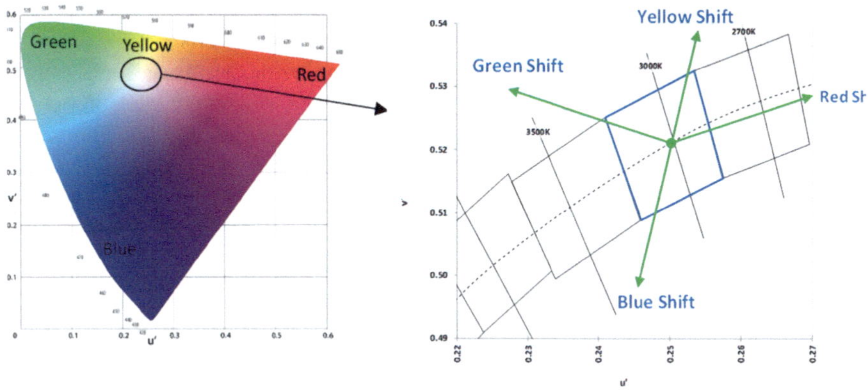

Fig. 8.10 Illustration of color shifts in the CIE 1976 color space. Any change in the spectral proportion of any colors results in a chromaticity shift [14, Reproduced with permission]

Table 8.2 Summary of main chromaticity shift mechanisms that can occur in LED lamps and luminaires [3]

Shift direction	Reason
Blue shift	– Oxidation of the molding compound in PLCC (plastic leaded chip carrier) or tarnishing of exposed metal surfaces such as lead frames and reflectors – Top-to-bottom fractures of the binder in the phosphor-binder layer, resulting in blue photons bypassing the phosphor layer
Yellow shift	– Increase in phosphor quantum efficiency due to chemical changes or temperature decreases – Cracking or delamination of phosphor-binder layer, which can lengthen the path of blue photons through the phosphor – Discoloration/oxidation of the lenses – Discoloration of the reflector
Green shift	– Chemical changes in the phosphor such as oxidation that shift emission intensity to lower wavelengths – Reduction in red emissions such as those from a red LED
Red shift	– Shift in emission properties of direct red emitter

5 Harsh Environments

Recently, Yazdan Mehr et al. studied the degradation of silicone used in secondary optical designs combined with an LED package under harsh environment conditions [58]. The effects of UV 360 nm exposure or blue 450 nm light exposure, combined with high temperature (100 °C), water environment (35 and 100 °C), and saline 100 °C water environment on the color, and radiant power intensities of silicone samples were studied. They have shown that silicone samples have a great degree of stability against light exposures (both UV and blue light). It is shown that 3500 h of radiation at 100 °C did not result in any aging in the optical properties of silicone samples. Major adverse effects of harsh environment on silicone lens were exposure to saline and high-temperature environmental conditions.

Significant reduction in the transmission of samples in the visible light range, reduction in the maximum radiant power at 450 nm, and a change in the color chromaticity values are the main degradation of silicone samples under saline water exposure at 100 °C. Saline water exposure at 100 °C changed the also mechanical properties of silicone plates by making them more brittle.

Figure 8.12 shows the effects of exposure to a saline water on the FTIR spectra of silicone samples. Decrease in the transmission of degraded samples with exposure time in this case is quite significant.

This specifies that among various potential degrading factors in a harsh environment condition, Cl ions dissolved in water which shows the most negative contribution to the structural stability of silicone. Cl ions significantly accelerated the kinetics of degradation in silicone samples. Observing at the FTIR spectra, it can be concluded that chemical degradation was associated with breaking Si–O–Si bonds and decomposition of methyl groups attached to silicone atoms. Several papers also reported parallel results concerning hydrothermal stability of cross-linked silicone rubber [28–31] (Fig. 8.11).

A shear punch test was also conducted on as received and aged specimens with 3500 h of exposure to saline environment to study changes in the mechanical properties of the degraded specimens in a saline environment. Figure 8.13 compares force–displacement curves of these two samples, illustrating that exposure to a saline water environment was accompanied with embrittlement and a decrease in the ductility of specimens.

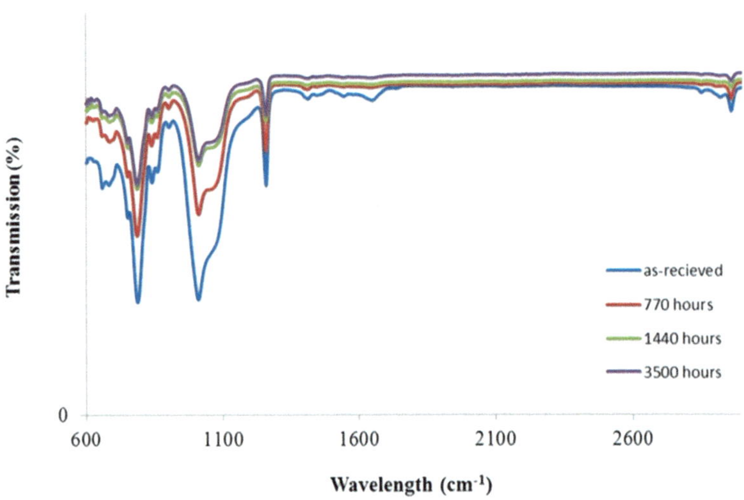

Fig. 8.11 FTIR-ATR spectra of aged Si in a salt batch at 100 °C

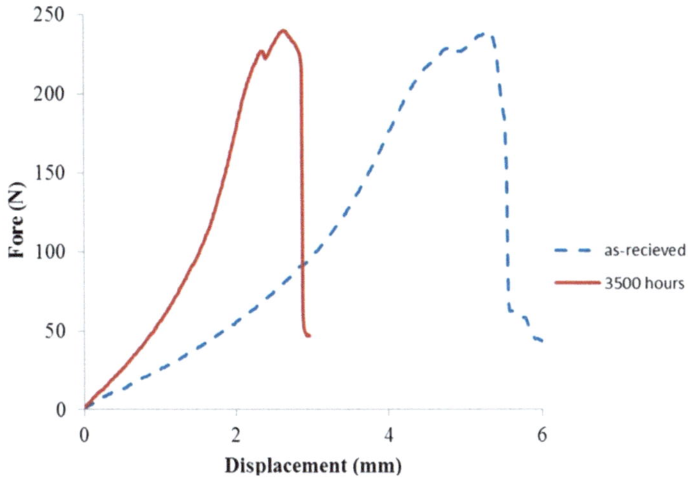

Fig. 8.12 Force displacement curve for aged silicone in a salt bath at 100 °C

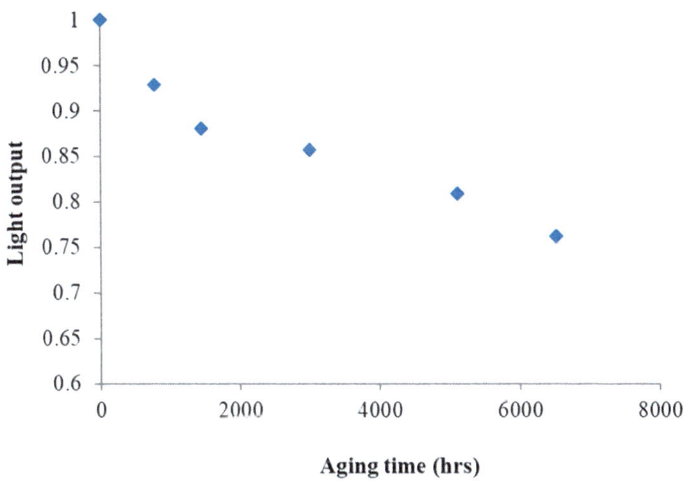

Fig. 8.13 Variation of maximum power intensity at 450 nm with exposure time under a saline water environment condition at 100 °C

Figure 8.13 illustrates the effects of aging time on the normalized maximum light output at 450 nm. Interestingly, there was a linear reduction of maximum light output with exposure time.

6 Reliability and High Accelerated Testing (HAST)

Degradation and failure of optical materials in LEDs are due to the operating under diverse stresses, including temperature, UV/Visible light, ionic contaminations, and humidity. Reliability of the LED-based products under different conditions in LED-based products are explained thoroughly in the literature [12, 35, 57]. In the high accelerated set up (HAST) [12, 35] the effects of both blue light intensity and the thermal stress on the lifetime and the kinetics of aging of phosphor plates used in LED-based products are well studied. Applying both thermal stress and light intensity in HAST set-up increases the rate of degradation. Three temperatures of 80, 100, and 120 °C were used, blue light is radiated on the sample with different light intensities of 825, 3300, and 13200 W/m2 and samples are aged up to 3000 h.

LED lifetime is measured by lumen maintenance, which is how the intensity of emitted light tends to weaken over time. Reliability models predict the lifetime of the LED under different stress.

In the HAST experiments, where the light intensity is also used as an extra acceleration factor, the Eyring relationship is an appropriate reliability model, as below [49]:

$$R = \gamma_0 (I)^n \exp(\frac{-E_a}{KT}) \tag{8.1}$$

where R is the reaction rate, $\gamma 0$ is the pre-exponential factor, I is the intensity of blue light, n is the constant factor, Ea is the activation energy (eV) of the degradation reaction, K is the Boltzmann gas constant (eV/K), and T is the absolute temperature (K) [57]. The radiation of light obviously accelerates the aging of remote phosphor by a factor 1.01, 1.2, and 1.9 for blue light radiations of 825, 3300, and 13200 W/m2, respectively [57]. The increase in acceleration factor by radiation of light indicates the effect of light intensity. It is also shown that the lifetime, defined as 30% lumen depreciation at 40 °C, is around 35 khrs, for the lowest energy power, which has almost the same lifetime as thermally aged phosphor. The lifetime of the phosphors with higher power energy is 25 khrs.

References

1. Holonyak Nick Jr, Bevaqua SF, Coherent light emission from GaAs(1-x)Px junctions, Applied Physics Letter 1962;1; 82-83
2. S. Nakamura, S. Pearton, G. Fasol, *The blue laser diode: The complete history* (Springer, 2000) chapter 2
3. US DOE, Adoption of Light-Emitting Diodes in Common Lighting Applications, prepared for U.S. Department of Energy, prepared by Navigant Consulting, Inc., July 2017.
4. Pattison, M., Bardsley, N., Hansen, M., Pattison, L., Schober, S., Stober, K., et al. (2017, September). Solid state lighting 2017 suggested research topics: Technology and market

context. Brodrick, J. (Ed.) Report prepared for the U.S. Department of Energy. https://energy.gov/sites/prod/files/2017/09/f37/ssl_supplement_suggested-topics_sep2017_0.pdf
5. W.D. Van Driel, X.J. Fan, *Solid state lighting reliability: Components to Systems* (Springer, 2012) ISBN 978-1-4614-3066-7; chapter 1
6. LED life for general lighting: definition of life, vol 1. Technical report, ASSIST recommendation, February 2005. 30, 138
7. M. Meneghini, L.R. Trevisanello, G. Meneghesso, Recent results on the degradation of white LEDs for lighting. J. Phys. D: Appl. Phys **43**, 354007 (2010) (11pp)
8. M. Yazdan Mehr, W.D. van Driel, H. Udono, G.Q. Zhang, Surface aspects of discolouration in Bisphenol A Polycarbonate (BPA-PC), used as lens in LED-based products. Opt. Mater **37**, 155–159 (2014)
9. W.D. van Driel, X.J. Fan, G.Q. Zhang, *Solid state lighting reliability: Components to Systems part II* (Springer, 2017) ISBN 978-3-319-58175-0 (Chapter 14)
10. W.D. van Driel, X.J. Fan, G.Q. Zhang, *Solid state lighting reliability: Components to Systems part II* (Springer, 2017) ISBN 978-3-319-58175-0 (Chapter 10)
11. Poplawski ME, Ledbetter M, Smith MA (2012). L-Prize: Stress testing of the Philips 60W replacement lamp entry. Available at http://www.lighting-prize.org/pdfs/lprize_60w-stress-testing.pdf
12. M. Yazdan Mehr, A. Bahrami, W.D. van Driel, X.J. Fan, J.L. Davis, G.Q. Zhang, Degradation of Optical Materials in Solid State Lighting Systems. International Materials Reviews **65** (2020)
13. S. Rosenfeld, Lighting art with LEDs at the Smithsonian American Art Museum (brief summary), in *2011 DOE Solid-State Lighting R&D Workshop*, (San Diego, CA, 2014) http://energy.gov/eere/ssl/transformationslighting-eighth-annual-solid-state-lighting-rd-workshop
14. U.S. Department of Energy, Solid State Lighting Technology Fact Sheet: LED Colour Stability, PNNL-SA-101828 (http://apps1.eere.energy.gov/buildings/publications/pdfs/ssl/colour-shift_factsheet.pdf)
15. M. Royer, R. Tuttle, S. Rosenfeld, N. Miller, *Colour Maintenance of LEDs in Laboratory and Field Applications, PNNL-22759* (Pacific Northwest National Laboratory, Richland, WA, 2013) http://apps1.eere.energy.gov/buildings/publications/pdfs/ssl/2013_gateway_colour-maintenance.pdf
16. W.G. Potter, *Epoxide Resins* (Springer, New York, NY, USA, 1970)
17. C.A. May, G.Y. Tanka, *Epoxy Resin Chemistry and Technology* (Marcel Dekker, New York, NY, USA, 1973)
18. Epoxy Resin Chemistry, Ed., R.S. Bauer, ACS Symposium Series No.114, American Chemical Society, Washington, DC, USA, 1979.
19. P. Jain, V. Choudhary, I.K. Varma, Effect of structure on thermal behaviour of epoxy resins. Europ Polym Journal **39**, 181–187 (2003)
20. X. Wang, Y. Hu, L. Song, W. Xing, H. Lu, P. Lv, G. Jie, Flame retardancy and thermal degradation mechanism of epoxy resin composites based on a DOPO substituted organophosphorus oligomer. Polymer **11**, 2435–2445 (2010)
21. A. Gu, G. Liang, Thermal degradation behaviour and kinetic analysis of epoxy/montmorillonite nanocomposites. Polymer Degradation and Stability **80**, 383–391 (2003)
22. S.J. Park, F.L. Jae, R. Lee, Thermal and mechanical properties of tetrafunctional epoxy resin toughened with epoxidized soybean oil. Mater Sci and Eng A **374**, 109–114 (2004)
23. Singh P, Tan C.M, Chang L.-B., Early Degradation of High Power Packaged LEDs Under Humid Conditions and its Recovery — Myth of Reliability Rejuvenation, 2015
24. W.D. Van Driel, X.J. Fan, *Solid state lighting reliability: Components to Systems* (Springer, 2012) ISBN 978-1-4614-3066-7; chapter 2
25. G. Strobel, *The physics of polymer, concept of understanding of their behavior* (Springer, 2006) ISBN 978-3-540-25278-8
26. A. Factor, Mechanism of Thermal and photo degradations of Bisphenol A Polycarbonate. Advance in Chemistry series **249**, 59–76 (1996)

27. I.B. Rufus, H. Shah, C.E. Hoyle, Identification of fluorescent products produced by the thermal treatment of bisphenol-A-based polycarbonate. J. of Applied Polymer Science **51**, 1549–1558 (1994)
28. D.T. Clark, H.S. Munro, Surface and bulk aspects of the natural and artificial photo-ageing of Bisphenol A polycarbonate as revealed by ESCA and difference UV spectroscopy. Polymer Degradation Stability **8**, 195–211 (1984)
29. G. Montaudo, C. Puglisi, Thermal-Decomposition Processes in Bisphenol-A Polycarbonate. Polymer Degradation Stability **37**, 91–96 (1992)
30. A. Davis, J.H. Golden, Thermal rearrangement of diphenyl carbonate. J. Chem. Soc. (B) **1**, 45–47 (1968)
31. I.C. Mcneill, A. Rincon, Degradation studies of some polyesters and polycarbonates – A. Bisphenol A polycarbonate. Polymer Degradation Stability **31**, 163–180 (1991)
32. L.H. Lee, Mechanism of Thermal Degradation of Phenolic Condensation Polymers. I. Studies on the Thermal Stability of Polycarbonate. Polym Sci A2 **5**, 1981–1989 (1964)
33. M. Yazdan Mehr, W.D. van Driel, K.M.B. Jansen, P. Deeben, G.Q. Zhang, Lifetime assessment of Bisphenol-A Polycarbonate (BPA-PC) plastic lens, used in LED-based products. Microelectronics Reliability **54**, 138–142 (2014)
34. M. Yazdan Mehr, W.D. van Driel, K.M.B. Jansen, P. Deeben, M. Boutelje, G.Q. Zhang, Photodegradation of bisphenol A polycarbonate under blue light radiation and its effect on optical properties. Optical Materials **35**, 504–508 (2013)
35. M. Diepens, P. Gijasman, Photodegradation of BPA. Polymer Degradation Stability **92**, 397–406 (2007)
36. Dow Corning Corporation, *Silicone chemistry overview* (Dow Corning Corporation, Midland, Michigan, 1997) https://www.dowcorning.com/content/publishedlit/01-4027-01.pdf
37. A.A. Miller, E.J. Lawton, J.S. Balwit, Effect of Chemical Structure of Vinyl Polymers on Crosslinking and Degradation by Radiation. J Polym Sci **14**, 503–504 (1954)
38. R.B. Fox, L.G. Isaacs, S. Stokes, Electron Spin Resonance Studies of Photodegradation in Poly(Methylmethacrylate). J Polymer Science A1 **1079** (1963)
39. M. Dole, *Radiation chemistry of macromolecules*, vol II (Academic Press, NY, London, 1973), p. 97
40. J.A. Moore, J.O. Choi, Radiation effects on polymers, in *American Chemical Society*, ed. by R. L. Clough, S. W. Shalaby, (1991), p. 158
41. B. Ranby, J.F. Rabek, *Photodegradation, photooxidation and photostabilization of polymers, principles and applications* (John Wiley & Sons, 1975), p. 153
42. W.B. Hardy, in *Developments in polymer photochemistry*, ed. by N. S. Allen, 3rd edn., (Applied Science Publication, London/New Jersey, 1980), p. 322
43. P. Schneider, C. Steitz, K.H. Schafer, C. Ziegler, Hot embossing of pyramidal micro-structures in PMMA for cell culture. Phys. Status Solidi A **206**, 501–507 (2009)
44. G. Palm, R.B. Dupaix, J. Castro, Large strain mechanical behavior of poly(methyl methacrylate) (PMMA) near the glass transition temperature. J. Eng. Mater. Technol. **128**, 559–563 (2006)
45. D.G. Legrand, J.T. Bendler, *Handbook of polycarbonate science and technology* (New York (NY), Marcel Dekker, Inc, 2000)
46. https://www.led-professional.com/resources-1/articles/led-and-lens-degradation-through-volatile-organic-compounds-by-cree
47. M. Yazdan Mehr, W.D. van Driel, K.M.B. Jansen, P. Deeben, G.Q. Zhang, Lifetime assessment of Bisphenol-A Polycarbonate (BPA-PC) plastic lens, used in LED-based products. Microelectronics Reliability **54**, 138–142 (2014)
48. W.D. van Driel, X.J. Fan, G.Q. Zhang, *Solid state lighting reliability: Components to Systems part II* (Springer, 2017) ISBN 978-3-319-58175-0 (Chapter 5)
49. M. Yazdan Mehr, W.D. van Driel, G.Q. Zhang, Reliability and Lifetime Prediction of Remote Phosphor Plates in Solid-State Lighting Applications Using Accelerated Degradation Testing. J. elect materi **45**, 444–452 (2015)

50. G. Lu, M. Yazdan Mehr, W.D. van Driel, X. Fan, J. Fan, G.Q. Zhang, Colour Shift Investigations for LED Secondary Optical Designs: Comparison between BPA-PC and PMMA. Optical Mater **45**, 37–41 (2015)
51. G. Lu, W.D. van Driel, X. Fan, M. Yazdan Mehr, J. Fan, K.M.B. Jansen, G.Q. Zhang, Colour shift and mechanism investigation on the PMMA diffuser used in LED-based luminaires. Optical Materials **54**, 282–287 (2016)
52. J. Huang, S. Dušan, K.S. Golubović, D. Yang, X. Li, X. Fan, G.Q. Zhang, Chapter 14 Degradation Mechanisms of Mid-power White-light LEDs. IEEE Transactions on Device and Materials Reliability **15**(2), 1 (2015)
53. P.J. Scully, S. Caulder, R. Bartlett, UV-laser photo-induced refractive index changes in poly methyl methacrylate and plastic optical fibres for application as sensors and devices. The Central Laser Facility Annual Report 1999–2000, 145–147 (2000)
54. Trevisanello L.R, Meneghini M, Mura G, Sanna C, Buso S, Spiazzi G, Vanzi M, Meneghesso G, Zanoni E, Thermal stability analysis of High Brightness LED during high temperature and electrical aging, SPIE 2007; 6669; 666913-[56] Davis JL, Mills KC, Johnson C, Yaga R, Bobashev G, Baldasaro N, (2017, May 31). System reliability model for solid-state lighting (SSL) luminaires. Contract DE-EE0005124. Final report prepared for the U.S. Department of Energy. Available at https://www.osti.gov/servlets/purl/1360770
55. M. Yazdan Mehr, W.D. van Driel, G.Q. Zhang, Reliability and Lifetime Prediction of Remote Phosphor Plates in Solid-State Lighting Applications Using Accelerated Degradation Testing. J. elect materi **45**, 444–452 (2015)
56. Davis JL, Young J, Royer M, U.S. Department of Energy CALiPER Report 20.5: Chromaticity Shift Modes of LED PAR 38 Lamps Operated in Steady-State Conditions, February 2016. Available: http://energy.gov/sites/prod/files/2016/03/f30/caliper_20-5_par38.pdf
57. Philips Lumileds (2012) IESNA LM-80 Test Report for Luxeon Products, Design Resource DR05-1
58. M. Yazdan Mehr, W.D. van Driel, F.D. Buyl, G.Q. Zhang, Study on the Degradation of Optical Silicone Exposed to Harsh Environments. Materials **11**, 1305 (2018)

Chapter 9
Degradation and Reliability of Organic Materials in Organic Light-Emitting Diodes (OLEDs)

Kelley Rountree, J. Lynn Davis, Karmann C. Riter, Jean Kim, Michelle McCombs, and Roger Pope

1 Introduction

Organic light-emitting diode (OLED) technologies have found significant use in mobile devices and have now become the dominant display technology for flagship smartphone and consumer products such as televisions and monitors [1]. OLEDs are attractive for these applications because the technology provides an excellent color gamut with relatively high luminance within a narrow viewing cone, properties that are derived from the light emission characteristics of the organic molecules used as light emitters. The increase in prominence of OLEDs in mobile and consumer products is coming at the expense of other white light-emitting technologies such as fluorescent lamps and light-emitting diode (LED) sources. Both fluorescent lamps and LEDs have historically served as backlights for active matrix displays that use white light combined with color filters to provide subtractive colors, but their use is in relative decline as OLEDs expand their foothold in mobile and consumer products [2, 3]. From a reliability standpoint, the expected longevity of mobile devices is modest, typically five years or less before the product becomes obsolete. This typical life expectancy serves to limit expectations for the long-term performance of an OLED display. Most of the reliability issues associated with OLED displays in mobile applications are tied to the failure of the thin-film transistors (TFTs, which are used to switch each pixel) or breakage of the display in mobile products [4].

The use of OLEDs as light sources for general illumination is still an emerging application that is not as well developed as OLED use in mobile and consumer products. In this application, a white OLED, typically made by combining red (R), green (G), and blue (B) emitters, is used to provide illumination for a task, a table,

K. Rountree · J. L. Davis (✉) · K. C. Riter · J. Kim · M. McCombs · R. Pope
RTI International, Research Triangle Park, Durham, NC, USA
e-mail: ldavis@rti.org

W. D. van Driel, M. Yazdan Mehr (eds.), *Reliability of Organic Compounds in Microelectronics and Optoelectronics*,
https://doi.org/10.1007/978-3-030-81576-9_9

or even a room. White OLED illumination is attractive because white OLEDs serve as a low luminance and diffuse lighting source with minimal glare and because they can be placed near the illuminated area. This type of light may be beneficial in some indoor lighting applications because of the potential for improved optical delivery efficiency [5].

There are some similarities between the use of OLEDs in mobile and consumer products and lighting applications. Some similarities include the basic structure of the OLED cell and the materials and manufacturing used to make the OLEDs. However, there are also many significant differences between OLEDs used in the two applications. For instance, OLEDs used in lighting applications do not require TFTs for switching, thereby greatly simplifying the manufacturing process. In addition, the illuminance requirements, operational power densities, and expected lifetime are significantly higher in a lighting application than in a mobile device.

OLEDs used in lighting applications must illuminate an area that is often one meter or more from the source. In achieving this level of illumination, the OLED can be expected to be operated at full power for up to 12 hours per day for 10 years or more. Therefore, it is not unusual to expect an OLED lighting product to operate at high current densities for 30,000 hours or more. In contrast, OLEDs used in mobile applications are often held within an arm's length of the human eye, thus requiring much lower luminance levels and lower operational power densities. For example, OLEDs used in mobile phone applications are only operated when the device is viewed, resulting in significantly lower usage time because the device is often in a standby mode. As a result, OLEDs in consumer products rarely experience 15,000 hours of use before being discarded. Therefore, from a reliability perspective, the use of OLEDs in an indoor lighting application is likely a greater test of reliability of the OLED materials and basic (i.e., non-TFT) construction methods than their use in mobile and consumer products.

Some of the current challenges of using OLED technologies in lighting applications are analogous to those experienced by LEDs during the early stages of technology development as LEDs progressed from display applications to consumer products to general lighting. Among these challenges are improving the luminous efficacy to compete with existing lighting technologies and meeting the expectations of long-term performance and reliability established by current lighting solutions [5]. However, today, LED lighting is an established technology with a significant market presence and is a major competitor of OLEDs. In short, LEDs have now set the standard for efficiency and reliability of a lighting product that OLEDs must strive to meet. There are potential benefits to OLED lighting that may not be easily addressed with inorganic LED technologies, thus making both of them potentially complementary lighting technologies.

This chapter will focus on the long-term reliability of OLEDs used in indoor lighting applications because this application is often a more stringent test of reliability than mobile and consumer product applications. In addition, the reliability of OLEDs is compared with inorganic LED technologies to provide a benchmark of OLED technologies relative to the current state-of-the-art lighting technology. To facilitate this comparison, accelerated stress testing and common parametric reliability properties such as luminous flux maintenance and chromaticity maintenance are used to measure the long-term stability of OLED products.

1.1 Emission Properties of OLEDs

OLEDs produce light through the radiation released by the combination of an electron-hole pair (i.e., exciton annihilation) on an organic molecule contained in an emitter layer of an OLED cell. The wavelength of the light emission is determined by the electronic structure of the emitting organic molecule, which is shown schematically in the Jablonski diagram shown as Fig. 9.1. Because both the electron and hole in an exciton pair have spin, these spins can either be paired to produce a singlet state (^1S) or unpaired to produce a triplet state (^3T). There is also a ground singlet state ($^1S_0$), a first excited singlet state ($^1S_1$), and a first triplet excited state ($^3T_1$) of the emitting molecule. There can also be higher order singlet and triplet excited states of the molecule (e.g., S_2, T_2); these states are omitted from Fig. 9.1 for brevity. The spins on the electron-hole pairs are random and, based on random probability, the spins are three times more likely to be unpaired than paired. As a result, 75% of the excitons on an organic molecule are in the ^3T state, and 25% are in the ^1S state [6].

Excitons in the singlet excited state can relax to the ground state (S_o) through either a fluorescence or a radiationless process known as internal conversion, during which the excited state dissipates energy through molecular vibrations instead of photon emission. The internal conversion process reduces the quantum yield of the

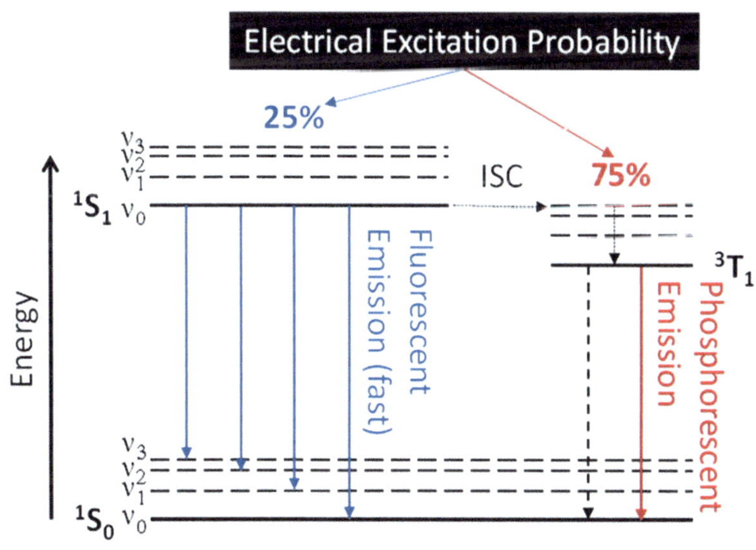

Fig. 9.1 Jablonski diagram of the electronic structure of an organic light-emitting molecule

state. In addition, singlet excitons can convert to the triplet state through another radiationless process known as intersystem crossing (ISC). The resulting triplet excited states can lose energy either through phosphorescence or internal conversion.

When a molecularly bound electron-hole pair recombines, energy is released either as a photon or through internal conversion. Photon emission occurs through the processes of fluorescence or phosphorescence. During fluorescence, the compound emits a photon when it relaxes from a singlet excited state to a singlet ground state (e.g., $S_1 \rightarrow S_0$). In contrast, during phosphorescence, the compound is in a triplet excited state and emits a photon to relax to the singlet ground state (e.g., $T_1 \rightarrow S_0$). Because a singlet-to-singlet transition is allowed by the selection rule of classical quantum mechanics, light emission in a fluorescence process is typically fast (on the order of nanoseconds). In contrast, light emission from a triplet state to a singlet state is forbidden by the selection rule of classical quantum mechanics; therefore, this process typically takes much longer (on the order of milliseconds to minutes). Unfortunately, only 25% of the excitons in organic light-emitting molecules are in a singlet state and can emit a photon by the allowed fluorescence transition. For organic molecules in which fluorescence is the only pathway to light emission, the triplet-to-singlet transition is forbidden, and energy relaxation from the triplet state typically occurs through an internal conversion process in which energy is dissipated through vibrational modes of the molecular lattice [7]. For such organic molecules in which singlet-to-singlet transition is the only pathway for light emission, the resulting internal quantum efficiency, defined as the number of photons emitted divided by the number of excitons created, is low because only 25% of the excitons are in the singlet state.

Although the quantum efficiency of organic molecules producing light through a fluorescence pathway is canonically limited to 25%, this quantum efficiency could be significantly increased if the phosphorescence states were also light-emitting on a short (<1 millisecond) timescale. Activating the light-emitting pathways of the triplet state can be achieved by capitalizing on the strong spin-orbit coupling of some heavy metal dopants (e.g., platinum, iridium) with the emitting organic molecule. Using this approach, much higher quantum efficiencies have been reported in OLEDs, and values approaching 100% have been reported in specially engineered devices because of contributions from the fluorescent and phosphorescent pathways [8, 9]. Such devices are called phosphorescent OLEDs (PhOLEDs). Although PhOLED devices exhibit higher efficiencies than fluorescent OLEDs, the lifetime of PhOLED emitters is generally shorter than that of fluorescent OLEDs, especially at short wavelengths (e.g., blue) [10]. As a result, commercial OLED devices often use PhOLED emitters for green and red light to capitalize on their higher efficiency, and they use fluorescent emitters for blue light to capitalize on its higher reliability [11].

In addition to the selection rule-based behavior previously described, each singlet and triplet state has an associated vibrational fine structure arising from the rotational, stretching, and bending modes of the emitting molecule. This vibrational fine structure, which is labeled as n_0 (ground state), n_1, ν_2, and ν_3 and is shown as dashed lines in Fig. 9.1, provides excited vibrational quantum levels that are displaced from the ground state by a small amount of energy. Although light emission most likely occurs from the lowest excited state (e.g., $^1S_1$, $^3T_1$) to the ground vibrational state (ν_0 of $^1S_0$), there is a probability that the emission will occur to one of the higher vibrational states (e.g., ν_1, ν_2, ν_3) of the ground state (i.e., $^1S_0$). This probability decreases as the energy of the vibrational state increases. This behavior imparts a vibrational fine structure on the electronic transitions that results in a broad emission profile from the organic molecule. Emission to the ground vibrational structure state is usually the most intense, followed by the higher order vibrational states. For a single-color emitting molecule, this behavior creates a broad emission profile of decreasing intensity, as shown in Fig. 9.2, instead of the relatively sharp emission lines found in inorganic LEDs.

1.2 OLED Device Structure

OLED device structures are formed by depositing successive layers of organic thin films on a substrate made from glass, plastic, or metal. The organic films range in thickness from 10 nanometers (nm) to 200 nm, and the deposition is accomplished by using one of several methods, including the evaporation of small molecules and various solution-based methods (e.g., printing) for polymers. An OLED stack includes multiple levels to support the creation and transport of holes and electrons to the active emission sites in the emissive layer (EML). These layers include a hole injection layer (HIL), a hole transport layer (HTL), a hole blocking layer (HBL), an electron injection layer (EIL), an electron transport layer (ETL), and an electron blocking layer (EBL). An anode (typically made from a transparent conductor such as indium tin oxide [ITO]) and a cathode (typically made from a low work function, reflective metal such as lithium, silver, or aluminum) are used as the source for holes and electrons, respectively. A list of the materials used in the different OLED layers is available [12]. Collectively, this structure creates a single electroluminescent (EL) cell, and an illustration of this structure is shown in Fig. 9.3.

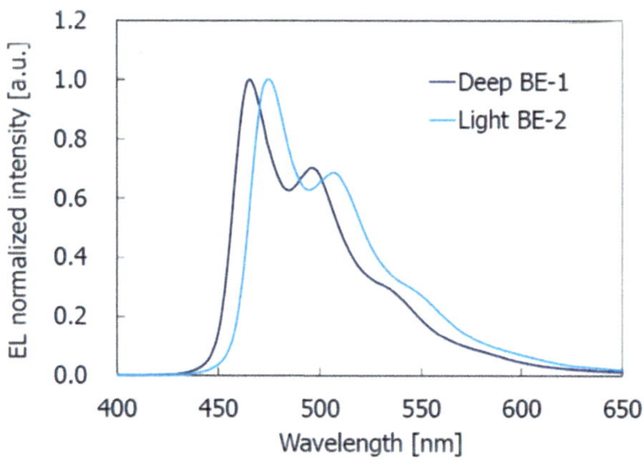

Fig. 9.2 Emission spectrum of two different OLED emitters [11]

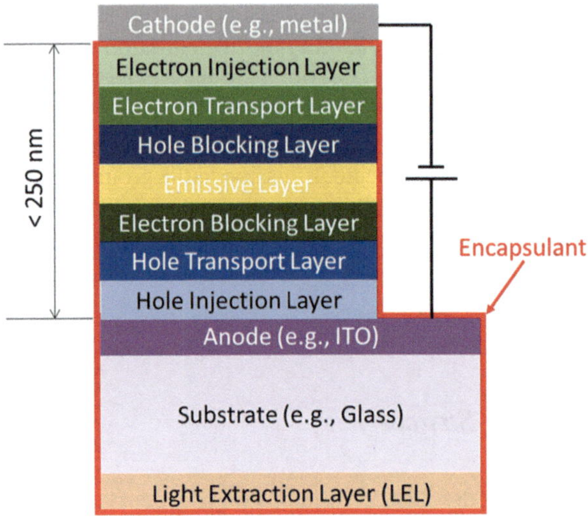

Fig. 9.3 Structure of an OLED device with a single EL unit containing one EML

During operation of the OLED shown in Fig. 9.3, electrons are supplied by the cathode. The EIL receives electrons from the cathode and injects them deeper into the device, and the ETL supports the transport of electrons to the EML. An HBL layer is often inserted between the ETL and EML layers to prevent the escape of holes from the EML, which may produce a non-emissive recombination of electrons and holes, thereby lowering device efficiency. In parallel, holes are supplied by the anode and travel through the HIL, HTL, and EBL driven by a repulsive electric potential at the anode. The electric potential required to operate this cell is typically on the order of 3–4 volts for a single OLED cell, depending on structure. The electrons and holes form excitons on the organic molecule in the EML, and the recombination of electrons and holes can produce photons as previously described. This structure produces broad light emissions at a color that is determined by the electronic structure of the EML, and multiple EMLs of different electronic structures are needed to make different colors of light (e.g., red, green, blue).

The lifetime of OLEDs is known to be sensitive to the current density. One approach to increasing the luminous flux of an OLED while simultaneously decreasing the current density is to increase the number of photons produced for each injected electron, which effectively increases the luminous flux produced by an OLED panel without increasing the current density. The single EL cell shown in Fig. 9.3 produces at most one photon for each injected electron. However, stacking multiple single EL cells, as shown in Fig. 9.4, allows a single electron to produce multiple photons. To build this structure, a charge generation layer (CGL) is needed to serve as an interconnect between the two EL units. The CGL generates both electrons and holes, has high optical transparency in the visible spectrum, and provides low series resistance [13]. This cell structure, which is called the tandem stack cell, consists of up to six separate EMLs that are connected in series within the cell

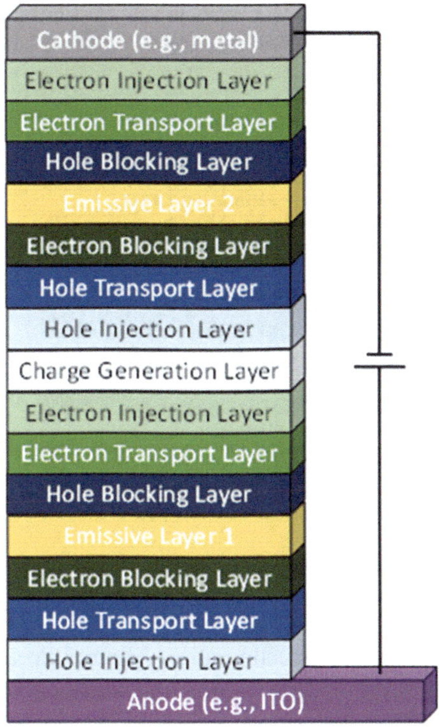

Fig. 9.4 An OLED device with the 2-cell tandem stack structure. The substrate and LEL are omitted for convenience

[14–16]. Examples of warm white and neutral white 6-cell tandem stack structures are shown in Fig. 9.5. Smaller tandem stack OLED cells are described in other publications [14, 15]. The HIL, HTL, HBL, EIL, ETL, and EBL are omitted from Fig. 9.5 for brevity, and only the EMLs and the CGLs are shown. In commercial tandem stack devices, the R and G organic emitters are often combined to form a broad emitter that appears to be amber in color because each is usually a PhOLED [14]. However, the blue EML is usually kept separate because this material is typically a fluorescent OLED due to lifetime concerns of blue emitters [16]. In a warm white tandem OLED, the 3-cell stacked tandem structure will usually have two combined amber (i.e., R + G) PhOLED layers and one blue emitter layer sandwiched between. As shown in Fig. 9.5, the warm white 6-cell tandem stack structure contains four amber PhOLED layers and two blue fluorescent layers, whereas the neutral white 6-cell tandem stack contains three amber PhOLED layers and three blue fluorescent layers.

The number of photons produced per electron injected into the device is proportional to the number of tandem EML layers. The use of the tandem stack OLED structure improves the electrical efficiency of the device because the photon yield is higher at the same current density. The most important benefit of the tandem stack structure is that lumen depreciation at a given illuminance (which depends on current density) is slower, thus the operational lifetime of the device increases. However,

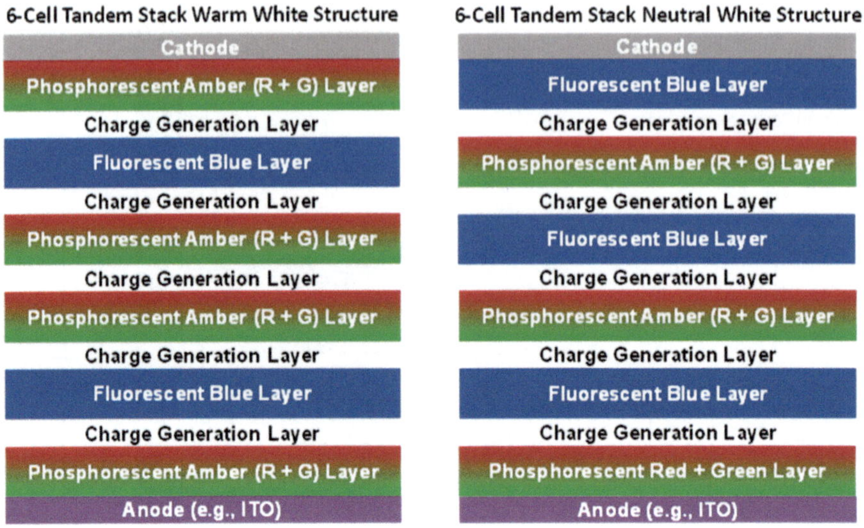

Fig. 9.5 Examples of the 6-cell tandem stack warm white (left) and neutral white (right) OLED structures

operating this cell requires a higher electrical potential than a single cell. In addition, the tandem cell structure increases device complexity and manufacturing costs. This chapter focuses on OLED devices with the 6-cell tandem stack structure because it generally produces the highest luminous flux at a given current density. Reliability information about 2-cell and 3-cell tandem stack structures is available in other publications [17–20].

1.3 Reliability Research for OLED Lighting Products

Overstress testing and accelerated life testing have been recognized as appropriate methods to identify failure modes and to study the reliability of lighting devices [21]. Although there have been over-stress testing methods developed for inorganic LEDs, the use of over-stress testing for OLEDs is less common. Therefore, there are no widely accepted accelerated stress tests (ASTs) for OLEDs. This fact can create confusion among lighting designers and skew the relative performance, both good and bad, of OLEDs compared with other lighting technologies. Standard testing procedures for OLED lighting products will ultimately need to comply with the existing testing infrastructure for LED lighting, which relies on environmental stresses of temperature, vibration, and power cycling. Existing accounts of overstress testing of OLEDs points to the use of mildly elevated ambient temperatures (e.g., 35 degrees Celsius [°C] to 65 °C) as being appropriate for ASTs [17–20, 22, 23].

In addition, the U.S. Department of Energy (DOE) has supported the development of OLED technologies through a series of technical reports, demonstration projects, and market analyses [24–27]. Although a general conclusion from DOE-sponsored studies of OLED lighting technology is that the cosine emission profile of OLED devices is beneficial for indoor lighting applications, OLED technologies have experienced issues with efficacy, reliability, driver performance, and initial costs, which are all potential market impediments requiring additional research.

2 Experimental and Analytic Methods

2.1 Devices Under Test

This chapter focuses on the latest generation of 6-cell tandem stack OLEDs because this device structure appears to provide a pathway to the high luminous efficacy required from an OLED used in an indoor lighting application. Findings for the lifetime performance of individual OLED panels at one of two different correlated color temperature (CCT) values, 3000 K (warm white) and 4000 K (neutral white), are reported here. Each warm white 6-cell tandem stack OLED panel consisted of four amber (i.e., R + G) PhOLED layers and two blue fluorescent emitter layers arranged as shown on the left side of Fig. 9.5. In contrast, the neutral white OLED panels contained three amber PhOLED layers and three blue fluorescent emitter layers, as shown on the right side of Fig. 9.5. These panels were tested individually and were not a part of a luminaire.

During all tests, the OLED panels were mounted on an aluminum base (for heat dissipation) and connected to a driver and a power supply to convert the alternating current (ac) mains to direct current (dc) power. The power supply converts the ac power into an intermediate dc voltage that was fed to the driver. Then, the driver converts the intermediate dc voltage to the final dc voltage used to power the OLED panel. In the tested configuration, each panel had an associated driver mounted on the aluminum base (Fig. 9.6). Each driver is associated with a specific OLED panel, but the external power supplies connected to the drivers were used interchangeably between panels during testing.

2.2 Stress Testing Methods

This chapter presents findings from room temperature operational life (RTOL) and ASTs (Table 9.1) involving temperature bakes at mildly elevated temperatures of 35 °C (35OL) and 45 °C (45OL). One temperature and humidity AST protocol, conducted at 65 °C and 90% relative humidity (6590), was also used during this study. Either a temperature oven or a temperature and humidity environmental

Fig. 9.6 Test configuration for the 6-cell tandem stack OLED devices examined during this study [20]

Table 9.1 Test methods used during this study

Test name	Test description
RTOL	Continuous operation at room temperature (nominally 25 °C) in ambient humidity
35OL	Continuous operation at a constant ambient temperature of 35 °C and ambient humidity
45OL	Continuous operation at a constant ambient temperature of 45 °C and ambient humidity
6590	Continuous operation at a constant temperature of 65 °C and relative humidity of 90%

chamber was used throughout. Humidity was not explicitly controlled when a temperature oven was used, and the ambient humidity was determined by the air handling system of the building in which this study was conducted. Because of these factors, the ambient humidity varied between 40% and 80%, depending on the time of year, for the RTOL test. The sample size at all conditions for the 6-cell tandem stack panels was three unless otherwise noted.

For all tests, the OLED devices were continually powered at the maximum output set by the manufacturer, and no attempts were made to power cycle the devices during testing. Operating at the maximum output should provide an indication of the luminous flux maintenance (LFM) of the devices. All OLED devices were operated with the drivers provided by the manufacturer, and these drivers were operated at their expected maximum output conditions (the power conditions were within the OLED panel manufacturer specifications) with no dimming signals applied to the product.

3 Measurement Methods

3.1 Luminous Flux

The spectral power distributions (SPDs) were taken in a calibrated 65-inch integrating sphere with each sample mounted in the center of the sphere (4π geometry), and data acquisition software was used to calculate quantities, such as luminous flux and chromaticity, from each SPD. Regular calibrations of the integrating sphere were performed by using a calibrated spectral flux standard that was traceable to standards from the National Institute of Standards and Technology (NIST). Background corrections were applied prior to calibration. Self-absorption corrections were made for all samples by using an auxiliary lamp mounted inside the sphere, which is in accordance with procedures in LM-79-19 [28].

3.2 Electrical Properties

Changes in the OLEDs' electrical properties (e.g., power consumption, power factor) were measured during photometric testing by using a plug monitor on the ac mains. A two-channel power analyzer was used when more accurate electrical measurements of voltage, current, power, power factor, and total harmonic distortion were required. Impedance (Z) measurements of the OLED panels were made at every time interval by using a hand-held inductance, capacitance, and resistance (LCR) meter. Phase angle (ψ) and impedance were measured at frequencies of 100 hertz (Hz); 1000 Hz; and 10,000 Hz.

4 Data Analysis Methods

4.1 Luminous Flux Maintenance

The Illuminating Engineering Society (IES) technical memorandum (TM)-28–14 is the established method for modeling and projecting the long-term LFM of LED lamps and luminaires [29]. This method uses data acquired according to the guidelines of LM-84 [30]. TM-28-14 provides a method for the projection of LFM for an LED lamp or a luminaire to future times by using data acquired at equal time intervals from three or more samples. The maximum projection time is determined by the sample size. If only three samples are used during testing, then projections of LFM cannot exceed more than three times the test duration (in hours). Currently, TM-28-14 is only approved for use with LED lamps and luminaires. The functional intent of OLED panels is similar to lamps and luminaires, so the TM-28-14 method may be extended to OLEDs given the caveat that its use in this way is not an approved standard at this time.

One procedure that has been widely used by researchers to model the degradation in luminous flux emitted by OLEDs over time is the double-exponential model [22], which consists of the two components in Eq. 1 as follows:

$$\Phi(t) = ae^{-\alpha_1 t} + be^{-\alpha_2 t} \tag{1}$$

In Eq. 1, a and b are constants determined by the initial conditions, α_1 is the decay rate of rapid initial processes, α_2 is the decay rate of long-term processes, and t is time. If the OLED test duration is sufficiently long, then the first exponential component of the double exponential model can be ignored, and the equation reduces to that used in TM-28-14. Consequently, measurement times of 6000 hours or longer may be preferred so that the OLED degradation model can be simplified to a single-exponential function.

Another approach to modeling the degradation of OLED emitters is to use the stretched exponential function [31] as presented in Eq. 2 as follows:

$$\Phi(t) = e^{\left(-\frac{t}{\tau_0}\right)^{\gamma}} \tag{2}$$

In Eq. 2, τ_0 is the characteristic time required for performance to degrade to 63% (i.e., 1–1/e) of the initial value, and γ is a parameter that characterizes the degradation rate. To use this model, the experimental time must be long enough for a reasonable value of τ_0 to be obtained, which places a significant burden on test duration especially if the LFM is high. As improvements in the LFM of OLED technologies are made, this type of model will be difficult to apply without extending test durations.

The parameters of the double exponential and stretched exponential models are difficult to compare. Because there are no standard methods currently accepted for modeling luminous flux in OLEDs and because the double exponential model reduces to a single exponential model at test durations commonly used in the lighting industry, the decision was made to modify the double-exponential model for OLEDs to make it align with the methods used in TM-28-14.[1] The TM-28-14 standard does not use data collected during early operation of the device, which can be equivalent to dropping the short-term component (i.e., $ae^{-\alpha_1 t}$) of Eq. 1 and using a single-exponential decay model to describe long-term behavior. In a single-exponential decay model such as the one used in TM-28-14, the ratio of the luminous flux at time t, $\Phi(t)$, to the initial luminous flux, Φ_0, can be expressed as shown in Eq. 3 as follows:

[1] Note: TM-28-14 is not currently approved for use with OLED data; therefore, the approach used here is referred as a "modified TM-28-14 method."

$$\Phi(t)/\Phi_0 = Be^{-\alpha t} \tag{3}$$

In Eq. 3, B is the pre-exponential factor, and α is the decay rate constant. Because TM-28-14 also uses Eq. 3 to model LFM and to project values at future times, comparisons of the α values of data derived from these measurements provide some relative measures of the OLED light source performance to that of inorganic LED technologies. Additional modifications in the TM-28-14 test method that were used in the Data Analysis Methods section presented herein have been discussed in another publication [20].

4.2 Photometric Analysis of 6-Cell Tandem Stack OLED Panels

The initial SPDs of the 6-cell tandem stack neutral white (i.e., 4000 K) and warm white (i.e., 3000 K) panels are presented in Fig. 9.7. The neutral white and warm white OLED panels have similar peak locations of their respective major emitters, but there are subtle differences in the peak shape of the emitter that produces light in the green-yellow region. This finding suggests that similar organic emitter

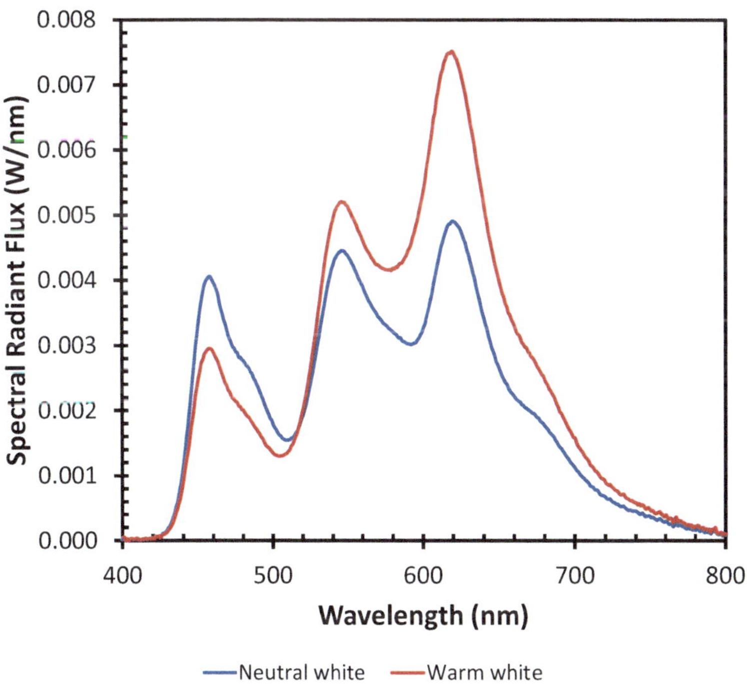

Fig. 9.7 Initial SPDs of the 6-cell tandem stack neutral white and warm white OLED panels [20]

chemistries are used in both warm white and neutral white devices under test (DUTs), but the relative concentration or layer thickness of the emitters is different.

4.3 Emission Spectra Deconvolution

Because the light emissions produced from the organic emitters in an OLED are broad, spectral deconvolution analysis is needed to determine individual emitters and to unravel how these emitters change over time. Because it is known that the stability of the organic emitters used in OLED devices are different, this information would be helpful in determining the factors that impact LFM and chromaticity maintenance. Unfortunately, the chemical identity of the emitters used in the tested OLED technologies is unknown because of the proprietary formulations used by the manufacturers. However, reasonable approximations of the spectrum and relative luminance of the organic emitters can be made by using minimal assumptions.

Given that the organic emitters are likely aromatic compounds, the emission spectra of the individual emitters contributing to the SPD of each white light OLED are anticipated to be asymmetric or skewed [32]. This structure likely reflects the vibrational fine structure shown in Fig. 9.1 and is further broadened by coupled roto-vibrational modes that occur in a small molecule and cause the emission lines to overlap. An empirical function [$f(s, A, \Delta p, p_o, p)$] to describe a skewed emission distribution (commonly called a skewed Gaussian) can be used following the method of Fraser and Suzuki [33] and is shown in Eq. 4 as follows.

$$f\left(s,A,\Delta p,p_0,p\right)=A*\exp\left(-\ln\left(2\right)\left(\ln\left(1+\frac{2s\left(p-p_0\right)}{\Delta p}\right)*\left(\frac{1}{s}\right)\right)^2\right) \tag{4}$$

In Eq. 4, A is the maximum radiant flux of the emitter, p_0 is the wavelength at which maximum emission occurs, and s is the asymmetry parameter. The s parameter is positive when the emission skews at wavelengths $p > p_0$ and is negative when the emission skews toward wavelengths $p < p_0$ (for an s near zero, the skewed distribution tends toward a symmetric Gaussian).

The linkage between Δp and the full width of the emission distribution at half-maximum radiant flux (w) is described in Eq. 5 as follows:

$$w=\Delta p\left(\frac{\sinh\left(s\right)}{s}\right) \tag{5}$$

For this analysis, the assumption was made that all peaks in the SPD of the tested OLED lighting products were produced by separate emitting compounds, and each peak was fit with a skewed Gaussian. Because of the broad spectrum of light produced by organic light-emitting compounds, the sum of two skewed Gaussian was

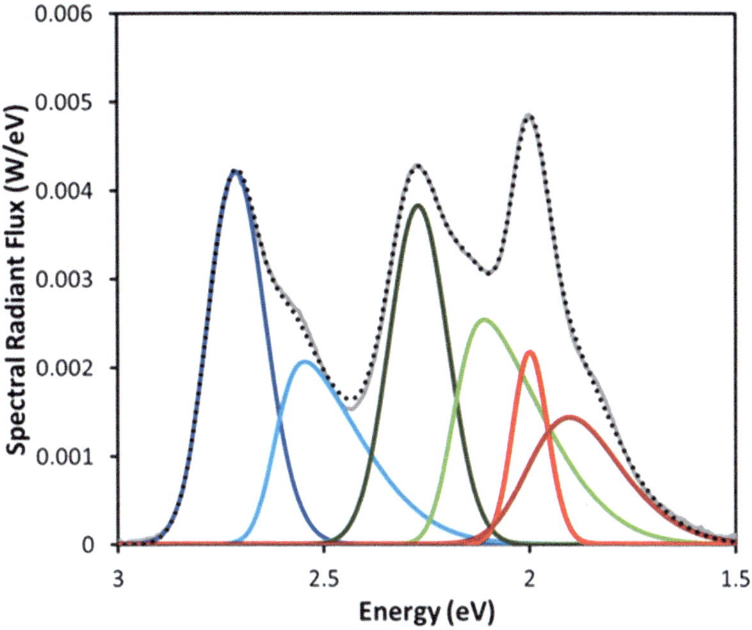

Fig. 9.8 Spectral emission deconvolution from a 6-cell tandem stack neutral white OLED panel with two skewed Gaussian used to model each organic emitter

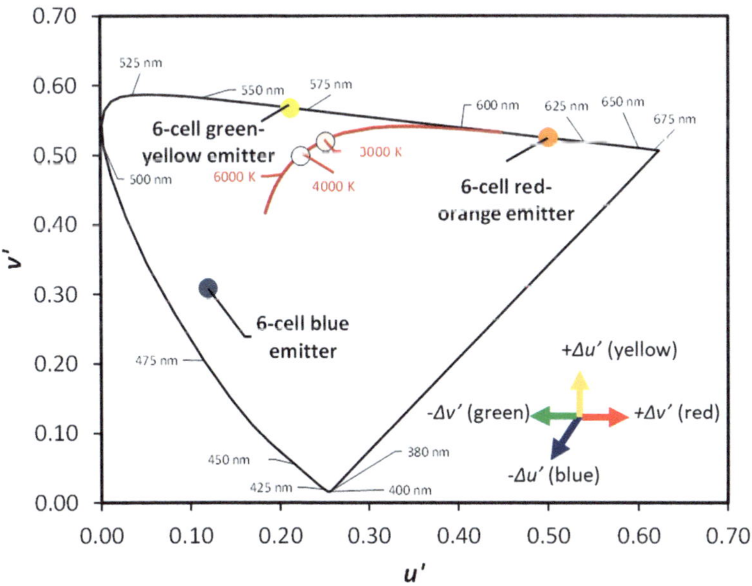

Fig. 9.9 Chromaticity points of the blue, green-yellow, and red-orange organic emitters in a 6-cell tandem stack OLED panel

often a better fit to the data for each emitting peak or region (e.g., blue, green-yellow, and red-orange emitters, as shown in Fig. 9.8). Although many factors play into the overall SPD produced by an OLED lighting device (e.g., organic emitters, dopants, diffusers, uniformity), the SPDs presented in this chapter were estimated as the sum of the individual emitters. In addition, the sum of squared errors was minimized through a non-linear regression analysis to complete the spectral deconvolution. Radiant power was estimated by using the trapezoid rule to approximate the definite integrals of the skewed Gaussian of the deconvoluted SPD. Radiant power of the individual organic emitters was subsequently calculated as the sum of the radiant power of their respective skewed Gaussian.

Deconvolution of the spectral emissions from an OLED panel allows a determination of the chromaticity points for the individual emitters by using the sum of their respective skewed Gaussians. For this analysis, the decision was made to combine the two curves for the blue emissions to obtain a single blue emitter chromaticity. A similar procedure was followed for the two phosphorescent red and green emitters. The chromaticities of the emitters for the 6-cell tandem stack panels calculated in this manner are shown in Fig. 9.9.

When starting from a typical white light chromaticity point on the blackbody locus (e.g., 3000 K; 4000 K), a shift in a generally blue direction involves moving toward the chromaticity point of the blue emitter used in the OLED. As shown in Fig. 9.9, this type of shift corresponds primarily to a reduction in the value of v' (i.e., $\Delta v' < 0$) with much less change in u'.[2] Likewise, from a typical white chromaticity point, a shift in the generally yellow direction involves moving toward the chromaticity point of the green-yellow emitter used in the OLEDs and corresponds to an increase in the value of v' (i.e., $\Delta v' > 0$) with much less change in u'. In a similar manner, when starting from a typical white chromaticity point, a shift in a generally red direction toward the chromaticity of the red-orange emitter used in the OLED increases the value of u' (i.e., $\Delta u' > 0$) with much less change in v'. A shift in a generally green direction corresponds to a decrease in the value of u' (i.e., $\Delta u' < 0$) with much less change in v'. Because chromaticity shifts can be thought of as changes in the relative contributions of the emitters, a shift in the generally green direction could be caused by a reduction in the intensity of the red-orange organic emitter, or combined changes in the blue, green-yellow, and red-orange emitters. These general changes in emitted light color are represented by the arrows in the lower right corner of Fig. 9.9.

[2] $\Delta u'$ represents the change in the u' chromaticity value from the initial (or reference) point to the value at the current time. Likewise, $\Delta v'$ represents the change in the v' chromaticity value from the initial (or reference) point to the current time.

Table 9.2 Initial performance of the population of warm white and neutral white OLEDs [18, 19]

Property	Warm White OLED	Neutral White OLED
Luminous flux	306 lumens	245 lumens
Luminance	8500 nits	7000 nits
Initial CCT	2954 K	3980 K
TM-30-20 color fidelity (R_f)[a]	85	84
TM-30-20 color gamut (R_g)[a]	98	95
Power consumption[b]	4.2 W	4.7 W
Luminous efficacy	73 lumens/W	52 lumens/W

[a]See reference [34] for the explanation of these metrics
[b]Operating the driver and power supply required an additional 1.8–2.0 W beyond the values shown here, which lowered the overall system efficiency

5 Results

5.1 *6-Cell Tandem Stack Panels*

The OLED panels used in this study were purchased directly from the OLED panel manufacturer in October 2018. At the time of purchase, these panels represented the state of the art for white light OLEDs. The panels were square in appearance (see Fig. 9.6) with the active area of the emitter surface being 105 centimeters squared (cm^2). Both warm white (nominal CCT of 3000 K) and neutral white (nominal CCT of 4000 K) panels were included in these tests. The different CCT values can be produced by using the same materials, but the relative proportions of blue fluorescent and amber (i.e., R + G) phosphorescent emitters are changed to achieve the different CCT values (see Fig. 9.5). The initial performance of the population of OLED panels included in this study is provided in Table 9.2.

5.2 *LFM of 6-Cell Tandem Stack OLED Panels*

Separate populations, consisting of three samples each, of warm white and neutral white 6-cell tandem stack OLED panels were tested in RTOL, 35OL, 45OL, and 6590. Each sample population in RTOL, 35OL, and 45OL was exposed to the AST conditions in increments of 500 hours up to 2000 hours of total exposure. After 2000 hours the exposure time duration was increased to 1000-hour increments for the remainder of the exposure test. For the 6590 sample population, the DUTs were exposed to the AST conditions in increments of 500 hours up to 4000 hours of total exposure, and the exposure duration was increased to 1000 hour increments thereafter. After each AST exposure increment, each sample population was removed from the test environment and its SPD measured in a calibrated integrating sphere. From the SPD, various performance parameters (e.g., luminous flux, LFM, chromaticity coordinates, CCT value) can be calculated by using standard methods.

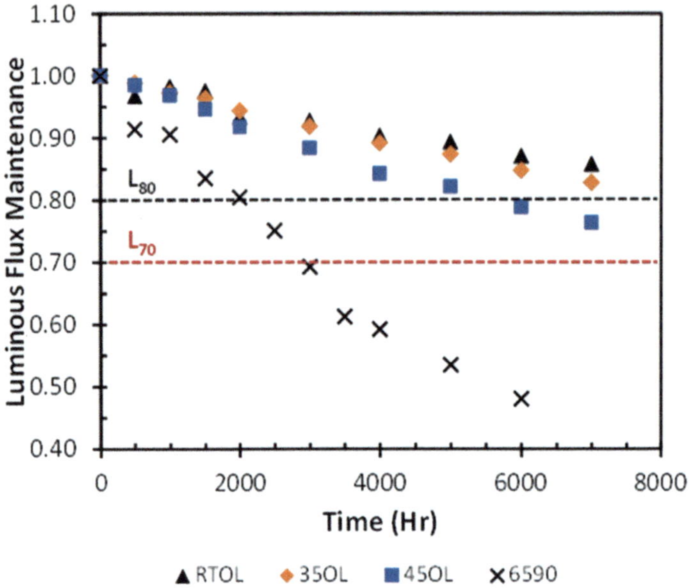

Fig. 9.10 Change in the LFM of the 6-cell tandem stack warm white OLED panels in different AST environments

The average changes in the luminous flux for the warm white OLED panels in the different AST environments are presented in Fig. 9.10. The luminous flux decreases at a steady rate in all AST environments, with the rate of decrease being higher at the higher test temperatures. The luminous flux remained higher than 0.80 (L_{80}) after 7000 hours of testing during RTOL and 35OL, and just dropped below 0.80 after 7000 hours during 45OL. In the very aggressive 6590 protocol, the luminous flux dropped below the 0.70 (i.e., L_{70}) threshold after 2940 hours of testing. This value was determined by using linear interpolation as required by TM-28-14 [29].

The changes in the luminous flux for the neutral white OLED panels in the different AST environments are presented in Fig. 9.11. The rate of luminous flux decline was lower for the neutral white OLED panel than the warm white OLED panel, although the same temperature dependence on the rate of decline was observed for both CCT values. Because of the slower decay rate, the luminous flux remained higher than 0.90 (L_{90}) after 7000 hours of testing during RTOL, 35OL, and 45OL. In the very aggressive 6590 protocol, the luminous flux did not drop below the 0.70 threshold until 4310 hours of testing. This value was determined by using linear interpolation, as required by TM-28-14 [29].

A close examination of Figs. 9.10 and 9.11 reveals that the LFM experienced two regions of decay for all samples: a fast, initial decay that usually leveled off between 1000 and 2000 hours, followed by another decay period after 2000 hours. This observation is consistent with a double-exponential model. For the first 2000 hours, the average LFM for RTOL, 35OL, and 45OL were very similar. The similarity in

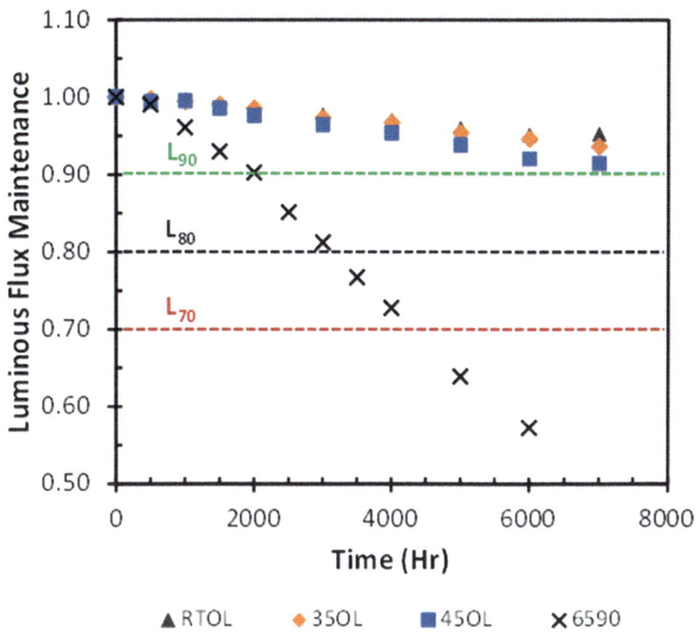

Fig. 9.11 Change in the LFM of the 6-cell tandem stack neutral white OLED panels in different AST environments

Table 9.3 Calculated models for luminous flux decay of the OLED panels in different ASTs by using the modified TM-28-14 method [20]

	Warm White OLED Panel			Neutral White OLED Panel		
Test condition	α	B	R	α	B	R
RTOL	1.8×10^{-5}	0.97	0.98	7.6×10^{-6}	1.0	0.95
35OL	2.6×10^{-5}	0.99	0.99	1.0×10^{-5}	1.0	0.99
45OL	3.7×10^{-5}	0.98	1.00	1.4×10^{-5}	1.0	0.99
6590	1.3×10^{-4}	1.0	1.00	1.1×10^{-4}	1.1	0.99

initial luminous flux decay could result from comparable levels of residual contaminants (e.g., water) present during the device fabrication process. After 2000 hours, the rate of luminous flux decay changed and exhibited a strong correlation with the AST protocols; lower LFM was observed for AST protocols with higher temperature stresses.

The LFM data were analyzed by using an exponential decay model, consistent with the recommendations of TM-28-14 [29]. During this modified TM-28-14 analysis, only data at 1000-hour intervals were used to avoid placing more statistical weight on data from the early portions of the test period when the test interval was 500 hours. The calculated models that used this modified TM-28-14 method for each test population are listed in condition in which the warm white α value was only 15% higher than the α value for the neutral white OLED panels.

Of particular significance is the finding that the α value for the neutral white DUTs under RTOL conditions was 7.6×10^{-6}—the lowest value observed in testing to date [20]. This value is better than the measured α value for earlier generations of OLED products, and it indicates a continual improvement in the technology. As expected, the calculated exponents for the decay process (i.e., the decay rate constant values α) were lower for the neutral white OLED panels than for the warm white OLED panels, often by a factor of approximately two. The only exception to this finding was during the highly accelerated 6590 condition in which the warm white α value was only 15% higher than the α value for the neutral white OLED panels. (Table 9.3)

The rapid decline in luminous flux from both OLEDs in the 6590 condition may be because of the combined effects of temperature and moisture. The increase in α values for the test populations during RTOL, 35OL, and 45OL provide a clear indication that temperature plays a major role in the rate of luminous flux decline in these samples. As evidenced in the panels operated at 6590, moisture ingress also likely played a role because darkening (i.e., no light emission) was observed at the edges of the emitting area. Although the calculated α value in 6590 is low compared to inorganic LEDs, the values reported here far exceed the performance of earlier generation OLED panels that were measured in a temperature and humidity environment of 75 °C and 75% relative humidity [17]. This finding is suggestive of improved encapsulation technologies for the OLEDs that significantly increase the lifetime in humid environments.

Because only the 6590 test conditions produced a sufficient drop in LFM for L_{70} to be determined experimentally, the test times necessary for the other panels to drop below the L_{70} threshold must be projected by using the modified TM-28-14 method. Of course, because of the sample size, projections of L_{70} are subject to the three times rule (i.e., the time to reach L_{70} cannot be projected past three times the actual test duration). The projected times to L_{70} are presented in Table 9.4, along with the experimentally derived values for DUTs during 6590. In general, the time to L_{70} is longer for the neutral white OLED panels than for the warm white OLED

Table 9.4 Average time to reach L_{70} for the 6-cell tandem stack panels during different test conditions [20]

LightColor	Test	TestLength	Time to L_{70}	Method
Warm white	RTOL	7000 hours	18,200 hours	Modified TM-28-14 method projection
Neutral white	RTOL	7000 hours	> 21,000 hours[a]	Modified TM-28-14 method projection
Warm white	35OL	7000 hours	13,200 hours	Modified TM-28-14 method projection
Neutral white	35OL	7000 hours	> 21,000 hours[a]	Modified TM-28-14 method projection
Warm white	45OL	7000 hours	9150 hours	Modified TM-28-14 method projection
Neutral white	45OL	7000 hours	> 21,000 hours[b]	Modified TM-28-14 method projection
Warm white	6590	6000 hours	2936 hours	Linear interpolation of experiment
Neutral white	6590	6000 hours	4311 hours	Linear interpolation of experiment

[a]Limited by the three times total test duration limit in TM-28-14

[b]Only two samples completed 7000 hours of testing during 35OL, and the time to L_{70} was set to three times the test duration based on the performance of these two samples

panels, which suggests higher reliability. The projected times for the neutral white OLED panels to reach L_{70} exceeded the maximum value of 21,000 hours permitted by the three times rule for RTOL, 35OL, and 45OL, and higher values could not be accurately determined by using this method. However, an earlier generation of this OLED product had projected lifetimes in excess of 36,000 hours during RTOL testing (based on 12,000 hours of testing) [20]. Consequently, it can be speculated that the lifetime of the neutral white OLED panels will be significantly longer during RTOL and 35OL than the values calculated here, perhaps exceeding 50,000 hours. (Table 9.4)

5.3 Chromaticity Maintenance of the 6-Cell Tandem Stack OLEDs

Chromaticity maintenance is the ability of a light source to maintain a chromaticity point, and this parameter is the chromaticity analog of LFM. The average chromaticity shift is given by the vector quantity $\Delta u'v'$.[3] The values of $\Delta u'v'$ for the warm white OLED panels measured after exposure to the different AST environment are

[3] $\Delta u'v' = [(\Delta u')^2 + (\Delta v')^2]^{0.5}$, where $\Delta u'$ is the change in the u' chromaticity coordinate and $\Delta v'$ is the change in the v' chromaticity coordinate.

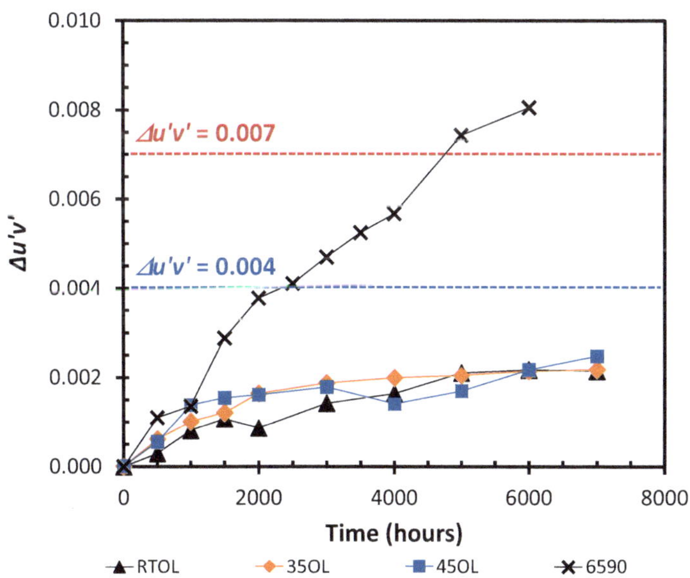

Fig. 9.12 Average chromaticity shift of the warm white 6-cell tandem stack OLED panel in the different ASTs

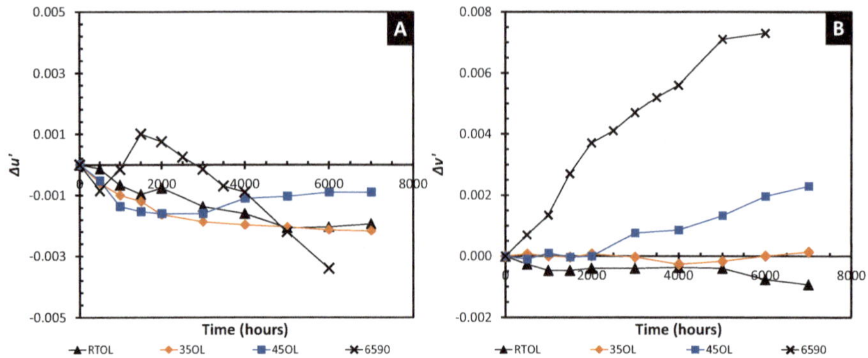

Fig. 9.13 Average change in (**a**) Δu′ and (**b**) Δv′ for the warm white 6-cell tandem stack OLED panels

presented in Fig. 9.12. This figure also contains lines indicating a $\Delta u'v'$ of 0.004 and 0.007, which are often taken as possible threshold limits for parametric failure because of an excessive chromaticity shift [21]. The only test condition that exceeded these thresholds was the 6590 environment during which the $\Delta u'v' = 0.004$ threshold was crossed at approximately 2340 hours and the $\Delta u'v' = 0.007$ threshold was crossed at approximately 5750 hours. None of the temperature-only ASTs (e.g., RTOL, 35OL, 45OL) approached the parametric failure thresholds during the 7000 hours of the test.

Additional details about the chromaticity shifts of the 6-cell tandem stack warm white OLED panels are provided in Fig. 9.13 as temporal plots of the change in $\Delta u'$ (Fig. 9.13A) and $\Delta v'$ (Fig. 9.13B). Based on an examination of the temporal changes in the chromaticity data (i.e., $\Delta u'v'$, $\Delta u'$, $\Delta v'$), there appears to be at least three main chromaticity shift processes occurring in some of the DUTs during the test period. These processes are most apparent in the 6590 data where: (1) there is an initial decrease in $\Delta u'$ until approximately 500 hours; (2) there is an increase in $\Delta v'$ accompanied by a smaller increase in $\Delta u'$ between 500 and 1500 hours; and (3) there is a steady decrease in $\Delta u'$ and a steady increase in $\Delta v'$ from 1500 hours until the 6590 exposure was stopped after 6000 hours. For both RTOL and 35OL conditions, only an initial change in chromaticity to a plateau value of $\Delta u'v' = 0.002$ was observed, and the chromaticity shift remained at that plateau until testing was terminated after 7000 hours. Samples in the 35OL test reach the $\Delta u'v' = 0.002$ plateau after approximately 3000 hours, whereas those in RTOL required longer (approximately 5000 hours) to reach this value. For the 45OL condition, there was an initial change in chromaticity to the $\Delta u'v' = 0.002$ plateau, driven primarily by a change in the $\Delta u'$ component, that lasted until approximately 3000 hours. This plateau was reached after only 1500 hours of testing in 45OL, faster than either RTOL or 35OL populations, and the reduction in time to reach the plateau as temperature increases suggests that this process is thermally activated. After 3000 hours of exposure to the 45OL conditions, the test population showed a steady increase in the chromaticity

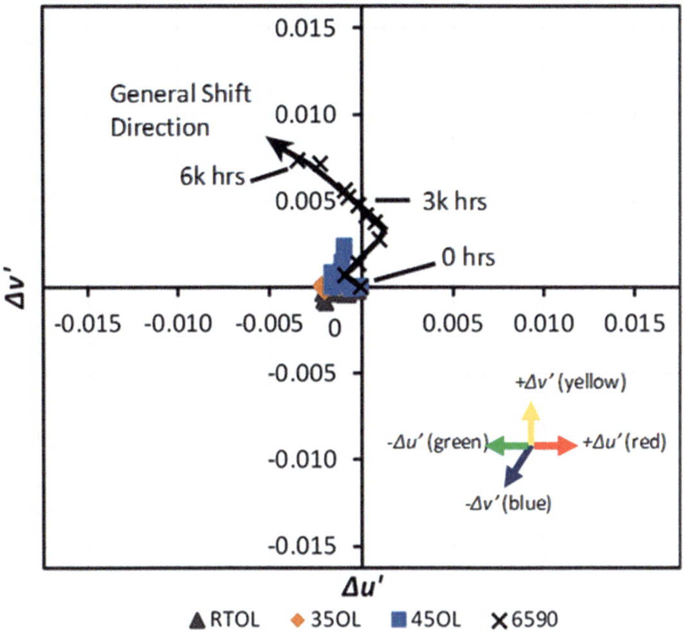

Fig. 9.14 Chromaticity shifts for 6-cell tandem stack warm white OLED panels in different AST protocols [20]

shift magnitude driven by an increase in both $\Delta u'$ and $\Delta v'$ over time. These findings suggest that only the first chromaticity process was observed during the test duration for the populations in RTOL and 35OL, whereas the first and second chromaticity shift processes were observed in 45OL. These findings are consistent with analytical models of chromaticity shifts, which demonstrated that initial chromaticity shifts often proceed rapidly to a limiting value that can be modeled with a bound exponential function. The initial shift is followed by longer lasting chromaticity shifts that are often responsible for parametric failures [35].

The direction of the net chromaticity shift for the 6-cell tandem stack OLED panels can be discerned by examining a plot of $\Delta v'$ versus $\Delta u'$ (see Fig. 9.14). The initial chromaticity shift for these devices is in the generally green direction ($\Delta u'$ is negative, and there is little change in $\Delta v'$). This is the first chromaticity shift that was previously mentioned, and it is the only chromaticity shift observed for samples in RTOL and 35OL during the test period. After the initial green shift, the chromaticity abruptly shifted in the generally yellow direction for the 45OL DUTs and in the yellow-red direction for the 6590 DUTs. It is possible that the processes responsible for these shifts require thermal activation and proceed slowly at temperatures below 45 °C, which would explain their absence in the RTOL and 35OL populations.

The SPD of each panel was deconvoluted into blue, green-yellow, and red-orange components (see Fig. 9.8) to determine the temporal behavior of the absolute radiant power of the organic emitters of the warm white OLED panels. The radiant

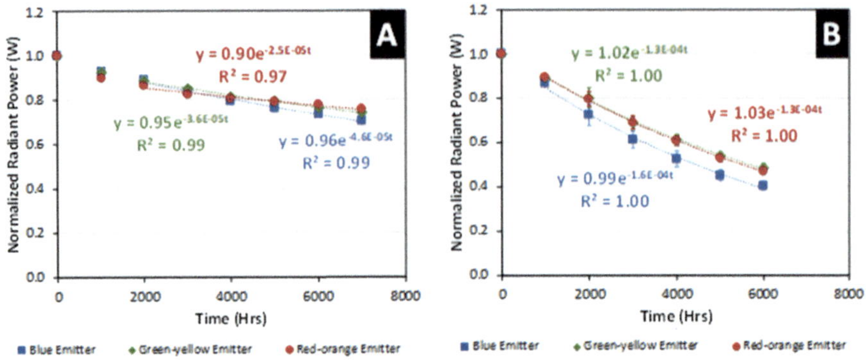

Fig. 9.15 Spectral deconvolution analysis of the warm white OLED panels at different times of AST exposure. (**a**) 45OL test population and (**b**) 6590 test population [20]

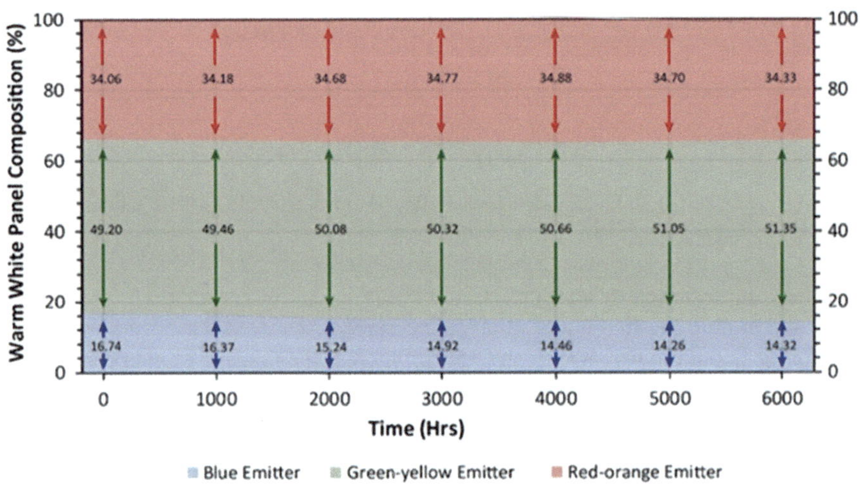

Fig. 9.16 Temporal dependence of the contribution of the different organic emitters to radiant power for the test population subjected to the 6590 test condition [20]

power calculated for each emitter was normalized and averaged. The temporal changes in normalized radiant power for the 45OL and 6590 panels are shown in Fig. 9.15a and Fig. 9.15b, respectively. By comparing the exponential values of the least square fits in Fig. 9.15, the blue emitter is shown to decay at the fastest rate, whereas the red-orange emitter and the green-yellow emitter decayed at similar rates during 6590.

Increases in the temperature and humidity of the experimental conditions in the 6590 environment resulted in decay rate increases for each emitter by approximately a factor of four, relative to 45OL. The contribution of each emitter of the 6-cell tandem stack warm white OLED panel in 6590 to the total radiant power (expressed as a percentage) is shown in Fig. 9.16. The relative contribution of the

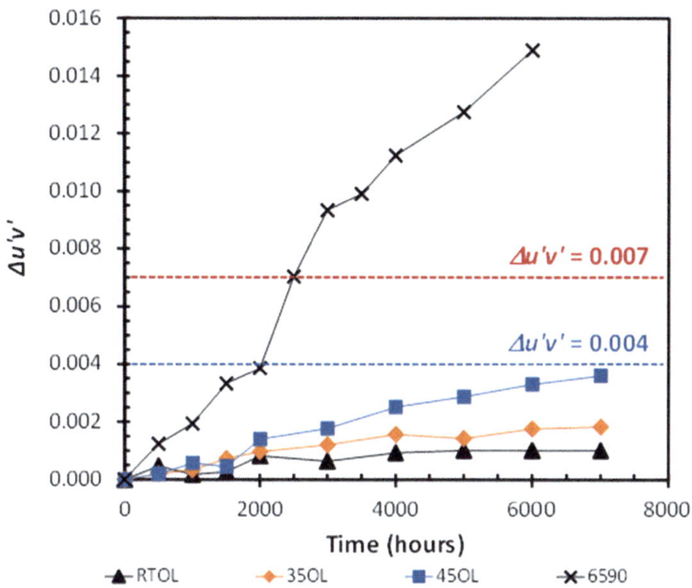

Fig. 9.17 Average chromaticity shift of the neutral white 6-cell tandem stack OLED panel in the different ASTs

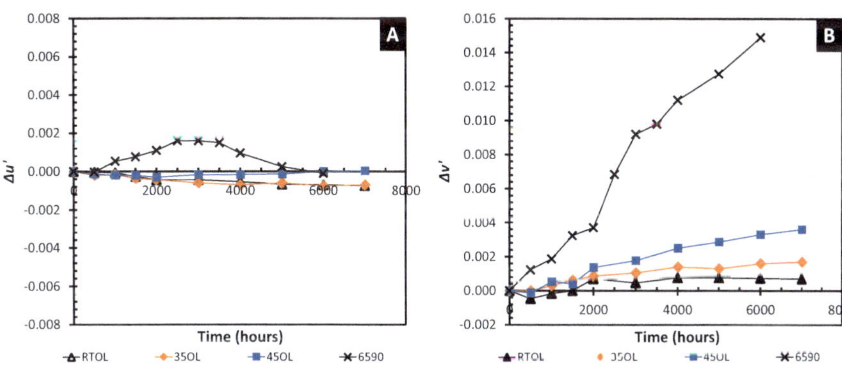

Fig. 9.18 Average change in (**a**) Δu′ and (**b**) Δv′ for the neutral white 6-cell tandem stack OLEDs in the ASTs

blue emitter decreased steadily through 6000 hours of testing, whereas the relative contributions of the green-yellow and red-orange emitters increased.

The neutral white 6-cell tandem stack OLED panels consist of three blue fluorescent emitters and three amber phosphorescent emitters (see Fig. 9.5). The average chromaticity shifts for the neutral white OLED panels measured after exposure to the different AST environments are presented in Fig. 9.17. This figure also contains lines indicating a $\Delta u'v'$ of 0.004 and 0.007, to denote the possible threshold limits for parametric failure because of excessive chromaticity shift [21]. The 6590 environment was the only test condition exceeding these thresholds, where the $\Delta u'v' = 0.004$ threshold was crossed at approximately 2020 hours, and the $\Delta u'v' = 0.007$ threshold was crossed at approximately 2490 hours. None of the temperature-only ASTs (e.g., RTOL, 35OL, 45OL) approached either parametric failure threshold during the 7000 hours of the test, which is consistent with a lower acceleration factor for these tests.

Additional details about the chromaticity shifts of the 6-cell tandem stack neutral white OLED panels are provided in Fig. 9.18 as temporal plots of the change in $\Delta u'$ (Fig. 9.18a) and $\Delta v'$ (Fig. 9.18b). Based on an examination of the temporal changes in the chromaticity data (i.e., $\Delta u'v'$, $\Delta u'$, $\Delta v'$), there appears to be at least three main chromaticity shift processes occurring during the test period, which is in agreement with the findings for the warm white OLED panels. These processes are most apparent in the data from the DUTs exposed to 6590, during which (1) there was minimal change in $\Delta u'$ and an increase in $\Delta v'$ through 500 hours of testing, (2) there was an increase in $\Delta v'$ accompanied by a smaller increase in $\Delta u'$ to a plateau value from 500 hours until at least 3500 hours of accumulated test time, and (3) there was a sharp increase in $\Delta v'$ from 2000 hours accompanied by a slow decrease in $\Delta u'$ (delayed until 3500 hours) that lasted until the 6590 exposure was stopped after 6000 hours. For both RTOL and 35OL conditions, only an initial change in chromaticity to a plateau value of $\Delta u'v'$ less than 0.001 to 0.002 was observed. Once this plateau was reached during RTOL and 35OL, the chromaticity remained at that value until testing was terminated after 7000 hours. For the 45OL condition, there was an initial change in chromaticity to $\Delta u'v' = 0.002$, driven primarily by a change in the $\Delta v'$ component, that lasted until approximately 3000 hours. This plateau was reached after only 2000 hours of testing during 45OL, faster than in either RTOL or 35OL, supporting that this process is temperature-activated. After 3000 hours of exposure to the 45OL conditions, the test population showed a steady increase in the chromaticity shift magnitude driven by an increase in $\Delta v'$ over time, with a minor contribution from $\Delta u'$. These findings suggest that only the first chromaticity process was observed during the test duration for the populations during RTOL and 35OL, whereas the first and second chromaticity shift processes were observed in 45OL. These findings are similar to those observed for the 6-cell tandem stack OLED panels, although the change in the $\Delta v'$ component was much larger, whereas the change in the $\Delta u'$ component was slightly smaller for the neutral white OLED panels compared with the warm white OLED panels.

The direction of the net chromaticity shift for the 6-cell tandem stack neutral white OLED panels can be discerned by examining a plot of $\Delta v'$ versus $\Delta u'$, which is given in Fig. 9.19. The chromaticity shift proceeded in the generally yellow direction (i.e., chromaticity changed primarily along the $+\Delta v'$ axis with minimal change along the $\Delta u'$ axis), and the magnitude of the shift increased with temperature. The same trends—only larger in magnitude—were found for the neutral white OLED panels operated in the 6590 conditions. This finding suggests that the same chromaticity shift mechanism is likely operating in all four AST conditions, but at different

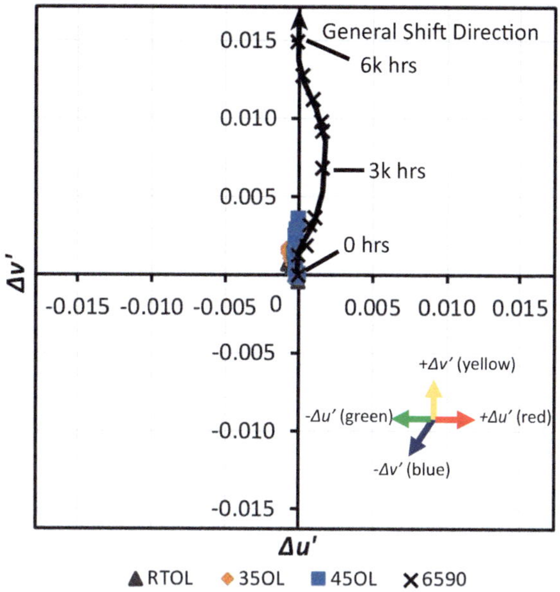

Fig. 9.19 Chromaticity shifts for 6-cell tandem stack neutral white OLED panel [20]

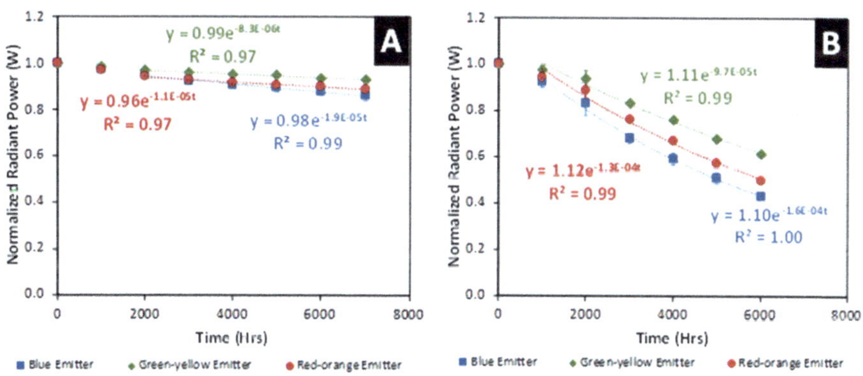

Fig. 9.20 Spectral deconvolution analysis of the neutral white OLED panels at different times of AST exposure. (**a**) 45OL test population and (**b**) 6590 test population [20]

rates (consistent with thermal activation). There was no evidence of a significant green shift, as observed for the warm white OLED panels.

A component analysis of the emission spectrum for each panel was conducted to determine the temporal absolute radiant power of the blue, green-yellow, and red-orange emitters of the 6-cell tandem stack neutral white OLED panels and to provide further insights regarding the cause of the chromaticity shift. The radiant power for each emitter was normalized and averaged, and the temporal changes in normalized radiant power for the 45OL and 6590 panels are shown in Fig. 9.20a, b, respectively. The blue emitter has the highest exponential component and decayed at the fastest rate. The green-yellow and red-orange emitters decayed at a slower rate, and their decay rates were lower than the net luminous flux decay rate presented in **condition in** which the warm white α value was only 15% higher than the α value for the neutral white OLED panels.

At the most aggressive testing conditions (i.e., 6590), the observed decay rate for each emitter was approximately one order of magnitude greater than the decay rate in 45OL, but the relative decay rate between emitters remained essentially the same, with the blue emitter decaying the fastest and the green emitter decaying the slowest. This finding is consistent with the same chromaticity shift occurring in all four ASTs, but the rate of chromaticity shift was higher during 6590 than during the other tests.

The contribution of each emitter of a 6590 neutral white 6-cell tandem stack OLED panel to the total radiant power (expressed as a percentage) is shown in Fig. 9.21. The neutral white OLED panels operated at 6590 experienced a steady decrease in the blue emitter's relative contribution to the emission spectrum and a relatively steady increase in the green-yellow emitter's relative contribution to the emission spectrum through 6000 hours. These changes are likely a significant contributor to the chromaticity shift in the yellow direction. The red-orange emitter contributed steadily to the emission spectra until 4000 hours, and then began to have a decreasing contribution. However, this decrease was smaller than that of the blue emitter. The overall magnitude of these changes in relative spectral composition explains the slight yellow-red shift during the first 3000 hours of testing in the otherwise strongly yellow-shifted data for the 6590 6-cell tandem stack neutral white OLED panels.

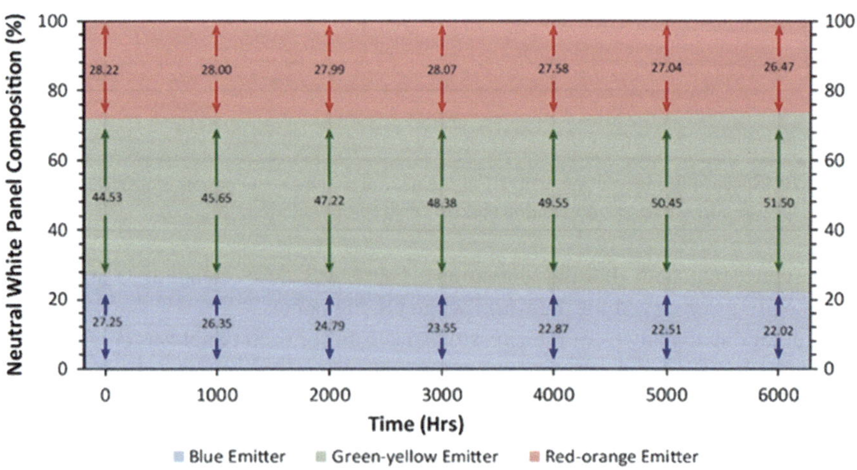

Fig. 9.21 Temporal dependence of the contribution of the different organic emitters to radiant power for the 6590 test population [20]

Table 9.5 Warm white OLED panel electrical data [20]

AST protocol	Voltage supplied to panel (V_{dc})	Current supplied to panel (I_{dc})	Power supplied to panel (W)
Control panel[a]	19.53 ± 0.29	0.217 ± 0.000	4.24 ± 0.06
RTOL (1500 hours)	19.45 ± 0.02	0.217 ± 0.001	4.23 ± 0.03
RTOL (7000 hours)	20.98 ± 0.31	0.217 ± 0.001	4.55 ± 0.08
35OL (1500 hours)	19.71 ± 0.05	0.218 ± 0.001	4.30 ± 0.02
35OL (7000 hours)	21.46 ± 0.19	0.218 ± 0.001	4.67 ± 0.04
45OL (1500 hours)	19.90 ± 0.22	0.217 ± 0.002	4.32 ± 0.09
45OL (7000 hours)	22.52 ± 0.42	0.217 ± 0.002	4.87 ± 0.13
6590 (1500 hours)	21.71 ± 0.23	0.230 ± 0.014	5.00 ± 0.26
6590 (6000 hours)[b]	26.98 ± 0.70	0.217 ± 0.001	5.86 ± 0.14

[a]Data are the average of the control panel taken over time
[b]Data are the averages of two panels

5.4 *Electrical Analysis of Neutral White OLED Panels*

Periodically, electrical characterization was performed with a power analyzer to determine the power supplied to the 6-cell tandem stack panels by their respective driver systems. The electrical drivers were pre-set to deliver 0.217 amperes (A) of constant current to the panels, and most changes occurred in the voltage and power levels for each panel. In extreme AST conditions (e.g., panels operated in 6590 conditions), the voltage required by the panels was larger than the driver could supply, so changes in current were also observed. The average electrical characteristics were calculated for the warm white (Table 9.5) and neutral white (Table 9.6) OLED panels by averaging the measured values for the three test panels in each AST environment. There were three notable exceptions to this approach as discussed in the footnotes to Table 9.5 and Table 9.6: the neutral white OLED panels at 35OL, the neutral white OLED panels at 6590, and the warm white OLED panels at 6590. In each of these exceptions, the average electrical data were calculated from the measured values for two panels because a neutral white OLED panel operated at 35OL failed by 3000 hours (electrical short) and a neutral white OLED panel operated at 6590 failed by 3000 hours (the panel still functioned, but a connection to a driver resulted in premature failure of the driver). Also, the heat sink backing of a warm white OLED panel operated in the 6590 environment was inadvertently water damaged by the test chamber; therefore, the panel was not considered for averaging.

Table 9.6 Neutral white OLED panel electrical data [20]

AST protocol	Voltage supplied to panel (Vdc)	Current supplied to panel (Idc)	Power supplied to panel (W)
Control panel[a]	21.52 ± 0.31	0.219 ± 0.001	4.72 ± 0.08
RTOL (1500 hours)	20.59 ± 0.70	0.217 ± 0.002	4.46 ± 0.19
RTOL (7000 hours)	21.38 ± 0.68	0.216 ± 0.002	4.63 ± 0.18
35OL (1500 hours)	21.24 ± 0.80	0.218 ± 0.002	4.63 ± 0.18
35OL (7000 hours)	21.98 ± 0.74	0.218 ± 0.001	4.79 ± 0.19
45OL (1500 hours)	21.50 ± 1.02	0.217 ± 0.002	4.68 ± 0.22
45OL (7000 hours)	22.52 ± 0.87	0.218 ± 0.001	4.97 ± 0.20
6590 (1500 hours)	22.50 ± 0.58	0.218 ± 0.001	4.91 ± 0.12
6590 (6000 hours)[b]	27.93 ± 0.23	0.200 ± 0.024	5.58 ± 0.72

[a]Data are the average of the control panel taken over time

[b]Data are the averages of the two panels. The driver used for one panel failed during the test, so the panel was powered with the driver from the other OLED panel to obtain these electrical measurements

Significant increases in voltage and power were observed for the warm white OLED panels as each AST progressed. For all AST conditions at the end of testing (6000 hours or 7000 hours), the warm white OLED panel voltage increases were found to be statistically significant from the control panel. For the data collected at the end of testing for the warm white OLED panels, the voltage increase was highest for the most aggressive testing condition (i.e., 6590), and the voltage increase was found to be second highest for the next most aggressive testing condition (i.e., 45OL). A statistical difference was not observed between the 35OL and RTOL warm white OLED panel voltages at the end of testing. For all AST conditions, the warm white OLED panel voltages increased in a statistically significant way between 1500 hours and the end of testing. The changes in power consumption and LFM resulted in a decrease in luminous efficacy of the warm white OLED lighting system (i.e., the OLED panel, driver, and wall-mounted power supply) from 50 lumens/watt (W) to 33 lumens/W after 7000 hours of 45OL. Although the performance of the OLED panels degraded, a significant contributor to the low luminous efficacy of these devices was the driver and power supply, which, when combined, had an efficiency of 65%. This low efficiency of the driver and power supply represents a potential opportunity for the improvement in the luminous efficacy of these OLED systems.

The neutral white OLED panels experienced much larger voltage variations than the warm white OLED panels; this issue may have been caused by the greater initial variation in impedance observed for the neutral white devices relative to the warm white devices. Specifically, the initial average impedance of the 12 test panels and control at 1000 Hz for neutral white DUTs was 548.7 ± 13.3 ohms (Ω) versus 558.8 ± 2.3 Ω for the same sample size of warm white OLED panels [20]. This impedance variation may be because of the slight differences in stack construction. Specifically, the warm white OLED panel has four amber phosphorescent layers and two blue fluorescent layers, and the neutral white OLED panel has three phosphorescent amber layers and three blue fluorescent layers. Throughout testing, the impedance values for both the neutral white and warm white OLED panels did not vary significantly from the control, indicating stable impedance values for the 6-cell tandem stack samples [20].

The larger variation in voltage across the neutral white samples continued throughout testing, and statistical differences between initial and final voltages at individual AST conditions were not observed as readily for the neutral white OLED panels due in part to the larger standard deviation values. To explain, increases in voltage and power were observed for the neutral white OLED panels in all AST conditions from the mildest (i.e., RTOL) to the most aggressive (i.e., 45OL and 6590). However, when compared with the control sample, no significant statistical difference was observed in the power or voltage supplied to panels that were tested at RTOL, 35OL, or 45OL. The lack of statistically significant differences likely reflects the larger standard deviation values of the neutral white DUTs compared with the warm white samples. However, there was a significant statistical increase (95% confidence level) in the power and voltages supplied to the 6590 panels relative to the control at 6000 hours, and the increases in voltage and power from 1500 hours to 6000 hours were also found to be significant for these panels. The increases in voltage and power at 6000 hours for the panels operated in 6590 conditions were accompanied by a decrease in current supplied to the panels (though the decrease was not significant). The decrease was likely the result of one of the panels requiring more voltage than the driver's designed maximum (28 V_{dc}) [20].

6 Discussion

The 6-cell tandem stack panels described herein represent the state-of-the-art OLED panels for illumination applications in the year 2020. These panels consist of either a warm white formulation containing four phosphorescent amber layers and two blue fluorescent layers (3000 K nominal CCT value) or a neutral white formulation containing three phosphorescent amber layers and three blue fluorescent layers (4000 K nominal CCT value). These layers are typically deposited via evaporation processes performed in glove boxes to minimize impurities. Light extraction layers (LELs) are added to the OLEDs to improve light extraction efficiency. In addition, an encapsulation process is used to protect the contents of the panel after manufacturing from contamination by moisture or other atmospheric components. The evaluations of the long-term performance of OLEDs discussed here not only reflect changes in the light-emitting organic materials, but also changes in the entire panel composition, including encapsulants and light-extraction layers.

In AST environments, the neutral white OLED panels exhibited better LFM but poorer luminous efficacy and chromaticity maintenance than the warm white OLED panels. The lower luminous efficacy of the neutral white OLED panels can be traced to the increased use of the blue fluorescent emitter because phosphorescent emitters have higher internal quantum efficiencies.

There are at least two mechanisms involved in the loss of luminous flux of these OLED panels. These mechanisms apply to operation in real-life environments and in the different AST environments. The first mechanism is the degradation of the organic emitters resulting in a reduction of light emission either through the loss of emission sites or a reduction in quantum efficiency. The calculated emission decay rates of the blue, green-yellow, and red-orange organic emitters in these OLED panels are presented in Fig. 9.15 (warm white) and Fig. 9.20 (neutral white). Although there are some differences in behavior among the different ASTs, the rates of emission decay, as given by the individual exponential decay rate constants (α), generally follow the order $\alpha_{blue} > \alpha_{red} > \alpha_{green}$ for both the warm white and neutral white OLED panels. Consequently, the emissions contained increasingly higher percentages of light from the green-yellow organic emitter as the AST progresses and the emissions become increasingly deficient in blue.

The second mechanism is the shorting of pixels creating dark spots. This shorting, which proceeds from the anode to the cathode, effectively renders the pixel non-functional from a light-generation standpoint. Fortunately, the anode and cathode of an OLED panel used in lighting applications are usually pixelated, so the entire panel does not go dark, just the area where the short occurred. An example of such a dark spot is shown in Fig. 9.22. There are several potential causes of these dark spots, including the presence of foreign impurities, pinholes, bubble formation, electromigration, and regions of high current density [17–19, 36].

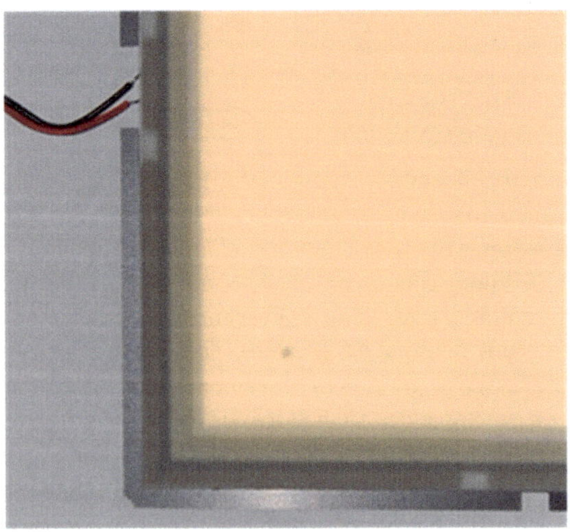

Fig. 9.22 A black spot formed on a warm white 6-cell tandem stack OLED panel during RTOL testing [18]

There was also a difference in the rate of chromaticity shift between the warm white OLED and neutral white OLED panels, with the warm white OLED panels generally exhibiting a lower rate in chromaticity change in all ASTs. For both warm white and neutral white panels, the dominant chromaticity change was in the yellow direction (i.e., $\Delta v' > \Delta u'$). The change in radiant flux of the individual organic emitters (see Fig. 9.16 and Fig. 9.21) suggests that this chromaticity change is caused by a relative increase in green-yellow emissions and a relative decrease in blue emissions. There are also some minor contributions from changes in the red-orange emissions. Because the neutral white OLED panel has a higher initial contribution from the blue organic emitter, the flux loss from this emitter has a greater impact on chromaticity shift at this CCT value than in the warm white OLED panels. Consequently, the rate of chromaticity shift can be expected to be higher in the neutral white OLED panel than in the warm white OLED panel, in agreement with the experimental observations.

The OLED panels also contain an LEL on the outside of the glass substrate to help out-couple light into the viewing space. LELs are typically made of polymers that are adhesively bound to the surface of the glass OLED panels. The encapsulant may also be a polymer material. The entire assembly, including the LEL, is encapsulated as shown in Fig. 9.5. One characteristic of polymer degradation over time is an increase in blue absorption [37]. It is unknown whether changes occurred in the LEL or encapsulant on these OLED panels. For the most part, the AST conditions were relatively mild, so minimal yellowing of the films can be expected. However, the 6590 environment is sufficiently aggressive that polymer yellowing may have occurred. If the polymer in the LEL or encapsulant is yellowing, then increased absorption of blue emissions would occur and the chromaticity would shift in a yellow direction, as was observed during the study.

The electrical properties of the 6-cell tandem stack OLED panels also changed during use in the different AST environments, indicating changes within the stack itself. As shown in Table 9.5 (warm white) and Table 9.6 (neutral white), the drive voltage required to operate the OLED panels increased with operational time. Furthermore, the rate of this increase approximately correlated to the degree of stress in the AST environment, with the mildest AST (i.e., RTOL) showing the smallest change from the initial drive voltage and the most aggressive AST (i.e., 6590) environment showing a much larger increase from the initial drive voltage. However, the impedance of these panels did not change significantly during any of the ASTs, which suggests that the higher voltages were necessary to compensate for increased trap states.

Because the output current of the drivers was fixed at 0.217 A, the power consumption of the panels increased with the drive voltage. The combined effect of decreasing luminous flux (because of emitter degradation and the appearance of dark spots) coupled with the rise in power consumption (which suggests an increase in trap states) has a negative impact on the parametric reliability of these OLED panels from the standpoint of luminous efficacy. The importance of these parameters in assessing the performance of OLED devices in lighting applications suggests that additional research is needed to overcome these limitations.

6.1 Comparison with Inorganic LEDs

Inorganic LEDs use a blue or violet LED that is covered with a phosphor to produce white light. This lighting source has found broad applicability in the lighting industry and is displacing fluorescent and incandescent sources in a variety of applications. The luminous flux maintenance of LED sources is typically evaluated according to the protocols established in LM-80-20 [38], and the data acquired using LM-80-20 methods are modeled with an exponential decay function to obtain a luminous flux decay rate constant α value [39].

The LM-80 data from more than 200 LED packages in different packaging platforms (e.g., high-brightness, medium-power, chip-on-board, chip-scale) were recently reviewed [40]. The analysis also examined a variety of operational junction temperatures ranging from 55 °C to 140 °C. The findings from this work can be best summarized by the α values found for inorganic LEDs under different operating conditions (see Fig. 9.23). In total, 96% of the α values in this study are less than 6×10^{-6}, which corresponds to an L_{70} time of nearly 60,000 hours [40].

Although the LM-80/TM-21 methods are for LED packages, TM-28-14 does allow for TM-21 data to be used to estimate the luminous flux maintenance of LED sources. Although the two methods are different, there are enough similarities to allow a comparison of α values from each method.

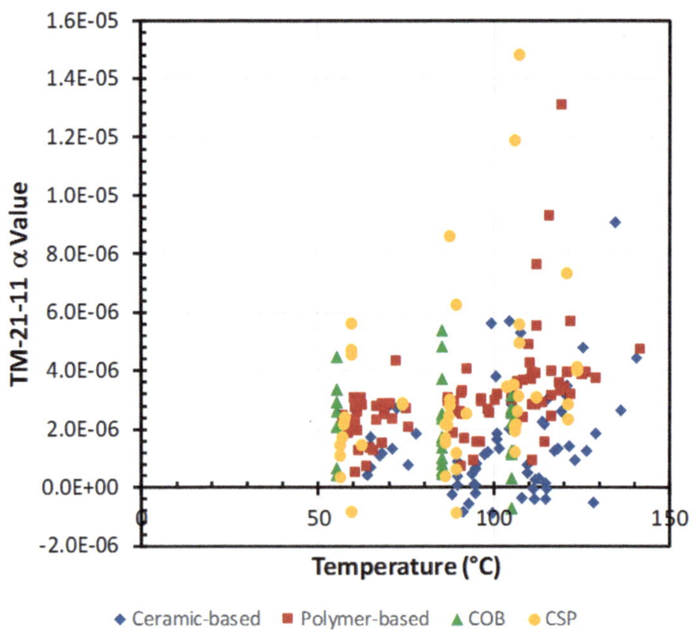

Fig. 9.23 Calculated TM-21-11 α values of inorganic LEDs as a function of temperature. The data are sorted by LED package platform [40]

The lowest α value calculated for an OLED panel was 7.6×10^{-6}, which was experimentally derived for the neutral white OLED population operated at room temperature (i.e., 25 °C). Although this value rates near the upper end of the α values computed for inorganic LEDs in the recent survey [40], it demonstrates that the performance gap between inorganic LEDs and OLEDs is narrowing. However, OLED performance is poorer at higher temperatures, in which the additional thermal energy is enough to promote an increase in the luminous flux decay. As a result, the recent improvements in OLED performance suggest that there are significant lighting applications of the technology provided that there are continued improvements in emitter stability, particularly at slightly elevated ambient temperatures.

7 Conclusions

OLED panels consist of stacked films of organic materials designed to control the transport and recombination of electrons and holes to produce visible light. The wavelength of the emitted light depends on the electronic structure of the emitting organic molecules and can range from blue to red. Electron-hole recombination can occur through a fluorescent (i.e., singlet-to-singlet transition) process or a phosphorescent (i.e., triplet-to-single transition) process, with random spin probabilities resulting in up to 25% of spin states being singlets and 75% being triplets. Phosphorescent emitters have higher efficiencies, but fluorescent emitters generally have higher stabilities.

During stress testing of commercially available, 6-cell tandem stack OLED panels at warm white (i.e., 3000 K) and neutral white (i.e., 4000 K) CCT values, the warm white OLED panels were found to exhibit higher initial luminous efficacies but lower LFM. Although similar chromaticity shift mechanisms appear to occur in warm white OLED and neutral white OLED panels, the rate of chromaticity shift was higher in the neutral white OLED panels than in the warm white OLED panels for the tested products.

Using a modified version of the LM-84-14 method, the luminous flux decay rate constants (α) measured in the different ASTs varied from 7.6×10^{-6} (neutral white OLED panels in RTOL, based on 7000 hours of testing) to 1.3×10^{-4} (warm white OLED panels in 6590, based on 6000 hours of testing). There is a clear dependence of α values on temperature, suggesting that the decay processes are thermally activated. The α values and chromaticity maintenance behavior of the neutral white OLED panels in mild AST conditions (e.g., RTOL, 35OL) suggest that the long lifetimes expected in indoor illumination applications are possible with neutral white OLED panels provided that proper precautions are taken to effectively dissipate heat. The LFMs of warm white OLED panels are slightly less. The results in temperature and humidity environments represent a significant improvement in the performance of OLED panels that is likely because of improved thermal management and encapsulation technologies [17, 20]. In particular, failure of these 6-cell tandem stack OLED panels in any test environment was solely through parametric failure mechanisms, and the abrupt "lights out" failures observed during early OLEDs were much less frequent.

Although the performance of OLED panels does not currently match that of inorganic LED devices from the standpoints of efficiency, LFM, and chromaticity maintenance, the technology does provide a low-glare, diffuse lighting source that has the potential to improve optical delivery efficiency. Although continued improvements in the shortcomings of OLED lighting technologies are being aggressively researched, by using standard lighting methods such as LM-84-14, the devices can now be projected to last many years in the field, which is a significant milestone for OLED technologies. This breakthrough indicates that OLEDs can meet the performance needs of indoor lighting and that the reliability of the technology has improved to the point that it will have a place among indoor lighting technologies.

Acknowledgments This document was prepared as an account of work sponsored by an agency of the United States Government. Neither the United States Government nor any agency thereof, nor any of their employees, makes any warranty, express or implied, or assumes any legal liability or responsibility for the accuracy, completeness, or usefulness of any information, apparatus, product, or process disclosed, or represents that its use would not infringe privately owned rights. Reference herein to any specific commercial product, process, or service by trade name, trademark, manufacturer, or otherwise does not necessarily constitute or imply its endorsement, recommendation, or favoring by the United States Government or any agency thereof. The views and opinions of authors expressed herein do not necessarily state or reflect those of the United States Government or any agency thereof.

Team KeyLogic's contributions to this work were funded by the National Energy Technology Laboratory under the Mission Execution and Strategic Analysis contract (DE-FE0025912) for support services.

References

1. OLED-Info. (2018). DSCC: The OLED market will grow 19% in 2019 to reach $31 billion in revenues. Retrieved from https://www.oled-info.com/dscc-oled-market-will-grow-19-2019-reach-31-billion-revenues
2. Smallwood, P. (2018, February 13–15). Global LED and lighting market overview. Presented at the 2018 Strategies in Light, Long Beach, CA
3. H.-W. Chen, J.-H. Lee, B.-Y. Lin, S. Chen, S.T. Wu, Liquid crystal display and organic light-emitting diode display: Present status and future perspectives. Light: Science & Applications **7**, 17168 (2018). https://doi.org/10.1038/lsa.2017.168
4. A. Laaperi, OLED lifetime issues from a mobile-phone-industry point of view. Journal of the Society for Information Display **16**(11), 1125–1130 (2008). https://doi.org/10.1889/JSID16.11.1125
5. DOE BTO Lighting R&D Program. *2019 Lighting R&D opportunities* (U.S. Department of Energy, Washington, DC, 2020 January). Retrieved from https://www.energy.gov/sites/prod/files/2020/01/f70/ssl-rd-opportunities2-jan2020.pdf
6. J. Miller, Building a better OLED. Physics Today Research and Technology. (2012, December). https://doi.org/10.1063/PT.4.0487

7. J.R. Lakowics, *Principles of fluorescence spectroscopy*, 3rd edn. (Springer, New York, NY, 2006)
8. M.A. Baldo, D.F. O'Brien, Y. You, A. Shoustikov, S. Sibley, M.E. Thompson, S.R. Forrest, Highly efficient phosphorescent emission from organic electroluminescent devices. Nature **395**, 151–154 (1998). https://doi.org/10.1038/25954
9. E. Longhi, L. De Cola, Iridium(III) complexes for OLED applications, in *Iridium(III) in optoelectronic and photonics applications*, ed. by E. Zysman-Colman, (Wiley, Hoboken, NJ, 2017). https://doi.org/10.1002/9781119007166.ch6
10. Forrest, S. R. (2017). Extending the lifetime of blue phosphorescent OLEDs. In Proceedings of the SPIE 10362 Organic Light Emitting Materials and Devices XXI. https://doi.org/10.1117/12.2276252
11. K. Kato, R. Iwasaki, T. Tsujimura, Over 130 lm/W all-phosphorescent white OLEDs for next generation lighting. Journal of Photopolymer Science and Technology **28**, 335–340 (2015). https://doi.org/10.2494/photopolymer.28.335
12. N.T. Kalyani, S.J. Dhoble, Novel materials for fabrication and encapsulation of OLEDs. Renewable and Sustainable Energy Reviews **44**, 319–347 (2015). https://doi.org/10.1016/j.rser.2014.11.070
13. Y.H. Deng, Q.-D. Ou, Q.-K. Wang, H.-X. Wei, Y.-Q. Li, S.-T. Lee, J.-X. Tang, The role of charge generation layers in the operational stability of tandem organic light-emitting diodes. Journal of Materials Chemistry C **2**, 1982–1989 (1982)
14. D. Lee, J. Spindler, M. Kondakova, A. Pleten, M. Boroson, 8–4: Invited paper: Amber OLED lighting technology development and application. Society for Information Display, International Symposium Digest of Technical Papers **48**(1), 91–94 (2017)
15. S. Jang, Y. Lee, M.C. Park, 44.1: Invited paper: OLED lighting for general lighting applications. Society for Information Display, Symposium Digest of Technical Papers **46**(1), 661–663 (2015, June)
16. J. Spindler, M. Kondakova, M. Boroson, J. Hammer, V. Gohri, M. Buchel, M. Ruske, et al., 24-2: Invited paper: High brightness OLED Lighting. Society for Information Display, International Symposium Digest of Technical Papers **47**(1), 294–297 (2016, May)
17. N.J. Miller, F.A. Leon, J.L. Davis, *CALiPER report 24: Photometric testing, laboratory teardowns, and accelerated lifetime testing of OLED luminaires* (U.S. Department of Energy, Washington, DC, 2017) Retrieved from https://www.energy.gov/sites/prod/files/2017/08/f35/caliper-24_oled-luminaires.pdf
18. J.L. Davis, K.J. Rountree, K.C. Mills, *Stress testing of organic light-emitting diode panels and luminaires* (U.S. Department of Energy, Washington, DC, 2018) Retrieved from https://www.energy.gov/sites/prod/files/2018/05/f51/ssl_oled-stresstest_2018.pdf. 2018, January
19. K.J. Rountree, J.L. Davis, K.C. Mills, J. Kim, M. McCombs, *Round 2 update of stress testing results of organic light-emitting diode panels and luminaires* (U.S. Department of Energy, Washington, DC, 2018) Retrieved from https://www.energy.gov/sites/prod/files/2019/03/f60/ssl_oled-stresstest_rd2_dec2018_0.pdf. 2018, December
20. K. Rountree, J.L. Davis, M. McCombs, K. Mills, J. Kim, Pope, R., *Round 3 update of stress testing results for organic light-emitting diode panels and luminaires* (U.S. Department of Energy, Washington, DC, 2020) Retrieved from https://www.energy.gov/sites/prod/files/2020/03/f73/ssl-oled-stresstest-rd3-mar2020.pdf
21. Next Generation Lighting Industry Alliance and the LED Systems Reliability Consortium, *LED Luminaire Lifetime: Recommendations for Testing and Reporting*, 3rd edn. (U.S. Department of Energy, Washington, DC, 2014) Retrieved from https://www.energy.gov/sites/prod/files/2015/01/f19/led_luminaire_lifetime_guide_sept2014.pdf
22. D.W. Kim, H. Oh, B.D. Youn, D. Kwon, Bivariate lifetime model for organic light-emitting diodes. IEEE Transactions on Industrial Electronics **64**(3), 2325–2334 (2017). https://doi.org/10.1109/TIE.2016.2623584

23. K. Sugimoto, T. Yoshioka, H. Kijima, H. Ohata, S. Miyaguchi, T. Tsutsui, N. Takada, et al., P-162: Accelerated lifetime testing of white OLED panels for lighting. Society for Information Display Symposium Digest of Technical Papers **47**(1), 1730–1733 (2016, May)
24. N.J. Miller, GATEWAY demonstration: OLED lighting in the offices of Aurora Lighting Design, Inc, in *Report number PNNL-SA-25337*, (U.S. Department of Energy, Washington, DC, 2016) Retrieved from https://www.energy.gov/sites/prod/files/2016/04/f30/2016_gateway_aurora-oled.pdf
25. N.J. Miller, F.A. Leon, *OLED lighting products: Capabilities, challenges, potential. Report Number PNNL-SA-25479* (U.S. Department of Energy, Washington, DC, 2016) Retrieved from https://www.energy.gov/sites/prod/files/2016/08/f33/ssl_oled-products_2016.pdf
26. N.J. Miller, F.A. Leon, J.L. Davis, *CALiPER report 24: Photometric testing, laboratory teardowns, and accelerated lifetime testing of OLED luminaires. Report number PNNL-SA-25932-Sup1* (U.S. Department of Energy, Washington, DC, 2017) Retrieved from https://www.energy.gov/sites/prod/files/2017/08/f35/caliper-24_oled-luminaires.pdf
27. N.J. Miller, GATEWAY demonstration: OLED lighting in the offices of DeJoy, Knauf & Blood, LLP, in *Report number PNNL-SA-26642*, (U.S. Department of Energy, Washington, DC, 2017) Retrieved from https://www.energy.gov/sites/prod/files/2017/08/f35/2017_gateway_dkb-oled.pdf
28. ANSI (American National Standards Institute) and IES (Illuminating Engineering Society), *ANSI/IES LM-79-19: Approved Method: Optical and Electrical Measurements of Solid-State Lighting Products* (IES, New York, NY, 2019)
29. IES (Illuminating Engineering Society), *IES TM-28-14: Projecting Long-Term Luminous Flux Maintenance of LED Lamps and Luminaires* (IES, New York, NY, 2014)
30. IES (Illuminating Engineering Society), *IES LM-84-14: Measuring Luminous Flux and Color Maintenance of LED Lamps, Light Engines, and Luminaires* (IES, New York, NY, 2014)
31. M. Ishii, Luminance decay mechanisms in organic-light-emitting diodes. R&D Review Toyota CRDL **38**, 55–60 (2003)
32. I.B. Berlman, *Handbook of fluorescence spectra of aromatic molecules*, 2nd edn. (Academic Press, New York, NY, 1971)
33. R.D.B. Fraser, E. Suzuki, Resolution of overlapping bands. Functions for simulating band shapes. Analytical Chemistry **41**, 37–39 (1969)
34. ANSI (American National Standards Institute) and IES (Illuminating Engineering Society), *ANSI/IES TM-30-18: IES Method for Evaluating Color Source Rendition* (IES, New York, NY, 2019)
35. J.L. Davis, K.C. Mills, G. Bobashev, K.J. Rountree, M. Lamvik, R. Yaga, C. Johnson, Understanding chromaticity shifts in LED devices through analytical models. Microelectronics Reliability **84**, 149–156 (2018). https://doi.org/10.1016/j.microrel.2018.03.023
36. M.M. Azrain, M.R. Mansor, S.H.S.M. Gadzullah, G. Omar, D. Sivakumar, L.M. Lim, M.N.A. Nordin, Analysis of mechanisms responsible for the formation of dark spots in organic light embittering diodes (OLEDs): A review. Synthetic Metals **236**, 160–175 (2018)
37. M.Y. Mehr, A. Bahrami, W.D. van Driel, X.J. Fan, J.L. Davis, G.Q. Zhang, Degradation of optical materials in solid-state lighting systems. International Materials Reviews **65**(2), 102–128 (2020). https://doi.org/10.1080/09506608.2019.1565716
38. ANSI (American National Standards Institute) and IES (Illuminating Engineering Society), *ANSI/IES LM-80-20 Approved Method: Measuring Luminous Flux and Color Maintenance of LED Packages* (Arrays and Modules, New York, 2015)
39. ANSI (American National Standards Institute) and IES (Illuminating Engineering Society). (2019). IES TM-21-19, Technical Memorandum: Projecting Long-Term Lumen, Photon, and Radiant Flux Maintenance of LED Light Sources. New York
40. L. Davis, M. Hansen, *Lumen and Chromaticity Maintenance Behavior of Light-Emitting Diode (LED) Packages Based on LM-80 Data* (U.S. Department of Energy, Washington, DC, 2020) Retrieved from https://www.energy.gov/sites/prod/files/2020/07/f76/rti_lm-80-white-leds_mar2020.pdf

Chapter 10
Artificial Intelligence and LED Degradation

J. Wei

1 Introduction to Artificial Intelligence

Artificial Intelligence is a field that studying computer algorithms to map input data and output behavior like a human expert. The AI algorithms can be very useful for modeling a complicated industrial system with numerous variables. The AI system can be built by modeling the relationship of the input features and output variables without human intervention. Nowadays, the AI system has greatly elevated the traditional industry through automation. Besides, as the industry continues shifting to the digital and cloud-based market, the availability of digital data will boost AI development so that it can drive many aspects of the industry including the manufacturing and the reliability for the next industry 4.0. For example, it will reduce the cost of system maintenance by predicted failures using the AI. This online monitoring system can be completely built with machine learning algorithms on top of the big-data platform. So far, there are many successful AI applications such as the public health system, industrial manufacturing, and biotech industry. In the hospital, imaging systems have integrated AI engines for the tasks including medical image enhancement, image segmentation, object recognition, and classification. In industrial, many manufacturers found the solutions for object defects detection, material degradation identification, and failure classification. Recently, the deep learning-based methods are very promising in many commercial applications including autonomous driving systems, natural language processing, oncology, genetic engineering, robotics, and surveillance system. In the AI field, it contains many computational algorithms that can be generally classified as unsupervised learning vs supervised learning. In this chapter, we will first introduce several traditional machine learning algorithms for predicting LED degradation using structure life

J. Wei (✉)
Auburn University, Auburn, AL, USA
e-mail: jzw0025@auburn.edu

W. D. van Driel, M. Yazdan Mehr (eds.), *Reliability of Organic Compounds in Microelectronics and Optoelectronics*,
https://doi.org/10.1007/978-3-030-81576-9_10

data. Then, we will focus on Deep Learning, a breakthrough technique in recent years that can be applied for some potential applications in the LED system.

2 Unsupervised Learning Algorithms

2.1 Bayesian Regression

The Bayesian Regression has been widely used in the statistical regression analysis, and in this chapter, we introduce the Bayesian Regression algorithm for tracking and estimating the Remaining Useful Life of electronics. For the basic Bayesian Regression problem, we firstly introduce the posterior probability that is given by the condition probability:

$$p(C_k|X) = \frac{p(X|C_k)p(C_k)}{P(X)}$$

C_k: Category k
X: Samples

The $p(C_k|X)$ is the normalized posterior that is same definition as it in the classification problem. The $p(C_k)$ is the prior distribution. In the problem of regression, we must use the probability $P(X)$ by the summation over the different groups. The summation definition shows as the following equation:

$$P(X) = \Sigma p(X|C_k)p(C_k)$$

Because the Bayesian Linear Regression applies quadratic model, the matrix presentation has been naturally introduced to simplify the problem. Suppose the real-valued input variable column vector is the form of the following:

$$X = [x_1 \ x_2 \ \dots \ x_n]^T$$

$\mathbf{x_i}$: input data

The real-valued predict target column vector is the following:

$$T = [t_1 \ \ t_2 \ \ \dots \ \ t_n]^T$$

In this chapter, we can fit the data with fourth-order polynomial function with its synaptic weights. The basis function 1 × M matrix can be shown as the following:

$$\Phi(x) = [1 \ \ x \ \ x^2 \ \ \dots \ \ x^M]^T$$

The synaptic weights M × 1column matrix is:

$$W = \begin{bmatrix} w_0 \\ w_1 \\ w_2 \\ \cdots \\ w_M \end{bmatrix}$$

Therefore, the estimation matrix can be calculated as the following matrix multiplication:

$$y_i = W^T \cdot \Phi(x_i)$$

Then, the estimation column matrix can be represented as:

$$\mathrm{Y} = \begin{bmatrix} y_1 & y_2 & \cdots & y_n \end{bmatrix}^T$$

In the Bayesian curve-fitting problem, the likelihood function takes the Gaussian probability density equation, which is the condition probability over the input variables and synaptic weights as well as noise precision parameter $\boldsymbol{\beta}$.

$$P(t|,x|,W|,\beta) = N\left(t|,W^T \cdot \Phi(x_i)|,\beta^{-1}I\right)$$

Then, we take the product of all the likelihood functions over the input variables:

$$P(T|,X|,W|,\beta) = \prod_{i=1}^{n} N\left(t|,W^T \cdot \Phi(x_i)|,\beta^{-1}I\right)$$

Now, we can maximize the above equation with respect to the distribution precision parameter $\boldsymbol{\beta}$, which applies the Maximum Likelihood Estimation for the distribution variance:

$$\frac{1}{\beta_{ML}} = \frac{1}{N}\sum_{i=1}^{N}(y_i - t_i)^2 = \frac{1}{N}\cdot(Y - T)\cdot(Y - T)^T$$

N: Number of Samples

The above result gives the reciprocate variance that yields maximum probability of likelihood function. We can substitute the $\boldsymbol{\beta_{ML}}$ into $\boldsymbol{P(T|X, W, \beta)}$, which gives:

$$P(T|,X|,W|,\beta) = \prod_{i=1}^{n} N\left(t|,W^T \cdot \Phi(x_i)|,\beta_{ML}^{-1}\right)$$

Then, we introduce the Bayesian conjugate prior Gaussian probability in the regression as follows:

$$P(W|\alpha) = N(W|,M_0|,S_0) = N(W|,0|,\alpha^{-1}I)$$

α: the precision parameter of the distribution
M_0: Distribution Mean
S_0: Distribution Variance

Hence, we can write the Bayesian posterior function as:

$$\begin{aligned}
&P(W|,X|,T|,\alpha|,\beta) \propto P(T|,X|,W|,\beta) \cdot P(W|\alpha) \\
&= N(W|,M_N|,S_N) \\
&M_N = S_N\left(S_0^{-1}M_0 + \beta\Phi^T T\right) \\
&S_N^{-1} = S_0^{-1} + \beta\Phi^T\Phi
\end{aligned}$$

M_N: Posterior Mean
S_N: Posterior Variance

Once, we can also apply the sum and product rule of probability, thus, it can allow us to write the predictive distribution to be:

$$P(t|,\mathrm{T}|,\alpha|,\beta) = \int P(t|,W|,\beta) \cdot P(W|,T|,\alpha|,\beta)\,dW$$

The condition distribution $\boldsymbol{P(t|T,\alpha,\beta)}$ can be calculated out as the distribution and probability with its mean and variance depending on the variable '*x*'; Therefore, we can finally predict each output '*t*' at each time series input '*x*', which iterates to the beginning until the convergence of posterior distribution.

$$\begin{aligned}
&P(t|,x|,\mathrm{T}|,\alpha|,\beta) = \\
&N\left(t|,M_N^T\Phi(x)|,\beta^{-1} + \Phi^T(x)S_N\Phi(x)\right)
\end{aligned}$$

For LED life estimation, we use the Bayesian regression method to determine the Remaining Useful Life (RUL) for a specified LED lamp. In this testing, because the lumen degradation is much more distinct and quick than the color shifting, we utilize the Lumen Maintenance (LM) degradation as the main indicator of system decay, which by fitting the Lumen Maintenance degradation curve, so the future tendency can be predicted using this Bayesian curve fitting method.

First, it applies the linear Bayesian model (polynomial function) to track the degradation curve and predict the Remaining Useful Life (RUL) of the system. In the Bayesian linear regression method, the predicted Remaining Useful Life has been calculated at each measuring time. In addition, the true Remaining Useful Life can be checked and traced at the system failure time. Therefore, we can compare the predicted Remaining Useful Life with true Remaining Useful Life in the LED system to verify our predicting models and to pick up system white noise.

The Bayesian Regression can be adapted to the dynamic system. The prior distribution provides the information that we already know about the system, and the posterior is affected by the combined PDFs from the prior distribution and likelihood function. Therefore, the old information would be viewed as the prior affecting the posterior. Also, the last time posterior will become the next time prior. So, as the system time migrating forward, the continuous incoming new data would make the posterior distribution precisely. The following figure presents the evolution of Lumen Maintenance degradation and predicting curve. Since more data has been utilized into modeling the feature of decay, the predicting curve fits the experimental data nicely (Fig. 10.1).

2.2 Bayesian Classifier

In this chapter, we introduce the Bayesian Probabilistic Generative Models [1] for the classification problem. The goal of the Bayesian Classification is to find the update posterior distribution for each group given the continuous input data. From the Bayes' Theorem, we need to model the class-condition density $\boldsymbol{p}(\boldsymbol{X}|\boldsymbol{C}_k)$ that is assumed from the Gaussian distribution and the class-prior $\boldsymbol{p}(\boldsymbol{C}_k)$. Following by the

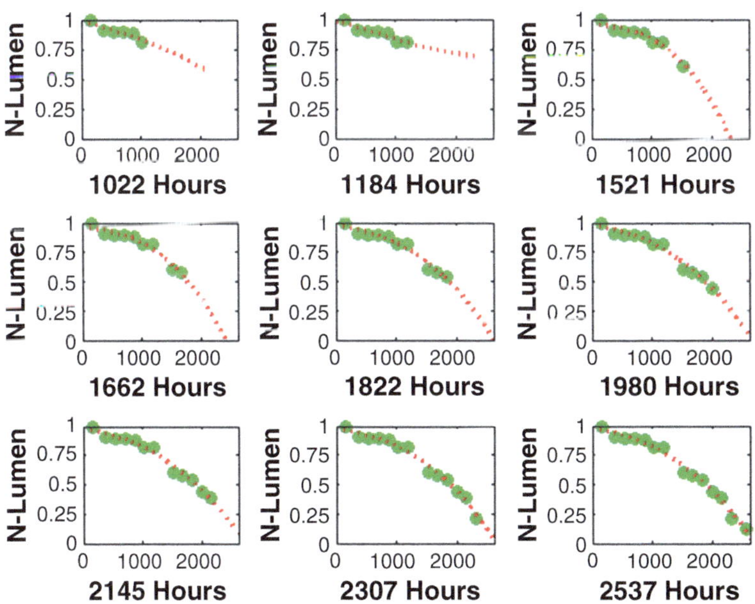

Fig. 10.1 Endoscope and instrument manipulators. (© IEEE 2007), Reprinted with permission

Gaussian model assumption, the Gaussian likelihood function for continuous input data shows below:

$$p(X|C_k) = \frac{1}{(2\pi)^{1/2}} \cdot \frac{1}{|\Sigma_k|} \cdot$$
$$\exp\left(-\frac{1}{2}(X-\mu_k)^T \Sigma_k^{-1} (X-\mu_k) \right)$$

The class prior $\boldsymbol{p(C_k)}$ utilizes the weighted group form as:

$$p(C_k) = N_k / N$$

$$p(C_i) / p(C_j) = N_i / N_j$$

The $\boldsymbol{N_k}$ is the number of samples in the 'k' group. The $\boldsymbol{N}$ is the total number of particles. The Weighted Posterior for classifying the 'k' group is the ratio between its posterior and the sum posterior from all the groups:

$$p(C_k|X) = \frac{p(X|C_k)p(C_k)}{\sum_j p(X|C_j)p(C_j)}$$

The activate function for the multiple-classification is defined as:

$$a_k = \ln p(X|C_k)p(C_k)$$

Typically, for the two-class problem, we introduce the Logistic Sigmoid activate function defined by:

$$p(C_1|X) = \frac{p(X|C_1)p(C_1)}{\sum_{j=1}^{2} p(X|C_j)p(C_j)} = \frac{1}{1+e^{-a}}$$

Where the coefficient 'a' is given by:

$$a = \ln \frac{p(X|C_1)p(C_1)}{p(X|C_2)p(C_2)}$$

The decision boundary can be calculated by equating the activate function or equating the posterior likelihood from above:

$$a_k = a_j$$

Then, we can take the logarithm for both of the Gaussian posterior distribution and find out the decision boundary as following quadratic form:

$$a_{kj} = X^T \cdot W_{kj}^1 \cdot X + W_{kj}^2 \cdot X + W_{kj}^3$$

Where the quadratic decision boundary coefficients are solved as:

$$W_{kj}^1 = -\frac{1}{2}\left(\Sigma_i^{-1} - \Sigma_j^{-1}\right)$$

$$W_{kj}^2 = \left(\Sigma_k^{-1} \cdot \mu_k - \Sigma_j^{-1} \cdot \mu_j\right)$$

$$W_{kj}^3 = -\frac{1}{2}\left(\mu_k^T \Sigma_k^{-1} \mu_k - \mu_j^T \Sigma_j^{-1} \mu_j\right) - \frac{1}{2}\left(\log\left|\Sigma_k\right| - \log\left|\Sigma_j\right|\right) + \log\frac{p(C_k)}{p(C_j)}$$

To solve the distribution coefficients, it is common and convenient to use the Maximum Likelihood Estimation (MLE) method to find out the parametric distribution that yields the maximum product of posterior. Therefore, we can solve this classification problem analytically and numerically. The product of the group's posterior can be written as:

$$\Pi\, p(X,C_i) = \Pi\, p(X|C_i)\, p(C_i) = \Pi \frac{N_i \cdot \mathrm{N}\left(X_i|, \mu_i|, \sum_i\right)}{N}$$

The overall group's posterior product can be thereby written as:

$$p\left(t, C_i, \mu_i, \sum_i\right) = \prod_{n-1}^{n} \left[\frac{N_i \cdot \mathrm{N}\left(X_i|, \mu_i|, \sum_i\right)}{N}\right]^{T_i}$$

where $\boldsymbol{T_i = 1}$ if $\boldsymbol{X_i \subset C_i}$.

Then, we can take the derivative of the above equation with respect to $\boldsymbol{C_i, \mu_i}$, $\sum_i$, and then we can solve those equations to get the value for them. The expression solutions are given directly by the following:

$$C_i = \frac{N_i}{N}$$

where $\boldsymbol{N_i}$ is the number of data in the i group, and N is the summation of all the numbers in each group. The Gaussian group's mean is given by the following:

$$\mu_i = \frac{1}{N_i} \sum_{n=1}^{N} T_i \cdot X_i$$

The Gaussian Covariance is given below:

$$\sum_i = \frac{1}{N_i} \sum_{n \subset C_i} (X_n - \mu_i) \cdot (X_n - \mu_i)^T$$

The synaptic weights have been used to characterize the distinguished classes. In the Bayesian classification method, the trained weights can be found as the mean value of the posterior distribution. The Probability Density Functions(PDFs) of two groups can be derived from synaptic weights as listed below:

$$F(x,y) = I^T W_2 I + W_1^T I + W_0$$

$$I = \begin{bmatrix} x \\ y \end{bmatrix}$$

$$W_2 = -\frac{1}{2} \Sigma_i^{-1}$$
$$W_1 = \Sigma_i^{-1} \mu_i$$
$$W_0 = -\frac{1}{2} \mu_i^T \Sigma_i^{-1} \mu_i - \frac{1}{2} \log|\Sigma_i| + \log(p(C_i))$$

We can expand the matrix expression of $\boldsymbol{F}(\boldsymbol{x},\boldsymbol{y})$ as the quadratic area function:

$$F(x,y) = W_2(1,1)x^2 + 2W_2(1,2)xy + W_2(2,2)y^2$$
$$+W_1(1,1)x + W_1(2,1)y + W_0$$

Once the Bayesian classifier has finished the training process, we can process to check the classification results. In the figure below, it shows the failure data for the L-prize lamp, and the initial degradation starts around the center at [1, 0.99], and goes to [0.4, 0.98]. It is clear to see that the degradation trend has been going from the top right corner to the bottom left corner. Also, the red dash line shows the failure threshold between the healthy L-prize data and failure L-prize data.

Consequently, the following figure provides us intuition about the class distributions: (1) Healthy Groups and (2) Failure Groups. The red PDF is corresponding to the failure of L-prize data while the green PDF is the healthy L-prize data. The overlapping area displays the transition failure area, and if the generated testing data under this area, it is more likely that the system will decay quickly to pass into failure class, and then the maintenance and replacement should be taken care of. Typically, we want this overlapping PDF region to be as smaller as possible to make a quick decision, which requires the variances of two PDFs to be smaller. This will

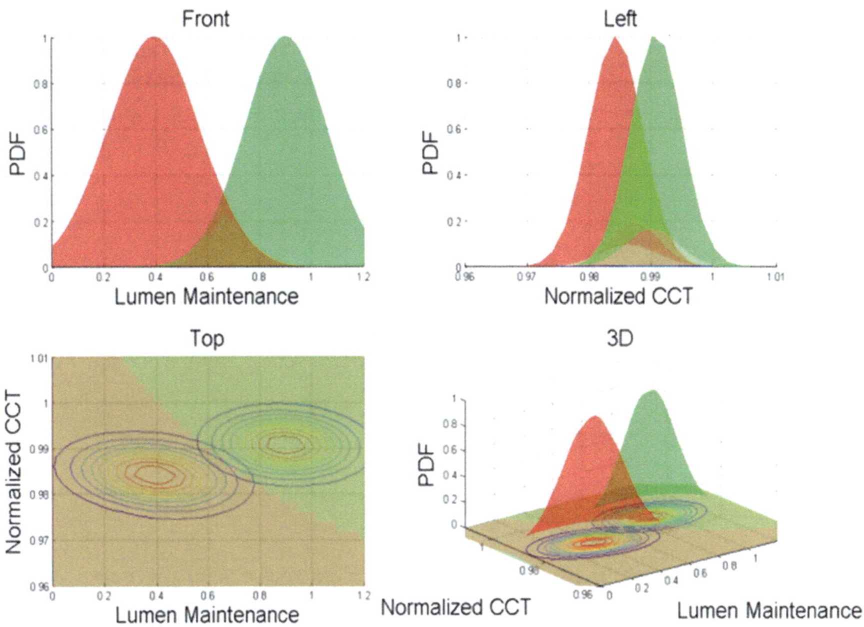

Fig. 10.2 Endoscope and instrument manipulators. (© IEEE 2007), Reprinted with permission

be achieved through the training data to be specialized and concentrated at the location we are interested in. Overall, the Bayesian unsupervised classifier is a powerful classification tool, because it does not need first to be systematically trained in the original model and there is no necessity to create a mapping between the input space and the output space. Therefore, any online data and incoming data can join into the unsupervised training process. The algorithm will learn from the prior distribution and decide which class the data belongs to (Fig. 10.2).

2.3 Kalman Filter and Extended Kalman Filter

The Kalman Filter (KF) is used for tracking the digital signal in the system, which has been widely used in space navigation, electronics control, and economics. Previously, the Kalman Filter has been used in the guidance and tracking applications [2]. The basic concept about the KF is to use the recursive and error correction algorithm to keep track of a signal in a dynamic system.

The Extended Kalman Filter (EKF) is built upon the KF algorithm, in which the nonlinear fundamental state matrix has been implemented to adapt to the advanced and nonlinear system. In its applications, system damage state estimation in the presence of measurement noise and process noise has been achieved using the Extended Kalman Filter (EKF).

The system state has been described in the state-space form using the measurement of the feature vector, the change of velocity in a feature vector, and the change of acceleration in the feature vector. The equivalent Extended Kalman Filter equation for state-space representation is in the presence of process noise and measurement noise is

$$\dot{x} = Fx + w$$

$$\dot{x} = f\left(x\right) + w$$

F is the system linear dynamic matrix; *f*(*x*) is a non-linear dynamic matrix; the *G* is the measurement matrix; $\boldsymbol{\mu}$ is the measurement vector; and $\boldsymbol{w}$ is the system white noise.

$$Z = H \cdot x + V$$

$$Z = h\left(x\right) + V$$

H is the measurement matrix, *z* is the measurement vector, *h(x)* is a measurement function, which is a nonlinear function of states, and v is a zero-mean random process described by the measurement noise matrix. The process noise can be calculated by taking the expectation of white noise:

$$\mathrm{Q} = \mathrm{E}\left[\mathrm{ww}^{\mathrm{T}}\right]$$

Similarly, the measurement noise matrix is derived from the measurement noise as follows:

$$\mathrm{R} = \mathrm{E}\left[\mathrm{vv}^{\mathrm{T}}\right]$$

Since the system-dynamics matrix '*F*' and measurement equations are nonlinear, a first-order approximation is used in the continuous Riccati equations for the systems dynamics matrix *F* and the measurement matrix *H*. It is expected that the progression of the failure is nonlinear, therefore, it needs to be linearized before it can be estimated. In the Extended Kalman Filter, the problem of linearization is addressed by calculating the Jacobian of the nonlinear function of states *f*(*x*) and the measurement function *h*(*x*) around the estimated state. System state at each future time has been computed based on the state space at predicting time point, system dynamics matrix, the control vector, control matrix, measurement matrix, measured vector, process noise, and measurement noise. The matrices are related to the nonlinear system and measurement equations according to:

$$F = \left.\frac{\partial f\left(x\right)}{\partial x}\right|_{x=\hat{x}}$$

$$H = \left.\frac{\partial h(x)}{\partial x}\right|_{x=\hat{x}}$$

From the linear dynamic equations, we can see that the world is linear as it assumed. All the problems can be solved linearly through the matrix approximation. Therefore, we can get the first and second derivatives from the linear matrix calculation.

$$\begin{bmatrix} \dot{x} \\ \ddot{x} \\ \dddot{x} \end{bmatrix} = F \cdot \begin{bmatrix} x \\ \dot{x} \\ \ddot{x} \end{bmatrix}$$

The system dynamic matrix for the *EKF* is:

$$F_{EKF} = \begin{bmatrix} \frac{\partial \dot{x}}{\partial x} & \frac{\partial \dot{x}}{\partial \dot{x}} & \frac{\partial \dot{x}}{\partial b} \\ \frac{\partial \ddot{x}}{\partial x} & \frac{\partial \ddot{x}}{\partial \dot{x}} & \frac{\partial \ddot{x}}{\partial b} \\ \frac{\partial \dot{b}}{\partial x} & \frac{\partial \dot{b}}{\partial \dot{x}} & \frac{\partial \dot{b}}{\partial b} \end{bmatrix}$$

The system dynamic matrix for the *KF* is:

$$F_{KF} = \begin{bmatrix} 0 & 1 & 0 \\ 0 & 0 & 1 \\ 0 & 0 & 0 \end{bmatrix}$$

We use this Jacobian matrix to linearize the problem; therefore, it can use similar KF updates. This is for the second-order system, and thus we can find the transfer function F to describe a certain system, and it is the key to find a fundamental matrix **Φ(t)**. In this chapter, for the EKF, we used the following exponential system model:

$$x = \alpha \cdot e^{\beta \cdot t}$$

The state vector is:

$$x_k = x \quad \dot{x} \quad \beta$$

This is an exponential function. The $\boldsymbol{\alpha}$ and $\boldsymbol{\beta}$ are two coefficients that are decided by different systems. The first derivation $\dot{\mathbf{x}}$ and the second derivation $\mathbf{x}$ are derived from the following equation:

$$x = \alpha \cdot e^{\beta \cdot t}$$

Therefore, the elements in the system dynamic matrix will be calculated as:

$$F_{EKF} = \begin{bmatrix} \beta & 1 & x \\ \beta^2 & \beta & 2\beta x \\ 0 & 0 & 0 \end{bmatrix}$$

Usually, the fundamental matrix $\boldsymbol{\Phi}(\boldsymbol{t})$ can be obtained from two ways: the first way, we can get it from Laplace Transform, simply as the following:

$$\Phi(t) = \ell\left[(SI - F)^{-1}\right]$$

where $\boldsymbol{\ell}^{-1}$ is the inverse of the Laplace Transform; the second way, known as the common way to find $\boldsymbol{\Phi}(\boldsymbol{t})$, deriving from the Tylor Series expansion:

$$\Phi(t) = I + FT + \frac{(FT)^2}{2!} + \frac{(FT)^3}{3!} + \cdots$$

Normally, we use the first two terms for representing the fundamental matrix $\boldsymbol{\Phi}(\boldsymbol{t})$, because simply adding more terms cannot improve the precision and convergence of the filter. The Fundamental Matrix in the Extended Kalman Filter is:

$$\Phi_{EKF}(T) \approx I + F_{EKF}T = \begin{bmatrix} 1+\beta T & T & xT \\ \beta^2 T & 1+\beta T & 2\beta \cdot xT \\ 0 & 0 & 1 \end{bmatrix}$$

The Fundamental Matrix in the Extended Kalman Filter is:

$$\Phi_{KF}(T) \approx I + F_{KF}T = \begin{bmatrix} 1 & T & 0.5Ts^2 \\ 0 & 1 & T \\ 0 & 0 & 1 \end{bmatrix}$$

In the Kalman Filter, the Fundamental Matrix will be directly used to update the estimation from the last estimation. Generally speaking, the process to find the optimal estimation can be expressed as following steps: First of all, we make the primary estimation, which should be approximate to the initial value in the dataset, and secondly, we can find the projection using the fundamental matrix **Φ**(t) and simply calculate as:

$$\bar{x} = \Phi(t) \cdot \hat{x}$$

For the KF, the projection can be represented by:

$$\begin{bmatrix} \Delta\dot{x} \\ \Delta\ddot{x} \\ \Delta\dddot{x} \end{bmatrix} = \begin{bmatrix} 0 & 1 & 0 \\ 0 & 0 & 1 \\ 0 & 0 & 0 \end{bmatrix} \begin{bmatrix} \Delta x \\ \Delta\dot{x} \\ \Delta\ddot{x} \end{bmatrix} + \begin{bmatrix} 0 \\ 0 \\ w \end{bmatrix}$$

For the EKF, the projection can be represented by:

$$\begin{bmatrix} \Delta\dot{x} \\ \Delta\ddot{x} \\ \Delta\dot{b} \end{bmatrix} = \begin{bmatrix} \frac{\Delta\dot{x}}{\Delta x} & \frac{\Delta\dot{x}}{\Delta\dot{x}} & \frac{\Delta\dot{x}}{\Delta b} \\ \frac{\Delta\ddot{x}}{\Delta x} & \frac{\Delta\ddot{x}}{\Delta\dot{x}} & \frac{\Delta\ddot{x}}{\Delta b} \\ \frac{\Delta\dot{b}}{\Delta x} & \frac{\Delta\dot{b}}{\Delta\dot{x}} & \frac{\Delta\dot{b}}{\Delta b} \end{bmatrix} \begin{bmatrix} \Delta x \\ \Delta\dot{x} \\ \Delta b \end{bmatrix} + \begin{bmatrix} 0 \\ 0 \\ w \end{bmatrix}$$

The next estimate vector could be obtained from the following equation:

$$\hat{x} = \bar{x} + K(Z - H \cdot \bar{x})$$

K: Kalman Gain
H: measurement matrix
Z: measurement

Each time we update the Kalman Gain (KG) and Covariance Matrix (CM), which minimizes the errors and makes optimal approximation during each step. Thus, the Kalman Gain mainly delivers information about how our estimation is close to the observation. The way to obtain Kalman Gain (K) is from three Riccati recursive equations:

$$M_k = \Phi_k P_k \Phi_k^{\ T} + Q_K$$

$$K = M_k H_t \left(H M_k H_t + R_k\right)^{-1}$$

$$P_k = (I - KH) M_k$$

In the above equations, $\boldsymbol{M}_k$ is the covariance matrix; $\boldsymbol{\Phi}$ is the fundamental matrix; $\boldsymbol{\Phi}_t$ is the transpose of the fundamental matrix; $\boldsymbol{P}_k$ is another covariance matrix that representing error according to the time; $\boldsymbol{Q}_k$ is the discrete process noise matrix, which is calculated as the following:

$$Q = \Phi_s \begin{bmatrix} 0 & 0 & 0 \\ 0 & 0 & 0 \\ 0 & 0 & 1 \end{bmatrix}$$

$$Q_k = \int_0^{Ts} \Phi(\tau) \cdot Q \cdot \Phi(\tau)^T \, d\tau$$

H is the unit measurement matrix, and $\boldsymbol{H}_t$ is the transpose of H; the K is the Kalman Gain; $\boldsymbol{R}_k$ represents the measurement noise of the different system. We notice that those three equations run in the recursions: for the initial covariance error $\boldsymbol{P}_0$, we can find a variance matrix $\boldsymbol{M}_k$ that represents the current error in time. Then, we use it in the second equation to find the Kalman Gain K. After that, we substitute the estimated Kalman Gain K into the third equation to update the covariance error $\boldsymbol{P}_k$ from the last time. Thus, we obtain the 'next' covariance error $\boldsymbol{P}_{k+1}$. Iteratively, as we go back to the first equation, we can update the covariance matrix $\boldsymbol{M}_k$ and Kalman Gain K. In the EKF, the Euler integration has been utilized instead of the KF's fundamental matrix. To get the estimation by projection, it can be found in the following two equations:

$$\overline{x}_k = \hat{x}_k + \overline{\dot{x}}_{k-1} T$$

$$\overline{\dot{x}}_k = \hat{\dot{x}}_k + \overline{\ddot{x}}_{k-1} T$$

We call the equation above is the update equations, $\boldsymbol{x}_k$ represents the projection from the last time k; $\hat{\dot{\boldsymbol{x}}}_{k-1}$ is the first derivative at the time (k-1); $\hat{\boldsymbol{x}}_k$ is the estimation at time k; $\hat{\dot{\boldsymbol{x}}}_k$ is the first derivative at time k; T is the sample time.

$$\hat{x}_k = \overline{x}_k + K_1 \left(Z - H \cdot \overline{x}_k \right)$$

$$\hat{\dot{x}}_k = \overline{\dot{x}}_k + K_2 \left(Z - H \cdot \overline{x}_k \right)$$

The above equations are the basic Extended Kalman Filter equations, which are to find the estimation and its 'velocity'. Also, in those equations, it uses the same three Riccati equations that are expressed in the Kalman Filter to obtain the Kalman Gain $\boldsymbol{K}_1$ and $\boldsymbol{K}_2$.

From the above, we can see that the Extended Kalman Filter has turned the nonlinear problem into a linear solution through the integration method. So at each step, the Extended Kalman Filter made a small integration, and if the integrated time is small, then the answer we get is becoming much closer to the real solution. However, the difficulty within the Extended Kalman Filter is to find the exact non-linear model to describe the system, which always contains many unknown coefficients. Therefore, the better we know about the test system, the better we can predict the system model in the Extended Kalman Filter, for example, the theories and functions in the situation of Electronics' failure. Therefore, the prediction of Remaining Useful Life (RUL) would be close to the real RUL in our PHM network.

2.4 LED Applications

In this chapter, we cover some applications of the KF/EKF and Bayesian Method in the LEDs system, and we firstly focus on the application of the Kalman Filter and Extended Kalman Filter Algorithm into the prediction of LED 70% Lumen Maintenance Life (L70), and the former prediction method, TM-21 calculator, can roughly make a prediction between average 20,000 and 36,000 hours for the bare LED's lifetime. The TM-21 calculator is a Curve-Fitting method basing on the Arrhenius Equation, which needs to calculate the Activation Energy, Pre-decay factor, and Decay Rates. Also, the TM21 is also overall prediction related to the knowledge of upper and lower temperature cases. However, the Activation Energy may not be obvious and easy to be obtained. Thus, we can introduce a substitutable prediction method: the KF/EKF prognostic method. It can dynamically update the prognostic L70 lifetime at each time of interest according to the system prognostic model. The KF/EKF algorithm bases on the identified system measurement noise and system process noise. The linear, polynomial, exponential stochastic models for predicting LEDs' life are discussed in this chapter, and each model has been incorporated into the KF/EKF algorithm to generate the L70 estimations. The Normal, Lognormal, and the Weibull distributions can be used to fit those estimations.

In the LEDs' Prognostic Health Management (PHM), we use the L70, which is the 70% Lumen Maintenance Output comparing with the 100% from the original as the failure criterion of LEDs. The RUL is estimated by the extrapolations at each evaluated point. The figure below shows the intuition of RUL for L70. In the plot the colorful line is a raw dataset from Philips LED LM80-08 Data, the dash green lines are KF/EKF fitting lines, and the dash green line is KF mean extrapolations for L70, and the red curve shows the LED life distribution from the last evaluated point (Fig. 10.3).

The Chromaticity u' and v' also could be tracked by the Kalman Filtering Algorithm. Therefore, we could gather the information into the reliability discussion of LEDs. Compared with lumen degradation, the chromaticity is very long-term degradation, but the change of delta u' and delta v' would result in shortening LEDs useful life in some color restricted situation. The dash blue line is the KF chromaticity tracking for u' and v' in 9000 hours operating hours. The figure below shows u' v' KF tracking in 2D 1976 CIE u' v' coordinate. The circle is ±0.06 confident interval; the circle center is the central point for 3000 K (Fig. 10.4).

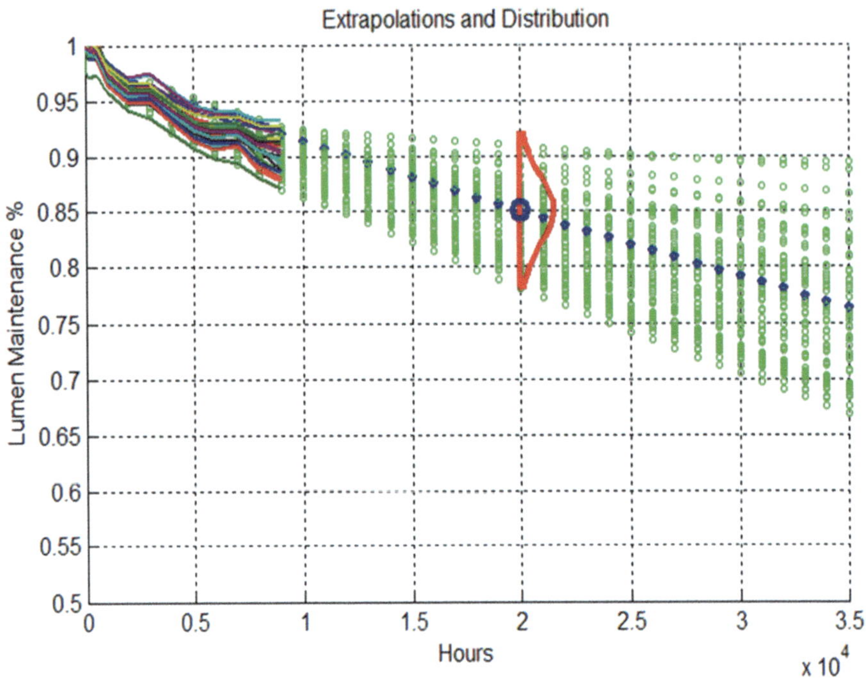

Fig. 10.3 Endoscope and instrument manipulators. (© IEEE 2007), Reprinted with permission

3 Supervised Learning and Artificial Neural Networks

Supervised learning is a learning algorithm that learns from inputs and labels by human experts. Usually, the algorithm can learn a set of parameters specified in a model during the training, in which the training procedure continuously reduces the loss function measuring errors between the model output and target labels. It takes advantage of the flexibility and can be adapted to any parametric and non-parametric distributions. Artificial Neural Networks (ANNs) are very basic and fundamental for constructing deep models.

The ANNs were first introduced as a perceptron algorithm that mimics a single neuron's activation in response to the input signal [3]. In the perceptron, the input signal is multiplied by a learnable weight and passes through a unit neuron with an activation that is like our neuron's response to a biological signal inside the neural cells. The activated output gives a quantitative value that can be compared with the target, from which an error is calculated. Then, the algorithm applies the error back-propagation to updates the weight so that the output can get closer to the target in the next iteration. This mechanism forms the basic idea and logic for the ANNs' prediction and training.

The ANNs' training relies on the error back-propagation algorithm, which was introduced and developed in the 1970s and 1980s [4]. Earlier this method was

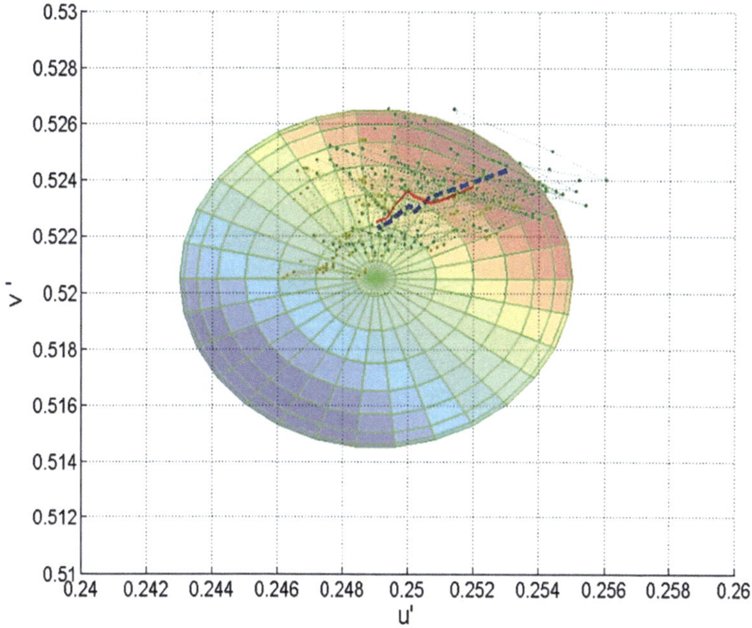

Fig. 10.4 Endoscope and instrument manipulators. (© IEEE 2007), Reprinted with permission

computational expensive when there are many neurons and layers in the network. Because all the weights in the network need to be updated at each training iteration, and the vectorization can save some computations but it is still a significantly high cost to use a large network.

In industrial applications, it is often impossible to find a universal function to solve and make an approximation to the problem. For example, when the system is growing bigger, its behavior with input and output can be very complicated, therefore, it is difficult to find a parameterized function to describe the behavior. However, using the ANN as a generalized model, the behavior of the system can be modeled precisely and trained using adjustable weights and training data. The availability of training data is not an issue and can be very selective, and choosing the right dataset for training the network can be very tricky and vital. In the Solid-State Lighting system, a set of quantitative internal and external variables, the internal variables such as current, voltage, lumen maintenance, and color-shifting are very important for modeling the system degradation, and the external variables can be collected from the experimental conditions such as temperature, humidity, time, and duration. Usually, when the system getting bigger and more complicated, the number of indicators, as the features in the system, can be as many as few hundreds.

Generally, the ANN can solve many kinds of issues, but major two categories and tasks: (1) Failure Mode Classification; (2) Failure Regression and Predictions. In a classification problem like Bayesian Classifier, a failure mode classifier can be trained with many datasets containing both healthy and failure data, and building such classifier requires a large amount of collected data and the data needed to be labeled with pre-identified methods. For example, different failure modes may require engineers to manually examine and classify. Once the labeled datasets are ready, the rest of the work is to choose a network, build the model, and testing.

Computational power has been improved dramatically during the past few years. Many data centers offer computational power to deploy deep models. Taking these advantages, industrial applications can be launched much faster and easier. The data center can collect big data from companies and users. With enormous industrial data, Deep Learning(DL) [5] can build a very robust model to handle the input from the different scenario.

There are many candidate networks for solving different engineering problems. Basically, there are shallow and deep neural networks. The shadow neural networks only have a limited number of hidden layers while deep neural networks can have many hidden layers that can be up to several hundred layers. Usually, the more hidden layer means the more computation and dataset that are required to train the network. Also, with more hidden layers, the network can be adapted to more complicated systems with complex behaviors and interactions.

In the deep neural networks, there are major two categories that have been widely used in the industrial AI applications, one is a general Deep Neural Network with fully connected layers and another is Convolutional Neural Network(CNN) [6]. Depending on the problem, we need to choose them separately.

The Convolutional Neural Network is widely used with applications related to computer vision, and in those applications, the target is usually very straight forward: (1) Object Detection and Recognition; (3) Instance Segmentation. The object detection deals with the target to identify many target objects in the images, for example, in the autonomous driving field, the targets can be one of the following: pedestrian, traffic light, car, and building [7]. The task initially returns the locations using a bounding box to locate the target and then put a label to identify its class. In the instance segmentation problem, the target is to identify the label of each pixel to form a region or a mask to segment the object from the other masks. This application is extremely useful in medical applications including the tumor or abnormal tissue segmentation from medical images, in which the size of the target can be calculated and evaluated precisely.

In the object detection and recognition, the CNN is used to extract the features from natural images, and the target objects are identified by the bounding boxes with labels in the images. Also, this model can be used for object classification. In the LED application, the failure images of LEDs can be classified and used for building the model, which would reduce the human labor and examination cost. For failure analysis of LEDs, the pixel level study about the failure features such as crack, delamination, and material degradation can be done by the instance segmentation network from which the size of crack and delamination and thickness of the degradation can be modeled and measured. Semantic Segmentation is widely used for a mask generation for a target region [8]. In the image, the mask can be labeled on each pixel. Usually, a binary mask is an example of foreground and background object detection. Multiple masks can also be labeled and generated. The semantic segmentation can be easily applied for material degradation detection. For example, it can detect the area that has obviously abnormal from the connected region. Similarly, it has already been applied to the automatic tumor segmentation for medical images.

The design of the networks contains several important aspects: first, designing for the inputs for the problem. We need to differentiate the problem to be a classification problem or regression problem. In the classification problem, the inputs are the features from the system as we mentioned before, the outputs are the class labels that human experts created for each input. In the regression problem, the inputs are the low-level features as those in the problems, the difference is that in the regression problem, the output is a metric with a numerical value and the error between it and the target is constantly minimized using a loss function. Once we have set our goal, we need to design a network for training. In general, we start with a network with enough hidden layers to ensure the capability of the network can match with a complex dataset, and later we can reduce the size of the network to improve its efficiency. If the problem is for the digital images, we need to apply the CNN, otherwise, a fully connected network will be adopted.

To design an appropriate network for the problem can be very tricky and requires some efforts and iterations. For example, the key is to deal with an overfitting issue with the model and data. The overfitting issue is a classical problem in statistics where the mismatches between the trained model and validation data. In deep learning, it is often the deep neural network that has more power on fitting given training data than under-fitting the data. It is important to use some regularization methods to control the performance of the model to avoid overfitting. Generally, the regularization can be achieved using some techniques like adding dropout layers, adding regularization terms of weights in the loss function, or adding more training data.

Once we finalized the structure of the network, we need to consider an appropriate loss function dealing with our problems. For the classification problem, it is often utilized as a sigmoid or softmax function for the output layers that will output a binary value or a one-hot vector for classes, and then the difference between the output and the target label can be calculated and minimized through the back propagation. For the classification problem, the loss function is often chosen as an error metric function like the Cross-Entropy function between the output and the target values.

After the design of the network, the next step is to train the network. There is a set of hyperparameters for training the network. Usually, the parameters that affect the training convergence of the network are the batch size and learning rate. The batch size is the size of training data feeding into the network at each step. The large batch size approximates the problem better but it may cause the training to diverge due to a large variance in the dataset. The learning rate affects the speed of training. A large learning rate would cause the oscillation of the loss and the network may never learn to approximate the goal. However, small learning causes the training slowly and may take a long time to reach the goal. It is often recommended to use start from a learning rate that does not cause the divergence and reduce the learning rate after the amount of training.

The training data and evaluation data are critical for developing a deep neural network, which prevents overfitting. The key is to split the available dataset into training, developing, and testing datasets, in which the training dataset is about 60–70% of the entire data while the developing and testing is about 30–40%. In the developing and testing dataset, usually we have 50% for developing and another 50% for testing, but it could have some overlaps between them. If the dataset is huge that contains millions of objects, we will use 98% of the entire data for training and 1% for developing, and 1% for testing. This will guarantee that the model is trained with more data with large variance and is able to generalize to the new data.

In the AI field, the data is the key to generating a high generality model for applications. Generally, in the industrial or medical field, it is very easy to acquire a big dataset. However, data labeling is very expensive in terms of human labor and time. To solve it, Data Augmentation is a technique for generating enormous data from limited labeling data. For example, computer graphics are widely used for image-based applications, and we can use computer graphics to simulate new images that are very similar to the data generated by imaging devices. Another way to generate an augmented dataset is to apply artifacts such as noises and backgrounds to the data. It has been proved in many applications, the data augmentation can extend the model's ability and generality for the problems. However, we need to pay attention to the scenario that several augmentations have been mixed, which could possibly shift the problem and increase the bias of the model.

In short, although there are many challenges in the development of modern AI techniques, the digital transformation in the industry has already achieved great success. The AI-based techniques have many advantages over the traditional model in terms of accuracy, efficiency, and robustness. Also, there are many potentials for those DL-based techniques to be applied for the LEDs material degradation and reliability analysis. The key to success for those applications is how we can seamlessly connect the existing AI techniques with valuable industrial datasets and efficiently deploy the models in the hardware. Overall, modern AI technology for our industrial applications is on the horizon.

References

1. C.M. Bishop, *Pattern Recognition and Machine Learning* (Springer, 2006)
2. H. Musoff, P. Zarchan, *Fundamentals of Kalman Filtering: A Practical Approach* (American Institute of Aeronautics and Astronautics, 2009)
3. F. Rosenblatt, The perceptron: a probabilistic model for information storage and organization in the brain. Psychol. Rev. **65**(6), 386 (1958)
4. D. E. Rumelhart, G. E. Hinton, R. J. Williams. Learning internal representations by error propagation. No. ICS-8506. (California Univ San Diego La Jolla Inst for Cognitive Science, 1985)
5. I. Goodfellow, *Yoshua Bengio, and Aaron Courville*. Deep learning (MIT press, 2016)
6. A. Krizhevsky, I. Sutskever, G.E. Hinton, *Imagenet Classification with Deep Convolutional Neural Networks* (Advances in neural information processing systems, 2012)
7. J. Redmon, et al. You only look once: Unified, real-time object detection. *Proceedings of the IEEE conference on computer vision and pattern recognition*, 2016
8. J. Long, E. Shelhamer, T. Darrell. Fully convolutional networks for semantic segmentation. *Proceedings of the IEEE conference on computer vision and pattern recognition*, 2015

Chapter 11
Degradation Analysis for Reliability of Optoelectronics

Cheng Qian, Zeyu Wu, Wei Chen, Jiajie Fan, Xi Yang, Yi Ren, Bo Sun, and Zili Wang

1 Introduction

Optoelectronics, or commonly known as light-emitting diodes (LEDs), are the devices which convert electricity into light with singlet or multiplet spectral power distributions (SPDs). They usually have a small but complicated structure by consisting of a number of components made by different materials, and exhibit therefore different failure features under a combination of temperature, humidity, driving current, and perhaps some contamination conditions [1–3]. Figure 11.1 illustrates a frequently used phosphor-converted white LED (pc-white LED) that comprises of serval parts such as LED chip, phosphors, encapsulant, bonding wire, die attach, lead frame, and housing [4]. Furtherly, the LED chip is divided by a stack of layers including the epi-layer, current spreading layer, chip substrate, chip electrodes, etc.; the phosphors are mixed with encapsulant to undertake an important mission that converts the blue light emitted from LED into yellow and red spectrum; the housing acts as a solid shell to prevent mechanical damages from external loads; the leadframe is embedded inside of the housing to connect the LED chip with the external circuit via bonding wires for electricity conductivity; the die attach layer with solder paste is placed to fix the LED chip on the leadframe, and also plays as a heat flow passage in between the LED chip and lead frame.

With the special electroluminescence performance, the light extraction efficiency of optoelectronics can be achieved up to 30% and more, which makes a great

C. Qian (✉) · Z. Wu · X. Yang · Y. Ren · B. Sun · Z. Wang
School of Reliability and Systems Engineering, Beihang University, Beijing, China
e-mail: cqian@buaa.edu.cn

W. Chen
Shenzhen Institute of Wide-Bandgap Semiconductors, Shenzhen, China

J. Fan
Academy for Engineering & Technology, Fudan University, Shanghai, China

W. D. van Driel, M. Yazdan Mehr (eds.), *Reliability of Organic Compounds in Microelectronics and Optoelectronics*,
https://doi.org/10.1007/978-3-030-81576-9_11

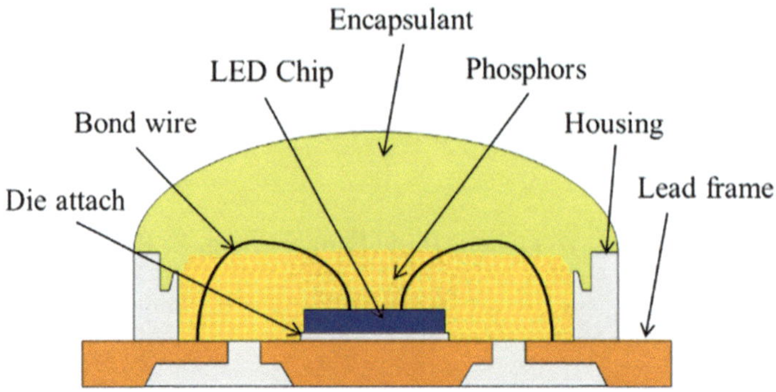

Fig. 11.1 Illustration of a pc-white LED package

advancement in general lighting applications compared to the traditional incandescent and fluorescent counterparts. However, drawback of the optoelectronics comes from their poor heat dissipation capacity. The remaining electricity energy turns to hard-to-be-dissipated heat to significantly raise the temperature of the device. In addition, high current density and improper packaging process are other important factors to create damages with high possibilities. In summary, these are the main causes leading to degradation failures, such as lumen depreciation, color shift, and Correlated Color Temperature (CCT) shift, in optoelectronics [5, 6].

Among the degradation failures, lumen depreciation has earned much more attention compared to the other two failure modes [7]. The lumen maintenance lifetime of an optoelectronic is generally defined by the time when its lumen maintenance depreciates to a certain degree p, indicated as L_p. According to a number of Solid-State Lighting standards published by Energy Star, Illuminating Engineering Society of North America (IESNA), International Electrotechnical Commission (IEC), and so on, a widely accepted consensus for the p value is 0.7, in correspondence to the L_{70} lifetime [8–13]. With modern technologies, the L_{70} lifetime of a well-manufactured optoelectronic device is considerably long and temperature-sensitive. It is usually higher than 25,000 hours at a temperature of 105 °C and could reach the magnitude of 100,000 hours at room temperature.

In addition to the lumen depreciation, color shift failure is an important reliability issue for the chromatic performance of optoelectronics. The color shift is used to qualify the gradual distortion change in SPDs of the optoelectronics. It reveals the deviation level of chromaticity coordinates from their initial values, as defined by Eq. (11.1).

$$du'v' = \sqrt{\left(u' - u_0'\right)^2 + \left(v' - v_0'\right)^2} \tag{11.1}$$

in which u' and v' are chromaticity coordinates in CIE 1976 color space, and u_0' and v_0' are the corresponding initial values respectively. As required by Energy Star standards, the CCT shift of the luminaires and lamps with the use of optoelectronics as light sources cannot exceed 0.007 for general lighting applications [9, 10]. In

Table 11.1 Nominal CCT categories

Nominal CCT (K)	Target CCT and tolerance (K)
2700	2725 ± 145
3000	3045 ± 175
3500	3465 ± 245
4000	3985 ± 275
4500	4503 ± 243
5000	5028 ± 283
5700	5665 ± 355
6500	6530 ± 510

some special circumstances with high color stability, a smaller threshold such as 0.004 or even 0.002 might be adopted for stricter requirements. Then the color maintenance lifetime of an optoelectronic can be defined by the time its $du'v'$ is out of the 0.007 (sometimes 0.004 or 0.002) threshold.

CCT failure is also a chromatic failure that affects the user comfort level in color exhibition. The CCT is defined by the most matchable temperature in degree Kelvin to the blackbody locus and used as a cold-to-warm appearance color indicator for optoelectronics. A couple of iso-temperature lines are preconstructed for the convenience of CCT categorization, based on the CIE 1976 color space. Table 11.1 lists the frequently used nominal CCTs from 2700 K to 6500 K. During the operation process, the CCT of an optoelectronic device gradually shifts. The CCT shift in a position direction means the color turns cold, while on the contrary means the color turns warm. The target and tolerance CCTs in respect to different nominal CCTs required by ENERGY STAR are provided in Table 11.1 as well [11]. An over-variation is regarded as CCT shift failure, and the CCT maintenance lifetime can be defined by the time as soon as the CCT is beyond the tolerance range around the corresponding target value.

In this Chap. 3 different types of models, i.e., the exponential decay model, Gamma process-based model, and spectral power distribution-based model, are introduced for degradation description of optical and chromatic performance of the optoelectronics, and also lifetime estimations on this basis, in the Sects. 15.2 to 15.4 respectively. In the end, the concluding remarks of these models are summarized in Sect. 15.5.

2 Exponential Decay Modeling

Among a number of commonly used empirical curve fitting models, the exponential function is the most widely accepted candidate for lumen depreciation prediction and lifetime estimation of optoelectronics [14]. The general equation of an exponential decay model is expressed as:

$$\Phi = \beta e^{-\alpha t} \tag{11.2}$$

where Φ is the lumen maintenance, β is the pre-factor, and α is the decay parameter. And the lumen maintenance lifetime L_p is then calculated by:

$$L_p = \frac{\ln(\beta / p)}{\alpha} \tag{11.3}$$

in which p is the lumen maintenance threshold to cause the lumen depreciation failure. Usually p is designated as 70% for the calculation of L_{70} lifetime, as specified in various standards.

By using Eq. (11.2), the degradation tendency of lumen maintenance of an optoelectronic device can be obtained from thousands of hours of test data by means of an extrapolation technique. A number of curve fitting algorithms such as Least Square, Maximum Likelihood and Non-linear Filter are possible options to get a high extrapolation accuracy, depending on the operating circumstances [15]. And L_{70} lifetimes of these optoelectronics can be then evaluated by extrapolating the lumen depreciation curve to meet the 70% lumen maintenance threshold by using Eq. (11.3). An illustration to obtain lumen depreciation trends of optoelectronics is shown in Fig. 11.2 where the lumen depreciation trends of two test samples under 55 °C and 85 °C respectively were curve fitted according to Eq. (11.2).

From Fig. 11.2, it is also found that temperature plays a significant role in lumen depreciation. As a result, the Arrhenius model shown in Eq. (11.4) can be applied to describe the temperature-dependent degradation rate for optoelectronics [2].

$$\alpha = Ae^{-\frac{E_a}{kT_s}} \tag{11.4}$$

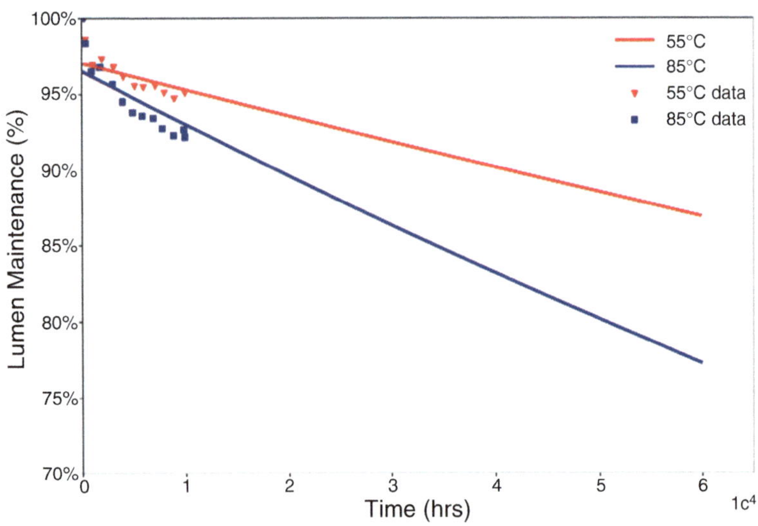

Fig. 11.2 Predictions of lumen depreciation by using an exponential decay model

in which, α is degradation rate, A and E_a are the pre-factor and activation energy respectively, k is the Boltzmann constant (i.e., 8.617385×10^{-5} eV/K), and T_s is the solder temperature in Kelvin. Because of the long-expected lifetime, it is difficult to test the lumen depreciation process of an optoelectronic device through its entire life span. Then the temperature-based Accelerated Degradation Test (ADT) has become a hot topic to replace the traditional operation tests at room temperature [16]. With the ADT data and exponential decay model, the lifetime of optoelectronics can be predicted with an acceleration factor AF calculated as:

$$\mathrm{AF} = \frac{\alpha|_{T_1}}{\alpha|_{T_2}} = \mathrm{e}^{-\frac{\mathrm{E_a}}{\mathrm{k}}\left(\frac{1}{T_2}-\frac{1}{T_1}\right)}$$

in which T_1, T_2 are the room and accelerated temperatures, and $\alpha|_{T_1}$, $\alpha|_{T_2}$ are the depreciation rates of the optoelectronics at room and accelerated temperatures, respectively.

There are two major solutions for temperature-based ADTs, which are Constant Stress Accelerated Degradation Test (CSADT) and Step Stress Accelerated Degradation Test (SSADT), to perform the optoelectronics' reliability evaluation and lifetime estimation [17, 18]. The difference in between these two tests is focused on the loading mode of test temperature. CSADT tests are those conducted by a constant high temperature throughout the tests. Whereas SSADT tests are those where the test temperature is step-wisely increased within a proper temperature range. Compared to CSADT tests, SSADT tests enable a higher lumen depreciation efficiency and comparable lifetime prediction accuracy only using less amount of samples and test resources, and for this reason are widely used for optoelectronics [19, 20]. As an illustration, Fig. 11.3 shows a comparison between the experimental results of lumen depreciation of optoelectronics under both SSADT and CSADT tests [21]. In the SSADT, the test samples were tested under 130 °C for the first 500 hours and then 140 °C for the last 500 hours. While in the CSADT, two batches of test samples were tested under 130 °C and 140 °C for 500 hours respectively. It can be seen that SSADT tests provide more significant degradation results on lumen maintenances within the same test period, compared to CSADT tests.

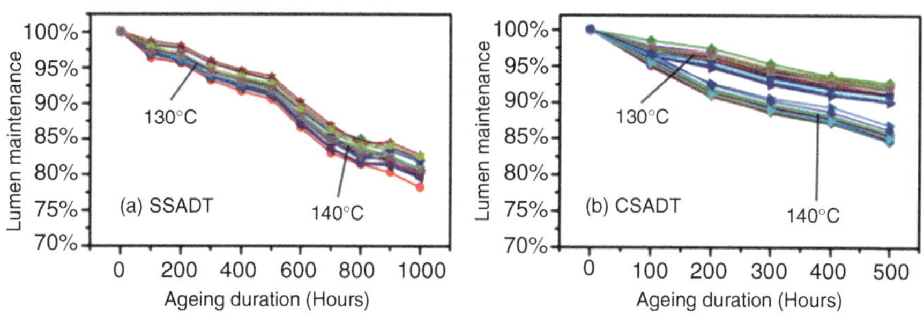

Fig. 11.3 Lumen maintenance degradation curves of optoelectronics by: (**a**) SSADT tests; (**b**) CSADT tests [21], with permission

For optoelectronics, the emitted lumen flux and intensity are proportional to the concentration of carriers in their p-n junctions [21]. As the concentration of carriers depends on the current density and junction temperature, lumen maintenance degradation curves of the optoelectronics are varied with their thermal and electrical stresses as well. Figure 11.4 shows the lumen depreciation curves of 1 W white LEDs submitted to different levels of thermal-electrical stresses [22]. It can be noticed that increase of the forward current significantly exacerbates the degradation of lumen maintenances of the optoelectronics under a junction temperature of 160 °C. This implies that ADTs driven by thermal-electrical stress can furtherly accelerate the lumen depreciation failure for these LEDs, compared to temperature-based ADTs.

When applied by both thermal and electrical stresses, the degradation rate of the optoelectronics can be calculated using an Eyring model:

$$\alpha = AI^{n} \exp\left(-\frac{E_a}{\mathrm{k}T_{\mathrm{j}}}\right) \tag{11.5}$$

where T_j is junction temperature of the optoelectronic device, I is the forward current, E_a and k are activation energy of the optoelectronic device and Boltzmann constant respectively, n and A are the model constants. By substituting Eq. (11.5) into Eq. (11.3), the lifetime of the optoelectronic device under a thermal-electrical stress can be established as:

$$L_{\mathrm{p}} = \Lambda \mathrm{I}^{-\mathrm{n}} \exp\left(\frac{E_a}{\mathrm{k}T_{\mathrm{j}}}\right), \Lambda = \frac{\ln(\beta/\mathrm{p})}{\mathrm{A}} \tag{11.6}$$

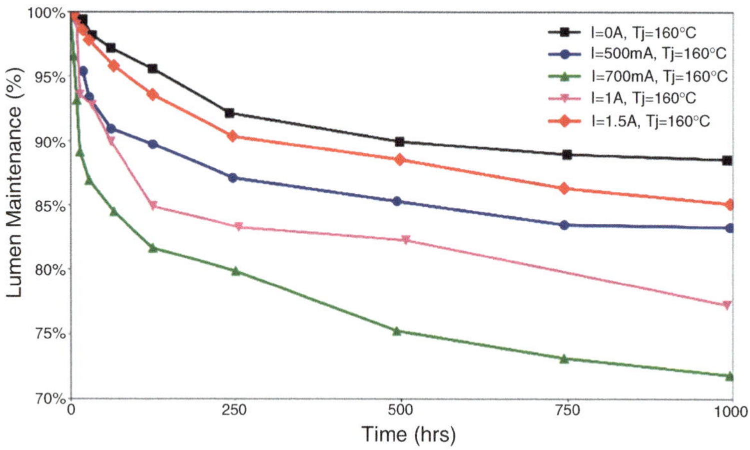

Fig. 11.4 Lumen depreciation curves of 1 W white LEDs under various levels of thermal-electrical stress [22], with permission

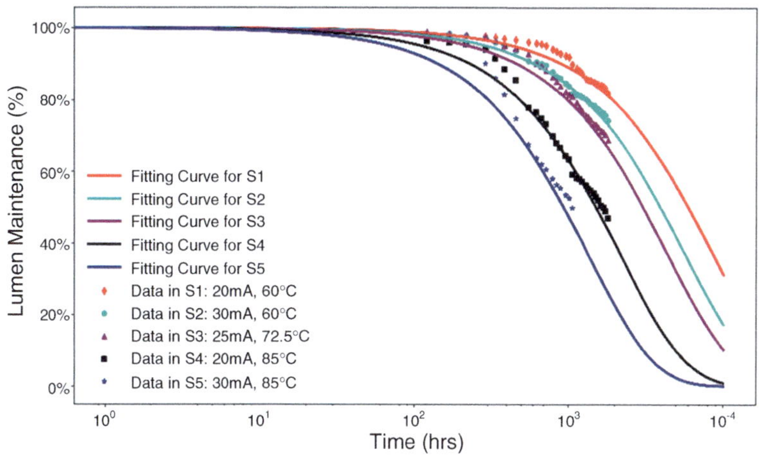

Fig. 11.5 Degradation curves of lumen maintenances of white LEDs under different stress levels [24], with permission

Figure 11.5 shows experimental measurements of lumen maintenances of 5 batches of white LEDs under different thermal and electrical stress levels, and the reasonable lumen degradation prediction curves by using Eq. (11.6) [23, 24]. It is also found that lumen depreciation failures of these test samples are more sensitive to thermal stress rather than electrical stress. This confirms that the temperature stress plays a primary role in the thermal-electrical-based ADTs.

Furthermore, considering effects of uncertainties originated from material defects, measurement error, turbulence of loading conditions, manufacturing tolerance, etc., lumen depreciation of the optoelectronics can be predicted by an expanded exponential decay model shown as:

$$\Phi(t) = M(t) + \varepsilon(t) \tag{11.7}$$

in which $\Phi(t)$ is the actual lumen depreciation path with random offsets, $M(t)$ is the theoretical lumen depreciation path calculated by Eq. (11.2), and $\varepsilon(t)$ is the stochastic error term resulted by those above-mentioned uncertainties. In this way, instead of deterministic values, statistical distributions can be obtained to achieve conservative predictions of L_{70} lifetimes of the optoelectronics. There have been two models (which are mixed effect model and stochastic process model) to incorporate the error distributions into Eq. (11.7) [25]. The mixed-effect model is developed by building up an analytical formula that mixes $M(t)$ and $\varepsilon(t)$ together [26, 27]. However, determination of the multivariate parameters in those models becomes quite expensive and complicated. By contrast, the stochastic process model is capable to reduce the computational complexity to a great extent since it only deals with the stochastic error prediction by using stochastic diffusion models such as the Gauss-Markov process, Wiener process, and Gamma process [25, 28–31].

3 Gamma Process-Based Modeling

Gamma process model has been firstly developed for predicting the random wear degradation process in 1970s by Abdel-Hameed [32]. Since then, it has been increasingly used in degradation analysis considering with stochastically measurement uncertainties in many applications [33–36]. And after nearly 50 years of development, it is believed that the Gamma process has been the most appropriate method for stochastic modeling of monotonic and gradual degradation processes that take place in a sequence of tiny increments over time [37–39]. And relevant studies in degradation prediction of the lumen and color maintenance are recently observed in prognostic investigations of the optoelectronics [40–42].

3.1 Methodology

Degradation of the optical performance variable, such as lumen maintenance, du′v′, and CCT, of the optoelectronics can be described by using a Gamma process, if its degradation (indicated by X) satisfies the following properties [30]:

1. X begins from 0, as described in below:

$$X(t=0)=0$$

2. X(t) has independent increments by a monotonical increasing (or decreasing) trend;
3. The increment (or decrement) of X in terms of time follows a Gamma distribution, as described in below.

$$\Delta X(t) \sim Ga(\alpha\Delta t,\beta)$$

where α and β are shape and scale parameters of the Gamma distribution. Then, the Probability Density Function (PDF) of *X*(*t*) can be derived as:

$$f(x|,\alpha|,\beta)=\frac{\beta^{\alpha(t)}}{\Gamma(\alpha(t))}\mathrm{x}^{\alpha(t)-1}\exp(-\beta x)\mathrm{I}_{(0,\infty)}(x) \tag{11.8}$$

where x is the measurements of X at a certain time t (i.e., $x = X(t)$), $I_{(0,\infty)}(x)$ is expressed as:

$$I_{(0,\infty)}(x)=\begin{cases}1, x\in(0,\infty)\\ 0, x\notin(0,\infty)\end{cases}$$

And $\Gamma(\alpha(t))$ is the Gamma function expressed as:

$$\Gamma(\alpha(t)) = \int_0^\infty e^{-x} x^{\alpha(t)-1} dx$$

For a Gamma process where $X = \{X(t), t \geq 0\}$, the expectation (mean) and variance values are given as:

$$E(X(t)) = \alpha(t)/\beta$$

$$Var(X(t)) = \alpha(t)/\beta^2$$

Empirical studies showed that the shape parameter α for an expected degradation over time-based on Gamma process is usually proportional to a Power law [43], namely:

$$\alpha(t) = ct^b \tag{11.9}$$

where b and c are fitting constants which are both larger than 0. Furthermore, the parameter b relies on the degradation process conditions and therefore can be determined by a rule of thumb. If the degradation of X depends only on the increment of time, it can be described by a stationary Gamma process with b equaling to 1. Then, estimating the parameters β and c becomes the main task of using Gamma process in prediction of the degradation of X, and can be achieved by the moments method (MM) and maximum likelihood estimation (MLE) method respectively [44].

Implementation of the MM method is briefly introduced by the following two steps [44]:

1. Suppose m is the number of test samples, and n is the number of measurements (total data) taken based on a uniform interval of time for each sample, then x_{ij} denotes the degradation readout at i^{th} time for the jth test sample. Then the degradation observation vector and matrix for each test sample can be denoted respectively as:

$$X_m = \left[x_{j1}, x_{j2}, \ldots, x_{jm}\right], X_{nm} = \begin{bmatrix} x_{11} & x_{12} & \cdots & x_{1n} \\ x_{21} & x_{22} & \cdots & x_{2n} \\ \vdots & \vdots & \ddots & \vdots \\ x_{m1} & x_{m2} & \cdots & x_{nm} \end{bmatrix}$$

2. For any arbitrary test sample, the estimates of β and c can be calculated as:

$$\frac{x_n}{\hat{\beta}}\left(1 - \frac{\sum_{i=1}^{n} \omega_i^2}{\left[\sum_{i=1}^{n} \omega_i\right]^2}\right) = \sum_{i=1}^{n}\left(\delta_i - \bar{\delta}\omega_i\right)^2$$

$$\overline{\delta} = \frac{x_n}{t_n} = \frac{\sum_{i=1}^{n} \delta_i}{\sum_{1}^{n} \omega_i} = \frac{\hat{c}}{\hat{\beta}}$$

where $\hat{\beta}$ and $\hat{c}$ are the estimated values of β and c respectively, x_i is the measurement of degraded X by the time t_i, $\delta_i = x_i - x_{i-1}$ is the degradation discrepancy by the time increment denoted by $\omega_i = t_i - t_{i-1}$, $\overline{\delta}$ equals to the ratio between x_n and t_n.

Alternatively, the estimates of β and c can also be determined from the following equations by using the MLE method [45]:

$$\hat{\beta} = \frac{\hat{c}t_n}{X_n}$$

$$\sum_{i=1}^{n} (t_i - t_{i-1}) \left\{ \frac{\partial \log \Gamma[\hat{c}(t_i - t_{i-1})]}{\partial [\hat{c}(t_i - t_{i-1})]} - \log(x_i - x_{i-1}) \right\} = t_n \log\left(\frac{\hat{c}t_n}{x_n}\right)$$

In practical applications, the MM method is usually comparatively easier to implement, whereas the MLE method provides higher estimation accuracies. Therefore, the MLE method has mostly the priority of use as long as the computational cost can be afforded.

3.2 Lumen Depreciation Modeling

Suppose a random variable L represents the lumen maintenance lifetime, $\{X(t), t \geq 0\}$ denotes time-dependent stochastic degradation process of lumen maintenance, and D denotes the failure threshold. Then, L can be described mathematically as $L = \inf\{t \mid X(t) \geq D, t \geq 0\}$, with its Cumulative Distribution Function (CDF) expressed as:

$$F_L(t) = P(L \leq t) = P(X(t) \geq D) = \int_{D}^{\infty} \frac{\beta^{\alpha(t)}}{\Gamma(\alpha(t))} X^{\alpha(t)-1} \exp(-\beta X) dX = \frac{\Gamma(\alpha(t), D\beta)}{\Gamma(\alpha(t))}$$

where $\Gamma(\alpha(t), D\beta)$ is an incomplete Gamma function expressed as:

$$\Gamma(\alpha(t), D\beta) = \int_{D\beta}^{\infty} \xi^{\alpha(t)-1} e^{\xi} d\xi$$

Then, the reliability function is presented as:

$$R(t) = 1 - F_L(t) = 1 - \frac{\Gamma(\alpha(t), D\beta)}{\Gamma(\alpha(t))}$$

And the Probability Density Function (PDF) of L can also be derived as follows:

$$\begin{aligned} f_L(t) &= \frac{\mathrm{dF_T}(t)}{\mathrm{dt}} = \frac{\alpha}{\Gamma(\alpha(t))}\int_0^{D\beta}\left[\ln(\xi) - \frac{\Gamma'(\alpha(t))}{\Gamma(\alpha(t))}\right]\xi^{\alpha(t)-1}e^{\xi}d\xi \\ &= \mathrm{deGamma}(\alpha(t), D\beta) \end{aligned} \tag{11.10}$$

Eq. (11.10) is also known as a de Gamma function (i.e., the logarithmic derivative of Gamma function), which analytical solution is in the form of generalized hypergeometric functions and therefore intractable to be achieved in engineering applications [46]. Due to this, the Birnbaum-Saunders (BS) distribution is implemented to reduce the calculation complexity, by converting the L distribution to a standard Normal distribution, expressed as [47]:

$$F_L(t) \approx \Phi\left[\frac{1}{h}\left(\sqrt{\frac{t}{k}} - \sqrt{\frac{k}{t}}\right)\right], t > 0 \tag{11.11}$$

where $h = \sqrt{D\beta}$, and $k = D\beta/c$. By differentiating both sides of Eq. (11.11), the PDF of L is then calculated as:

$$f_L(t) = \frac{\left[\left(\frac{k}{t}\right)^{\frac{1}{2}} + \left(\frac{k}{t}\right)^{\frac{3}{2}}\right]}{2\sqrt{2\pi}hk}\exp^{\left[-\frac{1}{2h^2}\left(\frac{t}{k} - 2 + \frac{k}{t}\right)\right]}, t > 0$$

And the reliability function is represented as:

$$R = 1 - F_L(t) = 1 - \Phi\left[\frac{1}{h}\left(\sqrt{\frac{t}{k}} - \sqrt{\frac{k}{t}}\right)\right], t > 0$$

The mean time to failure (MTTF) by the lumen depreciation failure of the optoelectronic device can be approximated as:

$$E(L) = \mathrm{MTTF} \cong k\left(1 + \frac{h^2}{2}\right) = \frac{D\beta}{c} + \frac{D\beta^2}{2c}$$

3.3 *CCT Shift Modeling*

In principle, the CCT shift modeling process is very similar to the lumen depreciation modeling process. By defining the CCT maintenance lifetime as L, $\{X(t), t \geq 0\}$ denotes time-dependent stochastic degradation process of du′v′, and ρ as the CCT shift failure threshold, the CDF function of L (i.e., $F_L(t)$) is expressed as:

$$F_L(t) = P(X(t) \geq \rho) = \frac{1}{\Gamma(\alpha(t))} \int_{\rho\beta}^{\infty} \xi^{\alpha(t)-1} e^{-\xi} d\xi = \frac{\Gamma(\alpha(t), \rho\beta)}{\Gamma(\alpha(t))}$$

where $\Gamma(\alpha(t), \rho\beta)$ is an incomplete Gamma function. Then, the reliability function is presented as:

$$R(t) = 1 - F_L(t) = 1 - \frac{\Gamma(\alpha(t), \rho\beta)}{\Gamma(\alpha(t))} \tag{11.12}$$

And the PDF function of L is presented as:

$$f_L(t) = \frac{dF_L(t)}{dt} = \frac{\alpha(t)}{\Gamma(\alpha(t))} \int_0^{\rho\beta} \left[\ln(\xi) - \frac{\Gamma'(\alpha(t))}{\Gamma(\alpha(t))} \right] \xi^{\alpha(t)-1} e^{-\xi} d\xi \tag{11.13}$$

Likewise, the BS method is employed for the solution of Eqs. (11.12, 11.13) by using the standard Normal distribution to approximate the L distribution, as given by:

$$F_L(t) \approx \Phi\left[\frac{1}{\nu}\left(\sqrt{\frac{t}{\omega}} - \frac{\omega}{t}\right)\right], t > 0$$

$$f_L(t) = \frac{1}{2\sqrt{2\pi}\nu\omega}\left[\left(\frac{\omega}{t}\right)^{\frac{1}{2}} + \left(\frac{\omega}{t}\right)^{\frac{3}{2}}\right] e^{\left[-\frac{1}{2\nu^2}\left(\frac{t}{\omega} - 2 + \frac{\omega}{t}\right)\right]}, t > 0.$$

in which $\Phi(\nu, \omega)$ is a standard normal distribution with: $\nu = \sqrt{\beta / \rho}$, $\omega = \rho/c\beta$. And finally, the MTTF by the CCT shift failure can be approximately calculated as:

$$\text{MTTF} \approx \omega\left(1 + \frac{\nu^2}{2}\right) = \frac{\rho}{c\beta} + \frac{1}{2c} \tag{11.14}$$

3.4 Case Study

Figure 11.6 displays CCT shift trends of a batch of 3000 K white LEDs under 120 °C and 0.5 A. In general, the CCTs of the test samples have declined over time with dynamic and random variations, despite the initial 2000 hours in which they are slightly increased as the result of seasoning effect, as plotted in both the subgraphs (a) and (b) of Fig. 11.611.6. Such a degradation tendency on CCTs is going to be described by the usage of the Gamma process.

Firstly, in order to remove the seasoning stage, the first 2000 hours data were excluded from the experimental measurements for each test sample. Then, with the

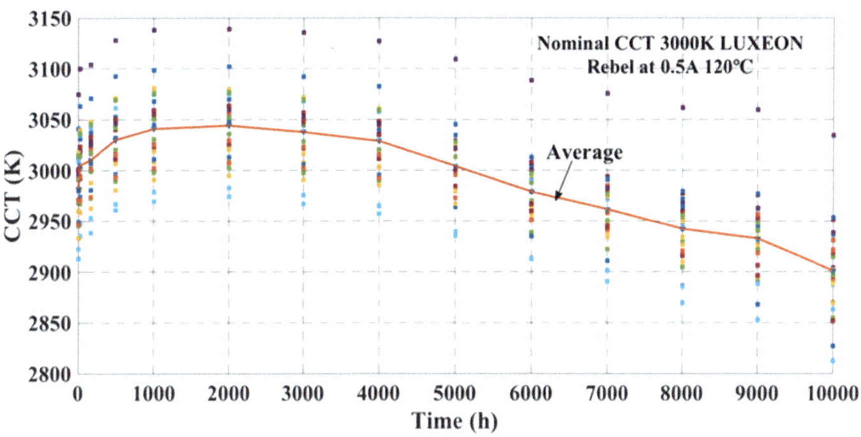

Fig. 11.6 CCT shift trends of a batch of white LEDs under 120 °C and 0.5A

remaining data, the parameters c and β (of Eq. (11.8) and Eq. (11.9) respectively) were extracted by using the MLE method. Figure 11.7 shows the PDF curves of the parameters c and β, both of which are supposed to follow Normal distributions with the mean values of 1.46×10^{-3} and 9.65 respectively. And the parameter ρ is determined as 175 according to the specifications in Table 11.1. Then, by substituting these parameters into Eq. (11.14), the MTTF of the test samples is calculated as 12,749 hours.

4 Spectral Power Distribution-Based Modeling

The SPD provides significant information of optical and chromatic performance for optoelectronics [48–50]. As a matter of fact, degradation failures such as the lumen depreciation, color shift, and CCT shift of an optoelectronic device are all linked by the degradation of its SPD curve which is the basis for calculating the photometric and colorimetric parameters such as lumen maintenance, chromaticity coordinates, Correlated Color Temperature (CCT), and Color Rendering Index (CRI) [51–53]. This provides the possibility of using SPD curves to investigate the failure mechanism [54, 55], transient optical performance [56], and long-term stability [57] of the optoelectronics.

4.1 SPD Modeling

As a mixed light, the entire SPD of an optoelectronic device is regarded as a superposition of multiple-coincident "bell-shaped" spectra, as shown in Eq. (11.15).

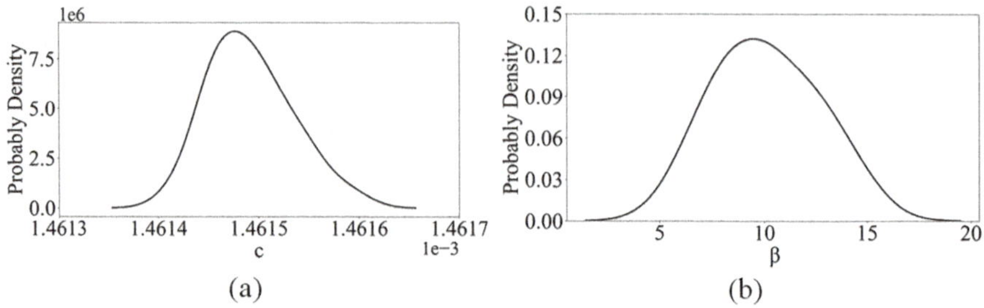

Fig. 11.7 Statistic plots of the Gamma distribution parameters (**a**): parameter c; (**b**): parameter β

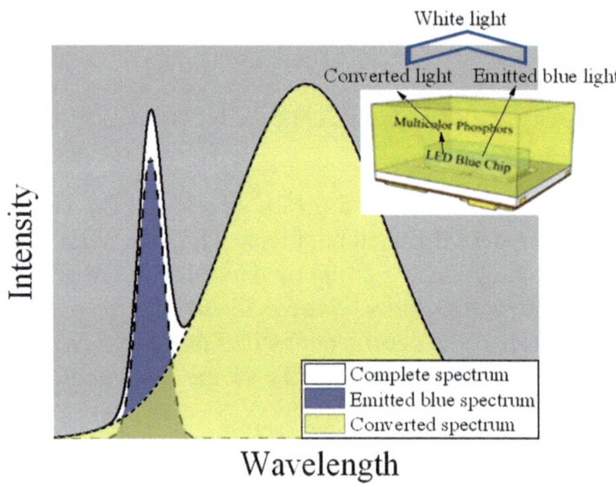

Fig. 11.8 Illustration of the entire SPD of a phosphor-converted white light-emitting diodes (WLED)

$$SPD_t(\lambda) = \sum_{i=1}^{n} SPD_i(\lambda) \tag{11.15}$$

in which λ is the wavelength, SPD_t and SPD_i indicate the entire SPD and its ith decomposed component of optoelectronics device respectively, and n is the total number of the decomposed SPDs. Figure 11.8 illustrates the entire SPD of a widely used phosphor-converted while light-emitting diode which is composed by multiple peaks: one located in the short wavelength region represents the blue light emitted by LED chip, the others are located in the long wavelength region with converted light from phosphors.

Candidate models for describing the decomposed SPD include the Gaussian function [54], Asymmetric Gaussian function [58], Asymmetric Double Sigmoidal (Asym2sig) function [59], Lorentzian function [60], etc. Equations of these above models are herein given as follows:

(1) Gaussian function

$$\mathrm{SPD} = \frac{a}{w\sqrt{\frac{\pi}{2}}} \mathrm{e}^{-2\left(\frac{\lambda-\lambda_c}{w}\right)^2} \tag{11.16}$$

in which a, λ_c, and w are the area, peak wavelength, and full width at half maxima of the light spectrum respectively.

(2) Asymmetric Gaussian Function

$$\mathrm{SPD} = \begin{cases} a\exp\left(-\frac{(\lambda-\lambda_c)^2}{w_1^2}\right) & x \le x_c \\ a\exp\left(-\frac{(\lambda-\lambda_c)^2}{w_2^2}\right) & x > x_c \end{cases}$$

in which parameters a, λ_c, w_1, and w_2 are the amplitude, peak wavelength, left and right FWHMs respectively.

(3) Asym2sig function

$$\mathrm{SPD} = a\,\frac{1 - \frac{1}{1+\exp\left\{-\left(\lambda-\lambda_c-\frac{w_1}{2}\right)/w_3\right\}}}{1+\exp\left\{-\left(\lambda-\lambda_c+\frac{w_1}{2}\right)/w_2\right\}} \tag{11.17}$$

in which parameters a, λ_c, w_1, w_2, and w_3 are the amplitude, peak wavelength, full width at half maximum (FWHM), variance of the low-energy, and high-energy sides respectively. In addition, Eq. (11.17) can be further simplified to Eq. (11.18), by ignoring the parameter w_1 as it is usually much lower than x_c in optoelectronics.

$$\mathrm{SPD} = a\frac{1 - \frac{1}{1+\exp\{-(\lambda-\lambda_c)/w_3\}}}{1+\exp\{-(\lambda-\lambda_c)/w_2\}} \tag{11.18}$$

(4) Lorentzian function:

$$\mathrm{SPD} = \frac{2a}{\pi} \times \frac{w}{4(x-\lambda_c)+w^2}$$

where a, λ_c, and w are the area, peak wavelength, and FWHM respectively.

The model accuracy of Eq. (11.15) depends on the number of decomposed SPDs it used. With the superposition of more decomposed SPDs, Eq. (11.15) can definitely provide a high fitting accuracy, but on the other hand suffer from tremendously increased model complexity. However, it is not convenient to investigate the evolution of SPD curves optoelectronics under degradation by an over-complicated model where each of the parameters influences the SPD deformation in its own way. Thus, it is important to find an optimized trade-off between model accuracy and complexity.

Once its SPD curve of the optoelectronics device is predicted, its optical and chromatic parameters such as lumen maintenance and chromaticity coordinates can be calculated by the following equations:

(1) Lumen maintenance:

$$\Phi = 683 \int_{380}^{780} \mathrm{SPD}(\lambda)\mathrm{V}(\lambda)d\lambda$$

where V(x) is the spectral luminous efficiency function as shown in Fig. 11.9.

(2) Chromaticity coordinate (x, y):

$$\begin{cases} x = \dfrac{X}{X+Y+Z} \\ y = \dfrac{Y}{X+Y+Z} \end{cases}$$

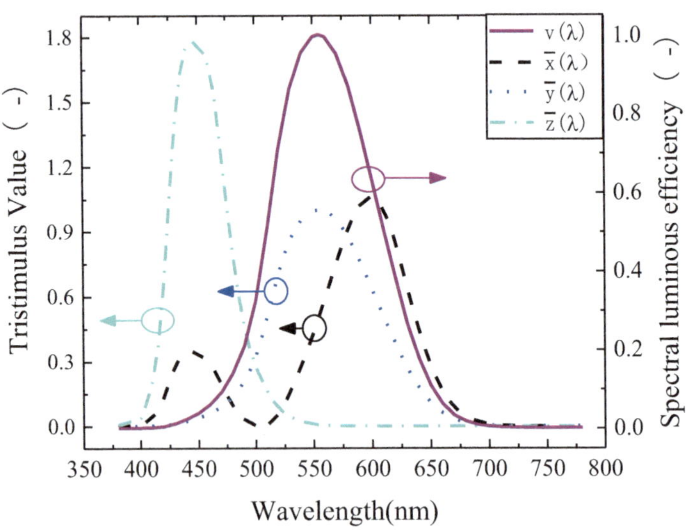

Fig. 11.9 Spectral luminous efficiency function for photopic vision and spectral color matching functions

$$\text{where}\begin{cases} X = \int_{380}^{780} \text{SPD}_{\text{LED}}(\lambda)\bar{x}(\lambda)\,d\lambda \\ Y = \int_{380}^{780} \text{SPD}_{\text{LED}}(\lambda)\bar{y}(\lambda)\,d\lambda \\ Z = \int_{380}^{780} \text{SPD}_{\text{LED}}(\lambda)\bar{z}(\lambda)\,d\lambda \end{cases}$$

where X, Y, and Z are the tristimulus values, $\bar{x}(\lambda)$, $\bar{y}(\lambda)$, $\bar{z}(\lambda)$ are the color matching functions as also shown in Fig. 11.9.

4.2 SPD-Based Failure Analysis

Figure 11.10 reveals the effects of degradation failures of the LED chip, phosphor, and silicone encapsulant on the SPD curve of a phosphor-converted white LED. It can be seen that the chip degradation results in an equal amount of reduction on both decomposed peaks. This is because that the phosphor-converted light is originally from the blue light emitted from the LED chip. Therefore, once the LED chip is degraded, both the final emitted blue light and phosphor-converted light of the LED are faded at the same time. Similarly, the phosphor degradation results in a significantly higher reduction on the phosphor light peak, but hardly affects the blue light peak. Whereas encapsulation silicone degradation results in more significant decrease on the blue peak than the phosphor converted light, because that silicone is always sensitive to a short wavelength light. In general, these above-mentioned failures will not occur individually during the aging process of a white LED, but appear jointly in some way to result in a complicated SPD depreciation for the white LED.

Next, the SPD-based failure analysis process for white LEDs is experimentally demonstrated via the investigations on degradation of 4 kinds of LED packages, which are CF2040 blue LEDs, CF2040 white LEDs, HL mid-power 3000 K white LEDs, and HL mid-power 6000 K white LEDs. The schematic structures of all test samples are shown in Fig. 11.11. For the purpose of comparison, the CF2040 white LED is equipped with the same type of chip with the CF2040 blue LED, whereas the HL mid-power 3000 K white LED and HL mid-power 6000 K white LED are with the same type of chip but different phosphors. The rated currents of the CF2040 blue LEDs, CF2040 white LEDs, HL mid-power 3000 K white LEDs, and HL mid-power 6000 K white LEDs are 200 mA, 200 mA, 60 mA and 60 mA respectively. A brief description of all these test samples is shown in Table 11.2, in which the sample quantity in each test is also provided. It is noted that the samples in groups #5 and #6 are exactly the same as those in groups #3 and #4 after a Stepped Stress Accelerated Degradation Test (SSADT).

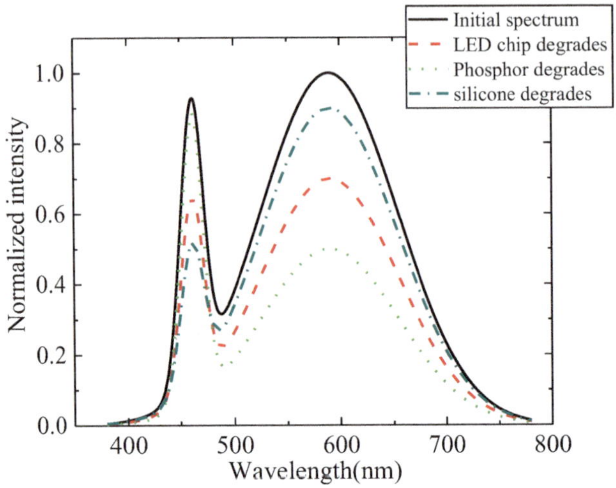

Fig. 11.10 Variation of the SPD curve a phosphor-converted white LED by degradation failures of its different components

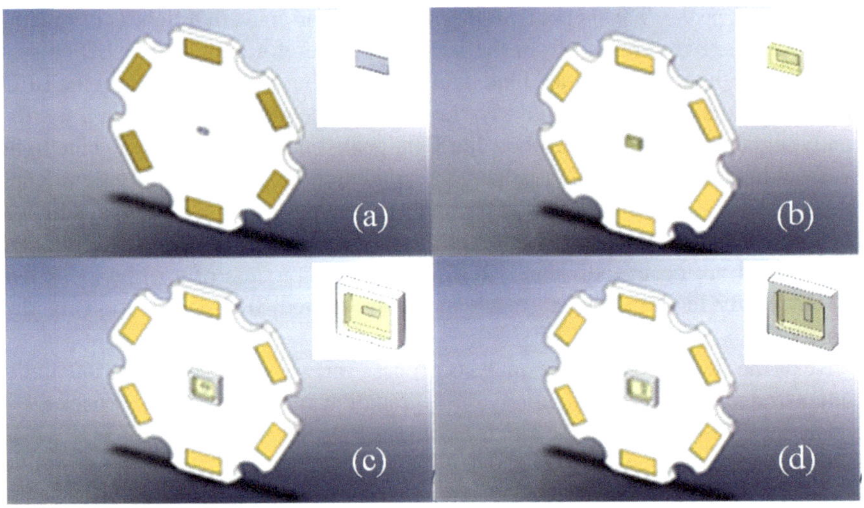

Fig. 11.11 Schematic structures of the test samples (**a**) CF2040 blue LED; (**b**) CF2040 white LED; (**c**) HL mid-power 3000 K white LED; (**d**) HL mid-power 6000 K white LED

Table 11.3 lists the test schemes for all samples under different driving currents and different case temperatures. A general description of the flow chart for all aging tests is shown in Fig. 11.12. It indicates that the experimental purpose is to quantitatively build the relationship between the optical parameter, such as lumen maintenance, with the driving current and junction temperature, via the analysis of the SPD curves obtained during the test period.

Table 11.2 Brief description of all test samples in different groups

Group ID	Brief description	Quantity
#1	CF2040 blue LED chip	5
#2	CF2040 white LED chip	5
#3	HL mid-power 3000 K white LED	15
#4	HL mid-power 6000 K white LED	15
#5	Samples in group #3 after SSADT accelerated tests	15
#6	Samples in group #4 after SSADT accelerated tests	15

Table 11.3 Summary of the test conditions

Group ID	Test scheme
#1 ~ #2	Driving currents: From 50 mA to 650 mA with an increment of 25 mA; Case temperatures: From 30 °C to 90 °C in an increment of 10 °C
#3 ~ #6	Driving currents: From 20 mA to 160 mA in an increment of 20 mA; Case temperatures: From 30 °C to 90 °C in an increment of 10 °C

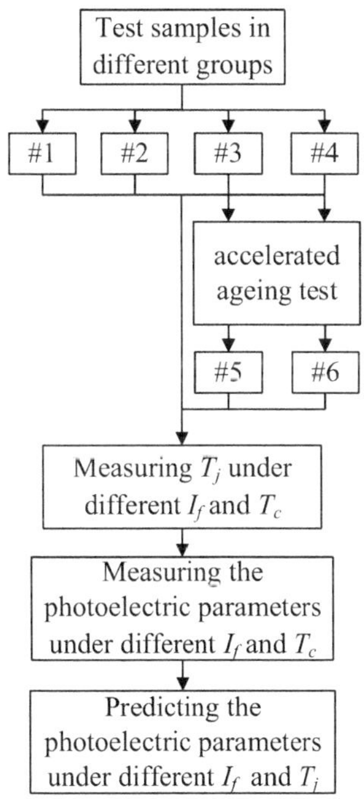

Fig. 11.12 Flow chart of the experimental process for all test samples

Figure 11.13 shows the comparison between the experimental and fitted SPD curves by Eq. (11.15) in which the Gaussian-based model (i.e., Eq. (11.16)) is applied for the description of decomposed SPDs of arbitrarily selected HL

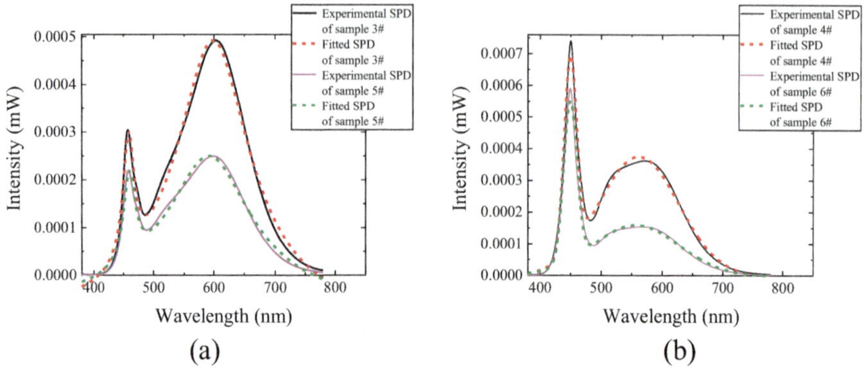

Fig. 11.13 Experimental and fitted SPD curves of the white LEDs before and after SSADT tests (**a**) HL mid-power 3000 K LED; (**b**) HL mid-power 6000KLED

Table 11.4 Comparison of decomposed SPD areas of the HL mid-power 3000 K and 6000 K white LEDs before and after SSADT tests

Sample	a_1	a_2
HL mid-power 3000 K WLED before SSADT test	0.00720	0.06884
HL mid-power 3000 K WLED after SSADT test	0.00517	0.03914
HL mid-power 6000 K WLED before SSADT test	0.01564	0.06218
HL mid-power 6000 K WLED after SSADT test	0.01363	0.02870

mid-power 3000 K and 6000 K white LEDs respectively. As shown in Fig. 11.13, the fitted SPD curves agree well with the experimental measurements for both white LEDs before and after SSADT tests.

Table 11.4 gives the extracted a_1 and a_2 parameters (which represent the areas of decomposed blue light peak and phosphor-converted light peak respectively) of the HL mid-power 3000 K and 6000 K white LEDs before and after SSADT tests. The reduction ratios of a_1 and a_2 parameters for both white LEDs in pristine state are calculated as 28.2% and 43.1%, whereas for both white LEDs after SSADTs are 12.9% and 53.8%, respectively. As the reduction rate on a_2 is much higher than that on a_1, it can be deduced that both the LED chip degradation and the phosphors degradation have occurred in HL mid-power 3000 K and 6000 K white LEDs. Furtherly, compared to the HL mid-power 3000 K white LED, the HL mid-power 6000 K white LED exhibits a more significant reduction on a_2 than a_1. This implies the phosphor degradation plays a more important role in the HL mid-power 6000 K white LED during the SSADT test, compared to the HL mid-power 3000 K white LED.

4.3 *Case Studies*

The optical and chromatic performance of an LED array does not equal the sum of individual LED packages' performances, as the thermal interactions between individual LED packages can't be ignored in the actual application [61, 62]. Thus,

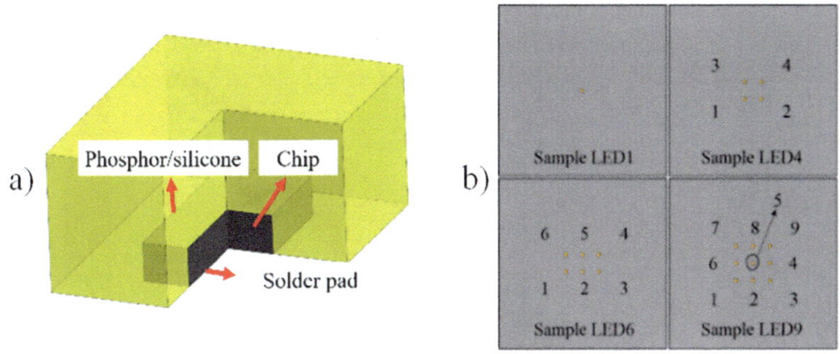

Fig. 11.14 Schematic diagram of (**a**) pc-WLED CSP and (**b**) LED arrays

Table 11.5 Experimental conditions on the test samples

	Description
Currents	From 40 mA to 280 mA with an interval of 20 mA
Temperatures	From 20 °C to 80 °C with an interval of 10 °C

optical and chromatic parameters of the LED array are not a simple accumulation by those of its constituent LEDs [63–65]. Nevertheless, the entire SPD curve of an LED array can be regarded as the superposition of the SPD curves of each individual LED inside of that array. This makes it possible to predict the optical and chromatic parameters of the LED array based on the SPD curves of its constituent LEDs. In this section, the SPD-based model is utilized to predict the lumen maintenance, CCT, and chromatic coordinates of a couple of LED arrays, which are illustrated in Fig. 11.14, via the SPD prediction of the individual LEDs in these arrays.

The operation conditions applied on these test samples are shown in Table 11.5. The implementation procedure is explained step by step as follows:

1. Case temperature and SPD curve measurements

Firstly, the case temperatures were measured by using IR camera for the test sample were measured under different operation conditions, as shown in Fig. 11.15. In the meantime, SPD curves of the test sample were also measured by using an integrating sphere spectrometer system, as shown in Fig. 11.16.

2. SPD decomposition

The SPD of the test sample is considered as a superposition of two decomposed SPDs corresponding to the blue light and phosphor-converted light peaks. Then by modeling the decomposed SPDs with the Gaussian model, Asym2sig model, and Lorentzian model respectively, Eq. (11.15) can be rewritten as Eq. (11.19) to Eq. (11.21).

(1) The Gaussian-based SPD model:

$$\mathrm{SPD} = a_1 \exp\left[-2\left(\frac{\lambda-\lambda_{c1}}{w_1}\right)^2\right] + a_2 \exp\left[-2\left(\frac{\lambda-\lambda_{c2}}{w_2}\right)^2\right] \quad (11.19)$$

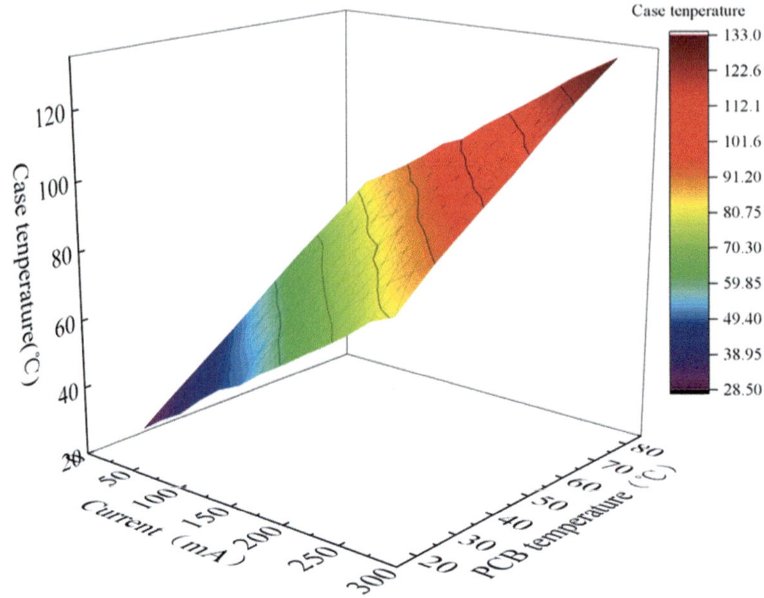

Fig. 11.15 Case temperature measurements of the test sample under different operation conditions

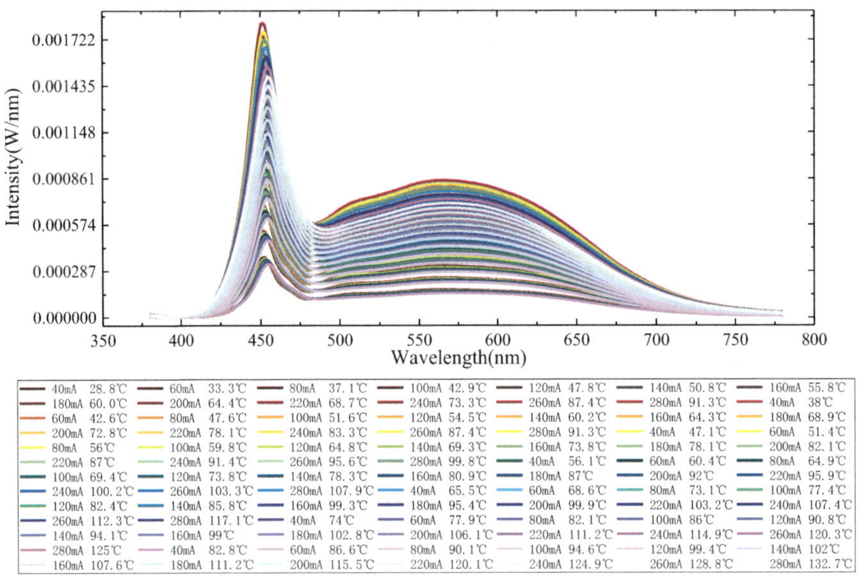

Fig. 11.16 SPD measurements of the test sample under different operation conditions

(2) The Asym2sig-based SPD model:

$$\mathrm{SPD} = b_1 \frac{1 - \dfrac{1}{1 + \exp\left[-(\lambda - \lambda_{c1})/v_1\right]}}{1 + \exp\left[-(\lambda - \lambda_{c1})/u_1\right]} + b_2 \frac{1 - \dfrac{1}{1 + \exp\left[-(\lambda - \lambda_{c2})/v_2\right]}}{1 + \exp\left[-(\lambda - \lambda_{c2})/u_2\right]} \tag{11.20}$$

(3) The Lorentzian-based SPD model:

$$\mathrm{SPD} = \frac{2a_1}{\pi} \times \frac{w_1}{4(\lambda - \lambda_{c1}) + w_1^2} + \frac{2a_2}{\pi} \times \frac{w_2}{4(\lambda - \lambda_{c2}) + w_2^2} \tag{11.21}$$

where λ is the wavelength; a_1, λ_{c1}, w_1, b_1, u_1, and v_1 are the area, peak wavelength, full width at half maximum (FWHM), the amplitude, variance of the low-energy, and variance of the high-energy sides of the decomposed blue light peak, respectively; a_2, λ_{c2}, w_2, b_2, u_2, and v_2 are the area, peak wavelength, full width at half maximum (FWHM), the amplitude, variance of the low-energy, and variance of the high-energy sides of the phosphor-converted light peak, respectively.

Next, all experimental SPD curves of the test sample shown in Fig. 11.16 were fitted by Eq. (11.19) to Eq. (11.21) for determining the best curve-fitting model for decomposed SPDs. As an example, Fig. 11.17 presents an arbitrary selected experimental SPD curve of the test sample (under the current of 120 mA with the case temperature of 64.8 °C) and fitted curves by using the SPD models based on

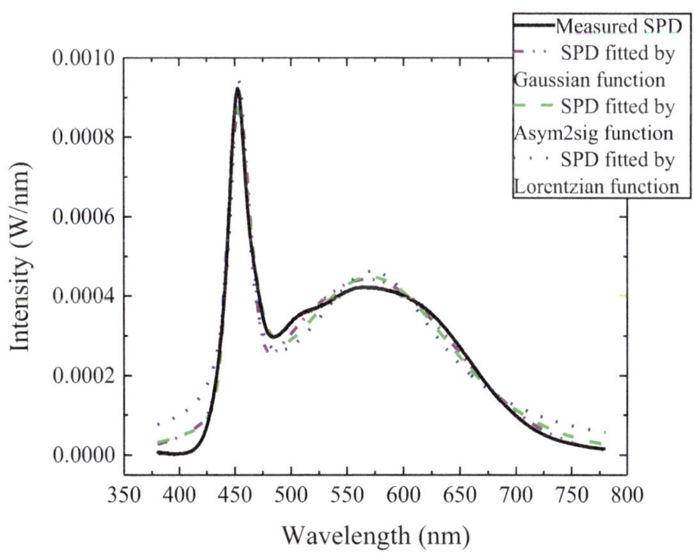

Fig. 11.17 Experimental measurements and SPD model predictions of the test sample the current of 120 mA with the case temperature of 64.8 °C

Table 11.6 Mean values and variances of $R2$ extracted from SPD predictions of the three SPD models

SPD model	Mean values of R^2	Variances of R^2
SPD model based on Gaussian functions	0.986061	0.003026
SPD model based on Asym2sig functions	0.986359	0.002553
SPD model based on Lorentzian functions	0.939333	0.006012

Gaussian function, Asym2sig function, and Lorentzian function respectively. Through the comparison of the R^2 values between the experimental and fitted SPD curves, it is found the Asym2sig and Gaussian functions are more suitable for description of the decomposed SPDs compared to the Lorentzian function. Furtherly, the mean values and variances of the R^2 values extracted from all SPD predictions by the three SPD models were calculated and compared in Table 11.6. It seems that the Asym2sig function works better than Gaussian function by virtue of the relatively higher mean value and smaller variance of the R^2 values. Therefore, SPD model based on Asym2sig functions was applied in the prediction of SPD curves of the test sample in the rest of this work.

3. Thermal-electrical-spectral relationship modeling

The 8 parameters of the SPD model based on Asym2sig functions were extracted from all experimental SPD curves and plotted as a function of the driving current (I) and case temperature (T_c) in Fig. 11.18. Such Thermal-Electrical-Spectral (TES) relationships shown in Fig. 11.18 can be described by a quadratic polynomial function as given in Eq. (11.22).

$$z(I,T) = p_1 + p_2 I + p_3 T + p_4 I^2 + p_5 IT + P_6 T^2 \tag{11.22}$$

in which z denotes each of the SPD model parameters, and p_1 to p_6 are the fitting constants. The fitted surfaces for each SPD model parameters are also plotted in Fig. 11.18, in which they are found in a good agreement with the extracted SPD parameters. In addition, all fitting constants of the quadratic polynomial functions for the total 8 SPD model parameters are listed in Table 11.7.

4. SPD prediction of the LED array

When the case temperatures of all LEDs in the LED array were obtained, the SPDs of these LEDs can be then predicted by the SPD model based on Asym2sig functions in which the model parameters were calculated from Eq. (11.22). In this study, the case temperatures of the 6 LEDs were 76.88 °C, 77.46 °C, 77.03 °C, 56.79 °C, 57. 46 °C, and 56.78 °C respectively. In this circumstance, the SPD model parameters for each LED in the LED array were calculated and listed in Table 11.8. With those parameters, the SPD curve of the LED array was predicted by the superposition of the SPD curves of all its constituent LEDs. The predicted SPD curve of the LED array shows a reasonable agreement with the experimental measurement, as shown in Fig. 11.19.

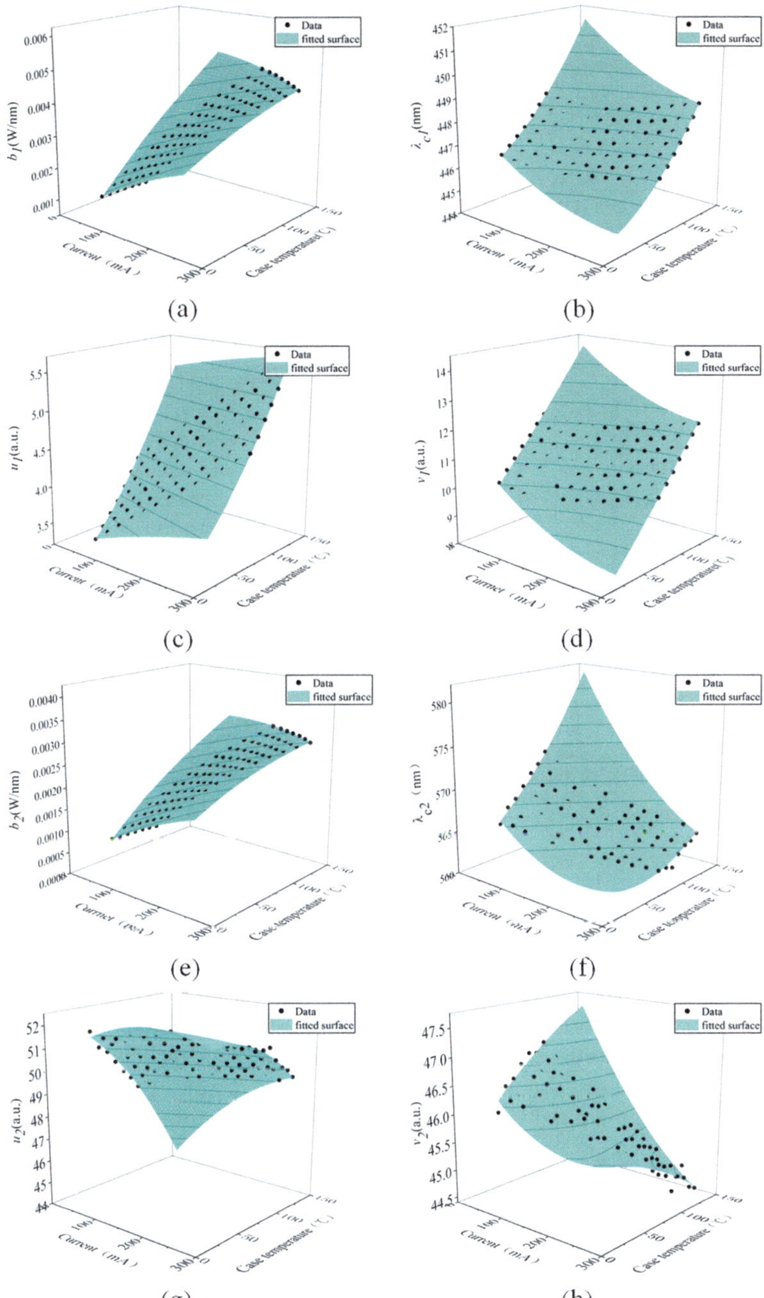

Fig. 11.18 Plots of 8 parameters of the SPD model based on Asym2sig functions as a function of the current and case temperature. (**a**) a1; (**b**) xc1; (**c**) u1; (**d**) v1; (**e**) a2; (**f**) xc2; (**g**) u2; (**h**) v2

Table 11.7 Fitting constants of the quadratic polynomial functions for the 8 SPD model parameters

SPD model parameter	P_1	P_2	P_3	P_4	P_5	P_6
a_1	−0.00015	2.61E-05	5.92E-06	−9.80E-09	−5.60E-08	−4.43E-08
x_{c1}	446.3	−0.0154	0.0175	2.85E-05	−4.28E-05	0.000210
u_1	2.723	0.002938	0.01163	−1.89E-06	−6.74E-06	5.57E-05
v_1	9.552	−0.01565	0.03463	2.57E-05	−7.94E-06	4.06E-05
a_2	−1.28E-05	1.79E-05	3.04E-06	−1.04E-08	−2.96E-08	−3.02E-08
x_{c2}	565.4	−0.05124	0.08277	0.000154	−0.00051	0.000545
u_2	45.9	−0.00757	0.02428	2.47E-05	−9.58E-05	−4.35E-05
v_2	51.87	0.01075	−0.02877	−4.49E-05	0.000177	−0.00025

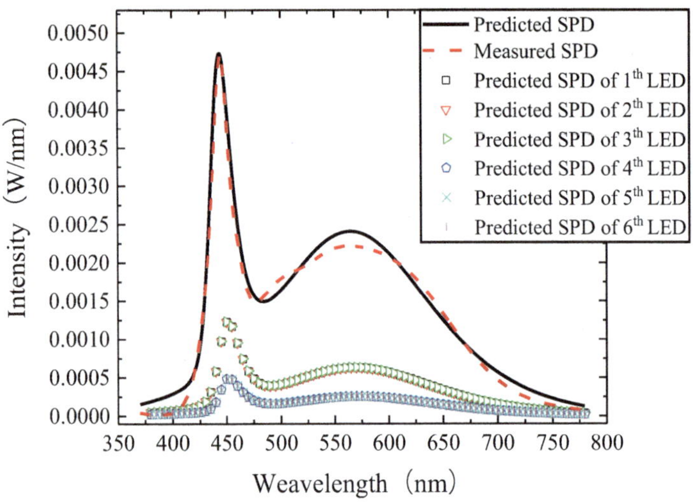

Fig. 11.19 Prediction of SPD curve of the LED array by the superposition of SPD curves of all its constitutive LEDs

To verify the proposed TES model (i.e., Eq. (11.22)), four kinds of test samples designed in Fig. 11.14 (b) operated under three different conditions to simulate the actual operation conditions. These experimental conditions include: (1) LEDs are driven all by the same current; (2) parts of LEDs are driven by the same current, while the others are not lightened; (3) LEDs are driven by different currents.

Case 1: LEDs Are Driven all by the Same Current In this case, all LEDs in an LED array are assumed to be lightened under the same current, to simulate the most common operation conditions in lighting applications. Three currents, i.e., 65 mA, 125 mA, and 185 mA, are applied to power-on all LEDs in the four kinds of test samples, as listed in Table 11.9. The predicted optical and chromatic parameters of these samples are shown in Fig. 11.20, which presents that the absolute prediction errors of lumen maintenance, CCT, chromaticity coordinate x and y are all lower than 5%.

Table 11.8 SPD model parameters of all LEDs in the LED array

LED	a_1	λ_{c1}	u_1	v_1	a_2	λ_{c2}	u_2	v_2
1st LED	0.00365	446.45	4.32	10.36	0.00252	563.61	45.62	51.07
2nd LED	0.00364	446.47	4.33	10.39	0.00251	563.65	45.62	51.05
3rd LED	0.00365	446.45	4.32	10.37	0.00252	563.62	45.62	51.06
4th LED	0.00138	447.00	3.71	10.78	0.00100	567.58	46.45	50.49
5th LED	0.00138	447.03	3.72	10.80	0.00100	567.65	46.46	50.46
6th LED	0.00138	447.00	3.71	10.78	0.00100	567.58	46.45	50.49

Table 11.9 Experimental conditions in Case 1

	Sample ID	Driving current
Condition 1	LED1	65mA
Condition 2	LED1	125mA
Condition 3	LED1	185mA
Condition 4	LED4	65mA
Condition 5	LED4	125mA
Condition 6	LED4	185mA
Condition 7	LED6	65mA
Condition 8	LED6	125mA
Condition 9	LED6	185mA
Condition 10	LED9	65mA
Condition 11	LED9	125mA
Condition 12	LED9	185mA

Case 2: Parts of LEDs Are Driven by the Same Current, while the Others Are Not Lightened In some situations, such as transferring the high beam to low beam for an automotive headlamp, LEDs in an array are not completely lightened. Table 11.10 lists a new test strategy including 7 experimental conditions in which only parts of LEDs are lightened by the same current. Figure 11.21 compares the difference between the experimental and prediction results of the optical and chromatic parameters of the LED arrays, and shows that the predicted parameters are in good agreement with the experimental obtained parameters within less than 4% errors.

Case 3: LEDs Are Driven by Different Currents To optimize the optical performance of an LED array, it is necessary to perform the current operation on each individual LED in that array. For simulating such a circumstance, another strategy on the experimental conditions is planned by distributing different driving currents to the LEDs inside an LED array, as listed in Table 11.11. And the predicted optical and chromatic parameters of the LED arrays are very close to the corresponding experimental obtained parameters within an error less than 3%, as plotted in Fig. 11.22.

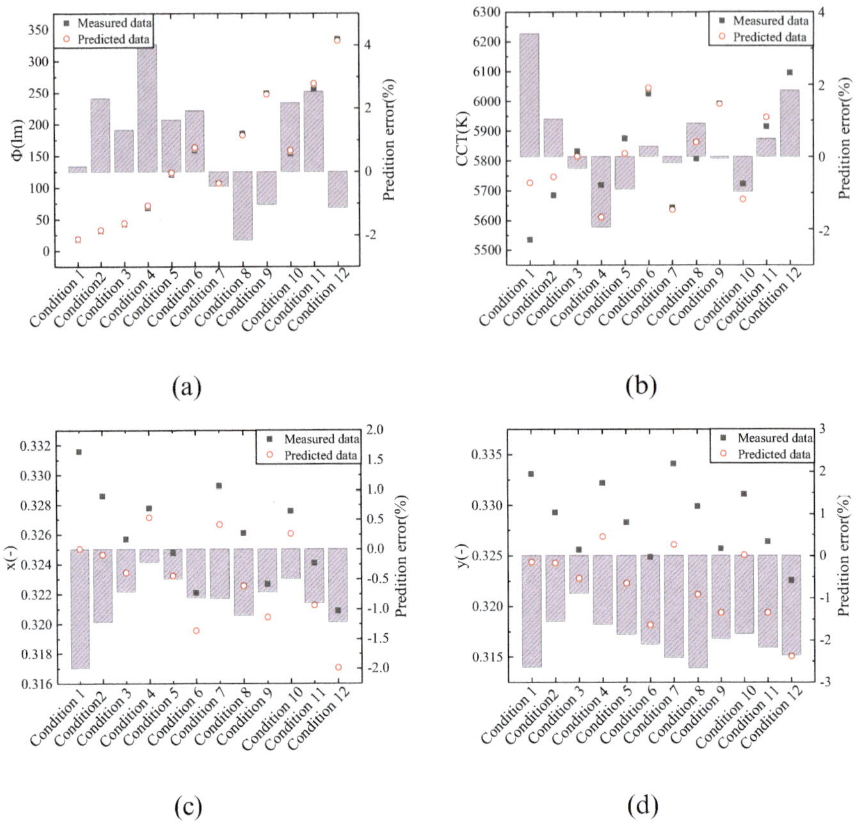

Fig. 11.20 Comparisons between the experimental and prediction results of the optical and chromatic parameters of the LED arrays in Case 1. (**a**) lumen maintenance; (**b**) CCT; (**c**) chromaticity coordinate x; (**d**) chromaticity coordinate y

Table 11.10 Experimental conditions in Case 2

	Sample ID	Driving current
Condition 1	LED4	1th, 2th LED under 125 mA
Condition 2	LED4	1th, 3th LED under
Condition 3	LED6	1th, 2th, 3th LED under
Condition 4	LED6	1th, 3th, 5th LED under
Condition 5	LED9	1th, 2th, 3th LED under
Condition 6	LED9	1th, 3th, 5th, 7th, 9th LED under
Condition 7	LED9	1th, 2th, 3th, 4th, 5th, 6th LED under

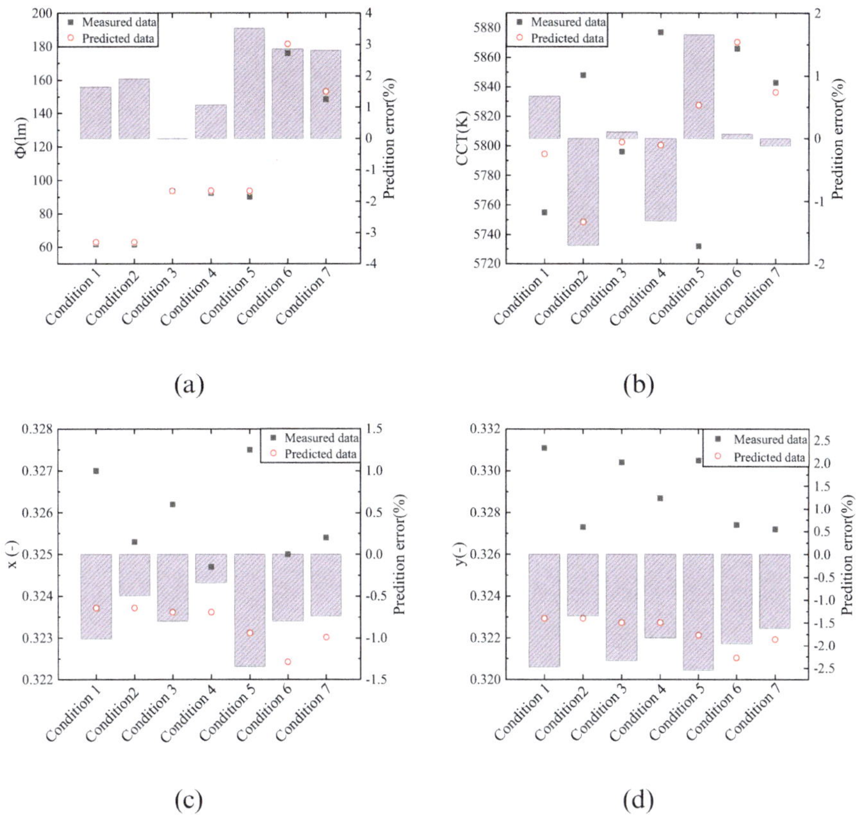

Fig. 11.21 Comparisons between the experimental and prediction results of the optical and chromatic parameters of the LED arrays in Case 2. (**a**) lumen maintenance; (**b**) CCT; (**c**) chromaticity coordinate x; (**d**) chromaticity coordinate y

5 Conclusions

The long-term reliability and lifetime estimation of optoelectronics has received a great concern to ensure that the optoelectronic products are in good conditions during their life cycle. In general, failures in optoelectronics are mainly caused by three failure modes, including the lumen depreciation, color shift, and CCT shift respectively. Depending on different using circumstances, the failure mode under the most consideration could be different, being investigated by using different estimation approaches. In this chapter, the experimental decay model, Gamma process-based model, and Spectral Power Distribution (SPD)-based model are introduced for coping with the degradation analysis of optoelectronics due to different failure causes. The features and applications of each above-mentioned estimation model are summarized as follows:

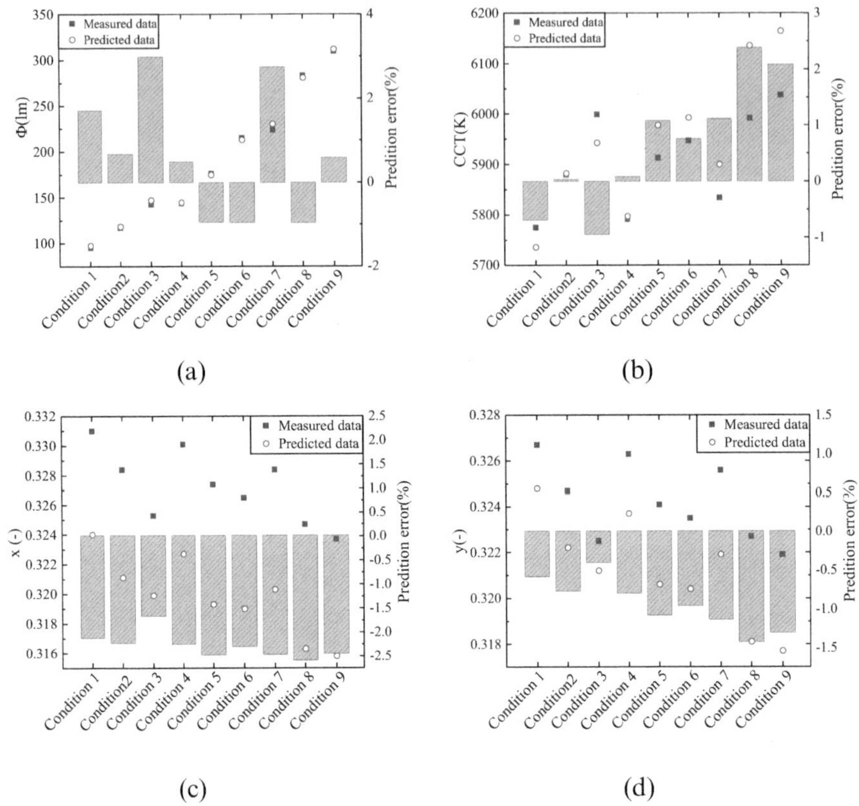

Fig. 11.22 Comparisons between the experimental and prediction results of the optical and chromatic parameters of the LED arrays in Case 3. (**a**) lumen maintenance; (**b**) CCT; (**c**) chromaticity coordinate x; (**d**) chromaticity coordinate y

Table 11.11 Experimental conditions in Case 2

	Sample ID	Driving current
Condition 1	LED4	1th, 2th with 125 mA; 3th, 4th with 65 mA
Condition 2	LED4	1th, 2th with 185 mA; 3th, 4th with 65 mA
Condition 3	LED4	1th, 2th with 185 mA; 3th, 4th with 125 mA
Condition 4	LED6	1th, 2th, 3th with 125 mA; 4th, 5th, 6th with 65 mA
Condition 5	LED6	1th, 2th, 3th with 185 mA; 4th, 5th, 6th with 65 mA
Condition 6	LED6	1th, 2th, 3th with 185 mA; 4th, 5th, 6th with 125 mA
Condition 7	LED9	1th, 2th, 3th, 7th, 8th, 9th with 125 mA; 4th, 5th, 6th with 65 mA
Condition 8	LED9	1th, 2th, 3th, 7th, 8th, 9th with 185 mA; 4th, 5th, 6th with 65 mA
Condition 9	LED9	1th, 2th, 3th, 7th, 8th, 9th with 185 mA; 4th, 5th, 6th with 125 mA

1. The exponential decay model has the simplest expression and most extensive application under a wide range of operating conditions. However, it can only deal with the lumen depreciation problem.
2. The Gamma process-based model can provide a more comprehensive estimation from a statistical viewpoint for all failure modes. However, the input data for modeling is strictly required in a monotonic increasing or decreasing trend.
3. The SPD-based model is capable to estimate the lumen depreciation, color shift, and CCT shift simultaneously. It also offers an accurate and convenient method to dynamically predict the optical and chromatic performances of LED array at real usages, only based on the SPD characteristics of individual LEDs in that array. However, to do so, it requires a complete set of SPD curves of optoelectronics throughout their test period, which are difficult to collect in practical situations.

Acknowledgment The authors would also like to acknowledge the support of China Association for Science and Technology (CAST) Program of International Collaboration Platform for Science and Technology Organizations in Belt and Road Countries (Grant No. 2020ZZGJB072014), the Fundamental Research Funds for the Central Universities (Grant No. YWF-20-BJ-J-909), and the National Natural Science Foundation of China (Grant No. 51805147).

References

1. S. Liu, X. Luo, *LED Packaging for Lighting Applications* (Wiley & Sons, 2011), pp. 1–32. https://doi.org/10.1002/9780470827857
2. W.D. Van Driel, X. Fan, *Solid State Lighting Reliability: Components to Systems. Vol. 1* (Springer, New York, 2012)
3. W.D. Van Driel, X. Fan, G.Q. Zhang, *Solid State Lighting Reliability Part 2: Components to Systems* (Springer, New York, 2017)
4. X. Cao et al., Defect generation in InGaN/GaN light-emitting diodes under forward and reverse electrical stresses. Microelectron. Reliab. **43**(12), 1987–1991 (2003)
5. M.-H. Chang et al., Light emitting diodes reliability review. Microelectron. Reliab. **52**(5), 762–782 (2012)
6. M. Meneghini et al., A review on the physical mechanisms that limit the reliability of GaN-based LEDs. IEEE Trans Electron Devices **57**(1), 108–118 (2009)
7. N. Narendran et al., Characterizing white LEDs for general illumination applications, in *Light-Emitting Diodes: Research, Manufacturing, and Applications IV*, (International Society for Optics and Photonics, 2000)
8. LM-80-08, I, *Approved Method for Measuring Lumen Maintenance or LED Light Sources* (Illuminating Engineering Society, New York, 2008)
9. E. Star, Program Requirements Product Specification for Luminaries (Light Fixtures), Eligibility Criteria-version 1.1. 2011, USA: ENERGY STAR
10. E. Star, Program Requirements–Product Specification for Lamps (Light Bulbs) Eligibility Criteria–Version 1.1. 2014, USA: ENERGY STAR
11. E. Star, Program requirements for solid state lighting luminaires. Eligibility Criteria-Version 1.3. 2010, USA: ENERGY STAR
12. IES-TM-21-11, *Projecting Long Term Lumen Maintenance of LED Light Sources* (Illuminating Engineering Society, New York, 2011)

13. I. Standard, ed. Self-ballasted LED-lamps for general lighting services. Performance requirements. 1.0 ed. 2009, International Electrotechnical Commission
14. H.A. Qiao, et al. A statistical method to analyze LED lumen depreciation and project useful LED product life. in IES Annual Conference. 2010. Wiley Toronto, Ontario
15. J. Fan, K.-C. Yung, M. Pecht, Prognostics of chromaticity state for phosphor-converted white light emitting diodes using an unscented Kalman filter approach. IEEE Trans. Device Mater. Reliab. **14**(1), 564–573 (2013)
16. S.H. Mohammadian, D. Aït-Kadi, F. Routhier, Quantitative accelerated degradation testing: Practical approaches. Reliab Eng Syst Saf **95**(2), 149–159 (2010)
17. J.L. Davis et al., *Insights into Accelerated Aging of SSL Luminaires. In LED-Based Illumination Systems* (International Society for Optics and Photonics, 2013)
18. C. Qian et al., An accelerated test method of luminous flux depreciation for LED luminaires and lamps. Reliab Eng Syst Saf **147**, 84–92 (2016)
19. M. Cai et al., Step-stress accelerated testing of high-power LED lamps based on subsystem isolation method. Microelectron. Reliab. **55**(9–10), 1784–1789 (2015)
20. F. Haghighi, S.J. Bae, Reliability estimation from linear degradation and failure time data with competing risks under a step-stress accelerated degradation test. IEEE Trans. Reliab. **64**(3), 960–971 (2015)
21. J. Huang et al., Lumen degradation modeling of white-light LEDs in step stress accelerated degradation test. Reliab Eng Syst Saf **154**, 152–159 (2016)
22. M. Dal Lago et al., Degradation mechanisms of high-power white LEDs activated by current and temperature. Microelectron. Reliab. **51**(9–11), 1742–1746 (2011)
23. F.-K. Wang, T.-P. Chu, Lifetime predictions of LED-based light bars by accelerated degradation test. Microelectron. Reliab. **52**(7), 1332–1336 (2012)
24. X. Yang et al., An alternative lifetime model for white light emitting diodes under thermal–electrical stresses. Materials **11**(5), 817 (2018)
25. S.-T. Tseng, C.-Y. Peng, Stochastic diffusion modeling of degradation data. J Data Sci **5**(3), 315–333 (2007)
26. W.Q. Meeker, L.A. Escobar, C.J. Lu, Accelerated degradation tests: Modeling and analysis. Technometrics **40**(2), 89–99 (1998)
27. S.J. Bae, P.H. Kvam, A nonlinear random-coefficients model for degradation testing. Technometrics **46**(4), 460–469 (2004)
28. K.A. Doksum, A. Hbyland, Models for variable-stress accelerated life testing experiments based on wener processes and the inverse gaussian distribution. Technometrics **34**(1), 74–82 (1992)
29. C.-C. Tsai, S.-T. Tseng, N. Balakrishnan, Optimal design for degradation tests based on gamma processes with random effects. IEEE Trans. Reliab. **61**(2), 604–613 (2012)
30. X. Wang, N. Balakrishnan, B. Guo, Residual life estimation based on a generalized wiener degradation process. Reliab Eng Syst Saf **124**, 13–23 (2014)
31. N. Narendran et al., What is useful life for white light LEDs? J. Illum. Eng. Soc. **30**(1), 57–67 (2001)
32. M. Abdel-Hameed, A gamma wear process. IEEE Trans. Reliab. **24**(2), 152–153 (1975)
33. J. Lawless, M. Crowder, Covariates and random effects in a gamma process model with application to degradation and failure. Lifetime Data Anal. **10**(3), 213–227 (2004)
34. C.J. Lu, W.O. Meeker, Using degradation measures to estimate a time-to-failure distribution. Technometrics **35**(2), 161–174 (1993)
35. C. Park, W.J. Padgett, Stochastic degradation models with several accelerating variables. IEEE Trans. Reliab. **55**(2), 379–390 (2006)
36. D. Liu et al., Bayesian model averaging based reliability analysis method for monotonic degradation dataset based on inverse Gaussian process and gamma process. Reliab. Eng. Syst. Saf. **180**, 25–38 (2018)
37. Z. Pan, Q. Sun, Optimal design for step-stress accelerated degradation test with multiple performance characteristics based on gamma processes. Commun. Stat. Simul. Comput. **43**(2), 298–314 (2014)

38. Z. Pan, N. Balakrishnan, Reliability modeling of degradation of products with multiple performance characteristics based on gamma processes. Reliab. Eng. Syst. Saf. **96**(8), 949–957 (2011)
39. S.-H. Park, J.-H. Kim, Application of gamma process model to estimate the lifetime of photovoltaic modules. Sol. Energy **147**, 390–398 (2017)
40. Z. Wu et al., A gamma process-based prognostics method for CCT shift of high-power white LEDs. IEEE Trans Electron Devices **65**(7), 2909–2916 (2018)
41. S.H. Park, J.H. Kim, Lifetime estimation of LED lamp using gamma process model. Microelectron. Reliab. **57**, 71–78 (2016)
42. S.T. Tseng, J. Tang, I.H. Ku, Determination of burn-in parameters and residual life for highly reliable products. Nav. Res. Logist. (NRL) **50**(1), 1–14 (2003)
43. J. Van Noortwijk, M. Pandey. A stochastic deterioration process for time-dependent reliability analysis. in Proceedings of the Eleventh IFIP WG. 2004
44. C.-Y. Peng, S.-T. Tseng, Mis-specification analysis of linear degradation models. IEEE Trans. Reliab. **58**(3), 444–455 (2009)
45. K.-C. Yung, B. Sun, X. Jiang, Prognostics-based qualification of high-power white LEDs using Lévy process approach. Mech. Syst. Signal Process. **82**, 206–216 (2017)
46. C. Park, W. Padgett, Accelerated degradation models for failure based on geometric Brownian motion and gamma processes. Lifetime Data Anal. **11**(4), 511–527 (2005)
47. A. Elwany, N. Gebraeel, Real-time estimation of mean remaining life using sensor-based degradation models. J. Manuf. Sci. Eng. **131**(5) (2009)
48. B.-M. Song, B. Han, Analytical/experimental hybrid approach based on spectral power distribution for quantitative degradation analysis of phosphor converted LED. IEEE Trans. Device Mater. Reliab. **14**(1), 365–374 (2013)
49. R. Supronowicz, I. Fryc. The LED spectral power distribution modelled by different functions-how spectral matching quality affected computed LED color parameters. In 2019 second Balkan junior conference on lighting (Balkan light junior). 2019. IEEE
50. H.-T. Chen et al., Chromatic, photometric and thermal modeling of LED systems with non-identical LED devices. IEEE Trans. Power Electron. **29**(12), 6636–6647 (2014)
51. H. Ye et al., Electrical–thermal–luminous–chromatic model of phosphor-converted white light-emitting diodes. Appl. Therm. Eng. **63**(2), 588–597 (2014)
52. S. Hui, Y. Qin, A general photo-electro-thermal theory for light emitting diode (LED) systems. IEEE Trans. Power Electron. **24**(8), 1967–1976 (2009)
53. G.B. Nair, H. Swart, S. Dhoble, A review on the advancements in phosphor-converted light emitting diodes (pc-LEDs): Phosphor synthesis, device fabrication and characterization. Prog. Mater. Sci. **109**, 100622 (2020)
54. W. Chen et al., Reliability assessment of light-emitting diode packages with both luminous flux response surface model and spectral power distribution method. IEEE Access **7**, 68495–68502 (2019)
55. J. Fan et al., Color shift failure prediction for phosphor-converted white LEDs by modeling features of spectral power distribution with a nonlinear filter approach. Materials **10**(7), 819 (2017)
56. J. Fan et al., Dynamic prediction of optical and chromatic performances for a light-emitting diode array based on a thermal-electrical-spectral model. Opt. Express **28**(9), 13921–13937 (2020)
57. C. Qian et al., Prediction of lumen depreciation and color shift for phosphor-converted white light-emitting diodes based on a spectral power distribution analysis method. IEEE Access **5**, 24054–24061 (2017)
58. G. He, H. Yan, Optimal spectra of the phosphor-coated white LEDs with excellent color rendering property and high luminous efficacy of radiation. Opt. Express **19**(3), 2519–2529 (2011)
59. L. Marsich et al., Raman spectroscopic study of bioactive silica-based glasses: The role of the alkali/alkali earth ratio on the non-bridging oxygen/bridging oxygen (NBO/BO) ratio. Spectroscopy **23**(3–4), 227–232 (2009)
60. J. Fan et al., Thermal/luminescence characterization and degradation mechanism analysis on phosphor-converted white LED chip scale packages. Microelectron. Reliab. **74**, 179–185 (2017)

61. B.-J. Huang, C.-W. Tang, Thermal–electrical–luminous model of multi-chip polychromatic LED luminaire. Appl. Therm. Eng. **29**(16), 3366–3373 (2009)
62. H.-L. Lu et al., Efficient measurement of thermal coupling effects on multichip light-emitting diodes. IEEE Trans. Power Electron. **32**(12), 9280–9292 (2017)
63. J.-M. Kang et al., Fabrication and thermal analysis of wafer-level light-emitting diode packages. IEEE Electron Device Lett. **29**(10), 1118–1120 (2008)
64. T. Treurniet, V. Lammens, Thermal management in color variable multi-chip LED modules. in Twenty-Second Annual IEEE Semiconductor Thermal Measurement And Management Symposium. 2006. IEEE
65. H. Zou, et al. Thermal characterization of multi-chip light emitting diodes with thermal resistance matrix. In 2017 14th China International Forum on Solid State Lighting: International Forum on Wide Bandgap Semiconductors China (SSLChina: IFWS). 2017. IEEE

Chapter 12
Reliability and Failure of Microelectronic Materials

A. Mavinkurve, R. T. H. Rongen, and M. van Soestbergen

1 Introduction

The use of microelectronics has exploded over the last few decades. The number of electronic components in mobile, consumer, industrial, and automotive applications has increased enormously. The drivers for this change have been miniaturization (e.g., hand-held devices), energy efficiency (e.g., the industrial and automotive industry), safety (automotive), connectivity and security (Internet of Things), and speed (e.g., data rate transfer in communication).

In addition to the number of applications, microelectronic components are designed and manufactured to withstand a host of different application conditions, also termed as "mission profiles" [1], which involve temperature (both static and dynamic), humidity, power-on time, vibrations, and shock. They are also being used in more stringent or harsh conditions, e.g., in commercial vehicles (e.g., trucks), electrical vehicles, and industrial applications. Figure 12.1 shows the diversity of semiconductor component envelopes, i.e., packages. Although relatively newer (and thinner) package technologies like (WL) CSP [(wafer-level) chip scale packages] and leadless packages like QFN (quad flat package no leads) have increased in usage across several years, more conventional packages like SO (small outline) and QFP (quad flat package) are still extensively used in more conventional applications where there are fewer space constraints, or also due to slower adoption of newer technologies.

Product quality is another aspect that is becoming more relevant in recent years. In addition to steadily more stringent reliability requirements that a semiconductor component should be capable of enduring during its lifetime (or after a given period of use in the application), recent focus has also shifted to improving quality at "zero

A. Mavinkurve (✉) · R. T. H. Rongen · M. van Soestbergen
NXP Semiconductors, Nijmegen, The Netherlands
e-mail: amar.mavinkurve@nxp.com

W. D. van Driel, M. Yazdan Mehr (eds.), *Reliability of Organic Compounds in Microelectronics and Optoelectronics*,
https://doi.org/10.1007/978-3-030-81576-9_12

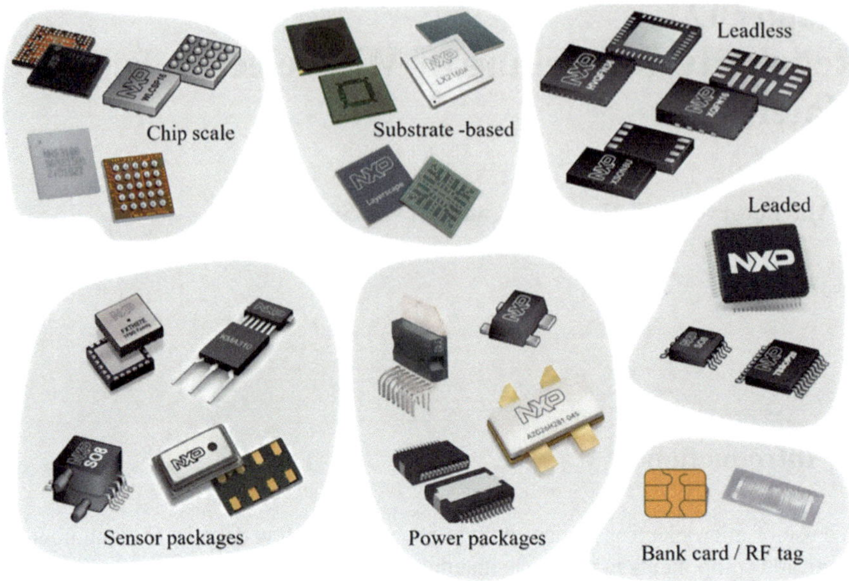

Fig. 12.1 Diversity of microelectronic packages for different applications

hour" and early life. This means that users demand ppb-level incidents after receiving semiconductor component during subsequent assembly at their facilities or downstream the supply chain at OEMs (original equipment manufacturers) or when operated by the end user in the final application: implying that no so-called early fails (e.g., due to extrinsic defects) are accepted. It is easy to understand why, on the one hand, OEMs differentiate themselves with new features and better quality. On the other hand, with the exponential growth of usage of electronics over the last decades, every single incident can multiply the risk of finding defects in the final application or product, be it a car, a mobile phone, or a base station. This can have a severe impact on the reputation of manufacturer as well as the semiconductor supplier, which is accompanied by large loss of financial resources to resolve the issue, as also potential business loss or even bankruptcy. With a large pressure on cost, whether it is manufacturing cost, material cost, or integral cost, manufacturers should make a conscious effort of not sacrificing quality in any way.

This chapter will go into aging-related reliability aspects of microelectronic devices. The first two sections will describe basics of microelectronic packages, the various types, how they are manufactured, and typical materials involved. Subsequent sections will describe aging mechanisms of different critical parts of the package: the encapsulant (e.g., the molding compound), the die attach, and the interconnect. The impact of environmental conditions like temperature and humidity will be described in detail; in addition to this, electrical conditions like applied bias levels becomes relevant when going into interconnect reliability.

2 The Basic Concept and Function of a Semiconductor Package

The essential role of a microelectronic package is to provide protection to the integrated circuit (IC) that is encapsulated within the package and also connect the device to the external environment, e.g., the printed circuit board (PCB) electrically, via the terminals, and thermally, via a heat sink. Packaging materials are designed and optimized to enable the device to withstand continuous or cyclic thermal stresses, protect it against intrusion from moisture and corrosion and contamination and corrosion promotors, and also provide robust electrical insulation between the conductive paths. Schematic images of a package and an example of an assembly process flow are shown in Fig. 12.2.

Traditionally, a package consists of a carrier (called a leadframe or substrate) on to which an IC is assembled as described below. The IC itself is an integration of many transistors connected via metal tracks and which is processed via subsequent diffusion/or deposition, mask, photolithographic, and etch steps on a carrier called wafer that is normally made of silicon. One wafer consists of typically hundreds or thousands of ICs. The assembly of the ICs into a package starts with separating the ICs from the wafer using a singulation process, which can be mechanical or a more advanced technique like laser singulation or a hybrid variant thereof. Next, the IC is secured on to the carrier using an adhesive that can be electrically and/or thermally conductive. The carrier can either be a metallic "leadframe" or an organic substrate. In the last metal layer of the IC that is covered with a scratch protection or passivation layer, larger areas are processed and opened up to allow for electrical connection. These areas are called bondpads and usually made of Al. One of the most common connections is simply by a wire, and a welding process is used for that, which is called wirebonding. The welds are created by transferring ultrasonic energy at the two contacts: the Al bondpad at the one and the leadframe or substrate at the other end of the wire. Au-based wires that were mainstream around 20 years ago have largely been replaced with Cu-based wires, mainly for cost reasons, with occasionally functionality/reliability benefits that will be discussed later in this chapter.

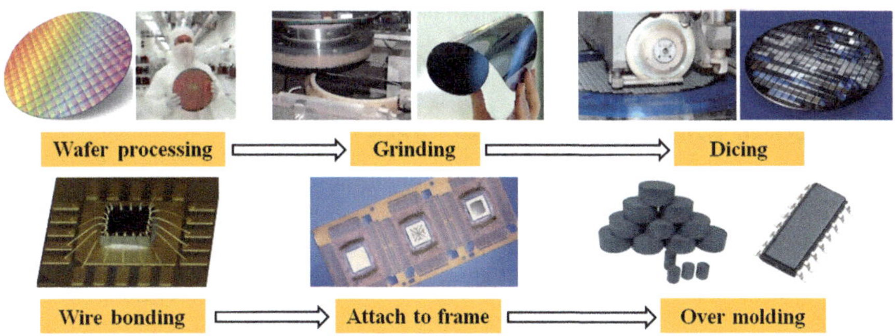

Fig. 12.2 Summary overview of a typical assembly process (wirebonded package)

Details on the wirebonding process have been described in a well-known reference book on the subject [2]. The next step in assembling ICs is an encapsulation process with epoxy molding compound (EMC) using transfer molding and mechanical singulation of the molded strips or panels into individual units or packages. This is sometimes followed by a process which shapes the terminals of the package and a plating process to coat the external terminals with Sn, for example, enabling connecting with solder on the PCB. In BGA packages (ball grid array), the external terminals involve solder balls.

In the more recently developed flipchip package, small Sn-based solder balls are placed on Al bondpads after covering them with a thin metal layer, so-called UBM ("under bump metallizations"). This is done on wafer level after which the ICs are singulated. The state of the art is to replace solder balls by smaller copper pillars which are covered with a layer of solder. During subsequent assembly steps, these "bumped" dice with balls or pillars are flipped and mounted on to the substrate on to which solder paste is printed. Then the carrier is passed through a "reflow" process at a temperature exceeding the melting point of the solder, enabling the formation of a connection between the bump and substrate via intermetallic compound (IMC) formation. The flipchip process is also followed by encapsulation, of which there are two possibilities. In the more conventional method, a "capillary underfill" material is applied via dispensing and essentially fills the volume below the die and in between the bumps. Finally, the conventional transfer molding process is used to seal the packages on the carrier. In more recent times, the application of underfill and encapsulation are integrated using a "molded underfill" in one process. The same material functions as underfill below the die, as well as an encapsulant around the die. Even in this case, the next step is the application of solder balls on the external terminals of the carrier followed by singulation of the units or packages from the panel. The final step before the product is packed is selection of the 100% functional components by electrical testing.

In some cases, this may involve testing at three temperature (room, hot, and cold), e.g., ICs for automotive electronics. In exceptional cases, every IC is operated for a short time at elevated temperature to screen out latent early fails before shipping, e.g., ICs for automotive safety applications. Finally, the assembled package is (dry)-packed and labelled according to its moisture sensitivity level [3] and shipped to the end user. Various package types are described in the next section.

3 Types of Microelectronic Packages

As shown in Fig. 12.1, microelectronic packages have evolved tremendously across the last few decades driven by several factors:

1. **Size/form factor**: Due to ongoing trend for miniaturization and integration in the electronics industry, microelectronic packages have been forced to reduce in size, form factor, and cost continuously. As the number of packages on a PCB

have increased, space constraints have led to packages becoming smaller and thinner while reducing the geometry (e.g., pitch) of the terminals.

2. **Assembly process**: Although the first package types were "through hole" packages, where connections were made on to the PCB via holes, mechanically as well as by individual soldering, mainstream packages these days are called "surface mount devices" or SMDs, where several components are connected in parallel using a soldering reflow process subsequently at both sides of the PCBs.
3. **Number of terminals ("I/Os")**: There is a large range in the number of packages terminals. Packages for discrete semiconductor components like diodes and transistors have low pin counts (< 8); IC packages have terminals that range from 8 to around even 1000. This is driven by the complexity of the IC devices that are assembled within the package.
4. **Wirebonded vs Flipchip**: As described in the previous section, broadly speaking, packages can be classified into wirebonded packages (older technology) and flipchip packages. Currently there is a drive toward increasing flipchip packaging to enhance electrical (RF) performance, better options for (3D) integration, enabling higher I/Os, etc.
5. **Multi-chip packages/advanced packages**: Conventional packages involve assembly of one single IC. In more state-of-the-art packages, ICs are integrated into a package by stacking, for example. This is among others driven by the fact that Moore's law leads to an enormous increase in cost and technology challenges to enable scaling down of features on the IC, and such multi-chip packages have enabled integrating functionality in a cost-efficient way [4]. Though multi-chip packages also fall within the advanced packages category, die stacking using wirebonding is known in the industry for at least 20 years. Details on some other most advanced packages are described [4], involving fan-out packages, (wafer-level) chip scale packages, package on package, etc. Multi-chip and these most advanced packages will not be in the scope of this chapter. However, aging mechanisms described for single-chip packages that are in scope will be equally relevant as similar organic materials are involved, but the interactions could be different due to design and type of chemistry involved.

Figure 12.3 gives an overview of some mainstream packages. Figure 12.4 depicts the construction of typical wirebonded and flipchip packages. Relevant materials are shown, which have completely different bulk characteristics and complex interfacial properties when they come into mutual contact within the microelectronic package. Successful integration involves a thorough understanding of the behavior of each individual material. Maintaining integrity of the interfaces is just as critical as that of each bulk material. The next section will go into more detail into the various packaging materials.

Fig. 12.3 Mainstream microelectronic packages (QFP, quad flat package; SO, small outline packages; HVQFN, heat sink very thin quad flat no leads; HVSON, heat sink very thin small outline no leads; BGA, ball grid array (T, thin; L. low profile; F, fine pitch); WLCSP, wafer-level chip scale packages

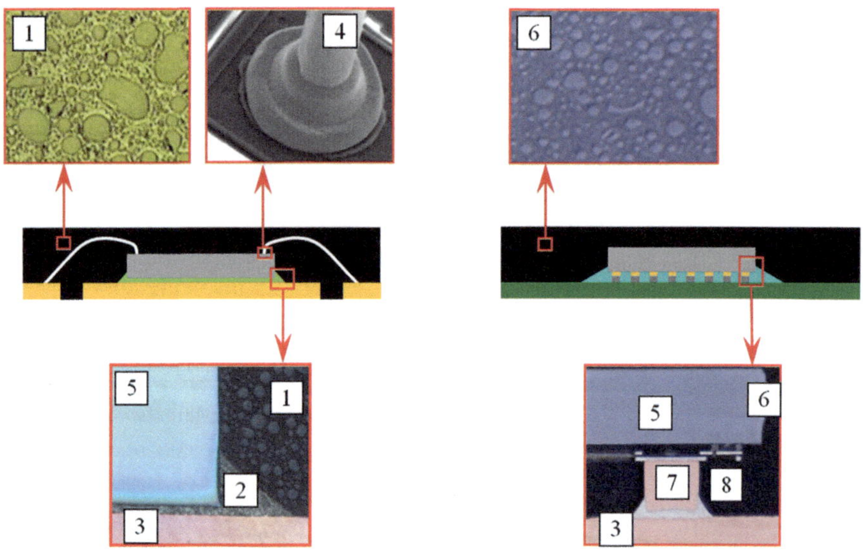

Fig. 12.4 Materials typically used in wirebonded package (left) and flipchip package (right); 1, epoxy molding compound; 2, die attach; 3, leadframe or substrate; 4, wirebonds; 5, die; 6, molded underfill material; 7, copper pillar with solder paste; 8, capillary underfill

4 Key Packaging Materials

Package materials can be split into several classes. A more detailed description on the characteristics will be given per class. In most cases, organic materials like polymers are thermosets (they cure or cross-link on heating) and are combined with inorganic fillers (like silica or glass or Ag) to enhance their mechanical properties while reducing moisture absorption.

1. **Carriers**: These are metal-based, generally Cu alloys (in some cases other alloys like Ni-Fe alloys are used) or organic-based PCB-like substrates.

   Leadframes are generally provided with a plated layer to enable connection via wirebonding or in some cases soldering. Examples are Ag plating or NiPdAu plating. In the former case, the Ag plating is selectively applied on specific areas, leaving the rest of the leadframe surface bare. In this case, a Sn-plating process is needed after encapsulation to enable soldering on to the PCB. In the latter case, the NiPdAu coating (also called "pre-plating") has the benefit of being solderable, so a separate "post"-Sn-plating step is not needed. In some cases, leadframes are provided with a roughening treatment to enhance adhesion to the encapsulant.

   Organic substrates are essentially a stack-up of glass fiber-reinforced bismaleimide-triazine-epoxy-based composites and plated copper layers to provide the routing between the connections to the die and to the outside terminals of the package. The outer layer consists of an insulation layer called soldermask, which is applied via a photolithographic process. The soldermask leaves openings for interconnections with solder balls on the outside terminals and bumps or wire connections on the inside terminals of the substrate in the package. Details of the manufacturing process have been described in the literature [5]. The technology is essentially along the same lines as PCB manufacturing, but substrate technology is more advanced due to smaller feature sizes, higher tolerances, lower contamination levels, the use of more advanced materials (e.g., low coefficient of thermal expansion (CTE), high glass transition temperature (Tg) materials), and so on. Organic substrates, by virtue of the Cu-based routing, enable a much higher I/O count than leadframes.

   The main disadvantage of organic substrates is its high cost (due to more expensive materials and manufacturing next to a dedicated supply chain). In more recent times, "hybrid" technologies called routable leadframes combine a more elaborate plated Cu routing with the use of a molding process to provide insulation and stability [4]. It has become more popular, because of substantial cost reduction. However, it cannot be used for very high pin counts.
2. **Die attach**: In most cases, dispensable paste materials are used to connect the die to the carrier. For products with stringent thermal requirements, metal-based die attach materials are used (e.g., high Pb solders or eutectic die attach layers). In recent times, the drive toward 100% Pb-free (note that high Pb solders fall under the exemption to RoHS legislation (Restriction of Hazardous Substances) has led to the development of alternative technologies like partial or complete Ag-sintered systems, though a few challenges remain (e.g., stress control) [6]. Mainstream organic die attach materials are generally filled with Ag flakes and consist of a whole variety of chemistries to impart and combine desired properties [7]. In more recent times, hybrid systems are employed, where chemistries are merged into a single formulation to achieve synergistic effects. For example,

epoxy-based systems enhance adhesion to various substrates but can be more rigid, whereas other chemistries like acrylate-based, bismaleimide-based and silicone-based chemistries can reduce moisture absorption and impart flexibility to the system [7]. Both conductive (generally Ag-filled) and nonconductive variants are in use. Nonconductive materials are generally silica-filled, but other types like Teflon-based fillers have also been used. In some cases, special fillers (like boron nitride) impart thermal conductivity but maintain electrical insulation of the die attach material. In addition to the resin, hardener and filler (Ag), several other components are added to the formulation in order to optimize other characteristics: adhesion promotors (to enhance interaction with specific metallic surfaces like Ag or Au), flexibilizers (to reduce the elastic moduli), reactive and nonreactive solvents (to reduce viscosity), etc.

Whereas conventional die attach materials were based on dispensable pastes, in more recent times, die attach films [4] have become more popular because they enable assembly of thinner dice as well as smaller clearances between dice, or between die to carrier or die to edge of the package, with the ultimate goal of miniaturization. Both conductive and nonconductive die attach films are commercially available. Unlike pastes that are dispensed on to the carrier before the die is placed on it, films are laminated on to the backside of the wafer, and the singulation process generally includes cutting through the die attach film as well. The most commonly available die attach films are supplied together with a dicing tape. A schematic image of the die attach film process is shown in Fig. 12.5.

3. **Encapsulants**: The most important type of encapsulants are epoxy molding compounds (EMCs). Although there are several other types of encapsulants that are used in microelectronic packaging, like glob top materials, dam and fill materials, and potting materials, EMCs are invariably based on epoxy chemistry consisting of resins, hardeners, fillers, and several additives and processed using transfer molding around 175 °C. Additives include catalysts, adhesion promotors, coupling agents, ion trappers, wax, coloring agents/carbon black, low-stress additives, and flame retardants, besides some proprietary ingredients to control warpage, reduce voids, etc. The hardeners are hydroxyl-based, and depending on the type and functionality of the resin and hardener, a range of cross-linking densities can be generated after cure, leading to a range of glass transition temperatures (Tgs). The filler used is generally fused silica, currently almost entirely spherical in shape (unlike flakes that were used in the past), with a maximum size of either ~55 um or 75 μm in mainstream materials. A key parameter in the

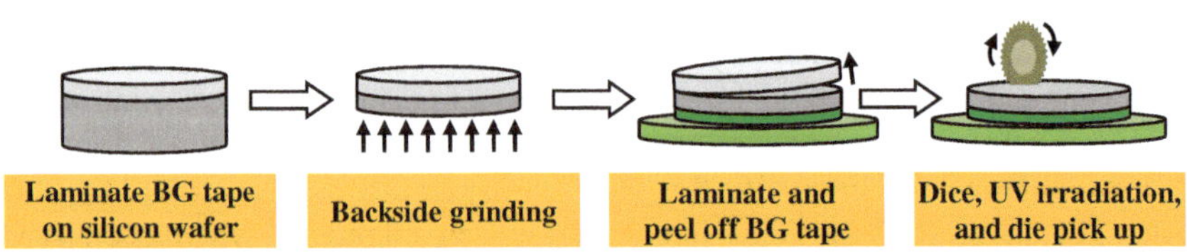

Fig. 12.5 Schematic flow of a die attach film combined with dicing tape (BG: back grind)

formulation of the EMC is the filler size distribution and filler content in the EMC, which determines key characteristics like the CTE, storage modulus as function of temperature, fracture toughness, moisture absorption, and flow properties. Proprietary size distributions are used by manufacturers to obtain higher filler loadings. With monodisperse systems, a maximum volume loading of 60% can be achieved. However, with special distributions, loadings up to around 80% by volume (around 90% by weight) can be obtained. Higher filler loadings are targeted to achieve lower moisture absorption or reduce thermomechanical stress due to a lower CTE mismatch with the silicon die. It is always challenging to obtain the right balance between various properties, e.g., flow properties to enable good processability, thermomechanical properties (to reduce or balance stress and package warpage), adhesion on various surfaces (but releasability from the mold), reduction of moisture absorption, good reliability (to overcome failure mechanisms like corrosion involving intermetal systems in the package), thermal and electrical stability, etc. A comprehensive reference on the challenges involved is described in a recent book [8].

4. **Underfill materials**: As mentioned in the second section of this chapter, there are generally two types of underfill materials: capillary underfill materials (CUF) and molded underfill materials (MUF). Molded underfills are essentially EMCs which contain finer silica fillers (sizes ranging from ~15 to 25 um). Capillary underfills consist of different chemistries. They can be epoxy-based, but the hardeners are amine-based or cyanate-ester chemistry and are essentially dispensable materials that also contain fine fillers and additives to enhance properties like flow and thermal stability. The filling process is driven by capillary forces due to the small gap between the substrate and die and is run at a somewhat higher temperature, in order to reduce viscosity. A very complete reference on this subject is a book on flipchip technologies [9].
5. **Interconnect materials (wire, solder bump, copper pillar)**: Metallic (alloy) wires are used to electrically connect the die to metallization on the leadframe or substrate using a process called wirebonding [2]. Conventional wires were based on Au (alloys). Commonly used dopants include Ag, Pt, Ca, and Pd; some reliability benefits were also demonstrated when using Pd coating [10]. Currently, Cu wire is being used in most applications, except in case no technical solutions exist to overcome the risk of damage during the bonding process of the more stiffer Cu wire compared to Au wire, or no acceptance of Cu wire to replace Au wire downstream of the supply chain. In addition to bare Cu, the most common type of Cu wire is coated wire (Pd coating with or without an Au flash layer). While wire suppliers have claimed reliability benefits of coated wires, the main driving factor for this wire type has been its improved processability in fine pitch applications, e.g., due to better bondability on the second bond (wedge bonding) and better free air ball control [11]. Extensive work has been done in the industry to understand the relevant degradation mechanisms evolving in the intermetallic systems involved: initially Au-Al and then Cu-Al, which is more susceptible. Details will be covered in subsequent sections.

In addition to this, Al wires with wedge-to-wedge bonding and Ag alloy wires are additional options in some power packagers or memory applications, respectively. The most important benefit of Ag-based wires above Cu-based wires is a lower risk of mechanical damage in the layer stack below the bondpads [12]. However, factors like lower cost benefit, higher corrosion sensitivity, and increasing maturity of Cu wire have prevented Ag alloy wire from becoming anything more than a niche area [12].

In flipchip applications, most of the solder ball materials used are lead-free, e.g., Sn-Ag, SAC305 (alloy of 96.5% Sn, 3%, and Ag, 0.5% Cu), In some cases, to enhance board-level performance, solder balls doped with bismuth have become popular [13]. In relatively recent time, plated copper pillars capped with Pb-free solder have become more common due to the drive toward finer pitch applications because of enhanced control of standoff height and joint diameter control [9]. Other benefits like improved thermal cycling performance and improved electromigration performance have also sometimes been reported [9].

Thermal interface materials, silicone-based gels, polymer-based repassivation or wafercoating layers are well-known and used in the industry but will not be covered in this chapter.

5 Material Characterization

This subsection will describe some typical properties of different materials and which characterization methods are used to determine them. In addition, to find out bulk properties of each of the individual materials like molding compounds, die attach, etc., it is equally important to investigate how they interact with each other within the package. Interfacial properties like adhesion are also relevant. The use of FE simulations (also called "virtual prototyping") is recommended to enable making predictions already in the design phase with respect to (hygro-) mechanical, thermal, or electrical performance of the package.

5.1 Thermomechanical Properties

Arguably, most of the failure mechanisms involving microelectronic packages are mechanical in nature. Hence, a thorough understanding of the thermomechanical properties of the various materials and their interactions is needed to enable a better prediction of reliability capability of the package. Unlike metals (like copper, gold) that are elastic until a specific strain limit beyond which they display plasticity, polymer-based materials like the molding compounds, die attach materials, underfill materials, coatings, etc. are viscoelastic in nature. At short timescales, they display more elastic behavior, so energy is stored when they are subjected to deformation. At longer timescales, they tend to be more "liquid-like" in nature, undergoing molecular rearrangements which allow for dissipation of energy. The reason for this is the molecular structure of the polymers, even when they form three-dimensional networks or when they are filled with inorganic materials like silica or metal. Depending on the timescale and magnitude of deformation applied to the system, the polymer network shows different responses, e.g., short-range interatomic deformations or long-range movements involving several segments of the polymer chain. A good reference on polymer viscoelasticity is a book by Aklonis et al. [14].

As shown in the previous sections, since microelectronic packages are made up of a combination of several materials, differences in CTE between them create stresses within each material or at the interfaces, depending on the moduli of the materials as the material passes through different temperatures during assembly, reflow soldering, and the product usage in applications or during reliability tests. This is because stress is a function of strain (which is coupled to CTE and temperature difference) and modulus. One of the ways to make a quick rough estimate of stresses is by using what is termed as "stress index' which is given in Eq. 1 below, so a function of temperature difference, difference in CTE ($\alpha(T)$) and temperature-dependent elastic modulus $E(T)$ [15]. The temperature range across which the integration is done can be between curing temperature of the relevant material and room temperature or the minimum temperature during temperature cycling reliability tests:

$$\sigma = \int E(T).\alpha(T)dT \tag{1}$$

Although this description does not consider several factors like the geometry of the system (design), time-dependent behavior, curing shrinkage, etc., it does sometimes provide good qualitative or even semiquantitative information when comparing materials. In many cases, elastic behavior is assumed in the organic materials when performing finite element simulations: this can sometimes be used to predict qualitative trends when comparing constructions or materials. However, to enable a better quantification of stresses, it is strongly advised to use a material model or constitutive relation that can describe the viscoelastic behavior of the material system, e.g., the generalized Maxwell model using a Prony series [16].

Another important factor to consider when estimating stress in a package is the presence of interfacial delamination. Microelectronic packages have several interfaces as shown previously. Delamination can be initiated when external loads or internal stresses exceed interfacial strength. Delamination at any interface can significantly increase stresses in the adjacent materials. This is partly due to creation of areas of stress concentration but also due to larger local deformations, which consequently increase the stress levels. Hence, it is essential to characterize the package in terms of delamination, as also quantify adhesion between different materials as function of temperature and moisture loading (e.g., EMC and die, EMC and leadframe or substrate, die attach and leadframe, etc.).

One of the topics covered extensively in the literature in relation to organic packaging materials is the glass transition temperature (Tg) [7]. All polymers display this second-order transition between what is called the glassy phase and the rubber plateau. In the glassy phase, only short-range molecular motions are possible, and the material is stiffer, whereas in the rubber plateau region, large-range movements are possible, and the material is soft and elastic, especially if the material is cross-linked to avoid chains from slipping over each other causing flow or creep (e.g., in amorphous thermoplastic materials). The Tg is not a single temperature but a relatively broad transition that can cover a window of several degrees of temperature, though in many cases a single value is reported in supplier datasheets. There are other definitions of the Tg, e.g., the onset of the transition rather than its midpoint. Furthermore, determination of the Tg can be done in several ways, and the value determined even for each of the methods depends on the rate of heating or cooling of the sample [7]. As mentioned, polymer behavior is time-dependent, and responses to heating and cooling are also time-dependent, e.g., faster cooling shifts the transition to lower temperatures, whereas slower cooling shifts the transition to higher temperatures (e.g., in a TMA). Another important fact is that in the presence of moisture, the Tg can drop by as much as 15 °C since absorbed water acts like a plasticizer [17].

A lot has been published on the Tg of packaging materials [18]. In many cases, writers have warned about crossing the Tg when performing reliability testing or in the final application. Strictly speaking, the material behaves in a different way above the Tg compared to below the Tg, so the concern is fully understandable. Some examples will be given in subsequent sections. However, the fact remains that in many cases, reliability tests are performed either above the Tg of some of the packaging materials (high-temperature storage life, (un)biased humidity tests), or the Tg is frequently passed during the reliability test (temperature cycling test). Even in real-life applications, usage temperatures can sometimes increase above the Tg for some or even all the packaging materials. In most cases, this does not pose any reliability issues, but in some cases, it can be of potential concern. Essentially, it is important to understand potential degradation mechanisms in relation to the properties of the material when crossing the Tg, which enables a better prediction of when it can have a significant influence. One should not reject materials purely based on their relatively low Tg. In some cases, low Tg materials are beneficial since they can enable reducing stress in the package [19].

Yet another important characteristic of organic materials is the "gel point" [7]. Almost all polymers used in microelectronic packaging are thermosets. Until the material is cured, its behavior is more "fluid-like" meaning mechanical energy is easily dissipated, which implies that stress is not really built up in the package. Once the material is fully cured, it behaves more "solid-like" even though viscoelastic in nature, thereby starting to store energy and build up stress. This is schematically shown in Fig. 12.6. The point at which the material transitions from a liquid to solid during cure is called the "gel point" since the material at that point forms a three-dimensional network (hence behaves like a gel) [7].

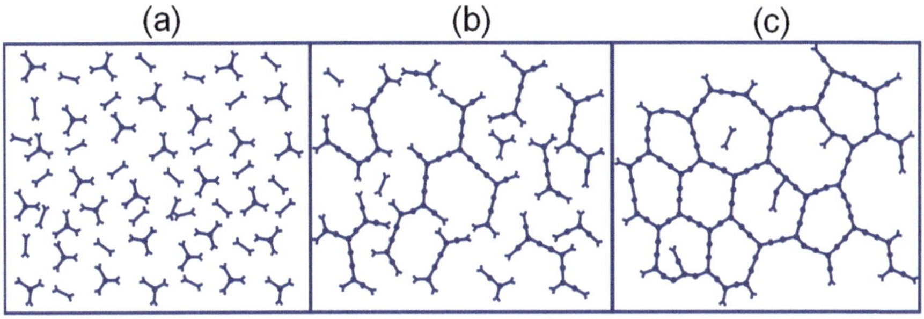

Fig. 12.6 Schematic description of gel point during the curing reaction: (**a**) monomers, (**b**) linear polymers, (**c**) gel (3D network)

The most commonly used techniques to characterize the materials will be described as follows [7].

5.1.1 DMA (Dynamic Mechanical Analyzer)

One of the most widely used instruments to characterize the thermomechanical behavior of organic packaging materials is a DMA. The basic concept involves applying a sinusoidal deformation (strain) to a sample and measuring the resulting stress as the temperature, deformation frequency, or both, are steadily increased from low to high values. Usually, the temperature ranges from – 65 °C to 300 °C, to cover the range of temperatures that the material will be subjected to during reliability tests or in the application (e.g., including the ~260°C reflow temperature). Due to the viscoelastic nature of the materials, there is a lag in response between stress and strain, called the phase lag (δ), which is between 0 and 90°. In this way the two components of the complex modulus – storage (G') and loss modulus (G") – as also the tan δ can be determined as function of time and frequency. There are different ways to stress the samples, e.g., three-point bending, single cantilever, dual cantilever, or in tensile mode. The method used depends on the stiffness range of the material to be tested.

Several characteristics of the material can be determined using a DMA: the most basic output is the storage (G'), loss modulus (G"), and tan δ as function of temperature. The peak of the tan δ is used to indicate the midpoint Tg of the glass transition window. This can be understood physically as the point where the maximum energy is dissipated in the system when the material transitions from the glassy state to the rubbery state due to large molecular deformations. The onset of the transition based on the storage modulus (G') can also be used to indicate the start of the glass transition. In addition to this, viscoelasticity can be characterized by performing measurements as function of frequency. This behavior can be phenomenologically described by what is called a generalized Maxwell model [16]. This model is used to derive a constitutive relation to describe the material behavior, called the Prony series (Eq. 2):

$$E(t) = E_R + \sum_{n=1}^{N} E_n \exp(-t/\tau_n) \tag{2}$$

where $E(t)$, viscoelastic modulus; E_R, elastic modulus for the rubbery state; E_n, elastic modulus; τ_n, relaxation times; and (t/τ_n), time-dependent viscous behavior. Shift factors are obtained using the Williams-Landel-Ferry (WLF) or Arrhenius relation. This is based on the time temperature superposition (TTS) principle [14], which postulates a relation between temperature and time dependence of the mechanical response. This can then be used to construct a Master curve. See Fig. 12.7. For additional reading on the subject, the reader is referred to two references [14, 16]. It is strongly recommended to use viscoelastic material data in finite element simulations involving packaging materials like EMCs, die attach, etc., assuming linear-elastic behavior will end up in inaccurate prediction of stress levels, even though the qualitative trend may still be accurate.

In addition to the influence of temperature, a special RH (relative humidity) accessory can be used to characterize the thermomechanical behavior under the influence of moisture uptake. A DMA can also be used to determine dimension changes (e.g., shrinkage) in a sample when subjected to heating, cooling, or moisture [20].

5.1.2 TMA (Thermomechanical Analysis)

TMA is used to measure thermal expansion of a sample when heated or cooled down by means of a probe that is placed on top of the sample to record the expansion of the sample. The most commonly used method for stiffer materials is called the expansion mode where a low force is applied on the probe. It can be used to determine coefficients of thermal expansion, by the simple relation shown in Eq. 3:

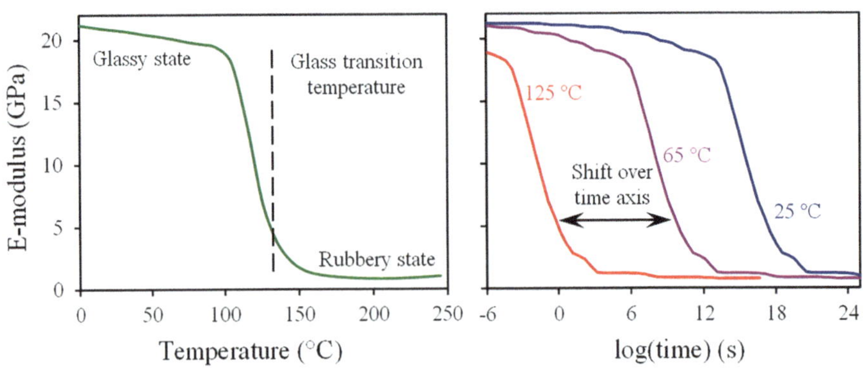

Fig. 12.7 Typical DMA measurement carried out as function of temperature (left) and the resulting curves plotted vs frequency including the horizontal shift factors (right). Both plots show the corresponding glassy region, the rubber plateau, and the glass transition

$$\alpha\left(C.T.E\right) = \Delta L / \left(L_0 * \Delta T\right) \tag{3}$$

where ΔL, change in length; L_0, original length; ΔT, temperature range. It can be determined from the slope of the displacement vs temperature curve (provided the initial length of the sample is recorded). The inflection point in the curve indicates the transition between the low-CTE region and high-CTE region, which is the Tg. For softer materials, the Tg can also be determined with TMA by using what is the penetration mode, where a somewhat higher force is applied on the probe/sample. As the material passes its Tg, it becomes softer and allows the probe to penetrate deeper into the sample, thereby enabling detection.

5.1.3 DSC (Differential Scanning Calorimetry)

DSC is used to determine thermal transitions in polymers like glass transition temperatures, melting temperatures (crystalline materials), reaction exotherms, etc. The working principle of the equipment involves measuring heat that is supplied to a sample and reference pan simultaneously to maintain the same temperature. When the sample undergoes a transition, the heat flow to the sample is changed compared to the reference accordingly and can hence be detected. Since packaging materials are generally (highly) filled with inorganic materials, transitions like the Tg are usually small and not easy to detect using a DSC. The most useful information that can be obtained for packaging materials is on their curing behavior. As a material cures, it undergoes a large exothermic transition that is clearly detectable with DSC using a dynamic scan: heat is liberated during the curing reaction. If cured material is rescanned, the transition is not visible; see Fig. 12.8. This can also be used to check whether a specific temperature profile is effective in curing the material.

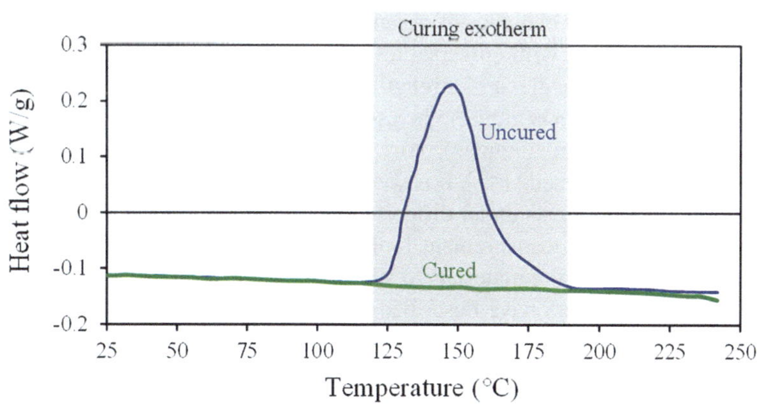

Fig. 12.8 Typical DSC measurement showing a curing exotherm of uncured material (blue) that disappears during a re-scan or measurement of a cured material

Another application of DSC is to determine the curing kinetics by using the approach as outlined by Kamal and Sourour [21]. By performing scans at different temperatures, the reaction rate can be determined as function of heating rate. Areas under the curves provide information on the conversion, e.g., the enthalpy of reaction.

5.1.4 TGA (Thermogravimetric Analysis)

TGA is used to characterize thermal stability of polymers, including packaging materials. The equipment measures weight loss (or increase) when a material is heated. The temperature range that is employed is typically between room temperature and about 900°°C. Measurements can be done under air or nitrogen. Key items to look for are the onset of decomposition (or temperature at which a certain level of weight loss takes place), as also what is called the ash content (which indicates the inorganic content in the material). Generally, the high-temperature exposure is carried out in air to enable complete removal of the organic fraction. Most organic materials show a large-scale decomposition within the range of 350–500 °C (even under nitrogen) due to chain scission. The thermal energy under those conditions is high enough to overcome the binding energy needed to break various bonds within the polymer. Beyond ~600 °C in air, the organic fractions in the material decompose completely. In addition to dynamic scans, isothermal scans can also be used to assess thermal stability at a specific temperature. A plot of differential weight loss (vs temperature or time) can be used to indicate at what point the maximum weight loss takes place.

5.1.5 Other Mechanical Tests

In addition to DMA, which is especially useful to characterize thermomechanical (especially viscoelastic) behavior of organic packaging materials, other techniques like tensile testing (e.g., with universal testing machines) can be used to determine properties like tensile or flexural strength of materials: both virgin materials and after aging. This equipment can also be used to characterize metal (alloys) like solder (e.g., using dogbone samples), wire, etc., in which case their visco-plastic behavior can be determined [141]. It is sometimes necessary to use a climate chamber to enable measurements under different exposure conditions.

Another very useful measurement is the determination of interfacial adhesion between organic packaging materials to substrates and leadframes or to the silicon-based die. Several papers have described methods to determine adhesion using a four-point bending method, e.g., dual cantilever beam method [22], to obtain a critical energy release rate, G_c, using cohesive zone modelling or an analytical equation. Alternatively, a J-integral approach (nonlinear energy release rate) has been used in some cases to analyze similar measurement data, combined with FE simulations. Here too it is advisable to perform measurements under different conditions, especially as function of temperature. More state-of-the-art equipment considers mode mixity when measuring adhesion in terms of a G_c [23], but even here, FE simulations are needed to enable determination of G_c. The advantage of these methods is that they can be used to quantify delamination issues by incorporating fracture mechanics (energetics) into FE simulations. This will enable prediction of whether or not delamination will progress as the stored energy in the system is released as the crack propagates.

Material suppliers often use button shear (room temperature or high temperature) to determine adhesion of molding compounds on various leadframe finishes (e.g., copper, silver-plating, etc.). It is powerful for comparing material combinations because it is a more qualitative method since factors like modulus differences at testing temperature, stress in the sample, etc. are not considered. However, some efforts have been made to quantify this method by using FE simulations [24], and it can provide useful information on adhesion properties between EMCs and different leadframe finishes.

5.2 *Chemical Properties*

In order to investigate failure mechanisms in relation to the chemical behavior of the material, chemical analysis techniques are extensively used to characterize organic packaging materials. One of the key items in microelectronic packaging is the ionic contamination levels. Trace levels of anions like halides (Cl^-, Br^-, etc.); cations/metals like Na^+, K^+, and Li^+; or acidic pH conditions (e.g., pH < 5) can be detrimental to package reliability, inducing corrosion, charge creep [25], and even ion ingression [26].

Hence one of the first tests used to screen materials is the determination of its ionic content. All material suppliers perform such tests usually. However, there is no common method to do this, though some mainstream suppliers use methods that are similar. The test results are dependent on several factors like the testing temperature and duration, sample preparation (dimensions), solid to liquid ratio during the test, etc. Hence it is essential to know the testing conditions or check the same method before comparing ionic contamination data. The most widely applied method for molding compounds is to perform an ultrapure water (UPW) extraction on mechanically ground samples at 121 °C for 24 h using a solid to liquid ratio of 1:10. Ion chromatography (IC) allows to determine the ion concentrations that are normalized to the content of the material in terms of ppm levels. The content of metals is sometimes analyzed with inductively coupled plasma mass spectrometry (ICP-MS).

While ionic extraction can be used to determine extractable ion content, a different technique is preferred to determine the total elemental content (the sum of extractable and bound fractions). This is also based on IC, but the material is initially combusted at very high temperatures (> 950 °C) in air, and the formed gases are led through UPW, followed by IC to determine the content of, e.g., halides and sulfur-based [126]. Even though it is usually the extractable ion content that poses most of the reliability risks, in some cases knowing the total level of halides or sulfur can be useful. An example will be shown in one of the subsequent sections.

For completeness, several other techniques are used as summarized below:

- FT-IR (Fourier Transform Infrared Spectroscopy): for identification of organic species (to investigate chemical changes) [27].
- GC-MS (gas chromatography-mass spectrometry): for identification of species being outgassed from the material at a specific temperature, enabling the determination of fingerprints of specific materials [28].
- Surface analysis techniques like Auger spectroscopy, XPS (X-ray photoelectron spectroscopy), ToF-SIMs (time-of-flight secondary ion mass spectrometry) for analysis of surfaces or interfaces [29]. Both surface spectra and depth profiling are carried out to enable identification and quantification of (near)-surface contamination. Whereas ToF-SIMS is the most sensitive technique (chemical information from the top nm), XPS and Auger are less sensitive (e.g., information from the top ~5 nm) but more quantitative.

Of course, the list of chemical analysis is far from complete, depending on the specific needs; other techniques like HPLC (high-performance liquid chromatography), GPC (gel phase chromatography), and combinations of techniques sometimes need to be used when resolving problems.

5.3 Moisture-Dependent Properties

There is a lot of published literature on moisture-related reliability failures in microelectronic devices, including methods to characterize materials like EMCs with respect to their behavior when exposed to a humid environment [32]. The most common assumption for moisture diffusion kinetics is that Fick's law holds, in 1 D denoted as

$$\partial C / \partial t = D . \partial^2 C / \partial x^2 \tag{4}$$

where C, moisture concentration; D, diffusion constant. Key parameters to describe a material are the values of D (which is temperature-dependent as

$$D(T) = D_0 \exp\left[E_{a1} / kT\right] \tag{5}$$

and the saturation concentration (C_{sat}) which also depends on temperature in the case of filled polymers like EMCs. The relation for C_{sat} as function of temperature is given as

$$C_{sat} = P_v S_0 \exp\left[E_{a2} / kT\right] \tag{6}$$

where P_v, vapor pressure around the sample; S_0, solubility parameter; D_0, pre-exponential diffusion term; E_{a1} *and* E_{a2}, activation energies; k, Boltzmann's constant; and T, absolute temperature. P_v depends on the relative humidity (RH) and P_{sat}: saturation pressure which is also a function of temperature:

$$P_v = R.H.P_{sat} \tag{7}$$

The diffusion kinetics can be determined by subjecting samples to various soak (T, RH) conditions and taking out samples at successive intervals to weigh them or by using the more sophisticated differential vapor sorption (DVS) technique or a gravimetric sorption analysis (GSA) [20]. Strictly speaking, although the above relations are good approximations of the moisture characteristics, the behavior is not fully Fickian in nature [30, 31]. The moisture level in EMCs does not saturate even after extended exposure to a specific soak condition; see an example shown in Fig. 12.9 [30]. This behavior has been reported by several authors, and various explanations have been provided, but the behavior is not fully understood. One potential mechanism is that as moisture is absorbed by the polymer network, it swells and undergoes molecular rearrangements, creating more free volume for additional moisture to be absorbed. Alternative models have been proposed to describe the behavior like the Langmuir and the dual-stage models [31].

In addition to this, other properties like the coefficient of moisture expansion (moisture-induced swelling) are conventionally determined by correlating TGA on one and TMA on a second sample, both soaked to achieve saturation. As soon as the samples dry, contraction will happen. TGA determines the corresponding level of moisture content, while the TMA monitors the dimensional change. The following equation holds

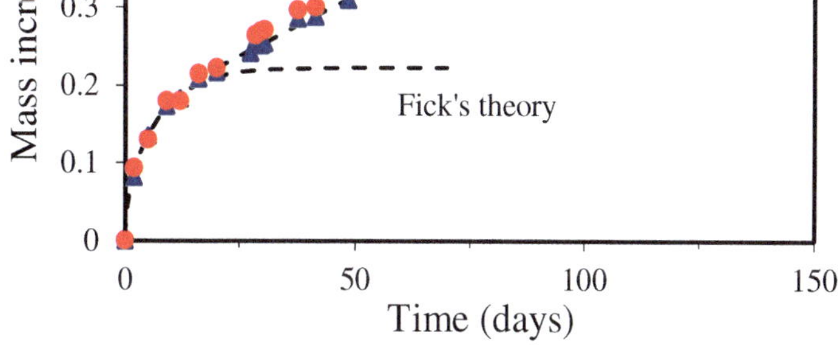

Fig. 12.9 Absorption of moisture by an EMC immersed in liquid water [30, with permission]

$$\varepsilon^{RH} = \beta.C \tag{8}$$

where ε^{RH} is the moisture-induced strain, β is the coefficient of moisture expansion (CME) [mm^3/mg], and C is the moisture concentration [mg/mm^3]. The method was first published by Wong et al. [32]. Other more sophisticated methods have also been described recently [20].

On a final note in this section, the reader is introduced to so-called moisture sensitivity levels (MSL), which is well-known in microelectronic packaging as shown in. Table 12.1 [33] shows an overview, and, as indicated, microelectronic packages with a greater moisture resistance (e.g., at reflow during PCB assembly by semiconductor users) have a higher-ranked MSL number (e.g., Level 1), also reflected in the standard soak requirements to which the package is subjected before performing the reflow assessment.

5.4 *Electrical Properties*

Since the function of a package is also to provide electrical protection to the device, electrical properties of the involved materials and construction are important and need to be characterized. Characteristics like dielectric constant and dissipation factor as function of frequency are needed for electrical package simulations (e.g., to determine dielectric/radio frequency (RF) losses). As there is an increasing drive toward RF, low loss packages, these properties need to be determined at much higher frequencies (GHz range) rather than only the MHz range that is usually provided by material suppliers. Typical ranges of these properties are dielectric constant(ε_r) ~ 3–4 and dissipation factor (tan δ) = 0.01–0.001. In addition, dielectric strength and breakdown voltage are also important to enable designing products that can withstand applied voltages and electrical fields under DC or AC conditions.

An important parameter in case of high-voltage insulation is called the Comparative Tracking Index (CTI), which reflects its ability to withstand breakdown under application of high voltage in the presence of external contamination.

Table 12.1 Moisture sensitivity levels in packages and soak time requirements [33]

Moisture sensitivity level	Floor life condition	Standard soak requirements	
		Time (h)	Condition
1	Unlimited (≤30°C/85%RH)	168 + 5/−0	85°C/85%RH
2	1 year (≤30°C/60%RH)	168 + 5/−0	85°C/60%RH
2a	4 weeks (≤30°C/60%RH)	696 + 5/−0	30°C/60%RH
3	168 h (≤30°C/60%RH)	192 + 5/−0	30°C/60%RH
4	72 h (≤30°C/60%RH)	96 + 2/−0	30°C/60%RH
5	48 h (≤30°C/60%RH)	72 + 2/−0	30°C/60%RH
5a	24 h (≤30°C/60%RH)	48 + 2/−0	30°C/60%RH
6	Time on label (TOL)	TOL	30°C/60%RH

The CTI voltage defines the material class and is used to design clearances between conductive parts on the outside of the package (creepage distance). More information can be found in the IEC standard [34].

Finally, one of the parameters that will be used to characterize EMCs is volume resistivity (VR), especially at high temperatures (e.g., after preconditioning). This parameter can be used to assess the behavior of EMCs under application of high electric fields [35]. More details will be shown subsequently.

To demonstrate the diversity in typical properties of various packaging materials, see Table 12.2.

6 Reliability and Qualification

6.1 *Introduction*

The definition of reliability is the probability that a product/package will perform a required function without failure under specified conditions for a stated period. Depending on the application, the expected lifetime of the product can be different. For example, the use life for a consumer application like a handheld device (few years) is less stringent compared to an automotive (15 years) or an infrastructure application (20 years). In addition, the conditions the products are subjected to during its use life are completely different from application to application and can even (periodically) change over time. This combined description of use life and application specific conditions is termed as the mission profile of a product [1]. Mission profiles typically consist of a combination of operational (power-on) time, steady-state temperatures, cyclic thermomechanical loading, humidity during standby, and power dissipation (self-heating).

Product quality refers to performance according to its specification at a specific moment: mechanical (including dimensional), electrical, and thermal. Hence reliability can be considered as product quality evolution over lifetime. Microelectronics are being used in steadily more applications, be it consumer, automotive, or

Table 12.2 Typical properties of key packaging materials (CTE: coefficient of thermal expansion)

	Silicon (die)	Cu (alloys)	EMC	DA	Underfill	Solder balls (Pb-free)
CTE1 (ppm/K)	~3	~17	~6–25	~20–50	25–35	22–23
CTE2 (ppm/K)	N/A	N/A	~18–75	~60–150	75–105	N/A
Tg (°C)	N/A	N/A	~100–225	~ – 100–150	~ 90–150	N/A
E modulus (GPa)	169	117	~10–30	~0.5–12	~8–12	~45–60
Shrinkage (%)	N/A	N/A	~ 0.1–0.3	~ 0.2–1	~0.2–0.5	N/A
Moisture absorption MSL1 (wt%)	0	0	~0.15–0.5	~0.1–1	~0.5–1	0

infrastructure, and consequently a lot of attention has been spent to improve on both the quality and reliability performance of microelectronic products and packages. The demands of even mobile phone makers have become more stringent lately. In addition to this, in the current world reliability is becoming even more important where a rapidly increasing number of devices are getting connected with each other to form a network. This is the so-called Internet of Things (IoT). IoT describes the network of physical objects – "things" – that are embedded with sensors, software, and other technologies for the purpose of connecting and exchanging data with other devices and systems over the Internet [36]. In a world becoming smarter and more automated, reliance on microelectronics and its performance will only become greater.

The reliability of an interconnected network is determined by the reliability of its weakest link. In cars these days, tens of electronics applications are present. Each application contains up to hundreds of components. This implies that a failure rate of single-digit ppm can lead to % level of failure in the car, which is totally unacceptable by any standards. In addition to this, a failure in automotive applications can be a matter of life and death, especially considering the increase of functional safety applications that rely on microelectronics [37]. One of the much talked about trends is autonomous driving where there is no margin for any error or defect. Hence it is not surprising that automotive manufacturers (original equipment manufacturers or OEMs) demand zero defects and zero incidents from their suppliers. In addition to this, applications are becoming more stringent from an environmental perspective. Microelectronic packages are finding their way into applications which are subjected to harsh conditions in terms of temperature, moisture, and chemical environment, e.g., automotive sensors, power devices, and infrastructure applications where the device is almost perpetually operational for 15–20 years.

Reliability over lifetime is characterized by what is called a bathtub curve [38]; see Fig. 12.10. Most standard reliability tests cover only useful life while not yet entering the wearout time domain characterized by the onset of failures due to intrinsic degradation mechanisms. Some automotive tests (like high-temperature operating life-early life failure rate/TOL-ELFR) do test products to cover the very first time domain with a potential risk on early fails (infant mortality). As state-of-the-art technologies are developed to become more robust from an intrinsic reliability point of view, in some cases reliability failures in the field (application) are extrinsic in nature, which means driven by defects or excursions (e.g., corrosion induced by particles) [39]. Hence, there have been intensive efforts to reduce the risk of such failures by taking precautionary measures like particle reduction programs in assembly [39]. Whereas wafer processing technology has always been several steps ahead in terms of defect control, assembly technology (including material manufacturing) is catching up in recent times, but there is still room for further improvement.

Competition between various OEMs and their microelectronic suppliers puts a very high pressure on achieving good quality and reliability next to a faster time to market for each new equipment generation. This may sometimes be contradictory because speeding up new product development can increase the risk of

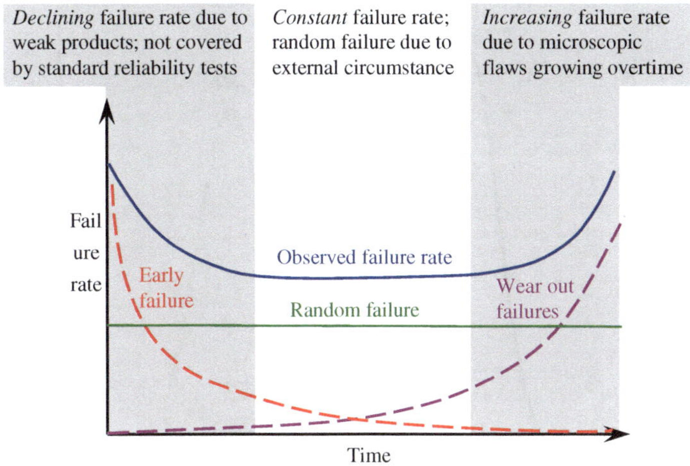

Fig. 12.10 The reliability bathtub curve

encountering unknown failure mechanisms: both extrinsic and intrinsic. In addition to this, multi-scale and multi-physics challenges are involved due to complex interactions between wafer technology (front end, on nm scale) and assembly (μm/mm scale), including multiple material interactions. Some examples of this will be given in the subsequent sections.

One of the approaches to improve reliability has been to increase robustness against the different failure mechanisms. To enable this, it is essential to understand the physics behind the mechanisms in detail so that the involved processing and material characteristics can be addressed. One of the ways to increase robustness is to improve process yields: higher yields usually also imply enhanced reliability performance [40].

A specific example is shown in Fig. 12.11: interfacial delamination or bulk fracture becomes a risk when (interfacial) *stress* approaches (interfacial) *strength*. This is of course a simplistic way to describe the complexities of fracture mechanics, when referring to interfaces, but allows us to explain the basic concept. Due to process and material variations, opposing quantities – stress and strength – always fall within certain distributions. If the distributions overlap to some extent, there will be a risk of failure occurrence. This can even happen in case of small changes. By shifting the distributions further apart, the overlap will reduce or even become extremely small and failure occurrence therewith unlikely, i.e., a certain robustness margin has been achieved [41]. One of the approaches to explore robustness margin is by building "corner lots" to determine the process or material specification window.

One of the key elements of product development is the reliability qualification process. This process aims to demonstrate by way of endurance testing that the technology and packaged product can meet the set requirements, including the mission profile. This is usually done by subjecting representative populations of products (multiple lots that are manufactured separately) to a set of acceleration and

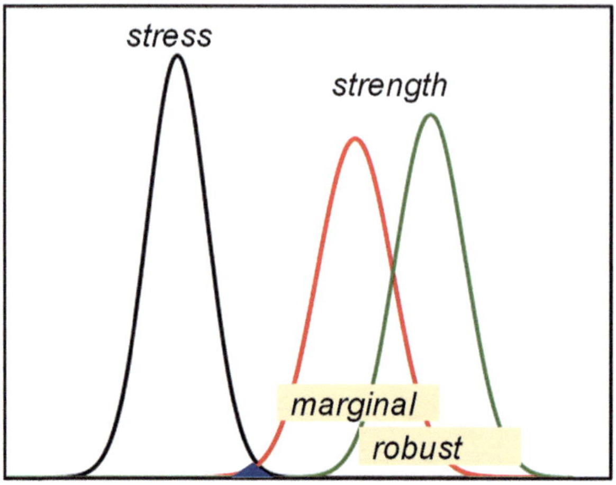

Fig. 12.11 Demonstration of robustness in relation to material interfaces

stress tests that aim to simulate the product lifetime and loading conditions within a much shorter period. Each test is defined to accelerate specific failure mechanisms by using environmental stimuli like temperature, relative humidity, and product-related stimuli like voltage and current. An overview of the most important test methods, conditions, and durations next to the acceleration models are shown in Table 12.3 [42].

Qualification methodology has changed significantly over the last 20 years [43]. Initially, stress-based qualifications were carried out by performing the specified tests with prescribed sample sizes. The qualification was classified "pass" when no electrical failures were observed by the end of these tests. All tests were executed on individual components. In the next generation, qualification approach moved to a more knowledge-based methodology with a deeper understanding of failure mechanisms and physics of failures. In addition to this, the concept of qualifying to meet specific component level mission profiles was introduced. And the test methods started to also include the interaction with the actual PCB (application board), called second-level reliability. Tests were extended beyond requirements (until fail) to assess robustness margins. Failures beyond readpoints covering the mission profile were allowed, if risks for actual failure were estimated low and beyond the useful life (refer to Fig. 12.10). Currently, while this trend continues, the concept of system reliability has come into the picture. This means a better awareness of the entire system is needed, with concepts like physics of degradation and built-in resilience, e.g., by redundancy and preventive replacement and/or repair.

Stress-based reliability qualification is generally done in three independently manufactured lots to capture process and material variations. The sample size per test per lot is 77; no electrical fails or 0/77 implies statistically that one can say with 90% confidence level and lots with a defect percentage above 3% will not be accepted [44]. These numbers are based on purely historical considerations and

Table 12.3 Overview of most commonly used package reliability tests, the test conditions for two typical applications, and the corresponding acceleration models [42]

Reliability test	Typical conditions	Test duration		Acceleration model [42]
		Consumer	Automotive	
High-temperature storage life (HTSL)	150 °C	1000 h	1000 h	Arrhenius
Highly accelerated stress test (HAST)	130 °C/85%R.H	96 h	96 h	Hallberg-Peck
Temperature humidity bias (THB)	85 °C/85%R.H	1000 h	1000 h	
Temperature cycling (TC)	−55 °C–150 °C	700 cycles	1000 cycles	Coffin-Manson

industry consensus. Knowledge-based reliability qualifications combined with the recent demand for higher reliability (speak larger robustness margins) call for extended testing, larger sample sizes, and physical analysis of the devices subjected to reliability tests, rather than rely on functional tests alone. For more details on failure statistics (e.g., the use of probability distribution functions) and approaches for risk assessment, the book by Meeker et al. is recommended [44].

Additionally, during package qualification, the use of what is called qualification by structural similarity is allowed [45]. If it can be shown based on historical data or FE simulations that certain geometrical features are more critical for a specific failure mechanisms (as a response to a reliability test), then reliability data from the more critical package (called reference package) can be used to cover the less critical package. This can save on costs and reduce the effort needed to qualify a new package in some cases. Of course, to claim structural similarity, certain strict rules need to be followed (e.g., applicability within certain package groups, the use of identical materials, etc.). Note: Some users may not accept this concept arguing with technical or nontechnical reasons.

On a final note in this section, the difference between a "failure" and "degradation" can be explained as follows. Failure in the case of a product takes place when it does not pass its electrical requirements or functionality. Degradation refers to a progressive physical (or chemical) mechanism that is initiated when the product is subjected to stress, which may or may not ultimately lead to failure.

6.2 *Material Qualification*

The scope of this chapter is the reliability of organic materials in microelectronic packaging. One of the phases during technology and subsequently product development is new material qualification. A new material (combination) is usually qualified for a specific reason, e.g., to improve product performance and/or reliability, to reduce cost, to improve manufacturability or productivity, for environmental reasons (e.g., conversion to lead-free), for standardization, or to enable a new packaging technology. Since the whole process takes effort, capacity, and time, it is highly advisable to qualify new materials with a broad applicability, considering future requirements. This also enables standardization and reduces diversity and cost in terms of material usage. Developing point solutions is not inefficient.

The process starts with defining a material profile in terms of required properties for the given application. It involves a paper study. Suppliers are approached to provide their recommended candidates. Next, generally a screening study is performed with several candidates involving the following steps successively narrowing them down to only one or two for the final qualification stage:

- Pre-screening (theoretical) based on the datasheet properties; material screening based on sustainability requirements (e.g., screening for banned or restricted substances) needs to be done as early as possible to avoid surprises at a later stage of the development.
- Material characterization as per the methods described in Sect. 4 (e.g., ionic content, thermomechanical properties, adhesion etc.).
- Use of finite element simulations (design optimization).
- Processability and process window evaluation, in some cases linked to the material characterization activity.
- Initial endurance studies using MSL testing and limited/preliminary reliability testing.
- If applicable, development of failure analysis (FA) techniques.

The use of risk analysis and mitigation techniques, like failure mode and effects analysis (FMEAs), is important in the initial stages of development. Historical experiences and lessons learned should be well documented and followed. In some cases, iterations are needed (e.g., minor material tweaks by the supplier) via a feedback loop before proceeding with the next steps. If the process is followed properly, it will lower the risk of running into unexpected failures during the qualification. However, no procedure is foolproof toward encountering unexpected issues during the final qualification, though it is always the endeavor to be "first time right." It reduces development time and the need to spend effort and capacity toward the end of the entire process.

6.3 *Failure Analysis*

Failure analysis (FA) is a process involving various techniques, to determine the root cause of product failure and gain more insight into the involved failure mechanisms. FA is relevant when dealing with "customer reject and returns" or reliability failures encountered during development or qualification. For more details, readers are referred to an excellent book [46]. Before FA is started, it is of vital importance to collect as much background information as possible (e.g., the product history, failure statistics, etc.). In microelectronics, it invariably starts with an electrical verification of the product to confirm the electrical failure signature. For this purpose, curve-tracing or micro-probing techniques can be used.

In some cases, electrical testing may indicate that a product has failed, but this does not necessarily mean the failure is valid. Terminal contact-related issues or tester-related issues need to be excluded. After it is confirmed that the fail is valid, in electrical terms (open, short, leakage, parametric shift or drift, etc.), then a number of nondestructive techniques are used to investigate the physical condition of the product, e.g., visual inspection using optical microscopy to look for external damage, X-ray to check for deviations in metal parts (e.g., wires, solder balls, particles), acoustical microscopy (AM) to check for potential delamination, and so on.

Based on insight gained from the nondestructive analysis, next steps are defined, including destructive techniques like the use of cross sections and decapsulation. Decapsulation is removal of the encapsulant chemically, by laser ablation or mechanically. Figure 12.12 shows a product after decapsulation, exposing the die. It is advisable to define the FA strategy carefully and step by step since once destructive analysis is started; there is no turning back, so valuable evidence that may contribute to the failure maybe lost or destroyed (e.g., particles).

State-of-the-art equipment like photoemission electron microscopy (PEM), optical beam-induced resistance change (OBIRCH), lock-in thermography (LIT), and magnetic imaging are used to localize faults like leakage, shorts, and hot spots, but some amount of mechanical preparation of the sample is needed in order to access internal circuitry, sometimes from the backside of the device. The challenge is to find the root cause of the observation: it could be die (device)-related (e.g., leakage due to internal gate oxide breakdown), package-related (e.g., open induced by a mechanical fracture in the interconnect induced), or a specific interaction between die and package (e.g., internal short due to passivation cracks induced by the encapsulant [47]. An additional challenge is the multi-scale nature of the device (nm scale within the die, μm/mm scale at package level).

It is of vital importance to understand the limitations and resolution of various techniques before drawing firm conclusions. Also, it cannot be stressed enough that it is always necessary to include a "known-good" reference.

During new material development, e.g., a new EMC, die attach, or interconnect material or even carrier (leadframe or substrate) technology, developing the needed FA techniques is to be considered. For example, to test interconnect integrity (e.g., using wire pull), it is necessary to chemically remove the EMC without damaging

Fig. 12.12 Example of a decapsulated device exposing the die and wirebonds

the wires. Hence decapsulation recipe development is needed during the material development [47, 48]. Also, it is essential to be able to analyze the new material using existing techniques or develop new techniques to do so (e.g., methods to analyze "fingerprints" of surface-sensitive materials like additives in die attach materials) [29].

7 Failure Mechanisms

Figure 12.13 shows an overview of failure mechanisms in microelectronic packaging based on an overview published by the ZVEI (Germany's Electrical Industry) Robustness Validation Knowledge Matrix [49]. About 30% of them are related to the package. This could be assembly- or package-related, e.g., cracks, corrosion, and migration; chip-package interactions, e.g., charge creep, corrosion, or intermetal shorts; or solder joint-related, e.g., migration, cracks, or corrosion. About 50% of these in turn are related to aging: either due to fatigue, corrosion, migration, or thermal degradation. Since the main theme of this book is degradation, the failure mechanisms described in this section will focus on degradation under steady-state loading, e.g., temperature, humidity, and bias. There has been a lot of work describing failure mechanisms under dynamic loading (e.g., during temperature cycling), but this will only be touched upon briefly. The section will be split into two main parts: bulk degradation and a section on interconnect degradation. Table 12.4 summarizes the failure mechanisms described in the subsequent sections.

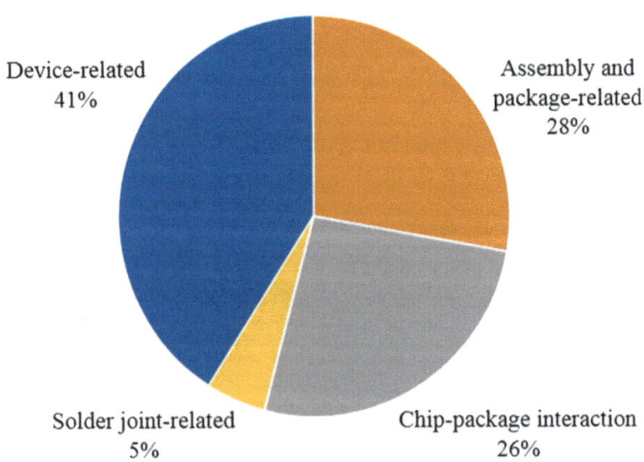

Fig. 12.13 Failure mechanisms in microelectronic devices (increase size and indicate %) [49, with permission]

Table 12.4 Overview of aging-related failure mechanisms described in the subsequent sections

Category	Subcategory	Section
1. Bulk	EMC	1(a)
	Die attach	1(b)
2. Interconnect reliability	Au wirebond	2(a) i
	Cu wirebond	2(a) ii
	Flipchip	2(b)
3. Others	For example, migration	3

7.1 Bulk Degradation

Most polymers used in microelectronic packaging, being organic in nature, are susceptible to thermal degradation or even hydrolytic degradation. Thermal degradation is by far the most important phenomenon since it will take place in the final application and can be assessed with accelerated life tests (reliability tests). A good reference on the thermal stability of polymers is a book by Compton [50].

Ambient temperature refers to the temperature of the environment surrounding the product during its application, whereas junction temperature refers to the temperature within the die or at the die surface. Applications (especially automotive) have increasingly more stringent requirements in terms of ambient temperatures. While Grade 0 applications require maximum operating ambient temperatures of 150 °C, short temperature transients up to about 200 °C are not uncommon. As microelectronic applications finds its way into other application like commercial vehicles (e.g., trucks), the ambient temperature can increase to temperatures approaching 200 °C. In addition to this, advanced RF power packaging (e.g., involving gallium nitride devices) should be able to sustain junction temperatures approaching even 250 °C for short durations.

During reliability testing, it is the aim to shorten the test duration by using temperature to accelerate potential failure mechanisms in the field. This assumes that the failure mechanism is accelerated based on chemical processes that follow the Arrhenius law [51]. On the one hand, this may seem an oversimplified way of looking at complex failure mechanisms that involve several competing processes. On the other hand, based on extensive empirical data collected over the several years, it so happens that the law does hold in many cases. However, one should always be cautious when assuming this to be a general rule. In some cases, the system may be over-accelerated (e.g., when testing above the Tg of the relevant packaging material), whereas in other cases, the behavior of the material at higher temperatures does not reflect its behavior under normal use conditions. This means that it is necessary to understand the temperature dependence of degradation mechanisms in more detail, to enable designing more robust materials for harsh conditions. Some examples will be shown in due course.

a. Epoxy Molding Compound (EMC) Degradation

This section will cover bulk degradation of EMCs under the influence of especially temperature and, to some extent, moisture. One of the first set of papers concerning the influence of the environmental humidity and temperature on bulk degradation of packaging materials, as also interconnect reliability, was from Teverovsky [52]. An extensive study was performed to evaluate weight loss (TGA) and changes in Tg and CTE (TMA) as functions of long-term aging at extreme temperatures until 225 °C. Weight loss data was shown to follow the Arrhenius law, with an activation energy of 1.45 eV. Tgs in general increased after long-term storage. CTE1 (in the glassy region) did not show a significant difference, whereas the CTE2 showed a decreasing trend after aging. Aging in air vs vacuum did not show any significant trend in the TMA measurement data, but differences in weight loss were evident, which indicated a clear impact of oxygen in the behavior under long-term storage. Notably no characterization was done using DMA at that time.

In addition to this, the impact of storage under various humidity conditions, including HAST at 130 °C and HAST at 150 °C, was performed showing non-Fickian diffusion behavior as was also reported by other authors before that [53], as also shown in Fig. 12.9. One of the more recent studies provide evidence that extended exposure to liquid water at high temperatures leads to chemical degradation of the epoxy material [30]. Hence after the initial weight increase due to moisture uptake, the weight decreases to below the initial weight; see Fig. 12.14. Additional characterization of the material under extreme conditions [54] also showed significant differences in behavior of the materials at higher temperatures in the presence of moisture, including evidence for chemical degradation. This implies that performing accelerated testing under these conditions can lead to nonrepresentative behavior and potential over-acceleration of failure mechanisms. More details on this will be provided in the section on interconnect reliability.

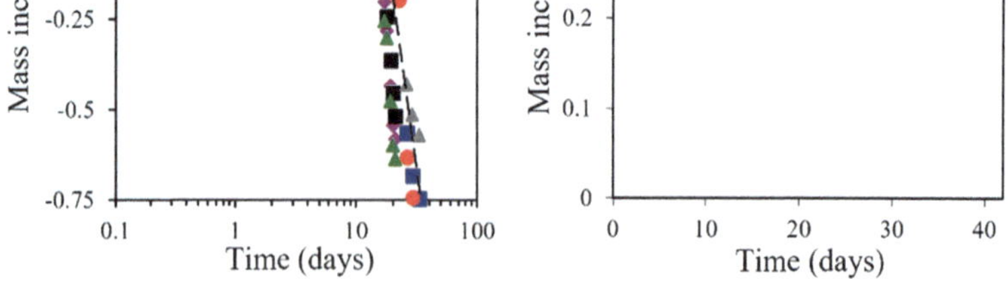

Fig. 12.14 Mass increase of biphenyl epoxy molding compound above glass transition at 130 °C: (**a**) for samples immerged in a water bath, and (**b**) for samples exposed to 85%RH. The dashed lines are guides to the eye [30, with permission]

In more recent times, Jansen et al. [55] studied the impact of aging in air on thermomechanical characteristics and microstructural changes in EMCs. Since DMA was used to characterize property changes, significant trends were observed. DMA is known to be far more sensitive toward capturing thermal transitions in materials. The glass transition region showed the most notable changes: a small shift of the initial Tg to higher temperatures and the formation of a second Tg and rubber plateau due to the formation of an oxidized layer on the outside of the EMC; see Fig. 12.15. In addition to this, significant shrinkage was observed after aging attributed to the fact that the oxidation leads to additional cross-linking. It is possible that a part of the shrinkage is also due to weight loss. The formation of the outer oxidized layer was found to follow a power law dependence. Subsequent work [56] performed in the same research group elaborated more on the potential degradation mechanism due to oxidation and proposed the use of a core-shell model to explain the thermomechanical behavior. One of the outcomes of this work is that the presence of oxygen (air) is necessary to induce such changes.

In any case, the use of FT-IR was subsequently used [57] to clearly demonstrate the impact of oxidation on the outer layer, whereas the inner layer was generally intact; see Fig. 12.16, supporting the fact that oxygen is an essential part of this aging behavior. Other workers around the same time performed extensive comparisons between various epoxy chemistries in molding compounds (and even underfill materials) [58] in relation to changes in thermomechanical properties and sensitivity to the formation of cracks in packages used for sensor applications [57, 59]. Like the work done by Vreugd et al. [56], a core-shell model was used to describe changes in several properties after thermal aging. Properties like moisture absorption, porosity, and changes in storage modulus and Tg, CTEs, and flexural strength as function of aging were determined and correlated to layer growth. The trends were also compared with package-level tests where onset of external cracks was the main response [57].

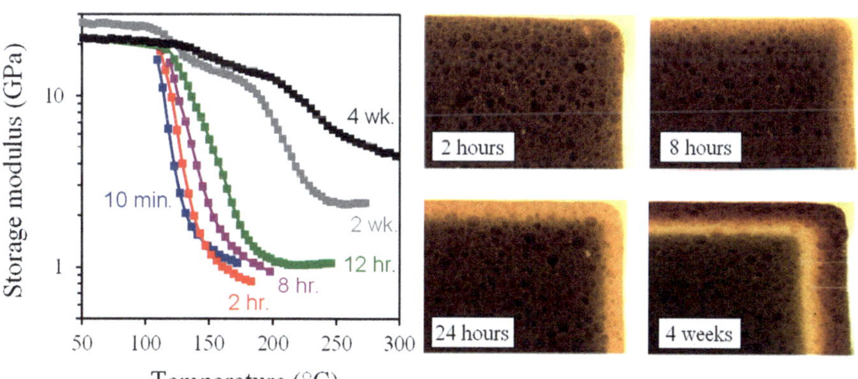

Fig. 12.15 (Left) DMA curves of a typical EMC after thermal aging at 175 °C. (Right), Fluorescence microscopy showing the oxidized layer after 24 h and 1 week at 175 °C [55, with permission]

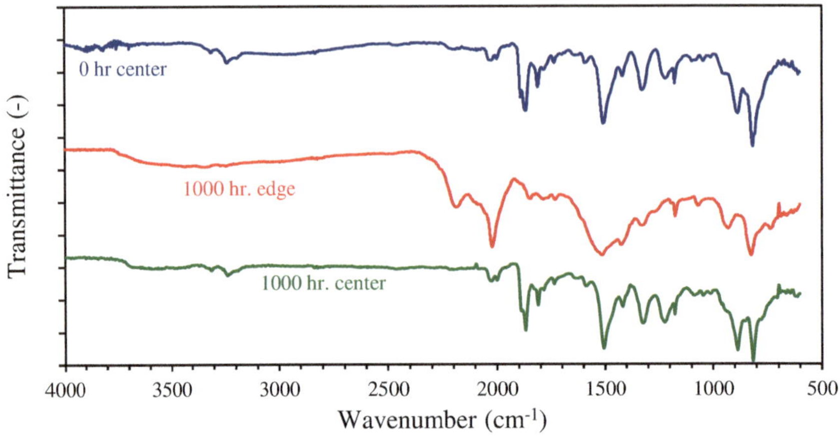

Fig. 12.16 An FT-IR spectrum of the oxidized outer edge (green), center (black), and unaged sample (red): after 1000 hrs aging in air at 200 °C (Class C). Main peaks seen changing as a result of oxidation, cross-linking, and decomposition are 1715 cm^{-1}, C=O ketone (increase); 1240 cm^{-1} epoxy/ether stretch (decrease); and 1600 cm^{-1}, C=C-C aromatic ring stretch (increase) [57, with permission]

The rate of growth of the oxidation layer at given temperature does not always correlate to how stable a material is at that temperature, although, as a rule, the faster the layer grows, the less stable the material is expected to be. For a given layer thickness at a given temperature and time duration, the mechanical properties of the specific material can be different depending on the influence of the oxidation reaction on the mechanical properties of the aged layer due to the competing responses of cross-linking and decomposition [57]. When comparing the various epoxy chemistries, it was also shown that a low Tg material is not necessarily worse in terms of thermal performance (and stability). Multi-aromatic systems (MAR) were found to be the most stable materials. Some high-Tg materials (e.g., ortho-cresol-novolak-based, or multifunctional epoxies) were found to be inferior to the MAR systems. To overcome disadvantages of low Tg materials in terms of electrical properties and high-Tg materials in terms of moisture absorption, suppliers have combined the two chemistries to obtain a good balance in properties.

Other workers [60] have investigated the impact of thermal aging effects on thin package warpage after reliability testing, related to shrinkage of the oxidized layer in an asymmetrical package construction. There have also been other groups working in this area [61, 62] where the focus has shifted toward obtaining better material models to enable predictions of product reliability using FE simulations.

Due to the strong focus toward flipchip applications for more extreme applications in recent times, similar work has been carried out on underfill materials, which are thermally less stable compared to EMCs [63].

Now for a small, but relevant, sidestep. Although the electrothermally induced parasitic gate leakage (GL) test as prescribed in the AEC-Q100 has been decommissioned since the last revision [64], failures were encountered in a few cases in the industry related to device sensitivity and EMC properties [65]. The failures were induced by charge that was trapped on the passivation layer after the GL test. One of the key properties of the EMC that determine sensitivity is the volume resistivity (VR), since the EMC conducts charges from the top of the package toward the passivation layer. The higher the VR, the more stable the material will be toward avoiding GL failures [66]. One of the methods to improve GL performance of a specific EMC was an extended post-mold cure (PMC) duration (temperature and time), which was also shown to increase the VR [66]. Revisiting the data considering the behavior of EMCs during thermal aging indicates that the increase in resistance is due to the chemical changes only in the outer oxidized layer. This means that the effective resistance of the EMC is made up of an in-series resistance of the oxidized layer and the intact inner layer, as follows:

$$VR_{shell} = \left(VR_{eff}\ x\, d - VR_{ref}\ x\, d_{core}\right) / \, TOL \tag{9}$$

where VR, volume resistivity; VR_{eff}, effective VR of entire sample; VR_{ref}, VR of reference material; d_{core}, thickness of unoxidized core; d, total sample thickness; and *TOL*, thickness of oxidized layer. Figure 12.17 shows the results: aging for only 200 h at 200 °C, results in significant changes in the volume resistivity measured at 150 °C in the shell layer (~100–200 um).

A full understanding of thermal aging effects at high temperatures is essential to enable better assessments of reliability capability. In case mechanical issues are observed during high-temperature testing (e.g., cracks), such an assessment can be used to evaluate the probability of such failures in the actual application. Eventually it comes down to assessing whether a given material can withstand the extreme

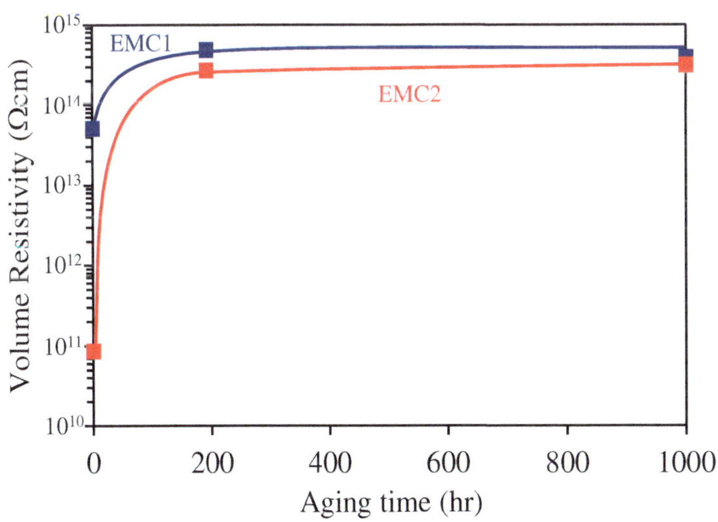

Fig. 12.17 Volume resistivity of the oxidized outer layer (shell) of two different EMCs after thermal aging at 200 °C

mission profile. To do this, one needs to understand all potential failure mechanisms that can take place at high temperature and the corresponding acceleration factors. In many cases it is assumed that the acceleration factors hold within the entire temperature range (test temperature and application temperature range). For high temperature-induced aging, it is often assumed to follow Arrhenius law, as mentioned previously [50]. For this purpose, one needs to define either a relevant physical parameter whose rate of change needs to be determined at various temperatures or define a failure criterion (e.g., appearance of package cracks) and assess the time to failure at different temperatures.

To determine the effective testing temperature and duration, an Arrhenius-weighted average temperature needs to be determined from the mission profile, if different temperatures are involved. Using the effective ambient temperature ($T_{amb, eff}$), the needed test duration is defined based on the same Arrhenius model. Based on the properties of the material, it is possible to predict whether the product will survive the relevant mission profile. It is, however, crucial to ensure that the behavior of the material can be described by an Arrhenius relation with a given activation energy within the relevant temperature range, so that the behavior of the material does not change significantly compared to what is projected beyond a certain temperature [67].

It is not easy to predict the behavior at lower temperature based on data collected under accelerated (higher temperature) conditions. In some cases, it is possible that the behavior is overestimated since material behavior can change at higher temperatures (e.g., > 200° C). A direct response that can be used to check this is crack occurrence in the material as a result of thermal aging. In some cases, it is worthwhile to confirm that the behavior at lower temperature (e.g., 125° C) is as what would be expected based on extrapolation to lower temperatures, even if it implies long-term testing. In the example shown in Fig. 12.18, EMC cracking was compared under various conditions ranging from 125 °C to 250 °C in a mainstream commercially available biphenyl-based EMC. Data from several different experiments was compared in terms of onset time to cracking, which was determined as the first readpoint showing cracks. This was done for an EMC known to be sensitive, so that response could be collected within a reasonable time. The data at 125 °C showed that almost 2 years testing duration was needed before a response was obtained. The dataset can be fitted to an Arrhenius relation with $E_a = 0.68$ eV. This implies that for this material, based on the mission profile, it would be quite reliable to predict the onset of cracking on the actual application within this temperature range. Note however that for this EMC, 125 °C is well into its Tg (transition), and it is quite probable that the behavior below Tg is over-predicted.

b. Die Attach Degradation

This section will cover bulk degradation of die attach materials under the influence of especially temperature. Much work has been published on die attach performance degradation due to thermomechanical loading after temperature cycling (TC) related to mechanical fatigue (cracks, delamination) [68, 69]. In contrary, the amount of work on thermal stability of die attach materials after high-temperature

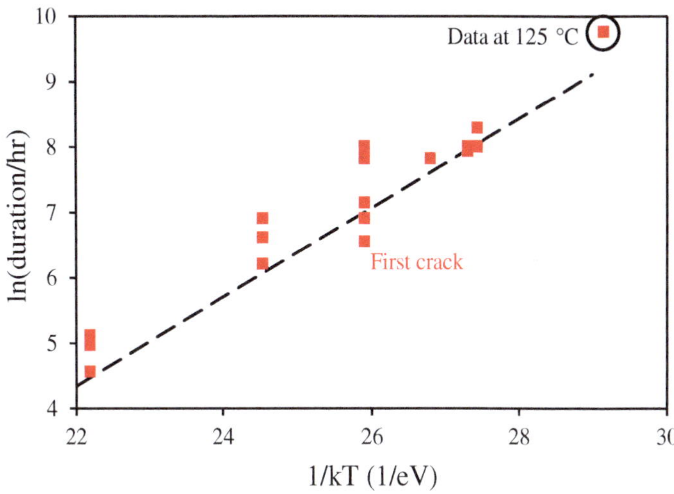

Fig. 12.18 EMC crack occurrence (first crack) data collected for several different experiments with the same compound across a very large temperature range, the encircled readpoint refers to the data collected at 125 °C, where cracks were initiated after nearly 2 years

exposure is limited. Although die attach materials are also polymer-based systems, they are less sensitive to degradation in over-molded systems due to a reduced access to air or oxygen from the outside of the package, since the EMC will act like a barrier. As mentioned in the previous section, the presence of oxygen or air is a prerequisite to initiate more severe thermal aging. There are a couple of cases where die attach thermal degradation can be relevant.

In the first case, in cavity packages, e.g., some microelectromechanical system (MEMS) packages, air can access the die attach as well, which can lead to changes in mechanical properties due to oxidation, hence an increase in elastic modulus due to oxidation-induced cross-linking or post-cure [70, 71]. In MEMS applications, where the low modulus of the die attach is important to induce low stress and prevent any drifts of the electromechanical response of the device, stability of the die attach is extremely important. Exposure of some die attach materials to air can lead to poor functionality of the MEMS, e.g., after reliability or in the application [72].

This has led to the development of materials that are more thermally stable or that retain their low modulus even after extended exposure to high temperatures (e.g., silicones) [73].

In some specific cases where the die or passive component attach is not exposed to air, in molded packages exposed to high-temperature storage under extreme conditions, changes in microstructure can be observed, as shown in Fig. 12.19. The silver flakes seem to undergo a kind of agglomeration, like sintering. This was found to be due to thermal degradation of the epoxy matrix in this specific case (even in the absence of air or oxygen) leading to a compressive stress on the Ag particles and hence to the agglomeration. This can be observed using TGA as shown in Fig. 12.20, where thermal degradation under N_2 is initiated below 200 °C for the relevant die

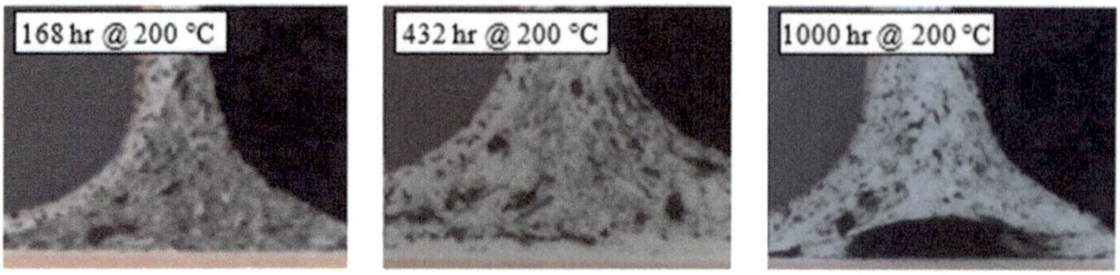

Fig. 12.19 Partial "sintering" of Ag flakes after thermal aging within a component attach despite surrounded by an over-molded package under an extreme condition of 200 °C, leaving behind large voids

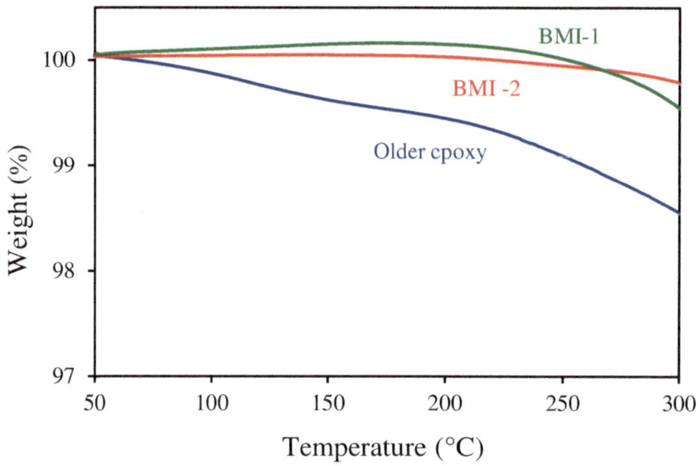

Fig. 12.20 Thermal stability (TGA data) of an older epoxy system used in the component die attach system shown in Fig. 12.18 compared to more stable bismaleimide (BMI) systems. The data was collected under N_2

attach material. It is remarkable that the degraded organic material in gaseous form can diffuse through the bulk. Hence, when selecting a die attach for a specific high-temperature application, the thermal stability of the material under N_2 should be checked, e.g., using TGA. For more extreme temperatures, other systems (e.g., Ag-sintered systems) may be more thermally stable, though other mechanisms like stress-induced cracking after TC need to be considered [74, 75].

7.2 Interfacial Degradation of Interconnects

This section will cover the interfacial degradation of interconnect materials (wirebonds and flipchip contacts) under the influence temperature, moisture, and bias. Interconnect technology (wirebonds, flipchip contacts) is what enables connecting the product/die to the carrier (leadframe or substrate), so are like lifelines for the product. The other equally critical electrical solder connections are between the carrier and the PCB. Reliability of both type of connections are equally important to enable product reliability and functionality in the final application. Package interconnect technology is usually referred to as "first level," whereas solder interconnections are referred to as "second level." A lot has been published on second-level reliability and will not be covered here in much detail. Second-level failure mechanisms to remember are fatigue cracking in the solder connection after board-level temperature cycling [76] and in some cases corrosion induced by contamination like sulfur [77].

Package interconnect technologies suffer from similar failure mechanisms: fatigue in the interconnect (either wirebonds or solder bumps/copper pillars) and corrosion. However, other mechanisms like Kirkendall voiding in bimetal systems (generally temperature-induced, or in some cases enhanced by electrical current in a specific direction) or electromigration can also lead to loss of integrity of the interconnect due to severe voiding as a result of exposure to more extreme temperature conditions. The focus of this section will be predominantly the impact of aging under steady-state temperature conditions, extended in some instances with humidity and bias or current.

Fatigue of wirebonds can take place either at the connection with the bondpad, so-called first bond, or at the other end on the carrier (substrate/leadframe), also known as second bond. The risk of fatigue fails depends on stresses induced by CTE mismatch between the different package material constituent, e.g., molding compound and die or leadframe, which is usually enhanced in the presence of local delamination [78].

a. Wirebond Degradation

i. Au Wire

Typical wire materials used for first level interconnects are aluminum (Al), gold (Au), and copper (Cu). The most commonly used wirebond material traditionally is Au wire (different purity levels indicated by 2 N, 3 N, 4 N, and even 5 N). 2 N means 99% pure, while 5 N implies 99.999% purity. In addition to this, the regularly used bondpad metallization is Al, containing low levels of other elements like Cu (to suppress electromigration). Whereas Al wire does not lead to intermetallic compound (IMC) formation on the die, the use of Au wire results in various Au-Al intermetallic compounds, as shown in Fig. 12.20. The process of wirebonding involves welding of the two (usually) dissimilar metals using heat, pressure, and ultrasonic energy. One of the keys to obtaining a good weld is the removal of the native oxide layer on the Al pad to enable an intimate contact between the pure metals, allowing for their interdiffusion to form initial IMC grains that will grow to larger IMC areas during subsequent wirebond process and other high-temperature assembly steps. A good book by Harman on the subject provides the reader with more background [2].

A lot of work has been published on the reliability of Au-Al. One of the mechanisms affecting the ballbond as mentioned previously is Kirkendall voiding [79]. Under high-temperature loading, extensive IMC growth can lead to the presence of voids due to the faster diffusion rate of Au in the Al compared to that of Al in the Au. Severe Kirkendall voiding can lead to loss of mechanical integrity of the IMC region, thereby sacrificing its reliability. Extensive IMC formation can also lead to mechanical issues since consumed Al is replaced by IMCs that occupy larger volumes compared to the elemental metals. This can also lead to damage under the pad, as was first reported by Clatterbaugh [82]. The role of current (density) and direction on Au-Al aging was investigated in detail by Krabbenborg [83] by using an in situ Static High Temperature Life setup. It was found that higher current densities accelerate this voiding due an electromigration-like effect where mass transport is accelerated by current flow. When current flows from the die into the wire, the voiding is accelerated. Opposite to the direction of the current flow, the flow of electrons can increase the diffusion rate of the Au into the Al [83].

As shown in Fig. 12.21, various IMC phases are formed during thermal aging of Au-Al interconnects: AuAl, Au_2Al, Au_5Al_2, and Au_4Al. The Au_2Al phase is called purple plague because of its typical purple appearance. The term is often mistakenly

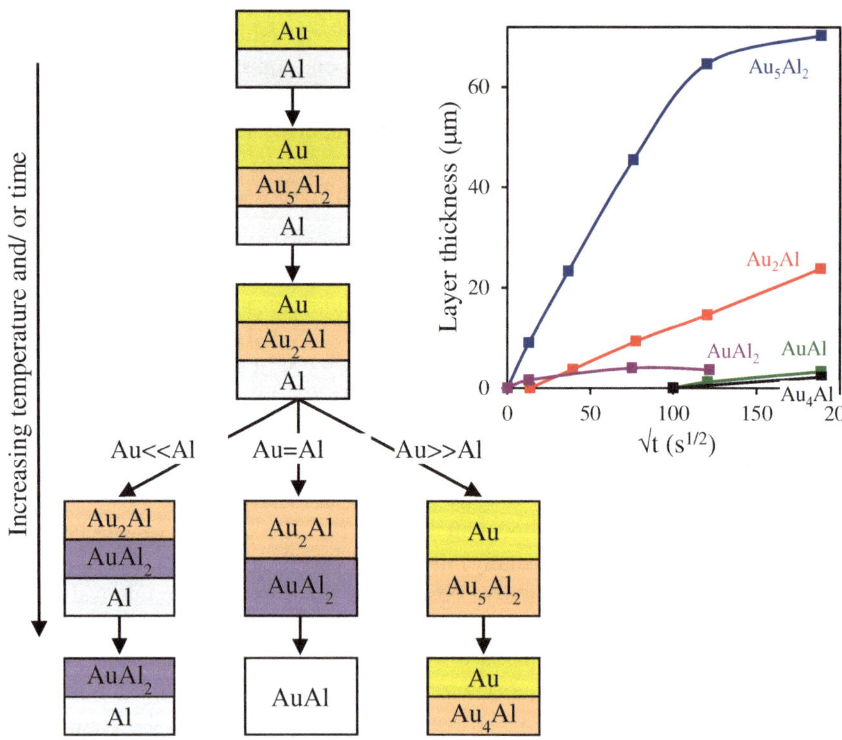

Fig. 12.21 Intermetallic compound (IMC) phases formed in Au-Al systems [80]. The graph [81] was collected by aging an Au-Al couple at 400 °C

used to describe a failure mechanism where impurities like halides can attack the Au_4Al phase leading to corrosion causing void formation [84, 85]. Horsting voids are especially observed when conventional EMCs containing brominated flame retardants are used. Bromide ions can react with the Au_4Al phase leading to AlBr3 formation at high temperature [84, 85] which is easily oxidized. The two mechanisms described above are shown in Fig. 12.22a and c.

Some workers have published studies on the impact of Al bondpad thickness [79] on thermal aging behavior, while others have studied the impact of dopants like Pd on high-temperature reliability of Au-Al interconnects, e.g., due to slowing down of Au-Al IMC growth rates [86]. One paper also describes a unique mechanism where metals can be transported along the wire from the leadframe to the ballbond under a galvanic driving force after extreme aging [87].

In addition to the impact of temperature (and current density), the impact of (biased humidity) is also well described in the literature. Whereas the main corrosion mechanism described in the past was Al bondpad corrosion under influence of halides [88], more recent literature described the corrosion of the intermetallics under the impact of halides [89]. In this case too, it is the Au_4Al phase that gets attacked by the halides under acidic conditions [89]. Depending on the level of (halide) concentration, either intermetallic corrosion (lower contamination level) or bondpad corrosion (higher contamination level) is initiated. This is also shown in Fig. 12.22b and c.

To impart robustness of Al-Au systems against both Kirkendall voiding and IMC or bondpad corrosion, the use of over pad metallization that is plated with Au is sometimes proposed [90]. This plated layer prevents the formation of sensitive intermetallics, as also protects the underlying Al pad from corrosion. However, the use of such technologies is not always free of issues like yield loss due to plating issues [91] and usually adds significant cost to wafer processing.

ii. Cu Wire

The most widely investigated subjects in wirebonding technology over the last 20 years are related to Cu wire. The drive toward Cu wire started with power packages (almost 25 years ago) where the cost benefit for converting from Au to Cu was the highest, due to the use of larger diameters (e.g., up to 50 and even 75 um wires). Despite the overall better reliability performance of thicker wires compared to thinner wires, whether in relation to second bond fatigue, or first bond corrosion and

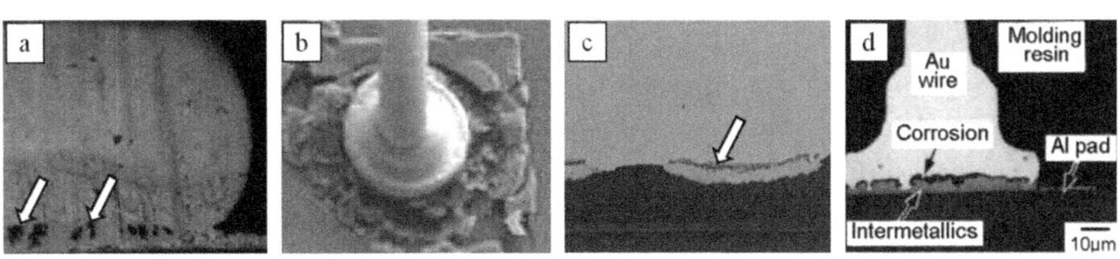

Fig. 12.22 Typical failure mechanisms observed in Au-Al systems (**a**) Kirkendall voiding, (**b**) Al bondpad corrosion, (**c**) intermetallic corrosion (right image from [84])

fatigue), the introduction of copper wire faced several challenges. Cu wire has some clear advantages compared to Au wire, like its lower electrical resistivity, better mechanical strength, higher stiffness, and slower IMC growth [89]. But there are also some drawbacks and therefore challenges which include the development of more robust wirebonding recipes that provide reproducibly good IMC contacts while limiting mechanical damage under the bondpads (e.g., cratering) [92]. Additional reliability challenges must be dealt with, like fatigue (after TC) and corrosion sensitivity of the Cu-Al system. Unlike failure mechanisms like fatigue or corrosion, which are generally wearout mechanisms, poor IMC contacts and damage under the pad can lead to early fails [92].

Not to be underestimated is extra demands on FA capability, e.g., the ability to decapsulate the package without affecting the wires, to enable mechanical integrity tests like wire pull and ball shear. In case wires are affected by decapping chemicals, this may have an impact the strength values measured [93]. The gradual implementation of this new technology to fine pitch and smaller wire diameter applications led to even more challenges. The development of wirebond recipes for recent, advanced wafer fab technologies became even more demanding (e.g., due to smaller metallization/bondpad openings) and due to reduced mechanical robustness of the bondpad layer stack-up [94]. Therefore, it constrains optimum transfer of ultrasonic energy toward IMC formation without damaging the structure below the pad or causing excessive deformation of the aluminum metallization layer ("Al splash"). New wirebonding concepts have been developed [95] that provide an optimum balance of good IMC coverage and no mechanical damage below the pad.

Several papers describe fatigue mechanisms involving Cu wire (e.g., after TC), especially involving the second bond or stitch/wedge [78]. Cu wire, due to its greater hardness compared to Au wire, has a lower fatigue resistance. Stresses induced in the package due to CTE mismatch between the EMC and leadframe or substrate can lead to cracks near the neckdown of the stitch or wedge; see Fig. 12.23a. One of the prerequisites for this failure mechanism is delamination between the EMC and leadframe [78]; see Fig. 12.23b. This is especially true for mainstream EMCs with higher silica contents, due to a larger mismatch in CTE with the generally Cu-based leadframes (CTE ~17 ppm/°C); see Fig. 12.23c. Delamination sensitivity is increased on Ag-plated areas or NiPdAu surfaces due to the inferior adhesion of EMCs on such surfaces. To overcome this, EMCs need to be modified (e.g., by increased adhesion promotor content, lower modulus), or leadframe surfaces need to be treated (e.g., via the use of roughening technologies).

Fatigue-induced fails of the first bond or ballbond (causing either ball lift or ball neck break) have also been described in literature [96, 97]. Ball lift is due to delamination on the die surface around the ballbond area. Ball neck break is a specific mechanism that occurs only when silicone-based coatings are used on die surfaces (to act as stress buffers): it requires complete coverage of the coating on the ballbond and part of the wire above to reduce stress as the gel generally delaminates from the EMC [97].

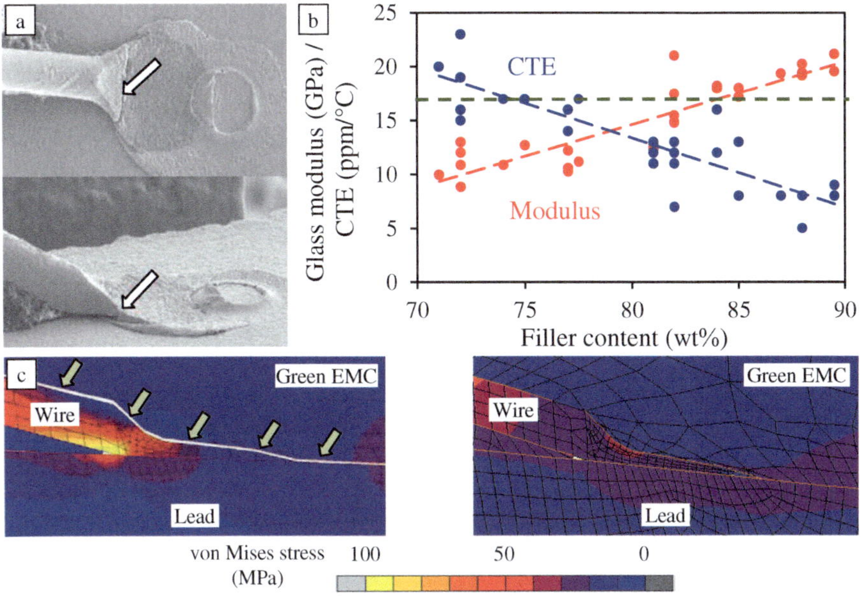

Fig. 12.23 (**a**) Example of a stitch with a fatigue crack (indicated by the arrow) after decapping a test vehicle that has been subjected to a thermal cycling test. (**b**) The glassy modulus and coefficient of thermal expansion as a function of filler content for several commercially available mold compounds; toward the left, conventional materials; toward the right, newer generation materials. The black dashed line refers to the CTE of copper (~ 17 ppm/°C). (**c**) Stress after cooling to −65 °C for the SO package showing the effect of delamination, indicated by the green arrows (left), and no delamination (right) [78]

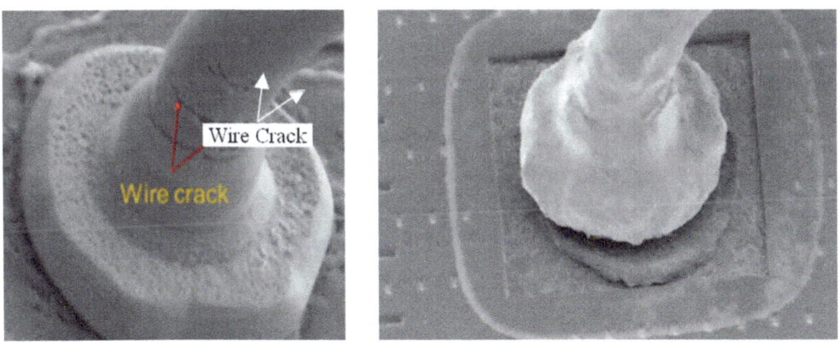

Fig. 12.24 Left, ball neck fatigue cracks after TC induced by delamination [96]; right, ball lift after TC induced by insufficient coverage by soft silicone-based gel [97]

In addition, since these silicones have a much higher CTE (~300 ppm/K) compared to EMCs (8–12 ppm/K), the ballbonds are subjected to large stresses during cycling; Fig. 12.24 shows two examples.

The most widely investigated aspect of Cu wire has been the corrosion sensitivity of its intermetallics formed with Al of the bondpad. Comparable to Au wire, the reliability of the Cu wire interconnect is determined by the properties of the intermetallics. A detailed investigation of the Cu-Al intermetallic formation kinetics was studied by Gubbels et al. [98], and various properties of the different IMC layers have been determined [98, 99]. Although a few papers describing the corrosion sensitivity of the Cu-Al system attributed the most sensitive phase to be the Cu-rich Cu_9Al_4 phase (like the Au_4Al phase in the Au-Al system) [89], new insight into the corrosion mechanism was proposed by Gubbels et al. [100]. They demonstrated an apparent interface-like corrosion which turned out to be oxidation of a very thin layer of the Al solid solution in the Cu ball, present adjacent to the IMC region. These studies are performed on a Cu-Al couple; see Fig. 12.25. Gubbels et al. [100] also showed that the presence of halides like Cl^- can accelerate the corrosion of the solid solution and further acceleration can be achieved by adding moisture.

This insight is based on the oxidation mechanism of Al alloys during high-temperature exposure as described by Plascencia [101], which was extended by Rongen et al. [102] by characterizing the intermetallic region in a wirebonded

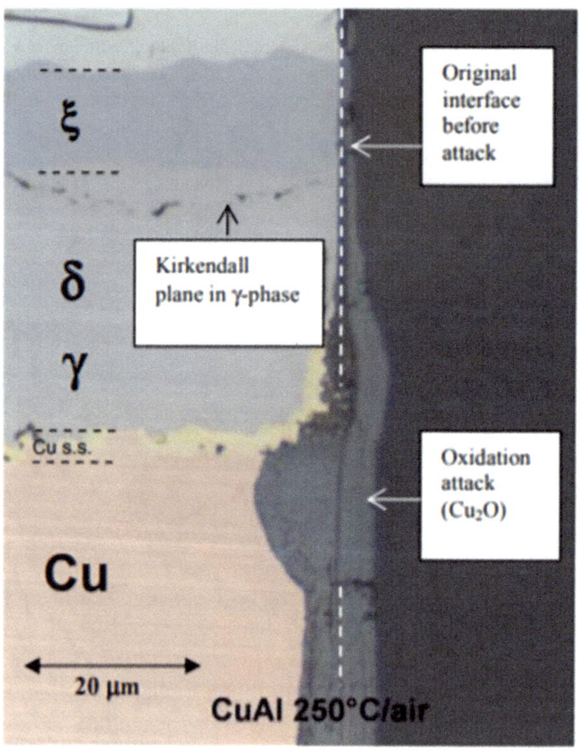

Fig. 12.25 Oxidation attack from the right due to heating at 250 °C in air for 150 hr. The yellow region is the Al-rich Cu solid solution [100]

sample subjected to extended aging (300 °C, 100 h) in clean air; see Fig. 12.26. The latter paper also describes the various stages of corrosion, depending on the level of catalytic ions in the system, as shown in Fig. 12.27. Furthermore, a clear dependence of degradation rate on EMC type was shown when the contacts were subjected to aging at higher temperatures. The initial degradation is characterized by low levels of corrosion catalysts in the IMC contact area. The corrosion, in cross section visible as separation of the bondball, is progressing from the outside toward the inside with the square root of time as shown in Fig. 12.28, indicating a diffusion-controlled mechanism.

An empirical model was proposed to describe this behavior as function of temperature and humidity as given in Eq. 10:

$$x = \sqrt{A \cdot \exp(a \cdot RH) \cdot \exp\left(-Ea/k_B T\right) \cdot t} \tag{10}$$

where A = pre-exponential constant (μm^2/hr); a, humidity coefficient (/%RH); Ea, activation energy (eV); k_B, Boltzmann constant (8.617×10^{-5} eV/K); and T, temperature (K). The activation energy for this growth is ~0.95 eV, and a key factor determining the growth rate at any given temperature is the pre-exponential factor D_0, which is a function of the chemistry of the EMC, e.g., depending on the level of halide, sulfur, and its mobility [103, 104]. The enablement of detailed studies of interfacial corrosion and the development of acceleration models were due to the development of a new method to determine the IMC area on reference and aged samples [105].

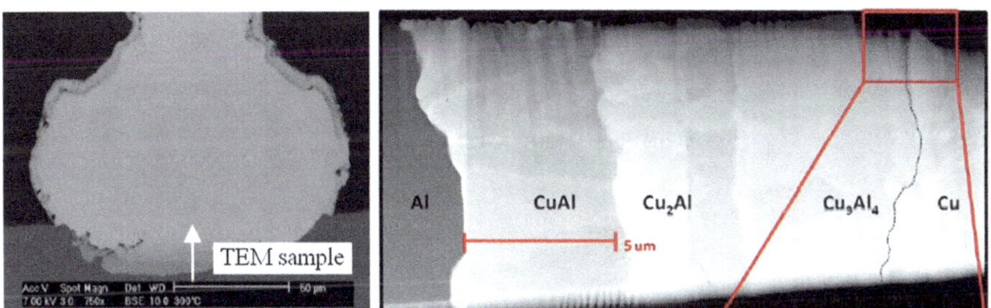

	At% @ location No.											
	CuAl	Cu_2Al	Cu_9Al_4	O_1	O_2	O_3	O_4	O_5	O_6	O_7	O_8	O_9
Cu	46	62	73	78	78	94	94	98	99	98	99	100
Al	54	38	27	22	22	6.1	6.2	2.3	0.9	2.0	1.4	0

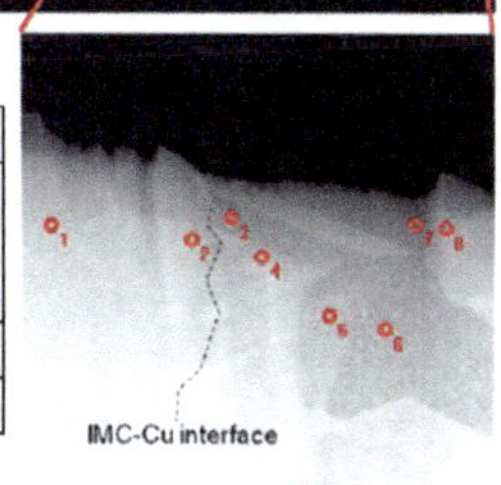

Fig. 12.26 Al solid solution adjacent to the Cu-IMC interface demonstrated by aging a wire-bonded sample for 300 °C for 100 h. Compositions of the various locations are shown in table below [102]

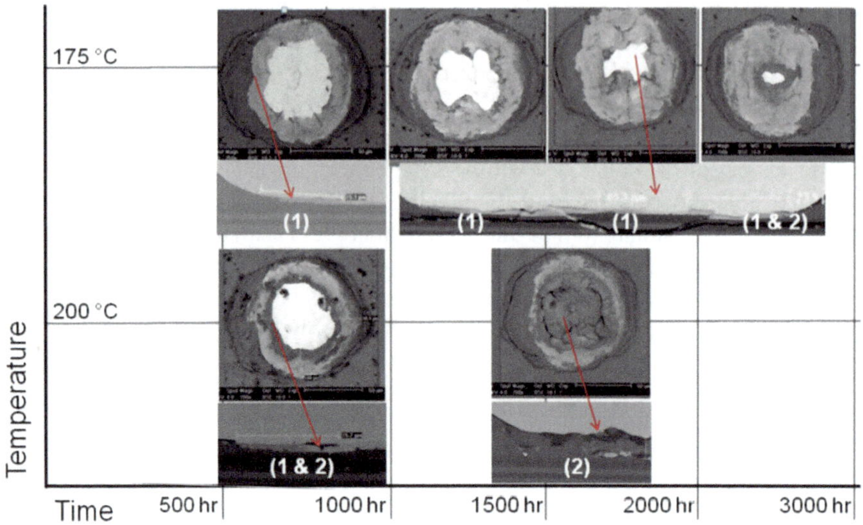

Fig. 12.27 Corrosion progression for aged Cu-Al ballbonds in time for different temperatures using planar analysis and cross sections; mechanism in brackets: (1) Cu-Al IMC interface corrosion, (2) Cu-Al IMC bulk corrosion [102]

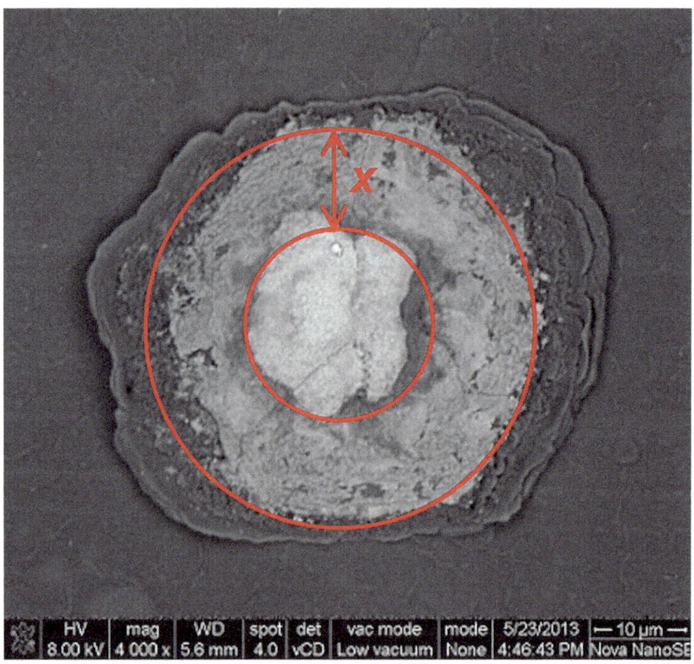

Fig. 12.28 Cu-Al interface corrosion progression according to the model given in Eq [8]

The subsequent stage when more and more catalysts arrive at the corrosion front is the onset of bulk IMC corrosion, as shown in Fig. 12.27. This is incidentally the first stage involving IMC corrosion in the Au-Al systems where interfacial corrosion does not occur. Finally, at the highest level of catalysts, bondpad corrosion takes place, which is also something that can be observed with Au wires. Note that while bulk IMC corrosion in long duration aging experiments indicates that the corrosion has progressed to a significant extent (e.g., due to higher levels of contamination being released by the system accumulated at the corrosion front), sudden or fast occurrence of bondpad corrosion usually implies there is a "too high" level of catalysts, generally due to extrinsic causes rather than background levels of, e.g., halides in the packaging materials [106].

Several papers have also described the key factors that can have an impact on especially biased HAST performance, to mention halide content (especially Cl^-) and pH [107]. The critical level of Cl^- to trigger corrosion was found to be dependent on the pH of the system, which is to be expected in view of early work on the corrosivity of Al metallization (bondpad corrosion) [108]. The reason for this is that at pH values between 4 and 9, Al is less likely to ionize due to formation of a passive oxide layer; see Fig. 12.29. Most epoxies undergo hydrolysis under high temperature and moisture loading, leading to the formation of organic anions, increasing the acidity of the system (hence low pH) [109].

Many studies are dedicated to improving biased HAST performance, including the role of higher biased voltages, e.g., > 5 V [110, 117]. The key factor determining corrosivity is the level of Cl^- ions and its mobility. Mainstream EMCs do not contain halides like Br^- nor F^-, but the presence of Cl^- is inevitable from the epoxy resin. In addition to reducing the Cl^- content of the EMCs by using high purity resins, the use of ion trappers to reduce mobility and enhance biased humidity performance has been reported [111]. It is remarkable that most literature still assumes that the corrosion involves the Cu_9Al_4 phase, which is a phase unlikely to have been

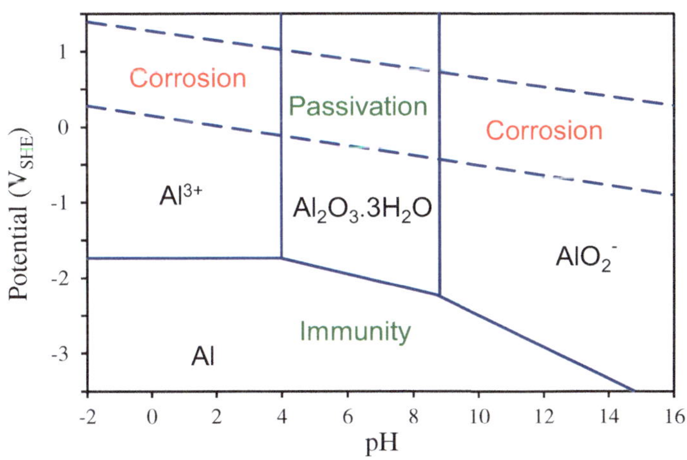

Fig. 12.29 Pourbaix diagram of Al [112, with permission]

formed after assembly or during HAST. In any case, the Cu-rich IMC phase or solid solution is the most sensitive for corrosion due to the highest anode to cathode ratio [113].

A generic equation for this mechanism can be described as below (cat: catalyst, e.g., Cl^-):

$$4Al + 3O_2 + cat + (+H_2O) -> 2Al_2O_3 + cat(+H_2O) \quad (11)$$

Some papers have also described the beneficial impact of Pd coating on biased HAST performance, due to lower corrosion sensitivity of the corresponding IMC phases. However, while the impact of Pd (coating) on based humidity performance is not always clear [111, 114], the impact of EMC properties is most critical in this regard [ref]. In these papers, however, it was shown that Pd is involved in IMC formation near the interface, which can change the electrochemistry at the interface [111, 114].

It is found that the HAST capability of EMCs, especially under higher bias voltage, can be pre-screened by the determination of their volume resistivity at the HAST condition (130 °C) subsequent to a moisture soak.

Based on previous work [115], it is proposed that the volume resistivity of an EMC is related to the mobility of ions (including Cl^-). A preliminary model is developed to compare the levels of Cl^- ions that can accumulate around the positively biased pad. In this work, the concept of an incubation time was introduced (as also observed during high-temperature tests): the time needed for the concentration of Cl^- ions is below the critical limit to initiate corrosion [116]; see Fig. 12.30.

Subsequent work shows that the incubation time is related to EMC properties like volume resistivity and not just the bias voltage level but the electric field induced by it between the biased pad and ground pad or metallization adjacent to it [117]. These studies also show how the HAST performance of EMCs can significantly be

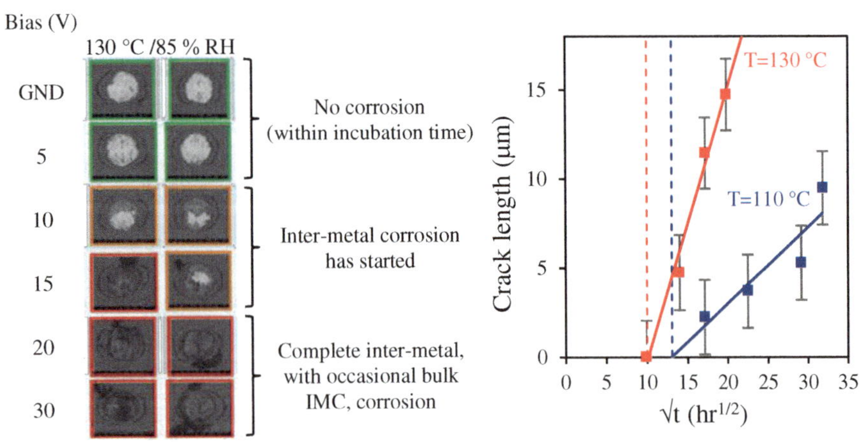

Fig. 12.30 Concept of incubation time during biased humidity tests [116, with permission]

improved by changing its formulation to achieve reduction of Cl^- level, increase of pH, and reduction of ion mobility. Other work used a multi-physics-based model corrosion of Cu wirebonded products under the impact of biased humidity [118].

However, one of the fundamental questions raised [117] is whether the classic acceleration model, the Peck model, still holds when defining appropriate test durations for biased humidity tests. It is shown that there are step changes in material properties across the Tg of the EMC like moisture absorption and volume resistivity. This would explain why the acceleration factor for high-Tg materials is somewhat closer to that defined by the Peck model, whereas there is an over-acceleration in low Tg materials. This over-acceleration of HAST vs THB is also shown in previous work on Cu-Al corrosion induced during HAST, where simulations are performed to compare moisture levels in the package as observed during accelerated tests with those in the actual application [119].

Other work on high-temperature aging using an in situ HTSL shows that it is possible to follow the progression of interfacial IMC corrosion by measuring the contact resistance of the ballbond contact as function of temperature for various EMCs [120]. Follow-up work using in situ SHTL shows that the failure mechanism does not depend on current (direction), in contrary to what is found in the case of Au wire. Enhanced corrosion in Cu-Al in the presence of current can be explained purely by Joule heating, i.e., temperature [121].

Another interesting subject of discussion is the role of S (sulfur). A few papers have described the role of S on initiating galvanic corrosion between Cu and Pd coating [122, 129]. This type of corrosion leads to voiding on the shoulders of the ballbond, especially after thermal aging under extreme conditions (> 175 °C) [122]. The same phenomenon is observed for stitch bond [123, 124], but in that study no specific reference has relation with sulfur in the EMC. Subsequent publications from wire suppliers include role of S though [125]. Figures 12.31 and 12.32 show some examples of this mechanism.

The impact of sulfur on the Cu-Al corrosion mechanism is less extensively covered in the literature. Some papers suggested there could be an impact of sulfur on Cu-Al interfacial corrosion [126] with Pd-coated wire. The paradoxical role of sulfur and interaction with the Cu-Al system, with both bare Cu and interactions with Pd coating, was described in a recent paper [127].

On the one hand, sulfur is an essential constituent of the most effective adhesion promotor systems to enhance adhesion on Ag plating. On the other hand, high levels of sulfur can catalyze Cu-Al interfacial corrosion. It was shown that sulfur is released at temperatures exceeding ~150 °C (which is above the Tg of the involved EMCs); see Fig. 12.33. Measurements are based on a special method used to determine the total content of sulfur present in the formulation after thermal aging [127].

The wire type also influences the corrosion sensitivity of the combination wire compound. For EMCs with low levels of total sulfur content (< 100 ppm), the difference between the wire types is not very significant. In case of intermediate levels of total sulfur content (between 100 and 200 ppm), Cu-Al IMC corrosion is retarded with Pd-coated wire compared to bare wire [127]. The reason for this is that sulfur-based species like H_2S can attack the Al solid solution in a similar way to what

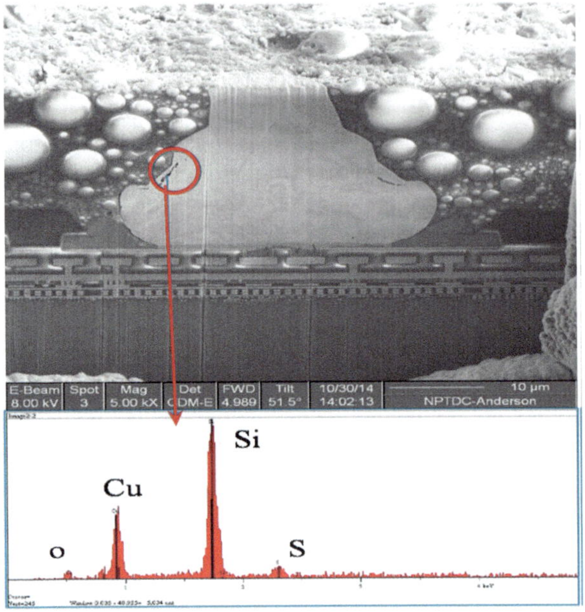

Fig. 12.31 Sulfur-induced galvanic corrosion of Pd-coated Cu wire at high temperature at the ballbond after175 °C, 1008 h [122]

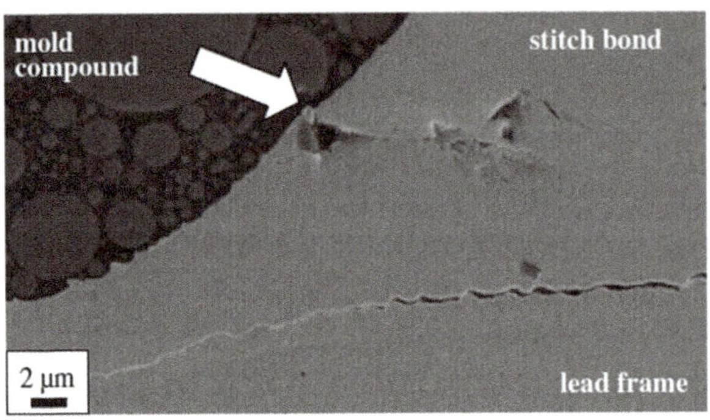

Fig. 12.32 Suspected sulfur-based galvanic corrosion of Pd-coated Cu wire at high temperature at the stitch after 175 °C, 2200 h [123, with permission]

occurs in the presence of halides. Evidence for the presence of sulfur was shown by Electron Spectroscopy for Chemical Analysis (ESCA). Pd is consuming sulfur in a kind of sacrificial reaction and by that suppressing the galvanic corrosion with the exposed Cu [127].

With higher levels of total sulfur content (>200 ppm), Pd-coated wire shows worse performance compared to bare Cu [128]. This is expected to be related to excessive voiding that allows for easier access of sulfur-based species to the Cu-Al

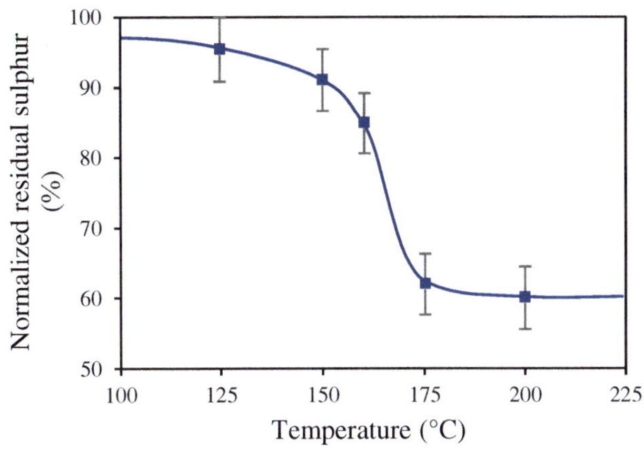

Fig. 12.33 Normalized concentration of residual sulfur as a function of temperature (measurement error is about 5%) [128].

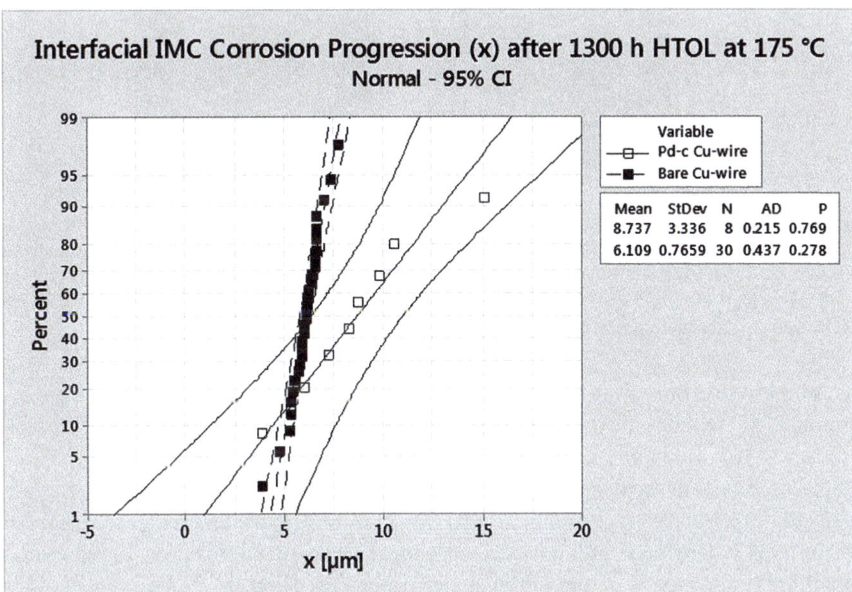

Fig. 12.34 IMC degradation, characterized by corrosion progression over distance x for Pd-coated Cu wire and bare Cu wire [128]

IMC region. In addition, the sulfur level is so high that it is not effectively "consumed" by the sacrificial reaction described above. In such cases, bare Cu is preferred for mission profiles that are extreme with respect to high-temperature exposure; see Fig. 12.34 [128]. Note, that the continuity of the Cu-Al IMC contact is crucial toward retarding the corrosion process.

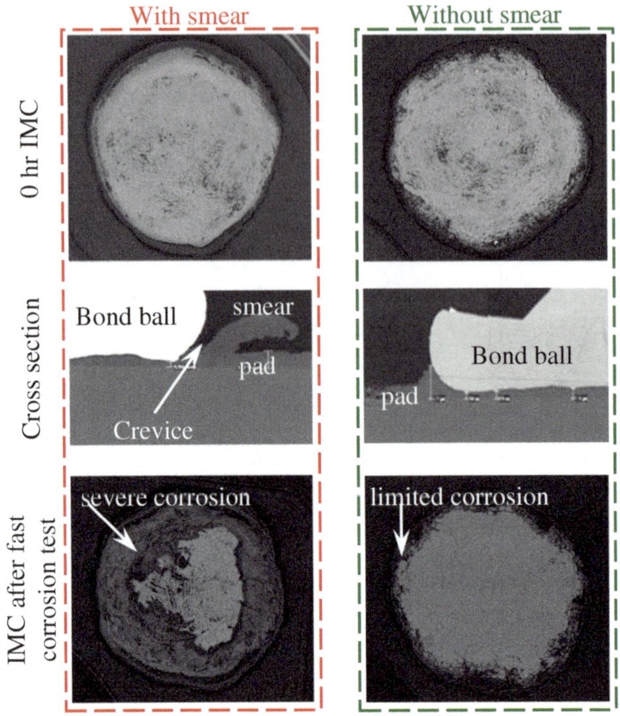

Fig. 12.35 Effect of smear and crevice on copper wire corrosion for products that were intentionally packaged with a non-suitable ("dirty") mold compound. The left column shows results for a bond process that gives significant smear and crevice, whereas the right column shows results for a process with no smear nor crevice. Both the 0 hr. and final IMC coverage after the fast corrosion test are given [131, with permission]

As mentioned previously, wire-bonding recipe concepts characterized by continuous IMC without the presence of a crevice will significantly retard the corrosion process [130], as shown in Fig. 12.35 below [131]. The electrochemical impact of the crevice was described by Soestbergen et al. [131]. The presence of a crevice leads to depletion of oxygen and a local decrease of pH, which creates a local corrosion cell where processes proceed at a faster rate (e.g., transport of Cl^- ions) which in return speeds up the initial stages of the corrosion.

Although the majority of published work on corrosion of the Cu-Al system refers to intrinsic reliability (wearout mechanisms), in more recent time, a number of papers have described extrinsic reliability cases (early life) or random events (use life). The impact of Cl-containing foreign material-like particles was described recently [132]. Penetration of Cl^- from the outside of the package transported by a combination of capillarity and electric fields along interfaces can also result in random fails [106]. A specific case of intrinsic behavior causing over-acceleration was due to hydrolytic decomposition of die attach films under HAST conditions [133]. In addition to corrosion of the ballbond (Cu-Al IMC) under moist conditions, extrinsic corrosion of the stitch was found in some cases, also related to delamination and penetration of corrosive contaminants from the outside of the package [134].

In addition to Au and Cu wires, other types of wires are also in use, e.g., Al wire [135] and Ag alloy wire [136]. Al wire is used especially in power and RF applications. The clear advantage is that no intermetallics are formed on the bondpad, making the system less sensitive to corrosion. For power packages, "heavy" Al wires and ribbons are used (meaning larger dimensions; ~0.2–0.5 mm range) to enhance current carrying capability. Extensive work has been published on the fatigue of Al wirebonds in power packages and methods to assess them [137]. Al wire also shows some interesting electromigration behavior, leading to a "bamboo-like" structure on the wire surface, as was reported by van der Wel et al. [138].

Most wires used in the microelectronic industry currently are Cu-based wires (>50%). The percentage of Au wire reduced over the years but is stabilizing and used for high reliability applications in automotive, military, and medical applications, where users are more conservative and averse to any risks and where costs matter to a lower degree. Although our understanding of failure mechanisms and improvement of robustness with Cu wire continues, the fact remains that Cu wire has its sensitivities (corrosion, bondpad damage, fatigue) that require special controls and monitoring during the manufacturing of parts with Cu wire. Automotive Electronics Council (AEC), a consortia of semiconductor suppliers and users, has defined special requirements for Cu wire, considering its sensitivities [139]. In more recent times, there is a push to using Ag alloy wires for specific applications (LED, memory) to reduce cost compared to Au wire. Ag alloy wire provides some advantages compared to Cu wire, like lower damage during wirebonding. However, corrosion sensitivity of Ag alloy wires remains to be a challenge [140].

b. Flipchip Degradation

With increasing interest in flipchip packaging (especially to enhance RF performance), the reliability performance of such interconnects has also received much attention recently. For fine pitch applications, solder bumps have been replaced by copper pillars with solder caps as mentioned earlier in this chapter.

The main topic of study has been fatigue resistance during TC and the role of materials like the capillary or molded underfill with respect to their thermomechanical properties. Several papers have described the detrimental effect of a low Tg material on bump fatigue [141]. When Tg is within the temperature swing of the TC experiment, the underfill offers a lower level of support to the bumps above Tg (lower modulus, high CTE), exposing them to higher stress, than below Tg (lower CTE). In addition, efforts have been made to predict the expected lifetime based on FE simulations. This requires reliable material models to describe the behavior of the bumps and intermetallics formed, especially creep compliance as function of temperature [142, 143], including the impact of thermal aging (during cycling) on changes in microstructure and impact on mechanical behavior of the bumps.

Next to fatigue performance of flipchip packages, challenges are faced during high-temperature exposure due to void formation. Voids can deteriorate the mechanical integrity of the joint and lead to electrical fails due to opens as well. This is generally triggered by IMC formation between Cu and Sn that causes volume shrinkage [144, 145]. However, interactions with plating chemicals [147] and thermomechanical stress [152] also need to be considered.

There are basically two types of voiding mechanisms. First, voiding near the metal – IMC interface can occur which is purely Kirkendall voiding (driven by IMC growth) [144, 145] and can become worse due to the presence of plating impurities acting as vacancy sinks [147]. Some papers have shown the presence of chemical species from the plating chemicals (e.g., sulfur) within the voids by using methods like TOF-SIMS [148]. This type of voiding can be suppressed by a proper choice of plating chemicals.

Another type of voiding is observed within the solder joint after exposure to higher temperatures (>150 °C generally); see Fig. 12.36. The main factor that drives this mechanism is also IMC growth, but the effect in this case is predominantly mechanical. IMC growth leads to volume shrinkage due to IMC formation (8.1% for Cu3Sn and 4.1% for Cu5Sn6 phase formation) [145], which leads to tensile stresses within the solder joint due to constraints imposed by the package. At higher temperature the lower yield stress or creep compliance of the Sn-based alloy [146] can lead to deformation within the solder and therewith to large voids. Since both mechanisms are a consequence of IMC formation, its activation energy is relevant to derive an acceleration factor. Values of about 0.8 to 1 eV have been given in the literature [149]. However, since the creep compliance is a nonlinear function of temperature, for the large voids described above it, it is very well possible that the failure mechanism at high temperature (e.g., >150 °C) is "over-accelerated."

Ways to overcome the large-scale voiding is by using barrier layers to slow down the IMC formation, e.g., a Ni barrier on the Cu [144, 150] or by controlling stress as was proposed in electromigration studies [151]. Suppressing this voiding mechanism in these ways will be key to enable flipchip packages to achieve more extreme mission profiles, e.g., AEC Grade 0 and beyond that. Another topic well covered in the literature is electromigration in bumps: enhanced void formation at higher current densities (which can accelerate IMC formation) [152]. More recent efforts have been to physically model the phenomenon based on self-diffusion, thermo-migration and stress migration [153].

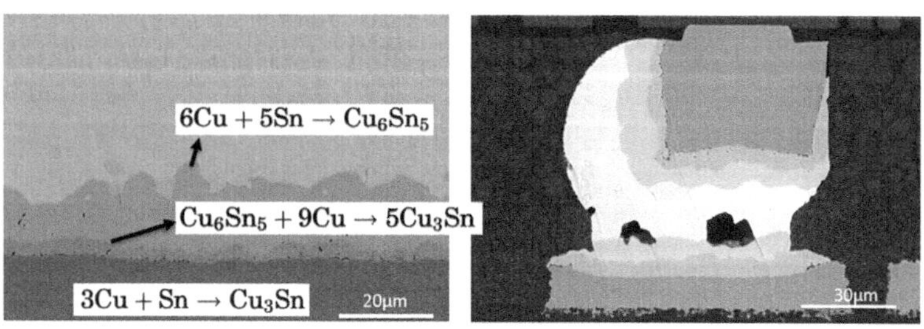

Fig. 12.36 Left, IMC growth at a Cu-Sn interface [144]. Right, larger voids as observed after aging at high temperatures (e.g., >150 °C)

On a final note in this section, one of the topics not well described in the literature is the corrosion of flipchip joints. Although the corresponding IMCs are less sensitive to corrosion compared to the wirebonded interconnects described in the previous section, there are some reports of corrosion of such Sn-alloy solder joints under the influence of halides [154]. It is expected that more studies on this topic will follow due to the potential use of flipchip packages in more demanding applications.

7.3 *Other Failure Mechanisms*

In addition to the degradation mechanism described in the sections above, there are several other mechanisms related to aging that have been covered in the literature. This section will only provide a brief description of such cases.

One of well-known mechanisms is metal migration, generally leading to shorts or leakage. In most cases, it appears for metals like Ag or Cu and is driven by the presence of a bias voltage (or galvanic potential) or an electrolyte (e.g., due to contamination generally in combination with moisture). One of the most common types encountered is Ag migration with dendrites growing from the cathode to the anode [155]. In most cases, other metals like Cu and even Au can also migrate under the appropriate conditions. Generally, delamination is a prerequisite that allows for such migration. Delamination provides the possibility for the formation of a bridging electrolyte (moisture), as also for penetration of contaminants from the outside of the package. In some cases, the source can be intrinsic to the package; one of the well-known cases was the migration of Ag driven by red phosphorus-based flame retardants in EMCs [156].

A special case of Cu migration driven by moisture and sulfur-based gases sometimes in combination with other oxidizing agents like chlorides is called creeping corrosion, as described by Vogel [77]. This is usually encountered after board assembly when exposed Cu is subjected to specific environmental conditions. Cu migration can also be observed in HVQFN packages due to interaction between exposed Cu and flux [157].

Another special mechanism encountered in high-voltage devices (e.g., > 400 V) is mobile ion ingression, specifically ions like Na + and K+ which are always present in very small quantities in the EMC [158]. During reliability tests like THB (85 °C/85%RH) or HTOL (e.g., Tj: 150 °C), under high bias voltages and consequently high electric fields, such mobile ion can penetrate into the back-end layers of the die, usually supported by some level of mechanical or passivation integrity damage. The mobile ions eventually reach transistors like NMOST transistors (negatively biased), causing electrical Vt shifts or leakages in the device. The properties of isolation layers used in such devices are extremely important to avoid any damage and to offer enough ion-gettering performance (e.g., the use of phosphorus-doped TEOS layers) [158].

Finally, it should be mentioned that in some cases, failure are encountered due to interactions with the execution method (e.g., the use of mechanical sockets during stressing of the product or during testing of the product after the stress test). Such mechanical loads can lead to cracks, corrosion [106], and even migration [157], which need careful risk assessment to judge the probability of the mechanism being of concern in the application. In general reliability execution and the test condition should always be optimized to avoid such issues or even over-acceleration.

8 Final Remarks

This chapter started with some basics of microelectronic packages, an overview of various types of packages, and the relevant materials involved. Next, a description of various ways to characterize such packaging materials, including their thermo-mechanical, chemical, moisture properties, and relation with stress and mechanical reliability, was given. This was followed by a basic introduction into package reliability, including (material) qualification methodology and failure analysis. The final part gave more details of relevant aging/failure mechanisms in microelectronic packaging: related to both bulk behavior (e.g., EMC aging) as also interconnect reliability with focus on (Cu) wirebond reliability since it is arguably the largest mainstream interconnect technology today. Key takeaways from this chapter are the following:

- Application areas for microelectronics are exponentially growing, driving a challenging mix of stringent requirements, high reliability, miniaturization, and time to market at low cost.
- Qualification of new materials and technologies takes a lot of time and effort; ideally most effort should be taken up front to streamline and speed up the process. Broad usage of new technologies and materials should be encouraged to reduce development cost, as also designing in robustness to improve reliability performance.
- There is a strong interaction between various materials in a package that can have a large impact on either bulk material reliability or interconnect reliability.
- Qualification should be based on better understanding of the physics, mechanics, and chemistry governing various degradation and failure mechanisms in order to develop better acceleration and prediction models or challenge conventional models.

Acknowledgments The authors are extremely grateful to several current and former NXP colleagues who have contributed to the work from NXP referred to in this chapter. In addition, the long-standing collaboration with material suppliers like Sumitomo Bakelite, Showa Denko, and Henkel is also kindly acknowledged.

References

1. M. Ciappa Microelectronics, Lifetime Prediction on the Base of Mission Profiles. Microelectron. Reliab. **45**, 1293–1298 (2005)
2. G. Harman, *Wire Bonding in Microelectronics: Materials, Processes, Reliability, and Yield*, 3rd edn. (McGraw-Hill Education, March 2010)
3. J-STD-020E, Joint IPC/Jedec standard for moisture/reflow sensitivity classification for non-hermetic surface-mount devices
4. D. Lu, C. P. Wong (eds.), *Materials for advanced packaging* (Springer International Publishing Switzerland 2009, 2017)
5. Substrate Manufacturing Technologies: Organic Packages and Interconnect Substrate, *Integrated Circuit Packaging, Assembly and Interconnections* (Springer, Boston, MA, 2007). https://doi.org/10.1007/0-387-33913-2_15
6. K. Sugiura et al., Thermal stability improvement of sintered Ag die-attach materials by addition of transition metal compound particles. Appl. Phys. Lett. **114**(161903) (2019)
7. Dr. Jeff Gottro, "Polymers in Electronic Packaging"
8. H. Sasajima et al., "New Development Trend of Epoxy Molding Compound for Encapsulating Semiconductor Chips" chapter 9 in [5] above
9. H.-M. Tong, Y.-S. Lai, C. P. Wong (eds.), *Advanced Flip Chip Packaging* (Springer, March 21 2013)
10. G. C. Liew, K. J. Lee and M. S. Aileen, Influence of mold compound type towards Palladium doped and copper doped 2N Au wire, 2014 IEEE 16th Electronics Packaging Technology Conference (EPTC), 2014, pp. 549–553, https://doi.org/10.1109/EPTC.2014.7028404
11. I. Singh et al, Pd-coated Cu wire bonding technology: Chip design, process optimization, production qualification and reliability test for high reliability semiconductor devices, 2012 IEEE 62nd Electronic Components and Technology Conference, San Diego, CA, USA, 2012, pp. 1089–1096, ECTC.2012
12. J. Xi, Evaluation of Ag Wire Reliability on Fine Pitch Wire Bonding. 2015 IEEE 65th Electronic Components and Technology Conference, San Diego, CA, USA, 2015, pp 1392 = 1396, ECTC-2015
13. R. Al Athamneh, S. Hamasha, Fatigue Behavior of SAC-Bi and SAC305 Solder Joints With Aging. IEEE Transactions on Components, Packaging and Manufacturing Technology **10**(4), 611–620 (April 2020). https://doi.org/10.1109/TCPMT.2019.2949719
14. Aklonis, John H. et al. "Introduction to polymer viscoelasticity." (1972)
15. X. Du, G. Xie, W. Tan, S. Qin, X. Cheng, *Stress reduction of epoxy molding compound and its effect on delamination* (2007 International Symposium on High Density packaging and Microsystem Integration, Shanghai, China, 2007), pp. 1 5. https://doi.org/10.1109/HDP.2007.4283560
16. J. Bergström, Ch 6 Linear Viscoelasticity in "Mechanics of Solid Polymers: Theory and Computational Modeling" pp. 309 351, Applied Science Publishers 2015
17. N. Sharp et al., Effects of water on epoxy cure kinetics and glass transition temperature utilizing molecular dynamics simulations. J. Polym. Sci. Part B Polymer Phys **55**(15) (2017)
18. A. Teverovsky, The significance of glass transition temperature of molding compounds for screening and reliability qualification of COTS PEMs, in *Proceedings of the 36th International Symposium on Microelectronics*, (IMAPS, Boston, MA, 2003, November 18–20 2003), pp. 483–488
19. K. Terada, T. Nejime, T. Ooyoshi, K. Kobayashi, K. Yamanaka, *Effect of solders, underfills and substrates on the reliability of flip-chip bonding of low-k semiconductor chips* (2010 IEEE CPMT Symposium Japan, Tokyo, Japan, 2010), pp. 1–4. https://doi.org/10.1109/CPMTSYMPJ.2010.5679666
20. Xiaosong Ma, "Fast qualification methods for Microelectronics Packages" PhD dissertation, Technical University Delft (2011)
21. M.R. Kamal, S. Sourour, Polymer Engineering and Science **12**, 1 (1973)

22. A. Xiao, H. Pape, G. Schlottig, Wunderle, Y. Y. Leung, K. M. B. Jansen, and L. J. Ernst. Procedure to determine interfacial toughness of emc-copper (oxide) interfaces. In Proceedings - 2010 11th International Conference on Electronic Packaging Technology and High Density Packaging, ICEPT-HDP 2010, pages 356–362, 2010
23. M. Sadiginha, "Failure and delamination in microelectronic packages", PhD dissertation, Technical University of Delft, 2013
24. N. Pflügler, et al, 2018 19th International Conference on Thermal, Mechanical and Multi-Physics Simulation and Experiments in Microelectronics and Microsystems (EuroSimE)
25. J.A. van der Pol, R.T.H. Rongen, H.J. Bruggers, Model and design rules for eliminating surface potential induced failures in high voltage integrated circuits. Microelectron. Reliab. **40**(8–10) (2000)
26. J.A. van der Pol, H.J. Gerritsen, R.T.H. Rongen, P.P.M.C. Groeneveld, P.W. Ragay, H.A. van den Hurk, Reliability issues in 650V high voltage bipolar-CMOS-DMOS integrated circuits. Microelectron. Reliab. **37**(10–11) (1997)
27. Y. Ng, W. Hooi, *FTIR method to distinguish mold compound types and identify the presence of Post-Mold-Cure (PMC) treatment on electronic packaging* (2007 9th Electronics Packaging Technology Conference, Singapore, 2007), pp. 652–656. https://doi.org/10.1109/EPTC.2007.4469789
28. W. Zhang, A. Fina, G. Ferraro, R. Yang, FTIR and GCMS analysis of epoxy resin decomposition products feeding the flame during UL 94 standard flammability test. Application to the understanding of the blowing-out effect in epoxy/polyhedral silsesquioxane formulations. J. Anal. Appl. Pyrolysis **135**, 271–280 (2018)
29. V. Dmitrovic, et al, "Surface Analysis and Material Characterization Techniques Used in Semiconductor Industry to Identify and Prevent Failures", Book Chapter in "Microelectronics Failure Analysis Desk Reference", Seventh edition, Ed: Tejinder Gandhi, 2019
30. M. van Soestbergen, A. Mavinkurve, Anomalous water absorption by epoxy molding compounds that are used for microelectronics packaging. J. Appl. Polym. Sci. (2014). https://doi.org/10.1002/APP.41192
31. M. Barink, A. Mavinkurve, J. Janssen, Microelectronics Reliability. Predicting non-Fickian moisture diffusion in EMCs for application in micro-electronic devices **62**, 45–49 (2016)
32. E.H. Wong, R. Rajoo, S.W. Koh, T.B. Lim, The Mechanics and Impact of Hygroscopic Swelling of Polymeric Materials in Electronic Packaging. ASME. J. Electron. Packag. June 2002 **124**(2), 122–126 (May 2, 2002). https://doi.org/10.1115/1.1461367
33. IPC/JEDEC standard J-STD-020E
34. IEC standard IEC60112
35. M. van Soestbergen et al., *Electrical characterization of plastic encapsulations using an alternative gate leakage test method* (2008 IEEE International Reliability Physics Symposium, Phoenix, AZ, USA, 2008), pp. 462–467. https://doi.org/10.1109/RELPHY.2008.4558929
36. M. Rouse, Internet of things (IoT). IOT Agenda. Retrieved 14 August 2019
37. M. Bellotti, R. Mariani, How future automotive functional safety requirements will impact microprocessors design. Microelectron. Reliab. **50**(9), 1320–1326 (September 2010)
38. G.A. Klutke, P.C. Kiessler, M.A. Wortman, A critical look at the bathtub curve. IEEE Transactions on Reliability **52**(1), 125–129 (2003. March 2003). https://doi.org/10.1109/TR.2002.804492
39. V. Mathew, E. Wikramanayake, S.F. Chopin, *Corrosion of copper wire bonded packages by chlorine containing foreign particles* (2020 IEEE 70th Electronic Components and Technology Conference (ECTC), Orlando, FL, USA, 2020), pp. 504–511. https://doi.org/10.1109/ECTC32862.2020.00086
40. F. Kuper et al., *Relation between yield and reliability of integrated circuits: experimental results and application to continuous early failure rate reduction programs* (Proceedings of International Reliability Physics Symposium, Dallas, TX, USA, 1996), pp. 17–21. https://doi.org/10.1109/RELPHY.1996.492055

41. M. Hamada, Using statistically designed experiments to improve reliability and to achieve robust reliability. IEEE Transactions on Reliability **44**(2), 206–215 (1995, June 1995). https://doi.org/10.1109/24.387372
42. JEDEC standard JEP122
43. M. Pecht, T. Shibutani, M. Kang, M. Hodkiewicz, E. Cripps, A fusion prognostics-based qualification test methodology for microelectronic products. Microelectron. Reliab. **63**, 320–324 (2016)
44. W.Q. Meeker, L.A. Escobar, *Statistical Methods for Reliability Data* (Wiley, July, 1998)
45. W.D. van Driel, A. Mavinkurve, M.A.J. van Gils, G.Q. Zhang, D.G. Yang, L.J. Ernst, *"Advanced structural similarity rules for the BGA package family," EuroSimE 2005* (Proceedings of the 6th International Conference on Thermal, Mechanical and Multi-Physics Simulation and Experiments in Micro-Electronics and Micro-Systems, 2005, Berlin, Germany, 2005), pp. 524–530. https://doi.org/10.1109/ESIME.2005.1502861
46. Microelectronics Failure Analysis Desk Reference, Seventh Edition, Editor: Tejinder Gandhi, Product code: 09111G, ISBN: 9781627082457
47. A. Mavinkurve, H. Cobussen, W.D. van Driel, L. Endrinal, M. van Dort, *Assembly - chip interactions leading to PPM-level failures in microelectronic packages* (2009 European Microelectronics and Packaging Conference, Rimini, Italy, 2009), pp. 1–8
48. S. Manoharan et al, "Decapsulation of Plastic Encapsulated Microelectronics with Copper Wire Bonds", Additional Conferences (Device Packaging HiTEC HiTEN & CICMT) · May 2016
49. ZVEI (German Electrical Industry) Robustness Validation Knowledge Matrix
50. T.R. Compton, "Thermal Stability of polymers", Smithers Rapra, 8 May 2012
51. JEDEC standard JESD91A, 8/01
52. A. Teverovsky, "Reliability of COTS PEMs", NASA Electronic Parts and Packaging Program (NEPP), October, 2005
53. E.H. Wong, K.C. Chan, T.B. Lim, T.F. Lam, Non-Fickian moisture properties characterisation and diffusion modeling for electronic packages, in *Proceedings of the 49th Electronic Components and Technology Conference, 1999*, (1999), pp. 302–306
54. A. Mavinkurve et al., *Moisture absorption by molding compounds under extreme conditions: impact on accelerated reliability tests* (ESREF, 2016)
55. Jansen, K.M.B. et al, "Thermal aging of molding compounds," Electronic Packaging Technology & High Density Packaging (ICEPT-HDP), 2010 11th International Conference, pp. 778–780, 16–19 Aug. 2010
56. J. van Vreugd et al, Eurosime, High Temperature Storage Influence on Molding Compound Properties, 2010 and PhD thesis Chapter 7, 2011
57. A. Mavinkurve, L. Goumans, J. Martens, *Epoxy molding compounds for high temperature applications* (2013 European Microelectronics Packaging Conference (EMPC), Grenoble, France, 2013), pp. 1–7
58. T. Braun et al., *Plastic packaging for high temperature applications* (International Conference and Exhibition on High Temperature Electronics Network (HiTEN), Oxford, 2009)
59. D. Yang, D. Liu, W.D. van Driel, H. Scholten, L. Goumans, R. Faria, *Advanced reliability study on high temperature automotive electronics* (2010 11th International Conference on Electronic Packaging Technology & High Density Packaging, Xi'an, China, 2010), pp. 1246–1249. https://doi.org/10.1109/ICEPT.2010.5582777
60. N. Lakhera, S. Shantaram, A.K. Singh, *Resolution of extreme warpage in ultra-thin molded array packages under High Temperature Storage Life* (2015 IEEE 65th Electronic Components and Technology Conference (ECTC)) 26-29 May 2015
61. B. Zhang et al., *Modelling of thermal aging of Moulding Compound by using an equivalent layer assumption* (2017 18th International Conference on Thermal, Mechanical and Multi-Physics Simulation and Experiments in Microelectronics and Microsystems (EuroSimE), Dresden, Germany, 2017), pp. 1–6. https://doi.org/10.1109/EuroSimE.2017.7926271

62. A. Inamdar et al., *Study of Thermal Aging Behavior of Epoxy Molding Compound for Applications in Harsh Environments* (2019 IEEE 69th Electronic Components and Technology Conference (ECTC), Las Vegas, NV, USA, 2019), pp. 811–818. https://doi.org/10.1109/ECTC.2019.00128
63. C. Lin, S. Islam, J.C. Suhling, P. Lall, *Material behavior changes in underfill encapsulants exposed to isothermal aging* (Thermal and Thermomechanical Proceedings 10th Intersociety Conference on Phenomena in Electronics Systems, 2006. ITHERM 2006, San Diego, CA, USA, 2006), pp. 1137–1146. https://doi.org/10.1109/ITHERM.2006.1645473
64. AEC-Q100 Rev H
65. R.D. Mosbarger, *Electro-Thermally Induced Parasitic Gate Leakage (GL)* (ISTFA '93: The 19th International Symposium for Testing & Failure Analysis, Los Angeles, California, USA, 15-19 November 1993)
66. R.D. Mosbarger, D.M. Hickey, *The Effects of Materials and Post-Mold Profiles on Plastic Encapsulated Integrated Circuits* (Proceedings of 1994 IEEE International Reliability Physics Symposium, 1994)
67. W. Schuddeboom, M. Wübbenhorst, Question marks to the extrapolation to lower temperatures in high temperature storage life (HTSL) testing in plastic encapsulated IC'S. Microelectron. Reliab. **36**(11–12) SPEC. ISS, 1935–1938 (1996)
68. Y. Liu et al., *"Impact of the die attach process on power & thermal cycling for a discrete style semiconductor package," EuroSimE 2005* (Proceedings of the 6th International Conference on Thermal, Mechanical and Multi-Physics Simulation and Experiments in Micro-Electronics and Micro-Systems, 2005, Berlin, Germany, 2005), pp. 221–226. https://doi.org/10.1109/ESIME.2005.1502805
69. O. Wittler, A.M. Nejadari, B. Michel, *Detection of degradation in die-attach materials by in-situ monitoring of thermal properties* (2009 11th Electronics Packaging Technology Conference, Singapore, 2009), pp. 153–157. https://doi.org/10.1109/EPTC.2009.5416558
70. H. Lee, R. Liu "MEMS die warpage during curing of die attach material", Proceedings of the ASME 2015 International Technical Conference and Exhibition on Packaging and Integration of Electronic and Photonic Microsystems InterPACK2015, July 6–9, 2015, San Francisco, California
71. M.A. Fard Sanei, L. Frisk, The effect of reflow process on the physical properties of die attach adhesives. Procedia Eng **168**, 1613–1616 (2016)
72. C. Lo, O.K. Yoke, C.C. Yong, L.B. Seong, *Assembly challenges for low modulus die attach material for MEMS devices* (2009 11th Electronics Packaging Technology Conference, Singapore, 2009), pp. 762–767. https://doi.org/10.1109/EPTC.2009.5416446
73. P. Peng, X.P. He, D. Lian Ming, Z. Wu, H.J. Yu, B. Peng, The Influence of Die-Attach Adhesives on the Packaging of MEMS Accelerometer. Key Engineer Material **645–646**, 533–537 (May 2015). https://doi.org/10.4028/www.scientific.net/kem.645-646.533
74. K. Sugiura, T. Iwashige, K. Tsuruta, C. Chen, S. Nagao, T. Sugahara, K. Suganuma, Appl. Phys. Lett. **114**, 161903 (2019)
75. H. Yan, P. Liang, Y. Mei, Z. Feng, Brief review of silver sinter-bonding processing for packaging high-temperature power devices. Chinese Journal of Electrical Engineering (3), 25–34 (6, Sept. 2020). https://doi.org/10.23919/CJEE.2020.000016
76. S. Radente, S. Borsini, Board level reliability, in *Memories in wireless systems. Signals and communication technology*, ed. by R. Micheloni, G. Campardo, P. Olivo, (Springer, Berlin, Heidelberg, 2008). https://doi.org/10.1007/978-3-540-79078-5_8
77. Vogel, Creeping corrosion of copper on printed circuit board assemblies. Microelectron. Reliab. **64**, 650–655 (September 2016)
78. M. van Soestbergen, A. Mavinkurve, S. Shantaram, J.J.M. Zaal, *Delamination-induced stitch crack of copper wires* (2017 18th International Conference on Thermal, Mechanical and Multi-Physics Simulation and Experiments in Microelectronics and Microsystems (EuroSimE), Dresden, Germany, 2017), pp. 1–4. https://doi.org/10.1109/EuroSimE.2017.7926294

79. M. Schneider-Ramelow, S. Schmitz, B. Schuch, W. Grubl, *Kirkendall voiding in Au ball bond interconnects on al chip metallization in the temperature range from 100–200°C after optimized intermetallic coverage* (2009 European Microelectronics and Packaging Conference, Rimini, Italy, 2009), pp. 1–6
80. G. Majni et al, J. Crystal Growth, Vol 47, 583–8, 1979, G Majni et al, J Appl. Physics, 52(6), 4047–52, 1981
81. E. Philofsky, *Purple Plague Revisited* (8th Reliability Physics Symposium, Las Vegas, NV, USA, 1970), pp. 177–185. https://doi.org/10.1109/IRPS.1970.362455
82. G.V. Clatterbaugh, H.K. Charles, The effect of high-temperature intermetallic growth on ball shear-induced cratering. IEEE Transactions on Components, Hybrids, and Manufacturing Technology **13**(1), 167–175 (March 1990). https://doi.org/10.1109/33.52866
83. B. Krabbenborg, *In-situ monitoring of bond degradation in power ICs under high current stress* (1998 IEEE International Reliability Physics Symposium Proceedings. 36th Annual (Cat. No.98CH36173), Reno, NV, USA, 1998), pp. 238–247. https://doi.org/10.1109/RELPHY.1998.670557
84. M. Khan, H. Fatemi, J. Romero, E. Delenia, *Effect of high thermal stability mold material on the gold-aluminum bond reliability in epoxy encapsulated VLSI devices* (26th Annual Proceedings Reliability Physics Symposium 1988, Monterey, CA, USA, 1988), pp. 40–49. https://doi.org/10.1109/RELPHY.1988.23424
85. T. Uno, K. Tatsumi, Thermal reliability of gold-aluminum bonds encapsulated in bi-phenyl epoxy resin. Microelectro. Rel **40**
86. J. Park et al., Interfacial Degradation Mechanism of Au/Al and Alloy/Al Bonds Under High Temperature Storage Test: Contamination, Epoxy Molding Compound, Wire and Bonding Strength. IEEE Transactions on Components and Packaging Technologies **30**(4), 731–744 (Dec. 2007). https://doi.org/10.1109/TCAPT.2007.906318
87. K.-C. Hsieh et al., Ag and Cu migration phenomena on wirebonding. J Electron Mater **29**(10), 1239–1242 (2000)
88. D. Peck. IEEE int. reliability physics symposium proceedings 1986, pp 44–50
89. C.L. Gan et al., Extended reliability of gold and copper ball bonds in microelectronic packaging. Gold Bull **46**, 103–115 (2013)
90. S. Qu et al., *Over pad metallization for high temperature interconnections* (2011 IEEE 61st Electronic Components and Technology Conference (ECTC), Lake Buena Vista, FL, USA, 2011), pp. 1496–1501. https://doi.org/10.1109/ECTC.2011.5898708
91. V. Mathew, T.A. Tran, Characterization of Over Pad Metallization (OPM) for High Temperature Reliability. International Symposium on Microelectronics (1), 001097–001104 (1 January 2012, 2012). https://doi.org/10.4071/isom-2012-THP32
92. R.T.H. Rongen et al., Cratering response method to study the effect of ultrasonic energy on Cu-wire bonding quality. Microelectron. Reliab. **51**, 1865–1868 (2011)
93. J. Tang et al., *Microwave induced plasma decapsulation of stressed and delaminated high pin-count copper wire bonded IC packages* (2013 IEEE 63rd Electronic Components and Technology Conference, Las Vegas, NV, USA, 2013), pp. 1717–1723. https://doi.org/10.1109/ECTC.2013.6575806
94. F. Zong, N. Zhou, J. Niu, Z. Sun, *Study on bond pad damage issue in bare Cu wire bonding on SMOS8MV wafer technology* (2015 16th International Conference on Electronic Packaging Technology (ICEPT), Changsha, China, 2015), pp. 108–113. https://doi.org/10.1109/ICEPT.2015.7236554
95. M.L. Farrugia, *A new copper wire ball bonding process methodology* (2013 IEEE 15th Electronics Packaging Technology Conference (EPTC 2013), Singapore, 2013), pp. 168–172. https://doi.org/10.1109/EPTC.2013.6745706
96. H. Fan, W. Chow, K. Lee, F. Wong, H. Chen, J. Wu, *Investigation of Cu wire neck crack under thermal cycling test* (2016 17th International Conference on Electronic Packaging Technology (ICEPT), Wuhan, China, 2016), pp. 1522–1525. https://doi.org/10.1109/ICEPT.2016.7583412

97. A. Mavinkurve et al., *Effective Stress Relief Without Jeopardizing Reliability in Overmolded Packages with Stress Buffers*, vol 2018 (2018 IEEE 68th Electronic Components and Technology Conference (ECTC), San Diego, CA, USA), pp. 1774–1780. https://doi.org/10.1109/ECTC.2018.00267
98. G.H.M. Gubbels, M.H.M. Kouters, O. O'Halloran, R. Rongen, *Growth and properties of intermetallics formed during thermal aging of Cu-Al ball bonds* (3rd Electronics System Integration Technology Conference ESTC, Berlin, Germany, 2010), pp. 1–6. https://doi.org/10.1109/ESTC.2010.5642821
99. M.H.M. Kouters, G.H.M. Gubbels, C.A. Yuan, *Characterization of intermetallic compounds in Cu-Al ball bonds: Mechanical properties, delamination strength and thermal conductivity* (2012 13th International Thermal, Mechanical and Multi-Physics Simulation and Experiments in Microelectronics and Microsystems, Cascais, Portugal, 2012), pp. 1/9–9/9. https://doi.org/10.1109/ESimE.2012.6191740
100. G.H.M. Gubbels, M.H.M. Kouters, O.D.S. Ferreira, *Characterization of intermetallic compounds in Cu-Al ball bonds: Thermo-mechanical properties, interface delamination and corrosion* (2012 4th Electronic System-Integration Technology Conference, Amsterdam, Netherlands, 2012), pp. 1–6. https://doi.org/10.1109/ESTC.2012.6542123
101. G. Plascencia, T.A. Utigard, High temperature oxidation mechanism of dilute copper aluminium alloys. Corros. Sci. **47**, 1149–1163 (2005)
102. R. Rongen, G.M. O'Halloran, A. Mavinkurve, L. Goumans, M. Farrugia, *Lifetime prediction of Cu-Al wire bonded contacts for different mould compounds* (2014 IEEE 64th Electronic Components and Technology Conference (ECTC), Orlando, FL, USA, 2014), pp. 411–418. https://doi.org/10.1109/ECTC.2014.6897318
103. A. Mavinkurve et al., *The Paradoxical Role of Sulphur in Molding Compounds: Influence on High Temperature Reliability of Cu-Al Wirebond Interconnects* (2017 IEEE 67th Electronic Components and Technology Conference (ECTC), Orlando, FL, USA, 2017), pp. 1171–1178. https://doi.org/10.1109/ECTC.2017.252
104. R.T.H. Rongen et al., *Package Material Selection Criteria for High Temperature Automotive Applications* (2019 IEEE 69th Electronic Components and Technology Conference (ECTC), Las Vegas, NV, USA, 2019), pp. 479–485. https://doi.org/10.1109/ECTC.2019.00079
105. G.M. O'Halloran et al., *Planar Analysis of Copper-Aluminium Intermetallics* (Proc. International Symposium for testing and Failure Analysis (ISTFA), San Jose, CA, Nov 3-7 2013), pp. 297–300
106. A. Mavinkurve et al., *Key takeaways and relevance of extrinsic corrosion mechanisms during extended biased HAST* (2020 IEEE 70th Electronic Components and Technology Conference (ECTC), Orlando, FL, USA, 2020), pp. 518–525. https://doi.org/10.1109/ECTC32862.2020.00088
107. P. Su et al., *An evaluation of effects of molding compound properties on reliability of Cu wire components* (2011 IEEE 61st Electronic Components and Technology Conference (ECTC), Lake Buena Vista, FL, USA, 2011), pp. 363–369. https://doi.org/10.1109/ECTC.2011.5898539
108. E.H. Hollingworth, H.Y. Hunsicker, *"Corrosion of aluminum and aluminum alloys." Metal Handbook*, vol 30 (American Society of Materials, 1990), pp. 583–609
109. V. Mathew et al., Copper Wirebond Compatibility with Organic and Inorganic Ions Present in Mold Compounds. J Microelectron & Electron Packag **11**, 70–74 (2014)
110. V. Mathew, and S. Chopin, Corrosion reliability of copper wirebond (CuWB) packages-Impact of voltage and corrosive ions from packaging materials IMAPS 2015 Orlando – 48th Annual International Symposium on Microelectronics – October 26–29, 2015
111. H. Abe et al., *Cu wire and Pd-Cu wire package reliability and molding compounds* (Proc. IEEE Electronic Components and Technology Conference (ECTC), San Diego, May 29–June 1 2012, 2012), pp. 1117–1123

112. J. Zhu and L. H. Hihara, Corrosion of continuous alumina-fibre reinforced Al- wt.% Cu-T6 metal-matrix composite in 3.15 wt.% NaCl solution. In: Corrosion Science, Vol. 52, No. 2, pp. 406–415. Pergamon – Elsevier Science Ltd. England (2010).
113. W. Qin, H. Anderson, T. Anderson, G. Chang and D. Barrientos, "Corrosion Mechanisms of Cu Wire Bonding on Al Pads," 2018 IEEE 68th Electronic Components and Technology Conference (ECTC), San Diego, CA, USA, 2018, pp. 1446–1454, doi: https://doi.org/10.1109/ECTC.2018.00221
114. T. Uno, T. Yamada, *Improving humidity bond reliability of copper bonding wires* (2010 Proceedings 60th Electronic Components and Technology Conference (ECTC), Las Vegas, NV, USA, 2010), pp. 1725–1732. https://doi.org/10.1109/ECTC.2010.5490741
115. M. van Soestbergen et al., Modified Poisson–Nernst–Planck theory for ion transport in polymeric electrolytes. J. Electrostat. **66**, 567–573 (2008)
116. A. Mavinkurve, R. Rongen, M. van Soestbergen, O. O'Halloran, L. Goumans, M.L. Farrugia, *"Biased humidity degradation model for Cu-Al wire bonded contacts based on molding compound properties," 2015 European Microelectronics Packaging Conference (EMPC)* (Friedrichshafen, Germany, 2015), pp. 1–10
117. A. Mavinkurve et al, "The How and why of Biased Humidity Tests with Copper Wire," 2019 IEEE 69th Electronic Components and Technology Conference (ECTC), Las Vegas, NV, USA, 2019, pp. 777–784, https://doi.org/10.1109/ECTC.2019.00123
118. P. Lall et al, "Multiphysics-Modeling of Corrosion in Copper-Aluminum Interconnects in High Humidity Environments", Electronic Components and Technology Conference, ECTC 65th 2015, pp. 1045–1056
119. J. J. M. Zaal, A. Mavinkurve, R. Rongen, J. H. J. Janssen and P. Drummen, "Over-acceleration of corrosion mechanisms during reliability testing: A method to relate biased HAST tests and application conditions for Cu wire products," 2014 IEEE 16th Electronics Packaging Technology Conference (EPTC), Singapore, 2014, pp. 426–431, doi: https://doi.org/10.1109/EPTC.2014.7028322
120. A. Mavinkurve et al., Copper wire interconnect reliability evaluation using in-situ High Temperature Storage Life (HTSL) tests. Microelectron. Reliab. **54**, 1661–1665 (2014)
121. R. Rongen, A. van IJzerloo, A. Mavinkurve, G.M. O'Halloran, *"Degradation of Cu-Al wire bonded contacts under high current and high temperature conditions using in-situ resistance monitoring," 2015 IEEE 65th Electronic Components and Technology Conference (ECTC)* (San Diego, CA, USA, 2015), pp. 1396–1402. https://doi.org/10.1109/ECTC.2015.7159780
122. C. Lee, V. Mathew, T. Tran, R. Ibrahim and P. Eu, "Copper Ball Voids for Pd-Cu Wires: Affecting Factors and Methods of Controlling," 2016 IEEE 66th Electronic Components and Technology Conference (ECTC), Las Vegas, NV, USA, 2016, pp. 606–613, doi: https://doi.org/10.1109/ECTC.2016.220
123. J.C. Krinke, D. Dragicevic, S. Leinert, E. Friess, J. Glueck, High temperature degradation of palladium coated copper bond wires. Microelectron Reliab **54**(9–10), 1995–1999 (2014)
124. R. Villa, C.M. Villa, C. Passagrilli, A. Mancaleoni, *Extended Lifetime Study of Cu and Cu-Pd Bonding Wires* (Automotive Electronics Council Reliability Workshop, AEC Workshop, 2015)
125. M. Eto, T. Uno, T. Haibara, R. Oishi, T. Yamada and T. Oyamada, "Thermal Bond Reliability of High Reliability New Palladium-Coated Copper Wire," 2017 IEEE 67th Electronic Components and Technology Conference (ECTC), Orlando, FL, USA, 2017, pp. 1297–1302, doi: https://doi.org/10.1109/ECTC.2017.65
126. M. Han et al, "Copper wire bond pad/IMC interfacial layer crack study during HTSL (high temperature storage life) test," 2016 IEEE 18th Electronics Packaging Technology Conference (EPTC), Singapore, 2016, pp. 797–800, doi: https://doi.org/10.1109/EPTC.2016.7861590
127. A. Mavinkurve et al, "The Paradoxical Role of Sulphur in Molding Compounds: Influence on High Temperature Reliability of Cu-Al Wirebond Interconnects," 2017 IEEE 67th Electronic Components and Technology Conference (ECTC), Orlando, FL, USA, 2017, pp. 1171–1178, https://doi.org/10.1109/ECTC.2017.252

128. R. T. H. Rongen et al., "Package Material Selection Criteria for High Temperature Automotive Applications," 2019 IEEE 69th Electronic Components and Technology Conference (ECTC), Las Vegas, NV, USA, pp. 479–485, https://doi.org/10.1109/ECTC.2019.00079
129. Tu Anh Tran, Chu-Chung (Stephen) Lee, Varughese Mathew; Reliability Performance of Palladium Coated Copper Wire in High Temperature Bake at Extreme Durations. International Symposium on Microelectronics 1 October 2015; 2015 (1): 000292–000297. https://doi.org/10.4071/isom-2015-WA52
130. M. L. Farrugia, "The influence of bond-pad smear crevices on IMC corrosion in Cu-Al bonds," 2016 IEEE 18th Electronics Packaging Technology Conference (EPTC), Singapore, 2016, pp. 120–122, https://doi.org/10.1109/EPTC.2016.7861456
131. M. Van Soestbergen, A. Mavinkurve, J. J. M. Zaal, G. M. O'Halloran, R. T. H. Rongen and M. L. Farrugia, Crevice Corrosion of Ball Bond Intermetallics of Cu and Ag Wire, 2016 IEEE 66th Electronic Components and Technology Conference (ECTC), Las Vegas, NV, USA, 2016, pp. 774–781, https://doi.org/10.1109/ECTC.2016.192
132. S. F Chopin, V. Mathew; Controlling Extrinsic Chloride Ions Effect on Copper Wirebond Reliability. International Symposium on Microelectronics 1 October 2019; 2019 (1): 000095–000099. https://doi.org/10.4071/2380-4505-2019.1.000095
133. H. Hsieh, L. Wang, W. Lei and C. Lin, "Cu wire development for thin QFN package using die attach film (DAF)," 2013 8th International Microsystems, Packaging, Assembly and Circuits Technology Conference (IMPACT), Taipei, Taiwan, 2013, pp. 302–305, https://doi.org/10.1109/IMPACT.2013.6706660
134. K. Turner; A Case Study of the Reliability of Copper Bond Wires In Plastic Encapsulated Integrated Circuits. International Symposium on Microelectronics 1 January 2014; 2014 (1): 000804–000809. https://doi.org/10.4071/isom-THP15
135. H. Zhang, H. Yang, M. Zhou, A.C. Tsui, Aluminum Ribbon Bonding Technology in a New Package of High Power and Thermal Performance, in *8th International Conference on Electronic Packaging Technology*, (Shanghai, China, 2007), pp. 1–3. https://doi.org/10.1109/ICEPT.2007.4441448
136. T. B. Wei, N. Y. Hua and W. K. Sheng, "Silver alloy wire for IC packaging solution," 36th International Electronics Manufacturing Technology Conference, Johor, Malaysia, 2014, pp. 1–4, https://doi.org/10.1109/IEMT.2014.7123077
137. L. Merkle et al, "Mechanical fatigue properties of heavy aluminium wire bonds for power applications," 2008 2nd Electronics System-Integration Technology Conference, Greenwich, UK, 2008, pp. 1363–1368, https://doi.org/10.1109/ESTC.2008.4684554
138. P. J. van der Wel, R. Otte, H. Roberts, F. de Bruijn, A. van Zuijlen and B. Merkus, "Electromigration behavior in aluminum wires for power base-station applications," 2017 IEEE International Reliability Physics Symposium (IRPS), Monterey, CA, USA, 2017, pp. PA-2.1-PA-2.5, doi: 10.1109/IRPS.2017.7936381.AEC-Q006 Rev A (2016)
139. AEC Q006 Rev A (July 2016)
140. Y. Chiu et al., *The corrosion behavior of Ag alloy wire bond on Al pad in molding compounds of various chlorine contents under biased-HAST, 2016 International Conference on Electronics Packaging (ICEP)* (Hokkaido, Japan, 2016), pp. 497–501. https://doi.org/10.1109/ICEP.2016.7486877
141. S. Shantaram, A.R. Nazmus Sakib, N. Lakhera, Coupled Thermal and Thermo-Mechanical Simulation for Flip-chip Component Level Copper Pillar Bump Fatigue, in *2018 17th IEEE Intersociety Conference on Thermal and Thermomechanical Phenomena in Electronic Systems (ITherm)*, (IEEE, 2018)
142. M. Mustafa, Z. Cai, J.C. Suhling, P. Lall, The effects of aging on the cyclic stress-strain behavior and hysteresis loop evolution of lead free solders, in *2011 IEEE 61st Electronic Components and Technology Conference (ECTC)*, (Lake Buena Vista, FL, USA, 2011), pp. 927–939. https://doi.org/10.1109/ECTC.2011.5898623
143. L. Wu, X. Han, C. Shao, F. Yao, W. Yang, Thermal fatigue modelling and simulation of flip chip component solder joints under cyclic thermal loading. Energies **12**, 2391 (2019). https://doi.org/10.3390/en12122391

144. F. Gao, *Effects of Co and Ni addition on reactive diffusion between Sn–3.5Ag solder and Cu during soldering and annealing* (Materials Science and Engineering, 2006)
145. C.C. Li, C.K. Chung, W.L. Shih, et al., Volume Shrinkage Induced by Interfacial Reaction in Micro-Ni/Sn/Ni Joints. Metall Mater Trans A **45**, 2343–2346 (2014). https://doi.org/10.1007/s11661-014-2263-8
146. "Mechanical properties of metals and alloys", in: W.F. Gale, T.C. Totemeier (Eds.), Smithells Metals Reference Book (Eighth Edition), Butterworth-Heinemann, Oxford, 2004, pp. 22-1–22-162
147. T. Laurila, V. Vuorinen, J.K. Kivilahti, Interfacial reactions between lead-free solders and common base materials. Mater Sci & Engineer: R: Report **49**(1), 1–60 (2005)
148. G. Ross, V. Vuorinen, M. Krause, S. Reissaus, M. Petzold, M. Paulasto-Kröckel, XRD and ToF-SIMS study of intermetallic void formation in Cu-Sn micro-connects. Microelectron. Reliab. **76–77**, 390–394, ISSN 0026-2714 (2017). https://doi.org/10.1016/j.microrel.2017.07.044
149. R. Roucou, A. Mavinkurve, R.T.H. Rongen, C. Bestory, Empirical Model for the Degradation of IMC in SACQ Solder Bumps during High Temperature Storage, in *2020 IEEE 70th Electronic Components and Technology Conference (ECTC)*, (Orlando, FL, USA, 2020), pp. 1502–1508. https://doi.org/10.1109/ECTC32862.2020.00237
150. C. Tang, W. Zhu, L. Wang, G. Li, A Study on the Joint Properties and Reliability of 25μm Cu/Ni/Sn-3.5Ag Bonding Process with Chip on Chip Thermal Compression Bonding, in *21st International Conference on Electronic Packaging Technology (ICEPT)*, (Guangzhou, China, 2020), pp. 1–4. https://doi.org/10.1109/ICEPT50128.2020.9202998
151. F. Su, Z. Zhang, Y. Wang, W. Li, Study of Electromigration-Induced Stress of Solder. ASME. J. Electron. Packag. June 1, 2015 **137**(2), 021006 (2015). https://doi.org/10.1115/1.4029463
152. L. Meinshausen, K. Weide-Zaage, B. Goldbeck, A. Moujbani, J. Kludt and H. Frémont, "Electromigration reliability of cylindrical Cu pillar SnAg3.0Cu0.5 bumps," 2014 15th International Conference on Thermal, Mechanical and Mulit-Physics Simulation and Experiments in Microelectronics and Microsystems (EuroSimE), Ghent, Belgium, 2014, pp. 1–6, https://doi.org/10.1109/EuroSimE.2014.6813775
153. Z. Cui, X. Fan, & G. Zhang (2020). Implementation of General Coupling Model of Electromigration in ANSYS in L. O'Conner, & H. Torres (Eds.), 2020 IEEE 70th Electronic Components and Technology Conference (ECTC) (pp. 1632–1637). [9159445] IEEE. https://doi.org/10.1109/ECTC32862.2020.00256
154. S.S. Ha, H. Kang, G.R. Kim, S. Pae, H. Lee, Effect of Corrosion on Mechanical Reliability of Sn-Ag Flip-Chip Solder Joints. Mater. Trans. **57**(11), 1966–1971, Released October 25, 2016, Online ISSN 1347-5320, Print ISSN 1345-9678 (2016). https://doi.org/10.2320/matertrans.M2016203
155. S. Yang, J. Wu, A. Christou, Initial stage of silver electrochemical migration degradation. Microelectron. Reliab. **46**(9–11), 1915–1921, ISSN 0026-2714 (2006). https://doi.org/10.1016/j.microrel.2006.07.080
156. Y. Deng and M. Pecht, "The story behind the red phosphorus mold compound device failures," 2005 International Symposium on Electronics Materials and Packaging, Tokyo, Japan, 2005, pp. 1–5 https://doi.org/10.1109/EMAP.2005.1598224
157. P. Dumoulin, J. -. Seurin and P. Marce, "Metal Migrations Outside the Package During Accelerated Life Tests," IEEE Transactions on Components, Hybrids, and Manufacturing Technology, vol. 5, no. 4, pp. 479–486, 1982, December 1982. https://doi.org/10.1109/TCHMT.1982.1135979
158. J.A. van der Pol, H.J. Gerritsen, R.T.H. Rongen, P.P.M.C. Groeneveld, P.W. Ragay, H.A. van den Hurk, Reliability issues in 650V high voltage bipolar-CMOS-DMOS integrated circuits. Microelectron. Reliab. **37**(10–11), 1723–1726, ISSN 0026-2714 (1997). https://doi.org/10.1016/S0026-2714(97)00148-0

Chapter 13
Degradation and Remaining Useful Life Prediction of Automotive Electronics

P. Gromala, A. Inamdar, A. Prisacaru, M. Dressler, and A. Kabakchiev

1 Functions of Epoxy-Based Thermosets in Automotive Electronics

Epoxy-based thermosets are widely used in automotive electronics to suit many different functions like sealing to environmental media, to improve thermal performance, to conduct or isolate electronically, and connect parts mechanically. These functions have to be fulfilled under various environmental loading and over an extensive duration. Loading can be vibrational loading (e.g., when driving on gravel or dirt roads or hitting curbs) or thermomechanical loading (due to temperature change and heat dissipation of electronic parts). Time durations required to fulfill these functions are in the range of minimum mileage, time in active mode, or usage time.

Figure 13.1 shows a schematic diagram of an engine control unit with different types of epoxy-based polymer used for encapsulation. The design and choice of material basis of the polymer determines the degree of reliability, as it translates the external and internal loads into stresses and specifies its strength – therefore it is called a design element (DE). In order to be able to evaluate and improve the reliability of the design element, the material behavior during its manufacturing and lifetime as well as the type of expected failure needs to be described correctly.

P. Gromala (✉) · A. Inamdar · A. Prisacaru · M. Dressler · A. Kabakchiev
Robert Bosch GmbH, Reutlingen, Germany
e-mail: PrzemyslawJakub.Gromala@de.bosch.com; a.s.inamdar-1@tudelft.nl; Alexandru.Prisacaru@de.bosch.com; Marc.Dressler@de.bosch.com; Alexander.Kabakchiev@de.bosch.com

W. D. van Driel, M. Yazdan Mehr (eds.), *Reliability of Organic Compounds in Microelectronics and Optoelectronics*,
https://doi.org/10.1007/978-3-030-81576-9_13

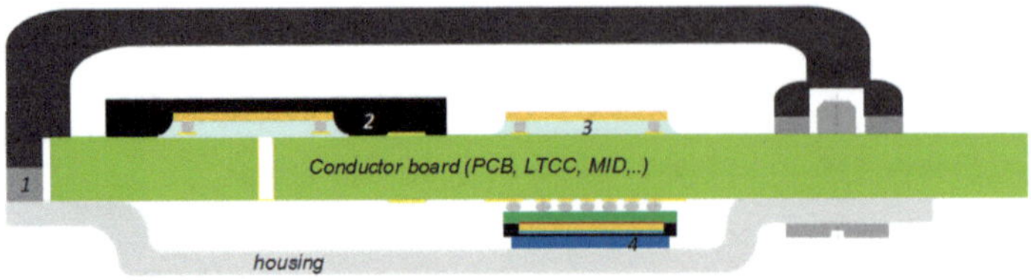

Fig. 13.1 Schematic view of engine control unit with different types of epoxy-based polymers – (*1*) sealant material to connect and seal housing parts (mostly silicone-based materials); (*2*) encapsulation material applied by potting, injection molding, or dispensing; (*3*) IC packaging like ICA, NCA, ACA, and capillary underfill; (*4*) thermal interface material (mostly silicone-based materials, not covered here)

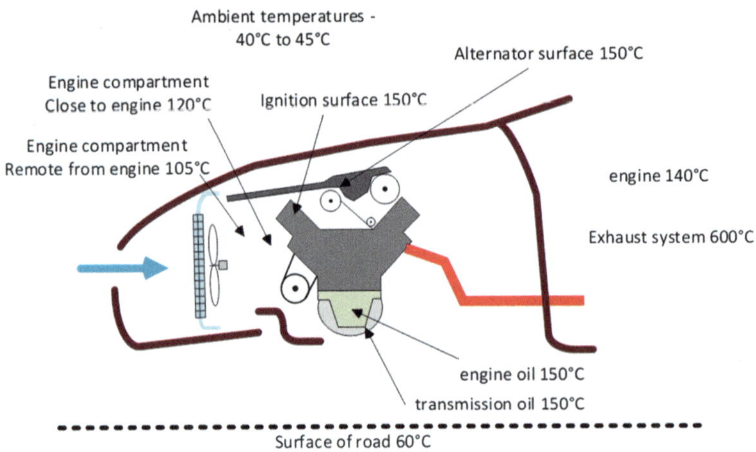

Fig. 13.2 Harsh environment considerations for epoxy-based polymers used in automotive applications

2 Environmental Requirements to Epoxy-Based Thermosets

Sensors and control units are being used at various mounting locations in a car, which lead to very different environmental conditions for the device. Among others, temperature, temperature cycling, humidity, and media are the main external loads, which need to be considered concerning reliability of polymers in sensors and control units.

Automotive electronics in the engine compartment face temperatures up to 150 °C when located close to the engine or transmission. For transmission control units and sensors, transmission oil as a media leads to additional loading due its corrosive behavior. For motor control units, the temperatures are going up to 105 °C depending on the location within the engine compartment. Figure 13.2 shows different high temperature zones within an automobile with a combustion engine. Such harsh environments are one of the key considerations for selecting epoxy-based polymers for automotive applications.

Sensors and control units which are located in the passenger compartment and trunk are exposed to overall lower temperatures but in general higher humidity loading. Control units for airbag are often placed in the center tunnel where humid and cold air from air-conditioning is acting on the used polymers. Temperatures are between −40 °C to 85 °C where the relative humidity can go up to 100%. Moreover, the polymers utilized for different functionalities, i.e., for different design elements, have to withstand these loads up to 15 years, 300000 kilometers, or up to 12000 hours of usage time.

3 Reliability Assessment of Polymer-Based Design Elements

A design element (DE) serves one main function as well as other sub-functions. If that function is not fulfilled anymore within the necessary usage time and loading conditions, the design element is defined as failed – it is not reliable. For example, the main function of a die attach material is to connect mechanically the chip to conductor board. Another function is to transfer heat produced by the chip to the conductor board. Once the thermal resistance is increased so much that the chip is heated up over its maximum temperature, the die attach material has failed its function.

Figure 13.3 shows the V-model of reliability, which is a representation of how the stress and strength are derived for a certain design element and failure mechanism. It can be split into two parts. The left part of the V is the load side and the right side the strength side. In the middle, the considered level is depicted, starting from the vehicle environment to the ECU to the component and finally to the design element of interest on the bottom of the V.

In general, risk assessment of a design element can be only done by comparing the stress and strength because they represent the relevant value for a certain failure mechanism [12–18, 19, 20].

One way to explain the V-diagram of reliability is to follow the workflow from the load side. The boxes stand for the loads and corresponding stresses at the specific level, whereas the arrows show the transfer functions.

The environment of the vehicle has certain conditions like temperature and relative humidity which might impact the stress at the design element. These conditions are transferred, shown by an arrow, to the ambient of the ECU. This is done mainly by measurements of the temperature at the ECU as a function of the ambient temperature of the vehicle. For example, the temperatures are different if the ECU is placed in the trunk or close to the engine, which means the transfer function is different from ambient to ECU. The ambient temperature of the ECU placed closely to the engine is dominated by the usage of the engine, whereas the temperature in the trunk is a strong function of the external temperatures of the vehicle and the exposure to the Sun.

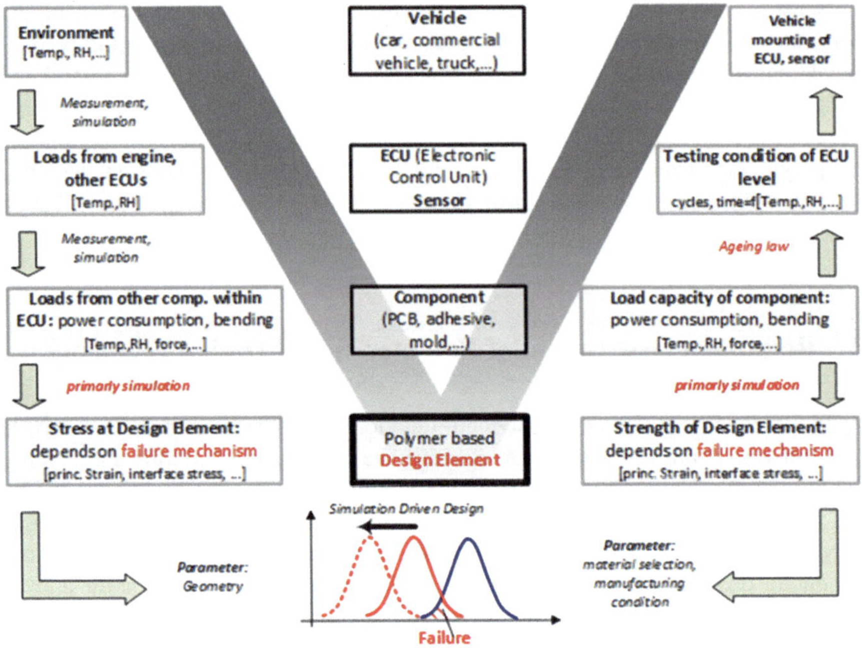

Fig. 13.3 V-model of reliability used to transfer load from a system (e.g., a car) to design elements (e.g., adhesive)

Once an ambient condition for the ECU is established, the loads from inside the ECU are considered. Depending on the specifics of a design element, inner loads are defined. Component self-heating as well as mechanical loading of the conductor board (e.g., printed circuit board (PCB)) during vibration or thermal cycling act as inner loads for a design element. For potted PCBs under temperature load, the inner load of the interface between potting material, component, and PCB is, on the one hand, the temperature; because it influences the mismatch of thermal expansion and therefore thermal strains. On the other hand, the bending of the PCB caused by, e.g., mechanical load causes additionally mechanical strain.

In order to derive the stresses on the design element which are finally directly connected to a failure mechanism, simulation can be used primarily. Direct measurements of stresses and strains or even displacements are not feasible without cutting and therefore destroying the specimen, e.g., inside an electronic package which is molded. This cutting process would change the complete stress state and therefore not represent the real situation.

In order to assess if a certain stress state is critical, which means failure occurs, the knowledge of strength is necessary. As for the stress, the strength can mostly not be directly measured, but deduced by simulation from load capacity tests on a component or a design element level. For example, the adhesive strength of a potting material on a PCB surface can be measured by static shear test. A critical shear force is obtained experimentally, but the failure occurs at a stress-concentration location like an edge or corner. By simulating the test and taking the force at failure as the loading condition, the local stress leading to the failure can be derived, which is nothing but the strength of the design element. Similar tests can be performed to obtain strength under different loading condition such as tension loading and other mixed modes, and even cyclic loading when it comes to thermal cycling or vibration. With that information, a risk assessment can be performed by comparing the stress and strength.

Additional care has to be taken when it comes to aging which can decrease the strength of a design element over lifetime. Thermal exposure over a longer period of time or higher temperature can increase the brittleness of epoxies. This can be described, e.g., by an Arrhenius-like aging law, where the activation energy E_a represents the value of how much an increase in temperature increases the aging process. Other aging mechanisms are used depending on the loading condition and physical or mechanical mechanism.

Knowledge of the aging mechanism is the key to derive field equivalent test conditions. As mentioned already, automotive electronics have often the requirement to survive 15 years of usage in a vehicle. Since no ECU can be tested for such long time periods during development, aging laws are used to design accelerated tests of shorter time periods. By assuming a certain field profile, e.g., temperature collective, a high temperature exposure, which induces similar magnitude of damage as expected under field operation, can be derived for the test.

In order to know the suitable stress and strength value to describe the failure as well as the aging law, the epoxy-based design element has to be investigated extensively. Otherwise, no precise risk assessment can be performed. In the case of epoxy-based design elements, the manufacturing process of the design elements impacts the stress as well as the strength level intensively.

4 The Use of Epoxy-Based Thermosets in Automotive Electronics

4.1 Epoxies Functioning as Electronic Housing

Due to the cost advantages, epoxy-based polymers are often used as housing material. This guarantees hermeticity of the integrated circuitry and provides mechanical stability. Many technologies for packaging electronic components are being used in order to ensure their reliability under harsh environments. In the following, some common direct packaging technologies are described.

4.1.1 Potting

The process of encapsulation comprises embodying the electronic component, e.g., a printed circuit board assembly (PCBA), with a liquid polymeric material, e.g., polyepoxides, and a subsequent curing step. An example of the potting process is illustrated in Fig. 13.4.

The process of curing is defined as the hardening by cross-linking of polymeric chains inside the material, forming solid polymeric of elastomeric bonds. The encapsulant compound may cure when, among others, exposed to an ultraviolet light, i.e., UV curing, or simply when exposed to room temperature air. One problem that may occur during curing is the formation of air-filled voids inside the resin, reducing the homogeneity of the packaging, which can add uncertainty to some parameters of interest, e.g., the coefficient of thermal expansion (CTE) of the packaging. To avoid the formation of voids, a vacuum chamber can be used during the curing. The encapsulation can be performed either in a pre-molded potting shell or directly in a mold.

The main reason for using encapsulation is to protect electronic components under harsh environments, e.g., moisture, electrostatic attraction of dust, corrosive atmospheres, chemicals, and service conditions, e.g., vibration or thermal or physical shock, which can cause loss of performance or failure. Protection is achieved through the mass of the resin surrounding the unit. The choice of the encapsulant is based on the intended application and must be done in such a way that ensures an adequate balance between mechanical strength, chemical resistance, dielectric insulation, and thermal conductivity of the packaging. Incautious choice of encapsulant material can lead to lower performance than expected or even to failure of the system.

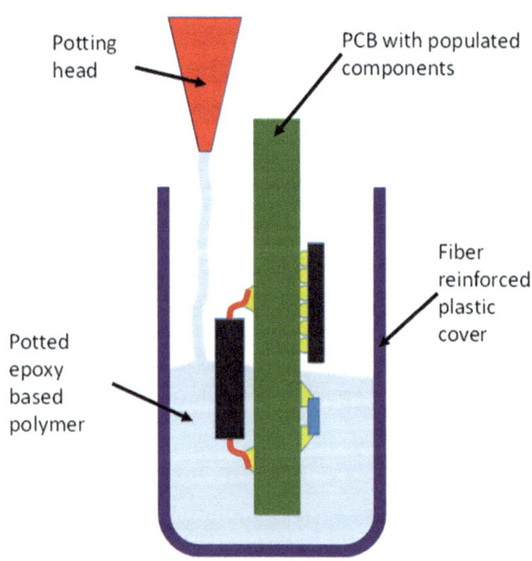

Fig. 13.4 Schematic view of the potting process

4.1.2 Transfer Molding

Transfer molding of electronic packages is performed in several steps, which are schematically shown in Fig. 13.5. First, a substrate with an integrated circuit is placed into the molding cavity and preheated for about 1 min. Parts can reach temperature up to 130 °C. Subsequently, the tool closes up to 1 mm below the final clamping position and the air is evacuated from the cavity. Afterward, the tool is clamped completely with a defined pressure. During that time, the substrate and circuitry rapidly heat up to temperatures of 160–170 °C. Resin pellets which are stored in a plunger cavity are melted. After 5–10 s, the transfer of the molding compound starts. The temperature of the molding compound is in the range of 145–165 °C. Time that is needed to fill the cavity is about 15–20 s.

In a next step of the process, the molding compound is cured. Here, the part is kept at 175–180 °C for 120 s. The molding compound shrinks during that time. This shrinkage is compensated by the movement of the tool, which is controlled by applying constant pressure. During that time the conversion level of the molding compound can reach up to 80–90%. After 2 min, the machine opens and the part can

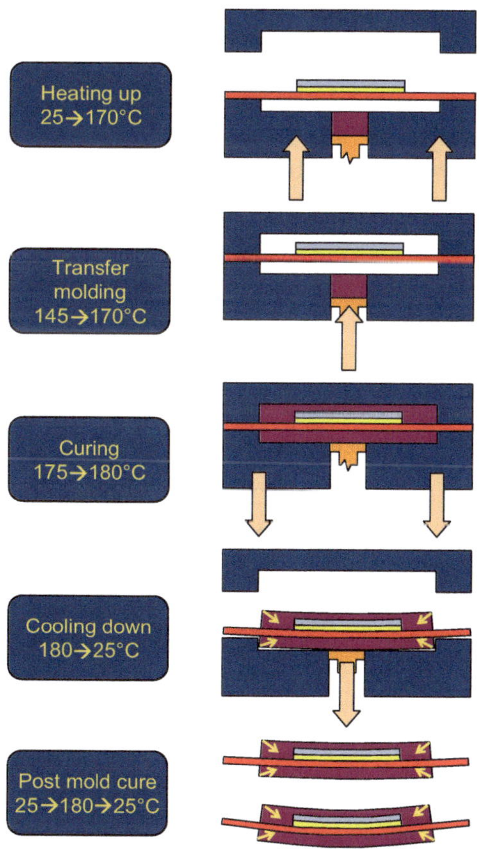

Fig. 13.5 Schematic view of transfer molding process

be removed. Already at that moment, the part warps with a magnitude depending on the conversion level. During cooling of the part to room temperature, the warpage changes, depending on the CTE mismatch.

During post mold cure (PMC), the part is stored at a temperature of about 175 °C for 4 h. Here, the remaining process of curing happens. After cooling down to room temperature, the part again deviates from flatness, which is the combination of chemical shrinkage after molding and PMC and the CTE mismatch between materials. After PMC, molding compound should not exhibit reactivity anymore.

4.1.3 Injection Molding

In the injection molding process, the thermosetting polymer is fed into a hopper which is connected to a coaxial screw-type plunger heated barrel, where the polymer is melted and forced through a nozzle and then into a mold. Once inside the mold, the molten polymer is cooled and removed from the mold by ejector pins. In Fig. 13.6, an injection molding machine is schematically depicted.

One advantage of the injection molding manufacturing process is the fully automated capability, which allows a very high rate of production. The process also allows the production of highly detailed and intricate parts.

The main disadvantage is the high tooling and maintenance costs, especially when working with thermosetting polymers. Overstaying of thermosetting polymers in the injection barrel can cause chemical cross-linking leading to the undesired seizing of the plunger.

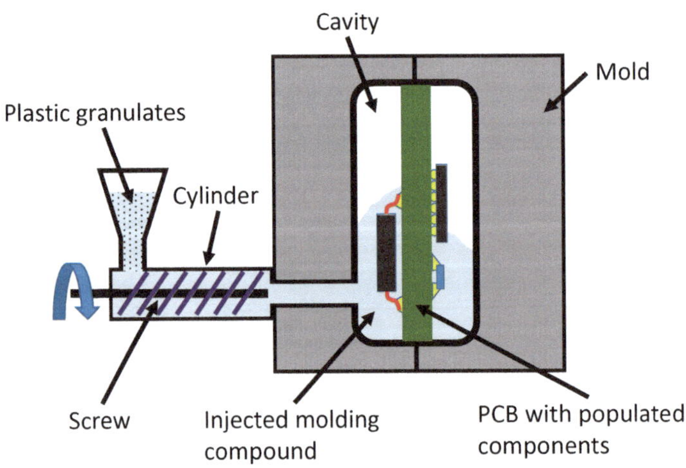

Fig. 13.6 Schematic view of injection molding machine

4.1.4 Polymers to Seal Housing Parts

Electronic control units are often mounted at locations where different kinds of media can damage the electronics. In the engine compartment, media like water from outside, salt water, washer fluid, etc. can act on the unit, whereas in the passenger compartment, liquid from beverages can lead to damage. In order to seal the top and bottom part of the housing, silicone-based sealant material is used.

Silicone has the advantage that it shows good adhesion to various substrates. Substrates like sheet metals, die-cast parts, or thermoplastic housing materials are used. For housing parts made of metal, a separate connector is used which is made of polyamide or other thermoplastic materials. Due to residues from the manufacturing steps, a cleaning process is mostly required to provide a suitable substrates surface. For housing material made of thermoplastics, an activation with oxygen is needed, which is applied by an atmospheric plasma treatment.

Silicone can withstand large deformations, which makes it suitable to deal with the thermal expansion of the different housing materials and housing material combinations. Additionally, silicone-based sealant material is inexpensive in material costs and can be easily applied. The curing mechanism can be thermal, humidity-based, UV-based, or a combination of these, where the curing condition and mechanism determines the stress-free condition. Often adhesion promoters are used in order to increase the adhesion strength.

Silicone materials are applied mostly using a time-pressure dispense unit, as shown in Fig. 13.7, where a certain dispense pattern is programmed. The dispense system follows the top or bottom housing contour, where also different component heights can be followed. The amount of material is controlled by the pressure and velocity of the needle movement. After dispensing, the other housing part is assembled, and the silicone is cured.

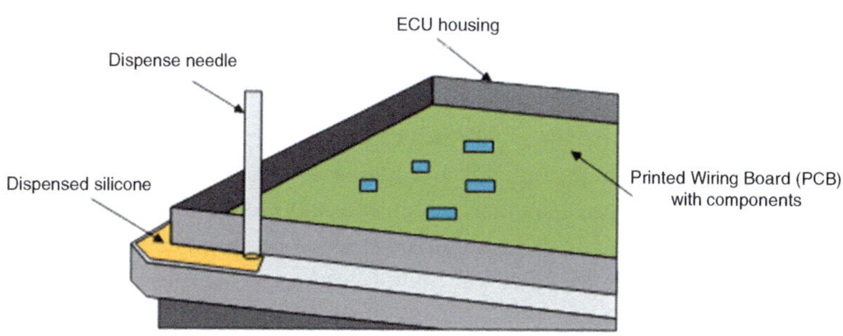

Fig. 13.7 Dispensing of polymer to connect housing, mostly silicones are used due to high tolerances of housing geometries

4.2 *Epoxies for Integrated Circuits (IC) Interconnections*

Since mechanical properties of epoxies can be adjusted easily by their constituent filler particles, they are also suitable to connect IC (integrated circuits, silicon chip, so called die) with a conductor board such as PCB, MID (molded interconnect device), or LTCC (low thermal coefficient ceramic). The epoxies provide directly or indirectly electrical conduction from chip to board. There are basically five different ways how epoxies are utilized and especially designed for these purposes.

4.2.1 Isotropic Conductive Adhesive (ICA) Interconnection for Flip Chips

On the one hand, there is the ICA (isotropic conductive adhesive) interconnection, where the filler particles provide the conduction directly, e.g., silver flakes. This interconnection is employed especially when balls or bumps made of gold, nickel, or copper are used. The chip with its bumps is dipped into the ICA and placed onto the board. Afterward it is cured at elevated temperatures. The ICA provides only the electrical function; the mechanical connection is secured by a capillary underfill also made of epoxy. The process is shown in Fig. 13.8, whereas Fig. 13.9 shows a cross-sectional cut of ICA interconnection.

4.2.2 Nonconductive Adhesive Interconnection for Flip Chips

On the other hand, there are nonconductive adhesive (NCA) interconnections where the adhesive provides a compression stress over lifetime between the bump respective ball and the conductor path on the board. This compression stress is established by the chemical cure shrinkage of the epoxy as well as by the thermal shrinkage, since the curing process is performed at high temperatures. The epoxy can be filled with NCA particles. The process flow is shown in Fig. 13.10. A cross-sectional cut of a gold bump on a MID substrate can be seen in Fig. 13.11.

4.2.3 Anisotropic Conductive Adhesive (ACA) Interconnection

The anisotropic conductive adhesive (ACA) interconnection is similar to a NCA interconnection, but single particles in the epoxy are used to provide an electrical connection between the pad between silicon die and conductor board. Such interconnection is indicated in Fig. 13.12. The number of particles is adjusted such that there is only a vertical conductance but no horizontal.

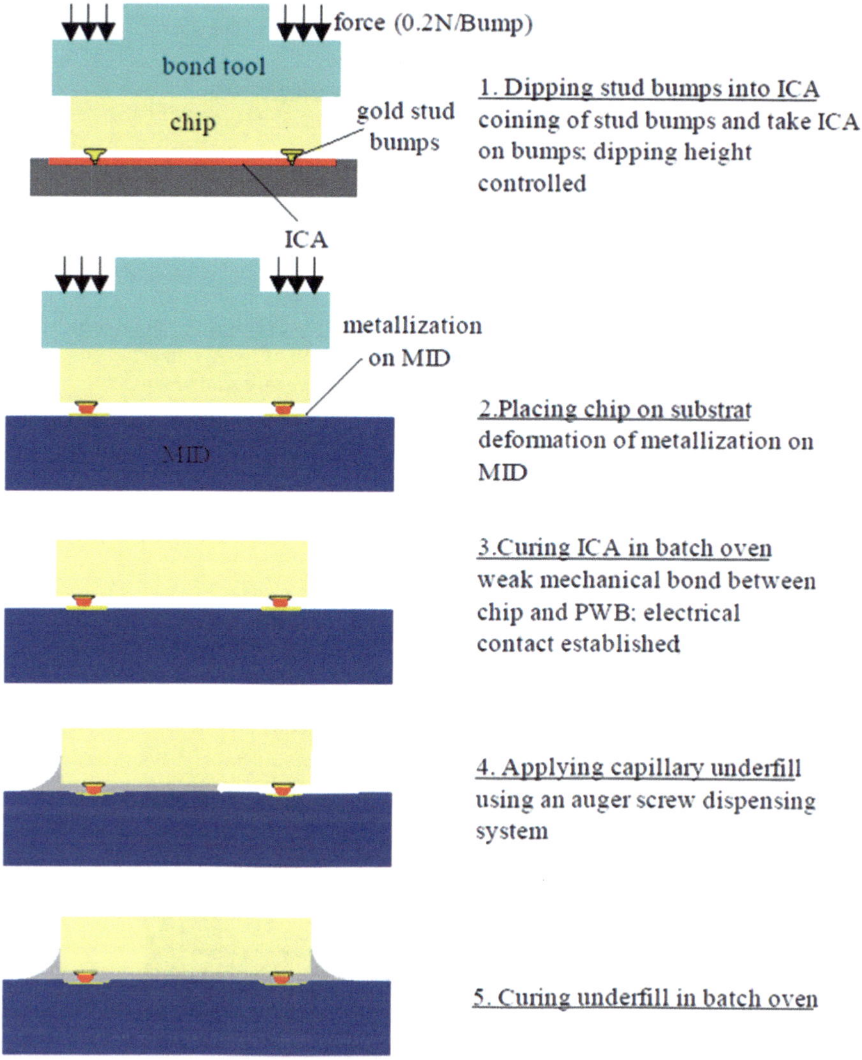

Fig. 13.8 Process flow of isotropic conductive adhesive (ICA) together with capillary underfill

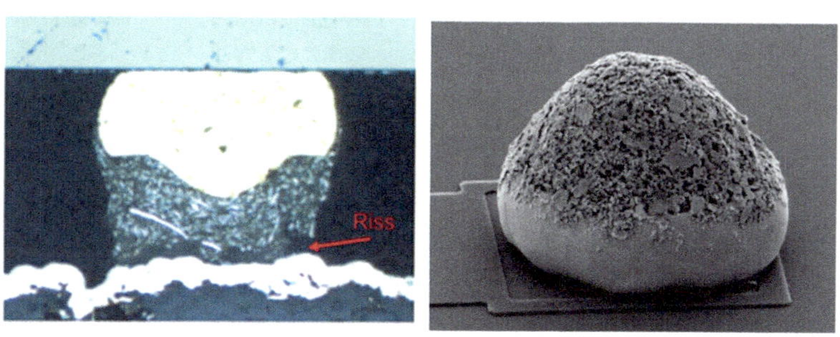

Fig. 13.9 Left, cross-sectional cut of a gold stud bump on a MID substrate using an ICA epoxy material and capillary underfill. The pictures show a crack caused by improper manufacturing parameters. Right, silver flakes of an ICA epoxy material on a gold stud bump

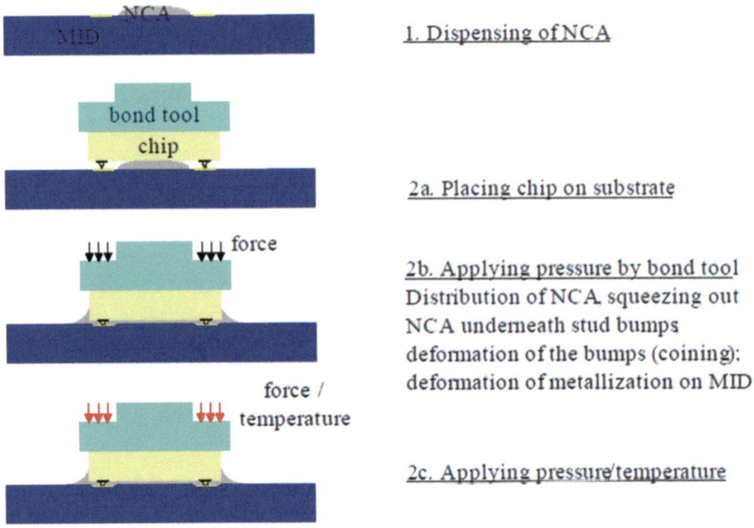

Fig. 13.10 Process flow of nonconductive adhesive (NCA) interconnection

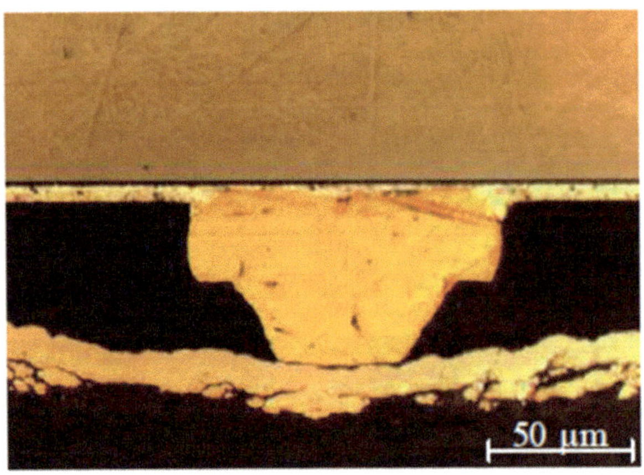

Fig. 13.11 Cross-sectional cut of a gold stud bump on a MID substrate using a NCA epoxy material

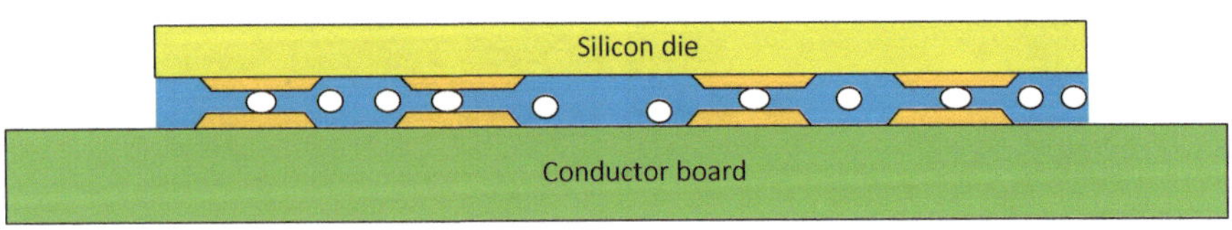

Fig. 13.12 ACA (anisotropic conductive adhesive) interconnection: conductive particles are compressed between the pads on the silicon die and the conductor board and lead to a vertical but no horizontal electrical connection. NCA (blue) builds up a compression stress due to thermal and chemical shrinkage to ensure constant low electrical resistance

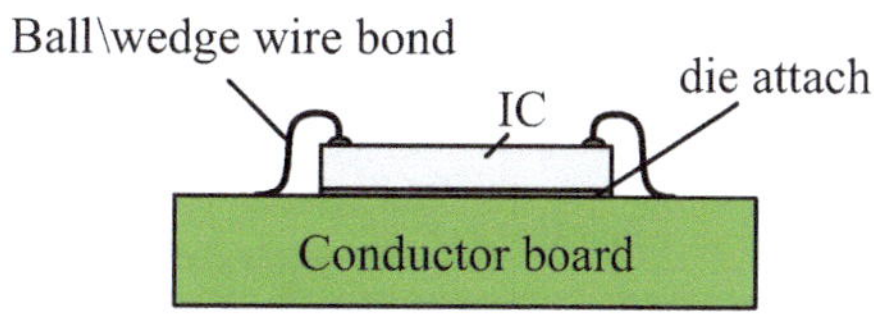

Fig. 13.13 Die attach adhesive: epoxy-based, mostly filled with silver particles, to connect mechanically and thermally the IC with the conductor board. The wire bonds are providing the electrical interconnection

4.2.4 Epoxy-Based Die Attach Adhesive for Silicon Dies with Wire Bonding

Another application how epoxies are used in chip interconnection is the usage as die attach material (Fig. 13.13). Die describes here the IC. The die attach adhesive is dispensed or transferred by stencil printing on the conductor board. The IC is placed into the die attach material and afterward it is thermally cured. The pads to be connected on the IC are on the top side. The electrical connection from IC to conductor board is hereby established by using thin wires made of gold or aluminum, so called wire bonds (mostly ultrasonic bonding). The die attach material has the function of mechanical and thermal connection. In most cases, the die attach materials are filled, similar to ICA, with silver flakes.

4.2.5 Capillary Underfill Dispensing to Protect Soldered or ICA Connected Bumps

The underfill process, simplified in Fig. 13.14, comprises dispensing a thermoset encapsulant on the bottom side of a silicon die or a ball grid array (BGA) device. The encapsulant is responsible for filling, sealing, covering or soaking the compound, protecting the fragile interconnects that are usually located in the underside of the electronic component, enhancing the connection strength of electrical contacts, and compensating for differences in thermal expansion rates of the two joining materials. For proper processing of the encapsulant, the substrate should be kept at an elevated temperature.

Underfill is typically recommended for applications where the electronic component must undergo high shock environments maintaining reliability.

5 Simulation-Driven Design

Development time is one of the major factors deciding about the success of any company. Especially in the automotive industry, in which product safety and reliability is crucial, development and qualification of a new ECU can take several

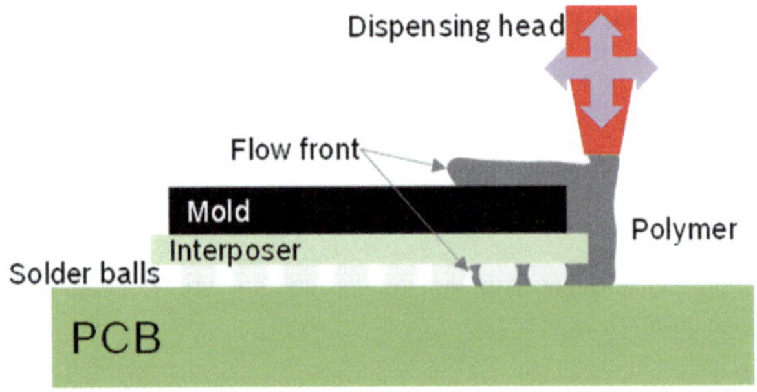

Fig. 13.14 Representation of the dispensing/underfill process

years. By utilizing simulation techniques even before the first prototype is manufactured or first product failure is observed during the testing phase, the true potential of numerical simulation in realizing validated designs is achieved. The obtained advantages in terms of reduced time, cost, and effective design are the essential prerequisites for any organization to remain competitive. The new age product development demands rolling out effective and efficient designs in short time span and reduced costs in the view of increased competition from market players. This requires the time conventionally needed for conceptualization and validation of new designs to be significantly reduced without having to compromise on the quality.

Conventionally, simulation is employed during verification and validation stages of product development, i.e., at the end of the design cycle. By making use of simulation in earlier stages, numerous design-analysis iterations can be performed in a timely manner, thereby zeroing in on the optimum variant economically. Acceleration of the design process of each new electronic control unit or IC packaging can be achieved by utilizing a simulation-driven design concept, e.g., to define the bill of materials before the first prototype is manufactured. The numerical models from simulation serve as good beginning for divergent activities such as sensitivity analysis and generation of concept candidates.

Figure 13.15 shows the direction of workflow process in the simulation-driven design methodology. It consists of four stages that are executed sequentially. These stages address the various concerns during module development, thereby resulting in an optimized ECU and IC packaging. They are described in this section.

5.1 *Virtual Design of Experiments (DoE)*

Virtual design of experiment (virtual DoE) is a technique recommended at the very early design stage to actively support and accelerate product development using numerical simulation. It augments simulation technique with statistical and stochastic methodologies. This combination of methods emphasizes on using simulation as means to execute the designed experiments to determine the correlation between

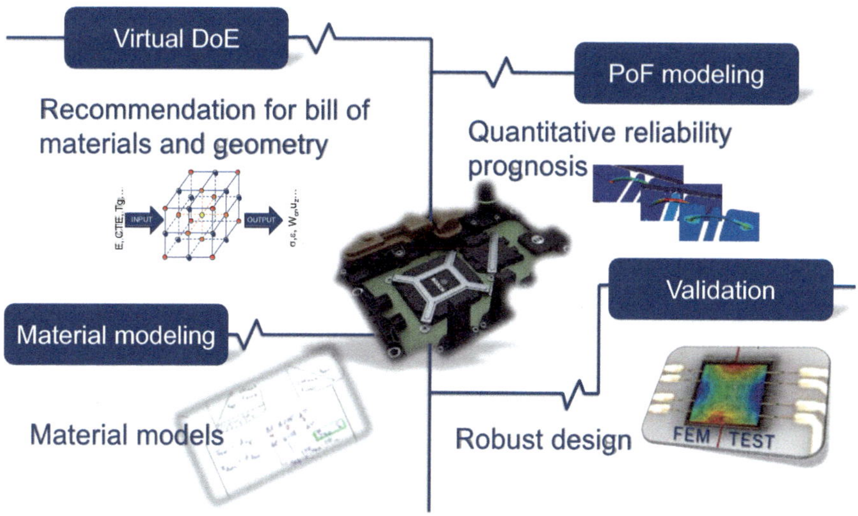

Fig. 13.15 Concept of simulation-driven design

input variables and output responses. It allows for virtual investigation of the design parameters, material properties, or process settings on a deformation and/or stress state. This stage helps in selection of optimum material type or geometric configuration for the chosen module. This concept is demonstrated for a selection of molding compound of an IC package. The important properties of molding compound that influences the stress state in the IC components are estimated.

In case of encapsulate materials such as molding compound, there are four parameters that influence the stress state:

- Modulus of elasticity
- Coefficient of thermal expansion (CTE)
- Glass transition temperature (T_g)
- Curing shrinkage

Each of these properties influences the warpage of the module and, hence, the internal stresses of the critical components such as epoxy-based molding compound, adhesive, wire bond, and solder. These properties are defined as factors/predictors in the design of experiment. The range for each property can be derived based on their availability in the market. Typical values can be found in Table 13.1.

Execution of the virtual DoE is divided into eight steps and requires using software that is capable to define statistically based design of experiment (DoE), perform regression analysis, and finally visualize generated response surface in a graphical manner. These steps are as follows:

(1) Define the parameters of interest that have an effect on response quantities.
(2) Select appropriate DoE model. For continuous variables it is recommended to opt for full factorial design. However, it can be computationally expensive, as it

Table 13.1 Parameters of epoxy-based molding compound available in the market

Parameter	Unit	Min	Max
Modulus of elasticity	[MPa]	10000	35000
Glass transition temperature (T_g)	[°C]	80	200
CTE below T_g	[ppm/K]	6	20
CTE above T_g	[ppm/K]	25	80
Curing shrinkage	[%]	0.0	0.5

generates all possible design trials; hence it is seldom used. In order to account for interaction between variables and their higher-order effects, it is necessary to choose a quadratic model which considers these effects. For continuous input parameters, either fractional factorial plan or central composite design can fulfill the requirements depending on the number of trial runs. In many engineering cases, classical plans such as fractional factorial or central composite cannot be applied. This can be due to the presence of, e.g., categorical parameters or constraints of the input parameters. In this case, it is recommended to use D-optimal plan.

(3) Create a model using selected DoE plan, i.e., to generate design trials. Keep the interaction between parameters and higher-order effects such as the quadratic terms shown in Equation (13.1).

$$\text{Output} = A + B * p_1 + C * p_2 + \ldots M * p_1 * p_2 + N * p_1 * p_3 + \ldots O * (p_1)^2 + P * (p_2)^2 \ldots \quad (13.1)$$

where:

$A, B, C, \ldots$ are the constants of the regression function
$p_1, p_2, p_3, \ldots$ are the input parameters

(4) Define the responses such as deformation, stress, strain, and energy. The response can be checked, e.g., in the enter material or in specific areas.
(5) Create a FEM model and execute simulation.
(6) Import results of the simulation into DoE software.

In the next step, regression analysis is conducted. The goal of the regression analysis is to create a mathematical function that describes the relationship between input parameters and one or more responses. This mathematical function has to be derived for each of the investigated response. This is usually achieved on the basis of the best fit of the design trial simulation results for each response. Finding the equation is done by minimizing the sum or squared values of residuals as shown in Fig. 13.16.

The quality of the regression model is assessed by:

(a) Checking the root mean square error of the model.
(b) Checking the residuals and search for any trends: Trends can indicate whether the model does adequately explain the variability of the data.

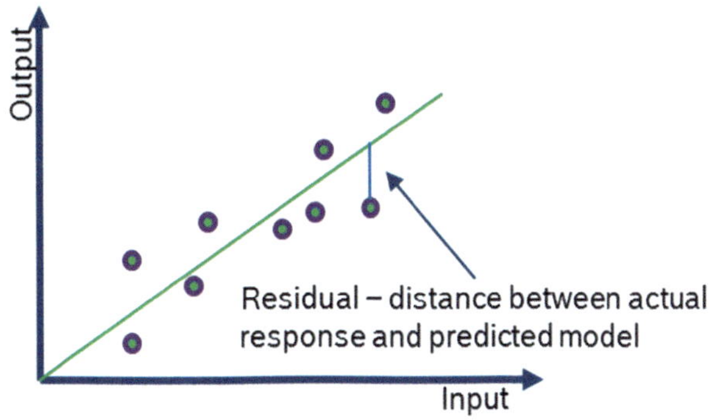

Fig. 13.16 Visualization of the linear regression function output = A+ B*(input)

(c) Excluding outliers.
(d) Checking and excluding input parameters that don't influence response.

In the last step, optimization of the input parameters is carried out. From the results of VDoE, it is possible to define value of properties of molding compound, at which the stress and/or deformation state will be minimal. This critical information forms the baseline for finalizing the material properties of molding compound by the product process team. They consult the material supplier and negotiate toward tailoring properties of the molding compound for specific product requirements. Figure 13.17 shows schematically process flow of virtual DoE used to select the optimal molding compound for designed IC package. Hence, the methodology supports selection of appropriate material within a range for a given application by combining finite element and statistical techniques.

5.1.1 Sensitivity Analysis of ASIC in QFN Housing

Variation in internal parameters associated with the QFN, i.e., its dimension and materials of its components, affects the stresses under the influence of external conditions such as temperature and/or mechanical loads. Hence, it is important to understand the effect of internal parameters on the local stresses generated inside the QFN package. This study is carried out with the help of sensitivity analysis using simulation-based virtual design of experiments (VDoE).

The virtual design of experiments in this study focuses on four regions of possible reliability concerns – (1) top surface of silicon die, (2) interface of EMC and leadframe, (3) bottom layer of adhesive between the silicon die and leadframe, and (4) bulk of solder joints. According to the relevant components of stress tensor for each of these regions, 12 most important output parameters are selected and presented in Table 13.4. Independent parameters or input parameters are broadly classified as material properties and geometrical dimensions of QFN. Individual virtual design of experiments was carried out using only material properties as independent

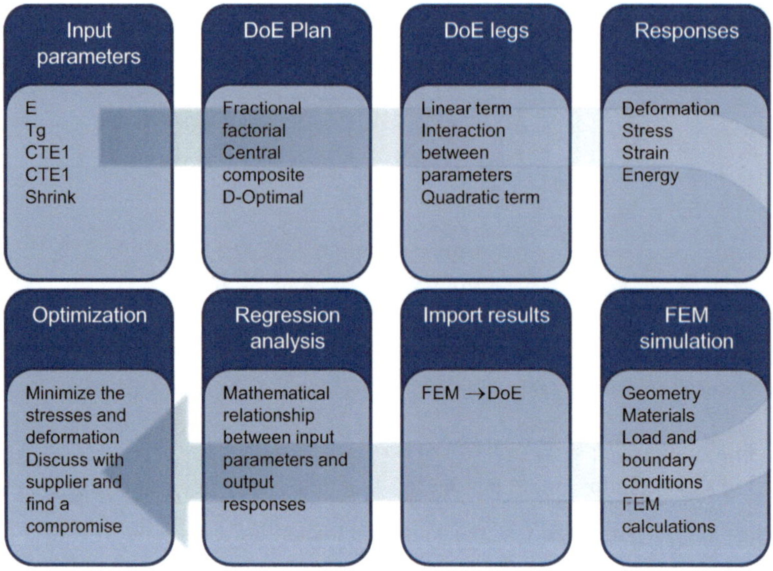

Fig. 13.17 Process flow of virtual DoE based on the selection of molding compound for IC package

variables (Material VDoE), using only geometrical parameters as independent variables (Geometry VDoE), and considering both as varying parameters (Combined VDoE).

Virtual design of experiments typically consists of large number of simulation-based experiments (in multiples of 100), utilizing different permutations and combinations of the input parameters, which are varied within a specified range. For this study, the finite element simulations were carried out in ANSYS Workbench environment with the optiSLang plug-in for sensitivity analysis.

The geometry comprises a single QFN package along with its assembly on a PCB with solder joints. Since the QFN has two planes of symmetry, a quarter model was prepared, which is shown in Fig. 13.18. This gives an advantage of significant reduction of the number of nodes. The fully parametric geometry was created in SpaceClaim environment using Python scripting. Table 13.2 includes the list of parameterized geometry variables of the QFN model along with their nominal values. They are also indicated in Fig. 13.19. The minimum and maximum values for each parameter (a to g) were set to +/−10% of respective nominal values. Since a quarter model is used, the position of the chip is kept fixed aligned to the center of the leadframe. Therefore, the parameters h and i are defined as constants in the sensitivity block of VDoE.

Linear elastic temperature-dependent material models were used for all materials within QFN package, in order to reduce the complexity of VDoE. A linear elastic orthotropic model was used for solder material, while the PCB was modeled as a single substrate material with equivalent linear elastic properties. Mechanical

Table 13.2 Geometry parameters of VDoE

Label	Parameter name	Nominal value
a	ChipHeight	0.28 mm
b	ChipWidth	3.55 mm
c	ChipLength	2.90 mm
d	ChipAttachHeight	0.03 mm
e	SolderStandOffAvgHeight	0.07 mm
f	ChipAttachMeniscusDie	0.10 mm
g	ChipAttachMeniscusLeadframe	0.10 mm
h	ChipPosOffsetX	0 mm
i	ChipPosOffsetY	0 mm

Table 13.3 Material parameters for VDoE

Nr	Parameter name	Nominal value
1	DieAttach_E1	7630 MPa
2	DieAttach_E2	176 MPa
3	DieAttach_Nu1	0.25
4	DieAttach_Nu2	0.45
5	DieAttach_CTE1	45 ppm/K
6	DieAttach_CTE2	169 ppm/K
7	DieAttach_Tg	–10 °C
8	EMC_E1	28140 MPa
9	EMC_E2	1000 MPa
10	EMC_Nu1	0.25
11	EMC_Nu2	0.45
12	EMC_CTE1	8.2 ppm/K
13	EMC_CTE2	32 ppm/K
14	EMC_Tg	130 °C

properties of the molding compound and adhesive materials were defined with the help of seven parameterized constants – E_1, E_2, ν_1, ν_2, CTE_1, CTE_2, and T_g. The nominal values of these parameterized properties are indicated in Table 13.3, as they are used as input parameters for Material VDoE. The range of variation for each parameter was defined by +/−10% of their nominal values.

Thermal condition has been defined using a single step drop from 180 °C to −40 °C with a cooling rate of 6 K/min. Since all of the material models are linear elastic, the cooling rate and number of steps do not affect the results. Therefore, the settings are optimized for minimal computation time. The reference temperature (T_{ref}) for all materials is set to 180 °C, and the results are evaluated at −40 °C. Recording of the output parameters, listed in Table 13.4, is automated using an APDL script using the specially defined named selections, viz., Chip-TopEval, DieAttachBottomEval, EMCLeadframeEval, and SolderJoints.

Table 13.4 Output parameters and coefficient of prognosis

Nr	Output parameter	CoP (%)
1	ChipTopEval_MAX_S1	85
2	ChipTopEval_MAX_SXXmSYY	95
3	ChipTopEval_MAX_SXY	96
4	ChipTopEval_MIN_S3	72
5	ChipTopEval_MIN_SXXmSYY	77
6	ChipTopEval_MIN_SXY	77
7	DieAttachBottom_MAX_S1	98
8	DieAttachBottom_MAX_SZ	91
9	EMCLeadFrame_MAX_S1	35
10	EMCLeadFrame_MAX_SZ	98
11	SolderJoints_MAX_S1	99
12	SolderJoints_MAX_SEQV	97

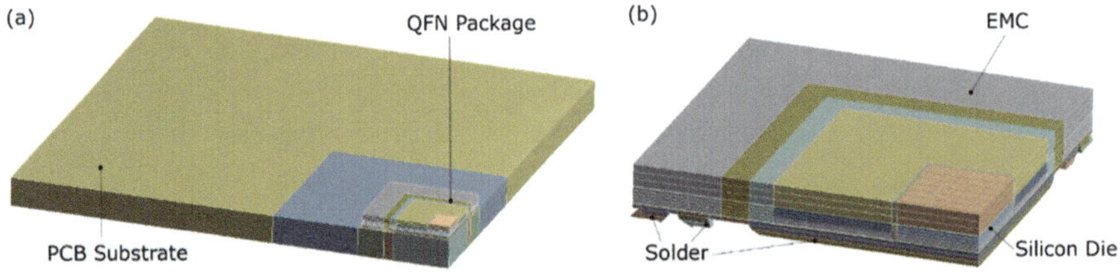

Fig. 13.18 Quarter model of (**a**) QFN-on-PCB assembly and (**b**) QFN package

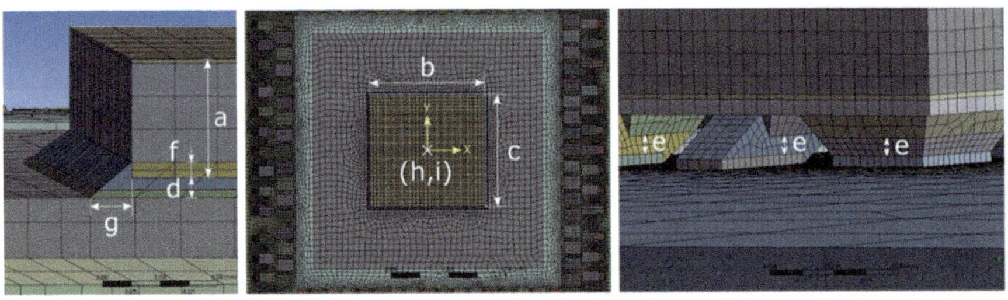

Fig. 13.19 Parameterized geometry of QFN on PCB assembly

Sensitivity analysis is carried out using the optiSLang software. There are two ways to use the sensitivity tool in optiSLang: one is by assigning ANSYS simulation as a solver in a sensitivity block within optiSLang environment and the other by integrating optiSLang sensitivity block into ANSYS Workbench environment. The second option has been utilized for this study, and the structure of the project is indicated in Fig. 13.20.

The structure shows a block for the parameter set, which contains all of the parameterized input parameters, geometrical and/or material, as well as the output

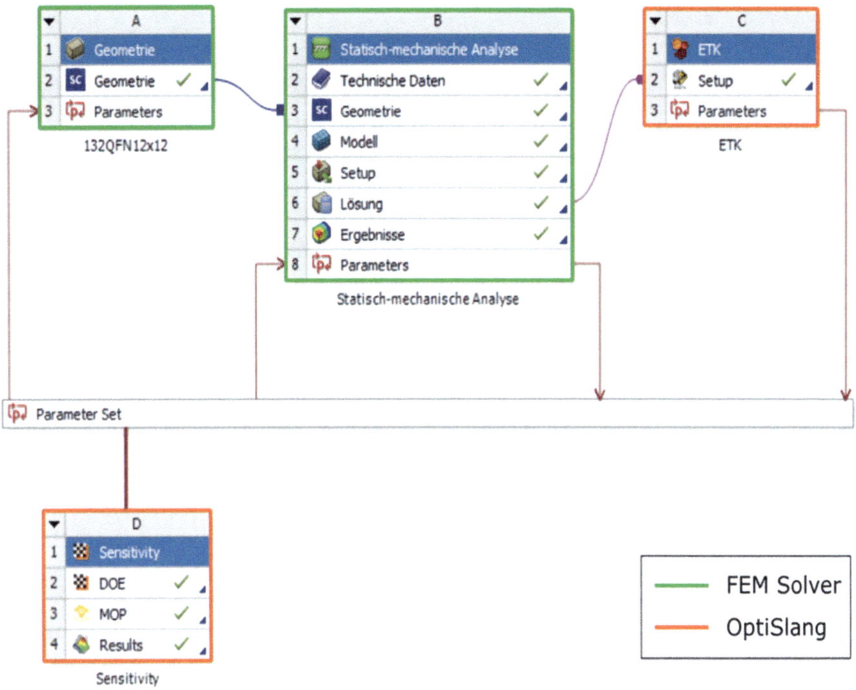

Fig. 13.20 Structure of VDoE

parameters extracted from evaluated results. The sensitivity block handles the input parameters. The nominal values and range of the input parameters are defined using this block.

Geometry parameters in Table 13.2 along with the material parameters in Table 13.3 contribute to the input settings used for the Combined VDoE, which utilizes Advanced Latin Hypercube Sampling algorithm. This algorithm selects 200 different combinations of input parameter values such that they are well-distributed inside the hypercube, which is defined by the lower and upper boundaries of each parameter.

At the beginning of each simulation of a VDoE, the sensitivity block creates a set of values for all input parameters and feeds it to the parameter set. These values are then picked up by the ANSYS solver for generating solution using finite element simulation. The extraction tool kit (ETK) block is used for the extraction of results or output parameters from the ANSYS solution file. The ETK tool allows a simple and convenient extraction and processing of simulation results for an optimization or stochastic analysis. The ETK reads the "solve.out" file after each simulation. Specific user-defined markers were created for the extraction of MAX and MIN values of the output parameters, mentioned in Table 13.4, at all of the four investigated regions.

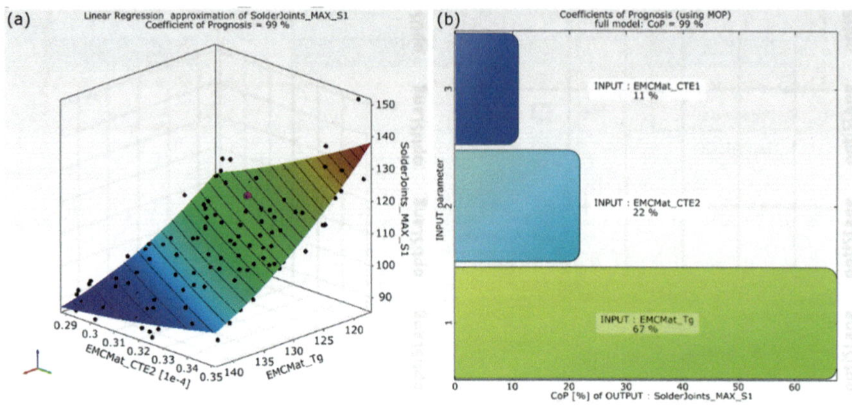

Fig. 13.21 Sensitivity of SolderJoints_MAX_S1: (**a**) metamodel; (**b**) breakup of COP

optiSLang extracts and deals with only values of the results, without their units. They are then passed on to and stored in the parameter set. The sensitivity block accesses the parameter set after all of the simulations are finished, for generating metamodel of optimal prognosis (MOP).

The sensitivity block generated a metamodel of optimal prognosis (MOP) based on the data of input and output parameters generated from the VDoE simulations. The metamodel is a mathematical model fit into the 200 design points obtained from VDoE, using internal algorithms of optiSLang. The fitted curve can be visualized for each output parameter against two input parameters at a time, an example of which is presented in Figure 13.21(a). Based on the quality of the fit, a parameter called coefficient of prognosis (COP) is assigned to each of the output parameters on the scale of percentage. For this study, the value of COP was further categorized into poor (COP <50), moderate ($50 \leq$ COP <70), and good (COP ≥ 70). The COP of a particular output parameter can be further visualized for the contributions of input parameters. Figure 13.21b shows the breakup of COP for output parameter SolderJoints_MAX_S1.

The COP matrix obtained from the final VDoE is presented in Fig. 13.22. Twelve most important parameters are considered. They are indicated on vertical axis and also listed in Table 13.4. The horizontal axis shows top 16 input parameters, which have most contribution to COP of output parameters. Good COP values for all output parameters, except one, are observed. EMCLeadFrame_MAX_S1 (σ_1), which depends on the normal as well as shear components of stresses, has only 35% COP. Since corresponding normal stress σ_{zz} shows a good COP of 98%, it was concluded that the shear component would be responsible for a poor COP of σ_1.

While analyzing the results, it is important to note the effect of nonassociated input parameter on a certain output parameter. For example, the stresses on the ChipTopEval layer are expected to be sensitive to the input parameters associated with the chip, such as ChipLength and ChipWidth. The more relevant result was a

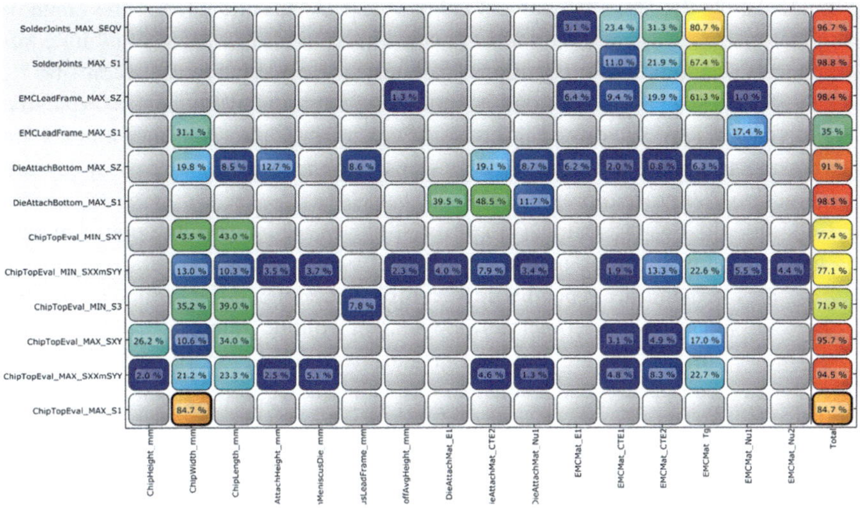

Fig. 13.22 COP matrix from VDoE

nonassociated parameter like EMC_Tg that affects the stresses at ChipTopEval. Similar observations were noted for other output parameters.

In conclusion, the evaluated stresses within the package are found to be most sensitive to the following five parameters – EMC_Tg, EMC_CTE1, EMC_CTE2, ChipLength, and ChipWidth. Hence, it is very important to characterize the material properties of EMC with highest possible accuracy. Similarly, it is crucial to control the dimensions of silicon die during its manufacturing process.

5.2 *Material Characterization*

Numerical simulations are one of the most important tools in predicting the reliability performance of automotive electronics. As mentioned in this chapter, most commonly epoxy-based thermosets are very often used as an encapsulate material instead of more expensive steel housing. In order to be able to quantitatively predict the stress state in the plastic, detailed material characterization and modeling is required. Material characterization of epoxy-based thermosets requires extensive testing [2, 8]. Tests are typically done for pristine as well aged material. In order to describe material behavior for predictive modeling, the following physical parameters are typically measured:

- Modulus of elasticity "E" – is a measure of stiffness of the material exposed. It describes the elastic properties of epoxy-based thermosets. In most of the cases, these physical properties are time and temperature dependent in the application range. It requires detailed characterization. In order to measure longitudinal modulus, the dynamic mechanical analysis (DMA) is conducted in the range of the temperature where material is expected to be used. In case of molding compounds for automotive applications, such tests are usually done between −60 °C and 250 °C. The harmonic load is applied by varying the frequency. Typically, material tests are performed in-house. It is very important that the samples are prepared following the same process steps, such as curing and post mold cure temperature, time, and pressure.

- Bulk modulus "K" – is a measure of resistivity of the material to applied hydrostatic pressure. It can be measured using pressure-volume-temperature apparatus. Measurements are typically done at different temperatures, which allows for the assessment of the temperature-dependent volume change of epoxy-based thermosets like molding compound or adhesives. This property is time dependent, so creep or relaxation test is recommended to get proper physical behavior.
- Coefficient of thermal expansion "CTE" – describes the change of the object size with a change of the temperature. Typically, it is measured using thermal mechanical analyzer (TMA). Another very common way is warpage measurement of a bi-material strip of the material, e.g., molding compound on copper substrate and FEM simulation. Using reverse engineering and optimization strategies, a CTE of molding compound can be optimized. This is one of the most important parameters that influence the deformation and the stress state in epoxy-based thermoset or at the interfaces.
- Cure kinetics – is investigated to obtain the conversion level as a function of time and temperature. Differential scanning calorimetry is used. Typically, at first, the uncured samples are subjected to a series of different heating rate tests (e.g., at 1, 2, 5, 10, and 15 °C/min), and the released reaction heat is recorded as a function of temperature. In the second stage, a series of isothermal cure experiments is performed as a consistency check.
- Tensile strength – New wide bandgap semiconductors and new die attach materials such as silver sintering lead to significantly higher strain rates and strain levels, to which epoxy-based thermosets were never exposed before. Epoxy-based thermosets such as molding compounds or adhesives now are used in the regime of nonlinear stress/strain [3]. Uniaxial tensile test until failure with different strain rates allows to get entire stress/stress curve. The experimental data is used to build nonlinear viscoplastic material models. These are capable of representing mechanical behavior under large strain regime. It is especially important for a new group of polymers used in the automotive industry, for example, power modules for hybrid and fully electric vehicles.
- Toughness – describes the ability of a material to absorb energy before fracture. Fracture parameters such as rupture strength and critical energy release rate are used to model crack initiation and propagation for thermo-mechanically driven cracks. Investigation is made for the bulk of the epoxy-based material, as well as for the interfaces between substrate and epoxy-based materials.

– Abovementioned properties allowed to build entire set of material constants and implement it as FEM material input deck. Material characterization and modeling is a first and mandatory step for quantitative FEM simulation, based on which reliability prediction can be done.

5.3 Physics of Failure Simulation

After selection of the epoxy-based thermoset materials based on the virtual DoE, and detailed characterization, the reliability of the product can be virtually tested by means of FEM numerical simulation. Here, to obtain quantitative results, still a couple of conditions must be fulfilled. The process flow of proposed physics of failure simulation is shown in Fig. 13.23. First of all a detailed geometry of all design elements has to be taken into account. The detailed geometry of any component is typically validated by cross sectioning, X-ray inspection, or etching of the epoxy-based molding compound. Next, all the material properties must be assessed as described before. Finally, the boundary and load conditions must be well defined.

For accurate prediction of thermomechanical behavior under complex loading conditions, such as active power condition, the power dissipation in all design elements is taken into account. Thus, the simulation workflow starts with the electrical domain. The result from this step is taken as an input into thermal domain to obtain the temperature distribution within the electronic control unit (ECU).

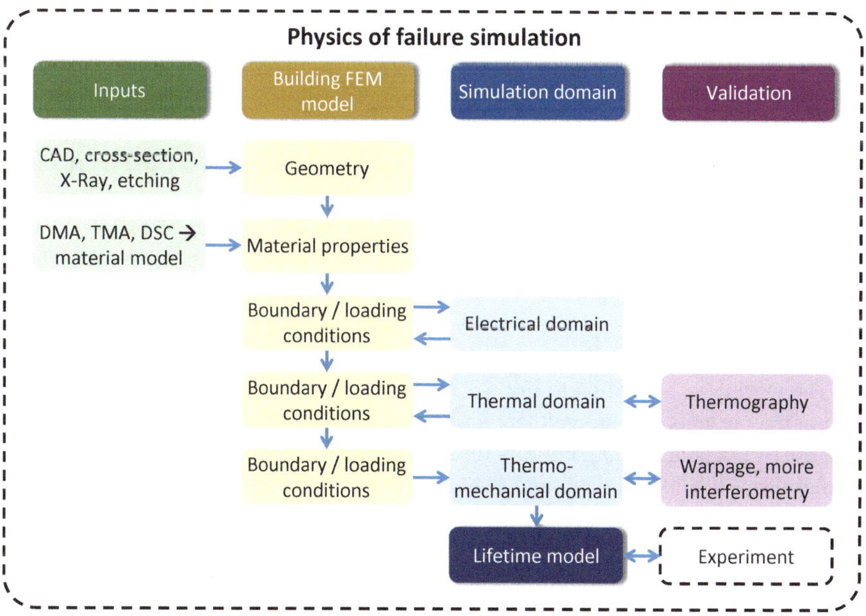

Fig. 13.23 Physics of failure simulation workflow

Next, the validated temperature distribution [5–7, 9] is used as an input to thermomechanical simulation. Based on the predictions of thermomechanical model, the stress/strain analysis of the ECU under real working conditions is performed. That enables a risk assessment for different failure modes. Additionally, a physics of failure-based lifetime model for a given failure can be built [21–27].

5.4 Validation

The last step of the simulation-driven design is validation. At this point, the virtually optimized ECU is built and tested. The experimental results are used to verify the numerical predictions. Depending on the simulation type, different experimental methods can be used. For validation of thermal models, the methods like thermography or temperature (e.g., using thermocouples) or thermal impedance measurements can be used. Also, a variety of methods is available to validate thermomechanical simulations. Most commonly used are optical methods that measure deformations, like digital image correlation for warpage measurements or moiré interferometry for in-plane displacement fields. Other methods successfully applied include stress and strain measurements [4, 10, 11].

6 Degradation and Remaining Useful Life Prediction

In the previous sections, different numerical and analytical models are predicting the degradation during lifetime. However, these predictions are rather statistical based on some a priori assumptions. Monitoring strategies can cope with this limitation and improve the existing methods. One example in the semiconductor packages is the online measurements of the mechanical stresses induced by the thermomechanical loads.

The coefficient of thermal expansion mismatch results in mechanical stresses between different material layers. Therefore, having the possibility to measure these stresses is an ideal situation for predicting the state of health of the package. These measurements are possible with CMOS (complementary metal-oxide semiconductor)-based stress sensor.

The workflow, according to Fig. 13.24, starts with the stress sensor data by measuring the currents and the voltage, based on which the stresses are calculated (Eq. (13.2)). Stress data is going initially through a preprocessing algorithm that removes outliers, smoothens the data, and extracts the relative stresses caused by the temperature difference. A damage index or a degradation estimation parameter is evaluated using either statistical models or machine learning models based on the stress data, followed by the Remaining Useful Life (RUL) estimation by extrapolating from the damage index or the degradation parameter.

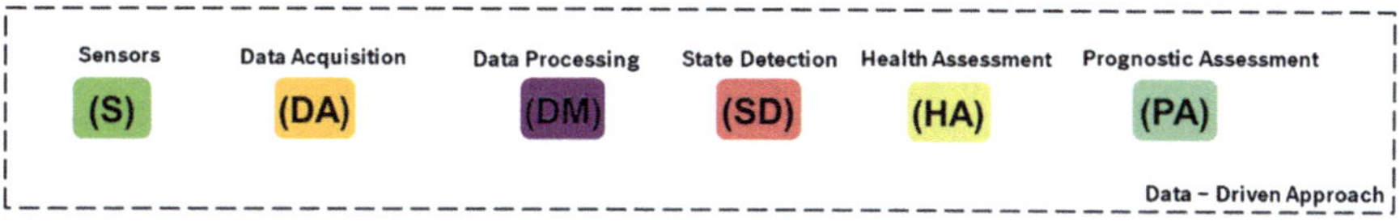

Fig. 13.24 Data-driven approach workflow

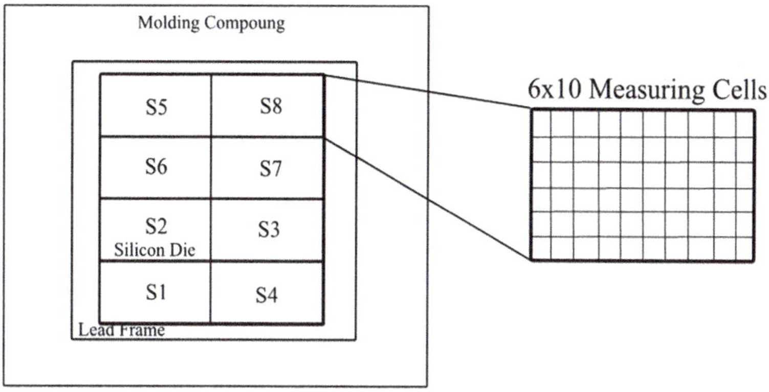

Fig. 13.25 Thin quad flat package stress sensor schematic

$$\sigma_{11} - \sigma_{22} = D(\sigma) \approx -\frac{2}{\pi_{44}^{p}(T)} \left(\frac{I_D|_{0^\circ} - I_D|_{90^\circ}}{I_D|_{0^\circ} + I_D|_{90^\circ}} \right) \tag{13.2}$$

The silicon-based stress sensor is packaged in a TQFP, resulting in replacing the functional die. The entire silicon die top surface consists of a 24 by 20 array of mechanical stress measuring cells, as depicted in Figure 13.25. These packages are mounted on a PCB to facilitate the measurement of the stress signal. The PCB pins are soldered with wires to a dedicated Acquisition Unit (AU). AU is required to steer the sensor and successfully record the data. Also, due to the necessity to cope with the large amount of data, size scaling, data remote access, computing power, and long running experiments, further improvements were required. Such system is depicted in Fig. 13.26.

The stress data is continuously recorded during Temperature Shock (TS) cycling of temperatures from −40 °C to 150 °C. The data from the package stress sensor is gathered with the AU by a cable connection, which is placed outside the chamber. Then the AU is sending the data to the Wi-Fi console, followed by the Raspberry Pi to collect this data from the console and save it. A simple graphical interface is used in Raspberry Pi to check the status of the experiments, whereas the stresses can be visualized in real time.

The raw data gathered on the experiment is composed of the high precision physical measures of voltages and currents from each individual cell present on the package and the times in which the measurements are being performed. By themselves,

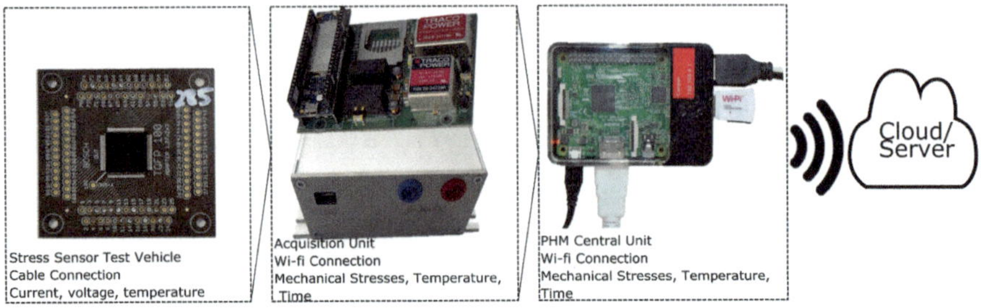

Fig. 13.26 Data flow during measurement

these measures give no valuable insight into the matter of this study. Therefore, data is processed, and only relative stress data is kept, as shown in Eq. (13.3).

For RUL estimations, a condition indicator is desired, whose change over time is observable and connected with the system degradation process in a reliable, measurable way. The RUL of the component is the expected life or usage time remaining before the component fails to fulfill the requirements. A prognostic metric is needed to incorporate the stress data of all cells. It is determined using the normal stress difference from all 480 cells of a sensor as:

$$\sigma^* = \frac{\sum_{j=1}^{N} \left| \left(\sigma_{xx} - \sigma_{yy}\right)_{T=-40^\circ C}^{j} - \left(\sigma_{xx} - \sigma_{yy}\right)_{T=150C}^{j} \right|}{N} \tag{13.3}$$

The evolution of such prognostic metric during TS cycles is shown in Figure 13.27. Sudden drops or sudden increases in stress indicate delamination initiation underneath that particular sensor. The stress rate decreases or increases as a result of the growing crack; since the crack length increases continuously, the stiffness of the interface decreases continuously as well.

Based on the SAM analysis, it is confirmed that the stress sensor is clearly able to detect crack initiation and propagation in a quantitative way. The prognostic metric stress evolution from the non-delaminated stage till the delaminated stage of one package is depicted in Fig. 13.27. It is clear that the healthy stage is given by almost constant stress behavior. Based on the SAM images and the prognostic metric values, three stages are identified:

- First stage is where the stress is constant. No delamination is observed in the SAM image.
- Second stage is where the stress drops in value with high rate. The SAM image confirms that the crack/delamination is initiated outside of the silicon die.
- Third stage is where the stress indicates an increase in value but in a low rate. The SAM image confirms the crack/delamination propagation underneath the silicon die.

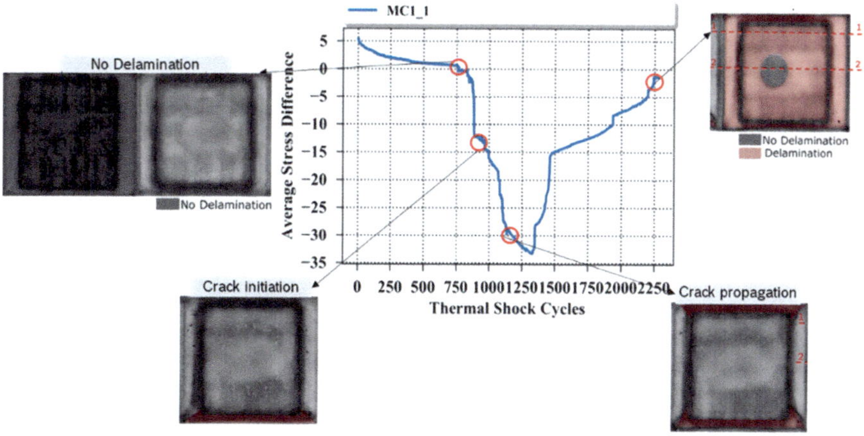

Fig. 13.27 Prognostic metric behavior from healthy until full delamination for module MC1_1

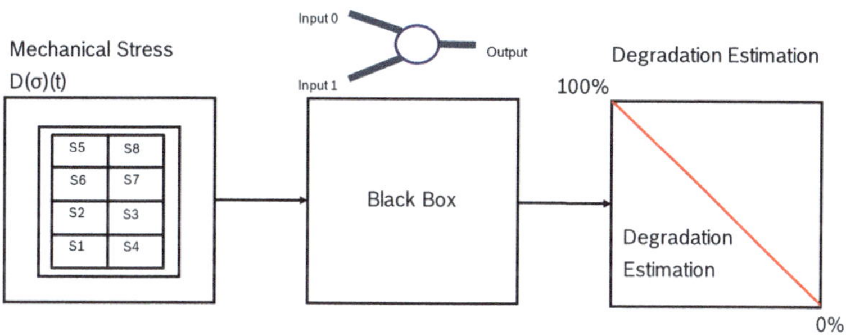

Fig. 13.28 Degradation prediction with black box concept

A black box concept is used to estimate the degradation of the package based on the stress values (Fig. 13.28). The stress values are considered as the input for any machine learning model (e.g., regression analysis), and the output is a linear degradation estimation. Different machine learning models such as artificial neural networks, support vector machine, and random forest can be used to model the degradation. In addition, different fusion or ensemble methods can be also implemented.

Neural network method is used for degradation estimation and prediction of the package based on incremental stress values. For the degradation estimation, the stress data (input) is transformed into a linear function from 0% to 100% (output). The result gives the degradation estimation of the package, from which the RUL can be extrapolated according to Eq. (13.4):

$$RUL_n = n \times \left[\frac{1}{\left(1 - \frac{dD_i}{D_0}\right)} \right]^{\alpha} \tag{13.4}$$

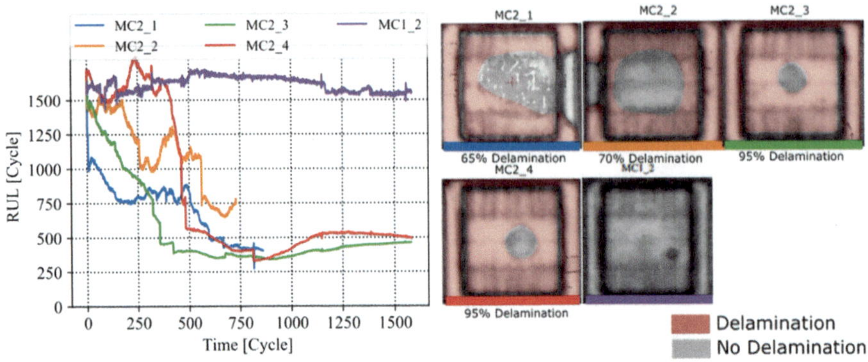

Fig. 13.29 Remaining useful life estimation and prediction. SAM images at the end of the experiment

where n is the number of cycle, D is the degradation predicted by machine learning algorithm, and α is the exponent fitted to the similar data.

RUL in cycles for all the testing modules are shown in Fig. 13.29. Comparing the results from the RUL prediction and the SAM images from Fig. 13.29, the following can be stated:

- For modules MC2_1 and MC2_2, the delamination is not complete. This is validated by the SAM images.
- For modules MC2_3 and MC2_4, the model predicts an almost full delamination close to few percentages. This is also confirmed by the SAM images.
- For MC1_2 no degradation is predicted. SAM image is showing no delamination.

In order to improve the RUL prediction, more data is needed to train the model for damage prediction. The more stress data is acquired, the more robust the model will be. In this study the similarity method to extrapolate the RUL is used. The end of life of the MC1_1 module is considered when the packages reach full delamination.

7 Conclusions

Epoxy-based thermosets in automotive electronics are used for many years now. Three major trends [1] in automotive sector, such as electrification, connectivity, and electrification, will significantly increase its amount. Especially electrification will introduce new scope of application of automotive electronics. Epoxy-based thermoset materials will be used more and more often as a housing to protect entire electronics systems from harsh environment and provide hermeticity. In the development process of such a complex system, we recommend the following steps to be completed:

- Materials selection of epoxy-based thermoset should be supported by virtual DoE techniques in order to obtain qualitatively reliable system at the very early design stage.
- Epoxy-based thermosets used in the automotive products should be characterized, and their behavior should be described by proper material models.
- The assembly process is required to include the effect of residual stresses during production process.
- Multi-domain loading conditions are required to describe the inhomogeneous temperature distribution during the temperature cycling. This is achieved by coupling electrothermal/thermomechanical domain and transfer results of each domain as an input load in the next simulation.
- Sub-modeling technique is required to model the stress state in a critical design element and estimate the lifetime of the component of interest.
- Results of physics of failure simulation combined with end of life experiments are used to create a lifetime model for known failure modes. Based on this reliability prediction is possible.
- Derivation of the aging laws also ensures suitable testing conditions to avoid over design as well as field failures.
- Certain failure modes, such as delamination in IC packaging, can be detected using data-driven approach, utilizing piezoresistive stress sensor.

Acknowledgments This work was funded by the Federal Government of Germany, Framework Program for Research and Innovation 2016–2020 Microelectronics – Innovation Drivers of Digitalization (Rahmenprogramm der Bundesregierung für Forschung und Innovation 2016–2020 Mikroelektronik aus Deutschland – Innovationstreiber der Digitalisierung) – Project smartSTAR (No. 16ES0965). This funding is gratefully appreciated. Many thanks to all members of the project team for the good cooperation.

References

1. J. Dokic, B. Müller, G. Meyer, European roadmap: Smart systems for automated driving, in *European Technology Platform for Smart System Integration*, (2015)
2. P. Gromala, B. Muthuraman, B. Ozturk, H. Yan, K.M.B. Jansen, L. Ernst, Simulation driven design of novel integrated circuits – Part 2: Constitutive material modeling of thermosets, in *Electronic Components and Technology Conference (ECTC), 2015 IEEE 65th*, (26–29 May 2015), pp. 1894–1900
3. P. Gromala, B. Muthuraman, B. Öztürk, K.M.B. Jansen, L. Ernst, Material characterization and nonlinear viscoelastic modelling of epoxy-based thermosets for automotive application, in *16th International Conference on Thermal, Mechanical and Multi-Physics Simulation and Experiments in Microelectronics and Microsystems, EuroSimE 2015*, (Budapest, Hungary, 19–22 April 2015)
4. D.S. Kim, B. Han, A. Yadur, P.J. Gromala, Electronic control package model calibration using moiré interferometry, in *Thermal, mechanical and multi-physics simulation and experiments in microelectronics and microsystems (EuroSimE), 2014 15th international conference on*, (7–9 April 2014), pp. 1–5

5. M. Riccio, A. Irace, G. Breglio, P. Spirito, V. Kosel, M. Glavanovics, A. Satka, Thermal simulation and ultrafast IR temperature mapping of a Smart Power Switch for automotive applications, in *Power Semiconductor Devices & IC's, 2009. ISPSD 2009. 21st International Symposium on*, (14–18 June 2009), pp. 200–203
6. X. Jorda, M. Vellvehi, X. Perpiny, J.L. Galvez, P. Godignon, Validation of dynamic thermal simulations of power assemblies using a thermal test chip, in *Thermal, Mechanical and Multi-Physics Simulation Experiments in Microelectronics and Micro-Systems, 2007. EuroSimE 2007. International Conference on*, (16–18 April 2007), pp. 1–6
7. L. Dupont, S. Lefebvre, Z. Khatir, FAUGIERE, J. C., Power cycling test circuit for thermal fatigue resistance analysis of solder joints in IGBT, in *10th European Conference on Power Electronics and Applications*, (January 2003), p. Sp
8. P. Gromala, J. Duerr, M. Dressler, K.M.B. Jansen, M. Hawryluk, J. de Vreugd, Comprehensive material characterization and method of its validation by means of FEM simulation, in *Thermal, Mechanical and Multi-Physics Simulation and Experiments in Microelectronics and Microsystems (EuroSimE), 2011 12th International Conference on*, (Moire, 18–20 April 2011), pp. 1/8–8/8
9. B. Wu, D.S. Kim, B. Han, A. Palczynska, P. Gromala, Thermal deformation analysis of automotive electronic control units subjected to passive and active thermal conditions, in *Thermal, Mechanical and Multi-Physics Simulation and Experiments in Microelectronics and Microsystems (EuroSimE), 2015 16th International Conference on*, (19–22 April 2015), pp. 1–6
10. P. Gromala, S. Fischer, T. Zoller, A. Andreescu, J. Duerr, M. Rapp, J. Wilde, Internal stress state measurements of the large molded electronic control units, in *14th International Conference on Thermal, Mechanical and Multi-Physics Simulation and Experiments in Microelectronics and Microsystems, EuroSimE 2013*, (Wroclaw, Poland, 15–17 April 2013)
11. N. Schafet, C. Lemm, U. Becker, H. Guttler, P. Schmid, Development of a submodel technique for the simulation of solder joint fatigue of electronic devices mounted within an assembled ECU, in *Thermal, Mechanical and Multi-Physics simulation and Experiments in Microelectronics and Microsystems, 2009. EuroSimE 2009. 10th International Conference on*, (26–29 April 2009), pp. 1–8
12. BS 5760 Part 5, Reliability of systems, equipment and components. Guide to failure modes, effects and criticality analysis (FMEA and FMECA). (1991)
13. A. Goel, R.J. Graves, Using failure mode effect analysis to increase electronic systems reliability, in *Electronics Technology, 30th International Spring Seminar on*, (9–13 May 2007), pp. 128–133
14. H. Ljazzar, M. Fischer, L. Grunske, M. Kuntz, F. Leitner-Fischer, S. Leue, Safety analysis of an airbag system using probabilistic FMEA and probabilistic counterexamples, in *Quantitative Evaluation of Systems, 2009. QEST'09. Sixth International Conference on the*, (13–16 September 2009), pp. 299–308
15. X.P. Zhu, L.B. Zhang, W. Liang, G.N. Shi, A quantitative comprehensive safety evaluation method for centrifugal compressors using FMEA-fuzzy operations, in *Instrumentation and Measurement, Sensor Network and Automation (IMSNA), 2013 2nd International Symposium on*, (23–24 December 2013), pp. 202–206
16. M. Hillenbrand, M. Heinz, N. Adler, J. Matheis, K.D. Muller-Glaser, Failure mode and effect analysis based on electric and electronic architectures of vehicles to support the safety lifecycle ISO/DIS 26262, in *Rapid System Prototyping (RSP), 2010 21st IEEE International Symposium on*, (8–11 June 2010), pp. 1–7
17. J. Gu, N. Vichare, T. Tracy, M. Pecht, Prognostics implementation methods for electronics, in *Annual Reliability and Maintainability Symposium, RAMS'07, Orlando*, (2007)
18. M. Pecht, Prognostics and health monitoring of electronics, in *International Conference on Electronic Materials and Packaging, EMAP 2006*, (Kowloon, 2006)
19. S. Mathew, D. Das, R. Rossenberger, M. Pecht, Failure mechanisms based prognostics, in *Prognostics and Health Management, 2008. PHM 2008. International Conference on*, (6–9 October 2008), pp. 1–6

20. S. Mathew, M. Alam, M. Pecht, Identification of failure mechanisms to enhance prognostic outcomes. J. Fail. Anal. Prev **12**(1), 66–73 (2012)
21. P. Gromala, A. Palczynska, B. Han, Prognostic approaches for the wirebond failure prediction in power semiconductors: A case study using DPAK package, in *Electronic Packaging Technology (ICEPT), 2015 16th International Conference on*, (11–14 August 2015), pp. 413–418
22. H.D. Solomon, Fatigue of 60/40 solder. Comp. Hybrids Manuf. Technol. IEEE Trans. **9**(4), 423–432 (1986)
23. W. Engelmaier, Fatigue life of leadless chip carrier solder joints during power cycling. Comp. Hybrids Manuf. Technol. IEEE Trans. **6**(3), 232–237 (1983)
24. A. Schubert, R. Dudek, E. Auerswald, A. Gollbardt, B. Michel, H. Reichl, Fatigue life models for SnAgCu and SnPb solder joints evaluated by experiments and simulation, in *Electronic Components and Technology Conference, 2003. Proceedings, 53rd*, (30–30 May 2003), pp. 603–610
25. A. Syed, Accumulated creep strain and energy density based thermal fatigue life prediction models for SnAgCu solder joints, in *Electronic Components and Technology Conference, 2004. Proceedings 54th*, vol. 1, (1–4 June 2004), pp. 737–746
26. P. Gromala, J. Reichelt, S. Rzepka, Accurate thermal cycle lifetime estimation for BGA memory components with lead-free solder joints, in *Thermal, Mechanical and Multi-Physics simulation and Experiments in Microelectronics and Microsystems, 2009. EuroSimE 2009. 10th International Conference on*, (26–29 April 2009), pp. 1–9
27. M. Dressler, H. Reichl, B. Wunderle, K.-F. Becker, *Reliability Study of Stud Bump Bonding Flip Chip Assemblies on Molded Interconnect Devices* (Ph.D. Thesis, Technical University of Berlin, 2010)

Chapter 14
Reliability and Degradation of Power Electronic Materials

R. Ross and G. Koopmans

1 Introduction to Reliability and Degradation

Power electronics (PE) obtain their properties from a constellation of materials. The performance of such devices depends on the properties of the applied materials and the quality of the assembly. This chapter describes reliability matters related to performance of materials and systems in R&D experiments and production. The focus is on the statistical description, failure data analytics, and inferences.

The chapter starts out with a description of the basic parts in PE and how their degradation can jeopardize the integrity of PE. Next the basics of relevant statistical analysis are discussed. This is followed by describing the acceleration due to enhanced stresses. The effect on the system lifetime by size, redundancy, and reparability is discussed next. Finally, it is discussed that a product batch may consist of various subpopulations (e.g., defected and sound products) and that various processes may occur simultaneously (e.g., random and wear-out failure) which leads to the phenomenon of mixed failure distributions. Some examples illustrate how to deal with such complicated situations that nevertheless occur frequently in testing.

A full review of the extensive literature that exists on reliability analysis is beyond the scope of this chapter. For a more extensive introduction the reader is referred to textbooks such as on lifetime data [7], accelerated testing [10], and reliability data [8]. The present approach also transfers practice on asset management in electric power supply [13] to power electronics. As a special topic the present

R. Ross (✉)
IWO, Ede, The Netherlands

TU Delft, Faculty EWI, Delft, The Netherlands
e-mail: r.ross@iwo.nl

G. Koopmans
IWO, Ede, The Netherlands
e-mail: g.koopmans@iwo.nl

W. D. van Driel, M. Yazdan Mehr (eds.), *Reliability of Organic Compounds in Microelectronics and Optoelectronics*,
https://doi.org/10.1007/978-3-030-81576-9_14

contribution also discusses the fact that multiple processes can lead to the degradation of devices, which leads to the concept of combined distributions. With accelerated aging the original distributions can all change, but probably at different rates. This is a noteworthy phenomenon which may help or hinder R&D testing.

2 Power Electronic Materials

The basic functionalities in PE devices are conduction, semi-conduction, insulation, and stress control. The performance and degradation of electrical materials in PE devices is described in industry handbooks and textbooks, e.g., [3], and relevant information can also be found in electrical materials handbooks, e.g., [4] and, e.g., IEEE transactions. The background is that phenomena that happen in high voltage engineering also appear to happen in miniaturized low voltage PE devices, because the electric fields are in the same range.

Most applied materials in increasing order of conductivity are aluminum, gold, copper, and silver. Other important properties of conductors are mechanical strength, thermal expansion, and chemical stability. Lack of conductivity can stem from design (insufficient cross section) or high Ohmic contact (e.g., fracture in soldered connection). Overheating may be caused by forced current (e.g., by inductances) through high impedances.

Semi-conducting materials owe their conduction to available charges and electric fields that advance or block current. This is the essential functionality of PE devices. Particularly during switching, electric losses occur that cause heating and may lead to thermal degradation of the device. Lack of cooling may lead to a thermal runaway that damages the insulating layers.

Insulating materials serve to keep charges separated. Such materials are often referred to as dielectric materials because many are polarized in an electric field. Characteristic dielectric properties are electric susceptibility χ_e and permittivity ε, of which the real part ε' reflects the capacitive properties and the imaginary part ε'' the conductive part (i.e., the leakage current through the insulator). The ratio $\varepsilon''/\varepsilon' = \tan\delta$ is the dielectric loss factor that quantifies the ratio of the leaked and the stored energy. Tan δ may increase with temperature which may contribute to a thermal runaway. The electric breakdown strength is the electric field at which an insulating material breaks down. A local transgression of the breakdown strength can lead to electrical treeing with high fields at the branch tips which may quickly lead to a full breakdown. The electric field may also exceed the strength in cracks and cavities (e.g., from delamination) which may first lead to partial discharging in such spaces and next to eroding the walls and consequently cause treeing and breakdown. Such phenomena are also quite common in high voltage (i.e., high electric field) engineering.

Finally, stress control materials are used to control the electric field in and at the surface of insulators in order to prevent exceeding the breakdown strength.

All the above-mentioned functionalities of the materials are essential to the performance of PE devices and malfunction can jeopardize the PE device. Not only the quality of the materials, but also the configuration and assembly are essential. As mentioned above, multiple processes can be expected to occur simultaneously.

3 Reliability

When multiple objects fail as a result of a given failure process, the individual failures will be distributed in time (or any other relevant variable X that reflects the object performance, but for clarity time is used as the representative variable). This distribution of failure events in time is often characteristic for the process. Even being one specific distribution in time, such a distribution is still represented by various basic functions: the cumulative failure distribution $F(x)$, the survival function $R(x)$, the distribution density $f(x)$, the hazard rate $h(x)$, and the cumulative hazard rate $H(x)$. A noteworthy distinction is made in statistics between the variable X and the values x it can adopt. A statement that the (variable) time X is smaller than time value x is written as: $X < x$.

3.1 *Basic Functions*

The description of the basic functions follows that of [13] maintaining the different meanings of distribution and probability function. A distribution describes how failure times are spread in time. Probability of failure describes the perceived likelihood of failure which depends on the available knowledge of individual objects. E.g., we may know how failure times overall tend to spread in time (the distribution), but if we also know that a specific test object, e.g., runs hot before it fails, we may be able to assign a higher probability of failure to those objects that feature a significant rise in temperature. The probablity of failure may differ per object, while the distribution remains the same for the group.

The cumulative distribution $F(x)$ describes the population part that failed at or before x (so $X \leq x$). The survival function $R(x)$ describes the population part that survived at least up to x (so $X > x$). The values of the function F range from 0 to 1 and those of R range from 1 to 0. $F(x)$ and $R(x)$ relate as:

$$R(x) = 1 - F(x) \tag{14.1}$$

The distribution density $f(x)$ describes the rate of failures:

$$f(x) = \frac{dF(x)}{dx} \tag{14.2}$$

The density is normalized, which means that its integral over the complete range of x equals 1. The mass function $f(x_i)$ has values at discrete values x_i. It is the discrete analog of the continuous density $f(x)$. The mass function is also normalized meaning that the sum of its values over all x_i equals 1.

The hazard rate $h(x)$ looks much like the density but only applies to the surviving population part $R(x)$.

$$h(x) = \frac{f(x)}{R(x)} = \frac{-d\ln\left(R(x)\right)}{dx} \tag{14.3}$$

Since $R \leq 1$ the hazard rate $h(x) \geq f(x)$. A discrete hazard rate is based on the mass function. The cumulative hazard rate $H(x)$ is the integral or sum of the hazard rates. A noteworthy relationship between R and H is:

$$R(x) = \exp\left(-H(x)\right) \quad \Leftrightarrow \quad H(x) = -\ln\left(R(x)\right) \tag{14.4}$$

If one of the fundamental functions F, R, f, h, or H is known, the other functions can be derived from that in principle.

Distributions are often reflected in measures like the mean time to failure and the variance (i.e., squared standard deviation σ) of failure times, which follow from the fundamental functions. The *mean* $\langle x \rangle$ follows from integrating x with $f(x)$:

$$x = \frac{\int_{x_{\min}}^{x_{\max}} x \cdot f(x)\mathrm{d}x}{\int_{x_{\min}}^{x_{\max}} f(x)\mathrm{d}x} \tag{14.5}$$

A non-trivial equation for the mean is also:

$$x = \int_0^1 x(F)\mathrm{d}F = \int_{x_{\min}}^{x_{\max}} R(x)\mathrm{d}x \tag{14.6}$$

The *variance* var(x) is found as (using the above equation for $\langle x \rangle$):

$$\mathrm{var}(x) = \sigma^2 = x^2 - x^2 = \frac{\int_{x_{\min}}^{x_{\max}} x^2 \cdot f(x)dx}{\int_{x_{\min}}^{x_{\max}} f(x)dx} - x^2 \tag{14.7}$$

The mean (and similarly variance) may also follow from the <u>complete</u> set of N values x_i ($i = 1,..,N$) as:

$$x = \frac{1}{N} \cdot \sum_{i=1}^{N} x_i \tag{14.8}$$

If only a *subset* of n observations is available with $n < N$, then this is a sample of which the *average* $\overline{x}$ and *sample variance* s^2 can be determined. This average and sample variance often serve as estimator of the mean and variance of the complete population.

3.2 Some Important Distribution Families

The previous sections discussed distributions as a concept. Various types of distribution families exist for various purposes. Here we will address four of those distributions: the Normal, Lognormal, Weibull, and Exponential distribution.

The Normal or Gaussian distribution is the best-known and possibly also the most misused distribution. The distribution has parameters mean μ and standard deviation σ. Because of the central limit theorem, the Normal distribution is particularly adequate (the asymptotic distribution) for describing the sums or averages of large data sets in the range (−∞, ∞). Many physical quantities are positive and do not fulfill this property. If the standard deviation σ is smaller than 3 or 4 times μ), the fraction $F(x)$ with $x < 0$ is small, but that remains to be checked. There is no analytical expression for the cumulative distribution F, but for the Normal distribution density f there is:

$$f(x;,\mu;,\sigma) = \frac{1}{\sigma\sqrt{2\pi}} \cdot \exp\left(-\frac{1}{2} \cdot \left(\frac{x-\mu}{\sigma}\right)^2\right) \tag{14.9}$$

Various approximations exist that enable numerical calculation of $F(x)$ to any desired precision ([18]). The mean x is equal to μ.

As mentioned, −∞ < x < ∞ and therefore x is not time in this case. The Lognormal distribution offers a way to fix the range issue. If x follows a Normal distribution, then $t = \exp(x)$ follows the so-called Lognormal distribution. It also has parameters μ and σ but they are the mean and standard deviation of $\ln(t)$, not of t. The Lognormal distribution is particularly useful for describing the sum or mean of large sets of log-positive t-values (if $t > 0$ then $-\infty < \log(t) = x < \infty$ for which the Normal distribution applies). It can also be applicable in its own right. Again, there is no analytical expression for the cumulative distribution, but the distribution density is:

$$f(t;,\mu;,\sigma) = \frac{1}{\sigma\sqrt{2\pi}} \cdot \exp\left(-\frac{1}{2} \cdot \left(\frac{\ln(t)-\mu}{\sigma}\right)^2\right) \tag{14.10}$$

The mean of the Lognormally distributed t is not equal to μ (which is the mean of the Normally distributed x). The mean θ of a Lognormal distribution is:

$$\theta_{\text{Lognormal}} = \exp\left(\mu + \frac{\sigma^2}{2}\right) \tag{14.11}$$

The Weibull distribution is the asymptotic distribution for the smallest of an increasingly large set of positive values t. While the Normal and Lognormal distribution focus on the center (i.e., the mean), the Weibull distribution focuses on the lower extreme such as the lowest breakdown time that characterizes the electric strength of a device. The Weibull distribution usually is applied with two parameters: namely a scale parameter α and a shape parameter β. Analytical expressions exist for all basic functions. The cumulative distribution function reads:

$$F(t;,\alpha;,\beta) = 1 - \exp\left(-\left(\frac{t}{\alpha}\right)^{\beta}\right) \tag{14.12}$$

The other functions follow from this as discussed before. A three-parameter version employs a threshold parameter δ as lower limit for *t*, in which case t in the equation above is replaced by $t-\delta$. A one-parameter version employs only the scale parameter α, and sets β to a fixed value by definition. The Weibull mean θ is:

$$\theta_{\text{Weibull}} = \alpha \cdot \Gamma\left(1 + \frac{1}{\beta}\right) \tag{14.13}$$

Here Γ is the Gamma distribution. If $\beta \geq 1$ then θ closely resembles α with a maximum deviation of −11.4% at $\beta = 2$. The ratio θ/α runs up fast with decreasing $\beta < 1$.

The exponential distribution can be regarded as such a one-parameter Weibull distribution with $\beta = 1$ per definition. In that case the scale parameter is often denoted as θ. This is the mean failure time, also called the characteristic time. Again, analytical expressions exist for all basic functions and the cumulative function reads:

$$F(t;\theta) = 1 - \exp\left(-\frac{t}{\theta}\right) \tag{14.14}$$

An important property of the Exponential distribution is its constant hazard rate $h(t) = h = 1/\theta$. Working devices having a constant hazard rate do not suffer from child disease, nor from wear-out. Failure is random in time ('bad luck') and often due to external impact like cosmic radiation or mechanical impact, if unrelated to the use and age of the device, etc. Another situation where the Exponential distribution applies is where maintenance is carried out to keep the hazard rate at a more or less constant level. This constant hazard rate implies an Exponential distribution.

3.3 Competition of Processes and Mixed Subpopulations

A batch of devices under test will often suffer from multiple processes. For instance, devices may be simultaneously exposed to child mortality processes (e.g., an ongoing cracking or chemical reaction originating from the production process, random failure (e.g., cosmic radiation), and wear (e.g., electrochemical degradation). Additionally, the batch of devices may be inhomogeneous if there are multiple subpopulations with distinct failure behavior. For instance, a certain subpopulation may contain defects from production that leads to failure behavior that is not found for the rest of the population. Competing processes and mixed populations lead to more complicated failure distributions.

As for competing processes, the functions F_{total}, etc., representing the total distribution are illustrated with the case of two competing processes A and B. A device only survives if it survives both A and B. From this principle follow:

$$R_{\text{total}} = R_A \cdot R_B \tag{14.15}$$

$$F_{\text{total}} = 1 - R_{\text{total}} = F_A + F_B - F_A \cdot F_B \tag{14.16}$$

$$f_{\text{total}} = \frac{\mathrm{d}F_{\text{total}}}{\mathrm{d}x} = f_A \cdot R_B + f_B \cdot R_A \tag{14.17}$$

$$h_{\text{total}} = h_A + h_B \tag{14.18}$$

$$H_{\text{total}} = H_A + H_B \tag{14.19}$$

It is noteworthy that items that fail by process A still bear information about process B. The reason is that the moment they fail at time t_A by process A, then it is also clear that they survived process B up to t_A. This finding should be taken into account with the data analytics for process B (see also the section on censored data).

In case of mixed subpopulations a similar set of equations can be derived. Let us assume each subpopulation is indexed with 'i' and forms a fraction p_i of the combined population. The sum of the fractions p_i amounts to 1. This leads to the following:

$$F_{\text{total}} = \sum_i p_i \cdot F_i \tag{14.20}$$

$$R_{\text{total}} = \sum_i p_i \cdot R_i \tag{14.21}$$

$$f_{\text{total}} = \sum_i p_i \cdot f_i \tag{14.22}$$

$$h_{\text{total}} = \frac{f_{\text{total}}}{R_{\text{total}}} = \frac{1}{R_{\text{total}}} \sum_i p_i \cdot f_i \tag{14.23}$$

$$H_{\text{total}} = \int \sum_i \left(p_i \cdot \frac{f_i}{R_{\text{total}}} \right) \mathrm{d}t \tag{14.24}$$

The hazard rates are less straightforward than in the case of competing processes.

As discussed, the total distribution can be deduced conveniently if the distributions of the competing processes and fractions of subpopulations are known. Unfortunately, starting from observed failure times without further knowledge, it may be much harder to disentangle distributions and reconstruct the competing processes and mixed subpopulations. Moreover, observation data generally scatter which can give the false impression of entangled distributions. Therefore, objective measures are needed to evaluate whether a study into competing processes and mixed subpopulations is justified.

However, plots may nevertheless reveal entangled distributions if they are sufficiently distinctive. As an illustration consider the case of two mixed subpopulations. Subpopulation 1 has a Weibull distribution with parameters $\alpha = 1$ and $\beta = 0.6$ (which has a character of a child mortality process as $\beta < 1$). Subpopulation 2 follows a Weibull distribution with parameters $\alpha = 20$ and $\beta = 3$ (which has a character of a wear-out process as $\beta > 1$). If these are mixed the resulting distribution functions can be found with Eqs. 14.18, 14.19, 14.20, 14.21, and 14.22. Figures 14.1 and 14.2 show respectively the hazard rate plot and distribution density plot. The figures show the original functions and the total function. The figures also show the observations that can be expected on average if the total population consists of $n_{total} = 13$ test objects ($n_1 = 4$ and $n_2 = 9$). Particularly the distribution density plot gives a clear indication of a mixed case.

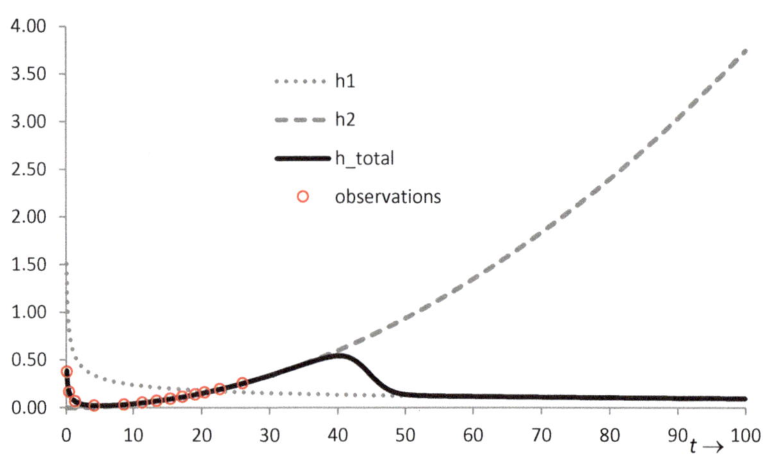

Fig. 14.1 Hazard rate plot of two entangled distributions due to a mixed population

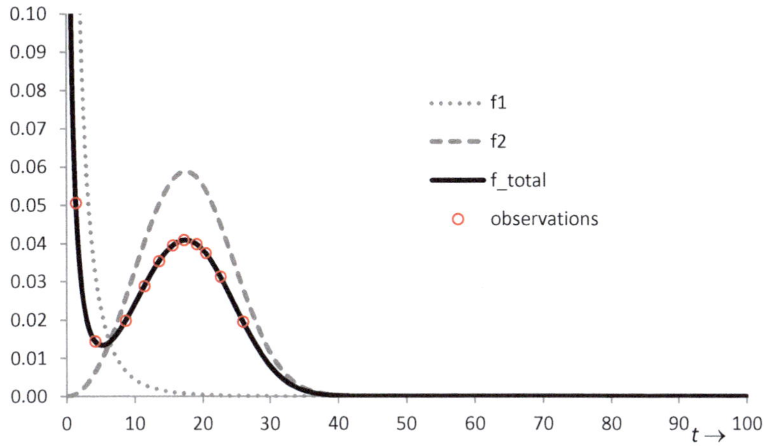

Fig. 14.2 Distribution density plot of two entangled distributions due to a mixed population

Table 14.1 Example composition of batch of devices. The total population consist of two subpopulations ('Normal' and 'Weak subpopulation') with multiple competing processes

Subpopulation	Normal	Weak subpopulation	Weibull parameters
Percentage	90%	10%	
Child mortality or early failures	Present	Present	$\alpha = 10{,}000$, $\beta = 0.005$
Random failure	Present	Present	$\alpha = 200$, $\beta = 1$
Wear-out	Present	Present	$\alpha = 25$, $\beta = 4$
Additional wear out	N.A.	Present	$\alpha = 0.3$, $\beta = 6$

3.4 Bath Tub Curves and Screening

Well-known examples of competing processes and mixed subpopulations are often found in so-called bath-tub curves. These are hazard rate graphs plotted as a function of time.

Table 14.1 shows an illustrative example of entangled distributions. The example is based on a case where typically three processes compete in each device, namely: child mortality or also called early failures; random failure; and wear-out failure. The early failures can be due to production errors or curing materials. With time this process loses importance. Wear-out can typically be due to degradation of insulating layers or other changing material properties in the long run. In between is an interval where, e.g., cosmic radiation or other often external influences deteriorate the devices. In this example this would be a set of competing processes that are typical for a 'normal' batch.

In this particular example 10% of the devices have been assigned an additional failure mechanism that causes fast wear-out and competes with the other three processes. This is a subpopulation in the total batch and turns the total batch into a mixed population.

Table 14.1 details the composition of the total. The rules in the previous section can be used to obtain the overall hazard rate for each subpopulation (cf. Eq. 14.18) and subsequently the total hazard rate (cf. Eq. 14.23).

Figure 14.3 shows this example of entangled distributions in a hazard rate plot. The entanglement is due to both competing processes and a mixed population. In various stages the underlying distributions dominate here. In a first stage a child mortality or also called 'early failure' process dominates. If such a phase is normal to produced batches of devices, a screening or burn-in process may be applied to prevent the delivery of early failing devices to customers. This process is normally done by applying enhanced stresses beyond rating which causes accelerating aging. This can be conceived as speeding up the consumption of operational life until a point that the hazard rate is reduced to acceptable levels.

A disadvantage of burn-in is that operational life of all devices is reduced. Therefore, the enhanced stress preferably targets the early failure mechanism specifically. Another approach is investigating whether the early failures can be prevented by improved production, after which the burn-in process may be shortened or even skipped, which translates in longer operational life of all devices.

Secondly, the plot shows a peak where the fast wear-out process of the weak subpopulation is dominant. This applies only to a part of the total batch and once this part is eradicated, the hazard rate drops. If batches normally contain such imperfection, a burn-in process may cause too much damage to the healthy part in case the mechanism cannot be targeted specifically. Other strategies may be developed to detect imperfect devices, e.g., by condition indicators such as the occurrence of hot spots, unusual leakage currents, partial discharge, etc.

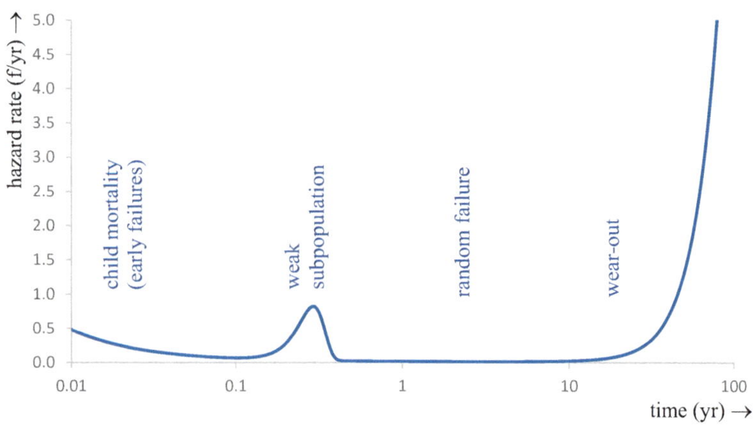

Fig. 14.3 A bath tub curve combined with subpopulation of weak devices based on Table 14.1. This is an example of entangled hazard rates based on competing processes (the bath tub) and a mixed population. The data here are illustrative and can be very different in other cases

The third stage is the desired operational mode where typically only random failures occur, i.e., the 'bad luck' region. Such failures may be due to effects that are unrelated to the age of the device, such as solar flames, lightning nearby, etc.

The final stage is the wear-out phase. Material and composition may deteriorate over time due to TEMA (thermal, electrical, mechanical, and ambient) processes. Knowledge of the timing of this wear-out is necessary for predicting product life, warranty, replacement planning, etc.

The same combination of distributions is shown in a reliability plot in Fig. 14.4. In this plot three types of screening are illustrated. The first is the full batch without screening which exhibits a considerable number of failures due to the child mortality process. The second graph shows the effect of screening for these early failures. The third graph shows the effect of extra screening until also the weak population is taken out. The resulting cumulative failure distribution $F_{\text{tot},s}(t)$ after screening until $t = t_s$ is obtained from:

$$F_{\text{tot},s}(t) = 1 - \frac{R_{\text{tot}}(t)}{R_{\text{tot}}(t_s)} \tag{14.25}$$

The screening or burn-in process has a great influence on the quality as perceived by customers. Whether or not the weak subpopulation is timely known will determine whether screening is applied and/or other measures are taken to prevent these low-quality devices to be delivered.

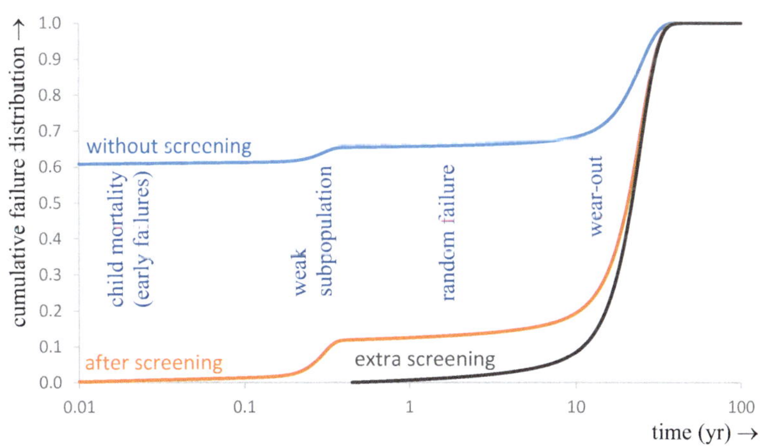

Fig. 14.4 Failure probability graph of the same population as in Fig. 14.3. In addition to the population without screening, also plots are added for screening (up to $t = 0.01$ year) to mitigate child mortality and extra screening (until $t = 0.45$ year) to also mitigate the weak population

4 Data Analytics

Failure data analysis is an important part of R&D and quality assurance. Quite straightforward is the analysis of results if all objects failed in the test and only one mechanism is active that is characterized by a single distribution. In general, many experiments and tests will not reveal all failure data to each failure mechanism. Such hidden data are called censored or suspended data and will be discussed first.

The two main categories of data analytics are graphical analysis and parameter estimation. Both have their distinct value. In the following these two are reviewed after discussing censored data.

4.1 Censored Data

As mentioned above, at the time of evaluation not all test results may be available, e.g., objects may survive a test for a certain time without a failure. This period of survival bears information about the failure distribution even if no failure is observed. Data that remain hidden are called censored or suspended data. Some causes of censored data are: withdrawal of test objects, disrupted tests, unequal starting times, and data from competing processes.

Test objects may be withdrawn from a test for various reasons like noticeably deviating and irrelevant aging conditions, or the data may be regarded unreliable and be discarded for that matter. Tests may also be terminated before all objects fail.

If the lifetime of devices in service is monitored, the tracked devices may not have been taken into operation at the same time. Some may have replaced failed devices from the same batch. This usually also leads to censored data.

If testing or aging in operation involves competing failure processes, then failure of an object by one process means it does not fail by other processes. For the process in question, it is a valid observation, but with respect to the other processes it is censored. Other objects will fail by other competing processes for which the observation is valid. Each failure principally yields a valid observation for one of the processes and is censored for the other competing processes. This implies that the data sets for each process can be significantly censored.

In summary, censoring is a very common phenomenon. It occurs not only in incomplete tests, but also in experiments with competing processes. Proper evaluation of test results requires analysis methods that can deal with censored data and are addressed in the following sections.

4.2 Graphical Analysis

Graphs are visuals that are informative about evolution and scatter of data. They are used to identify outliers, check the family to which the failure distribution might belong, check for entangled distributions, etc. In a graphical representation of observed failure times, it is customary to plot the observations along the horizontal axis. Along the vertical axis ranked theoretical values based on the assumed distribution are plotted. These are often means (or medians) of ranked values from transformed distribution functions like F, R, f, h, or H. These theoretical values are the so-called 'plotting positions' Z and are paired with the individual observations to form the plot. The purpose of the graph should determine the choice of the plotting positions.

Two major analysis families are non-parametric and parametric plots. Non-parametric plots show the distribution as a function of time without assumptions about the distribution. There are various methods. A rather straightforward technique is to plot the cumulative distribution, i.e., the failed fraction, $F(t)$ against time t. Assume a complete set of n failure times t_i ($i = 1,..,n$) is observed. The values for the corresponding fractions F_i are to be defined. Often the data set is treated as an n-sized sample randomly drawn from a (very) large N-sized population of devices. When repeatedly done, the mean fraction F_i of the ith test object can be proven to be:

$$F_i = \frac{i}{n+1} \tag{14.26}$$

A popular alternative for the mean is the median plotting position $F_{M,i}$. If this choice is made, the median plotting position is usually approximated with:

$$F_{M,i} = \frac{i-0.3}{n+0.4} \tag{14.27}$$

A typical example of a non-parametric plot of only the wear distribution in Table 14.1 is shown in Fig. 14.5. No assumptions have been made with respect to the type of the distribution like Lognormal or Weibull.

The parametric plots assume a type of distribution. The plotting position commonly depends on this type in such a way that the graph of the assumed distribution type becomes a straight line. This is very helpful for estimating parameters, for evaluating outliers, for extrapolating the graph to do predictions, etc. The challenge is to find a linear relationship between (functions of) the distribution function and the observed failure times.

For instance, for the Weibull distribution equation Eq. 14.11 can be rewritten as a linear relation between log(−ln(1−F)) and log(t). Other sources give the relation in terms of ln(−ln(1−F)) and ln(t). The choice in the present paper is to use 'ln' as the natural logarithm when taking the inverse of 'exp'. The choice for log or ln can be regarded a matter of taste. The expression for Z is:

$$Z = \log\left(-\ln\left(1-F\right)\right) = \beta \cdot \log\left(t\right) - \beta \cdot \log\left(\alpha\right) \tag{14.28}$$

The slope of such a graph is β and the intercept or initial value is −β•log(α). A graph is prepared by plotting the logarithm of observation t_i along the horizontal axis and

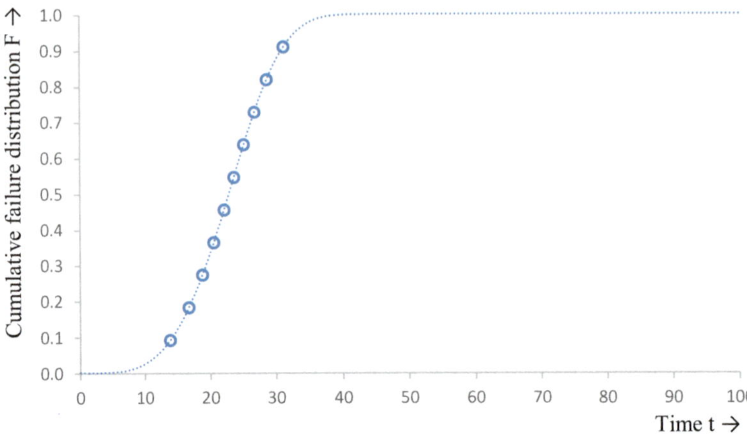

Fig. 14.5 Non-parametric plot with expected F_i plotting position of the wear distribution. The points have coordinates (t_i,‹F_i›). The dotted line would be the best fit

the plotting position can be chosen as the expected values ‹$Z_{i,n}$› and be plotted along the vertical axis. The exact expression for can be found with the Beta function, Sect. 6.2 in [1], as:

$$Z_{i,n} = \left[-\gamma + i \binom{n}{i} \sum_{j=0}^{i-1} \binom{i-1}{j} \frac{(-1)^{i-j} \cdot \ln(n-j)}{n-j} \right] \cdot \log e \tag{14.29}$$

A good approximation for finding ‹$Z_{i,n}$› is [5]:

$$Z_{i,n} \approx \log\left(-\ln\left(1 - \frac{i-0.44}{n+0.25} \right) \right) \tag{14.30}$$

Although Z-values are plotted along the vertical axis, normally the scale is displayed such that it shows F-values. As a result the scale is not linear anymore. An example is shown in Fig. 14.6. Here the same wear data as for Fig. 14.5 were used. From the graph the data can be fit quite well with a straight line, which supports the assumption that the wear-out process follows a Weibull distribution.

As mentioned previously, for various reasons data may be incomplete. If the first r of n observations are available, but the remaining n-r are missing, then this is called 'right censoring'. The failure times t_i with $i \leq r$ are known and $t_i \leq t_r$. Furthermore, for $i > r$ the times t_i are unknown as yet, but $t_i > t_r$. This typically happens if all objects entered the test simultaneously and the data are evaluated before the test is completed and/or the test is terminated while some objects survive as yet.

For plotting, the consequences are little. The plotting positions for n observations (censored or not) can be determined as above, but the last n-r observations are left open. Particularly parametric plots like Fig. 14.6 can still be conveniently produced.

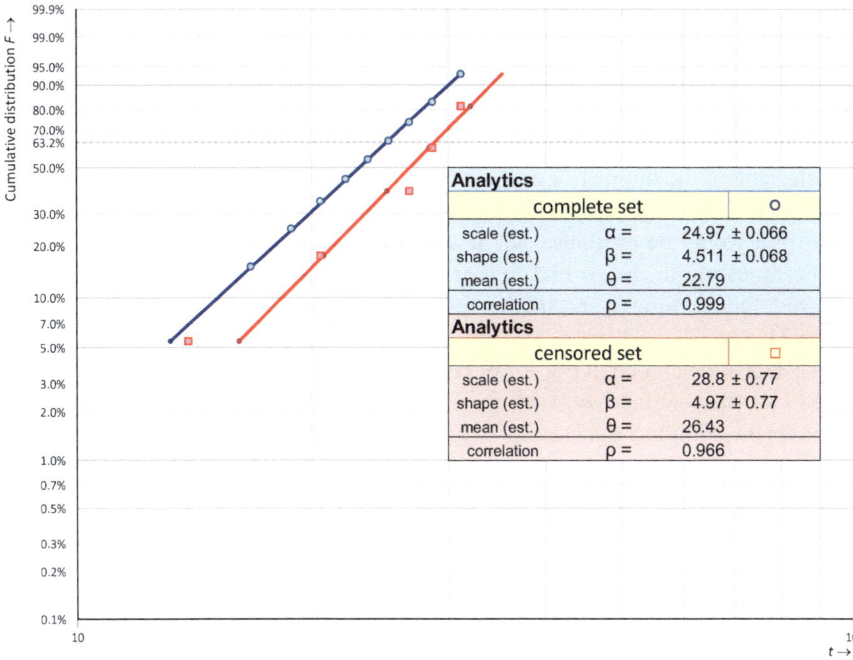

Fig. 14.6 Weibull plot employing a mean plotting position. The observations for the complete set are the same as in Table 14.2 and Fig. 14.5. The censored set is given in Table 14.3

However, in many other cases the censoring can be random: one of more failure times t_i are known, but the censored observations can be larger but also smaller than existing observations. This typically happens when a fleet of devices is taken into operation and failed devices are replaced by new devices. The new ones start their operational life later than other as yet surviving devices and they may fail at a younger or older age than the already failed devices. As a consequence, the ranking of the first r observations of n is no longer fixed. E.g., assume the items in Table 14.2 were each replaced after failure and the test was continued. At some moment, there may be as yet surviving objects with a shorter lifetime than, say, t_4. These surviving objects may fail before or after t_4, which implies that the presently fourth failure ultimately may be ranked higher like the sixth, tenth, or higher. The exact ultimate ranking remains hidden until the data are complete. However, there are methods that estimated the ultimate ranking based on actual observations and the lifetimes of the as yet surviving objects. A widely used method is 'adjusted ranking' that adjust the ranking indices i of observed data to adjusted ranking indices I, which no longer need to be integers. The adjusted ranks are calculated as [5]:

$$I(i) = I(i-1) + \frac{n+1-I(i-1)}{n+2-C_i} \tag{14.31}$$

Table 14.2 Observed breakdown times for Fig. 14.5

Index i	1	2	3	4	5	6	7	8	9	10
Time t_i (y)	13.9	16.7	18.8	20.5	22.1	23.6	25.1	26.7	28.6	31.1

C_i is the sum of the number of censored and uncensored observations including the failure i. $I(0)$ is 0 by definition. As an example, assume not all observations of Table 14.2 would be available, but at some time Table 14.3 would apply instead. Failure times are in the second row and surviving times without failure (i.e., the censored failure times) are in the third row. The adjusted rank follows from Eq. 14.31.

The data are included in Fig. 14.6. With data lacking, the fit is not necessarily the same as for the complete set. With increasing number of observations it approaches the fit for the complete data set. In some cases the set will never become complete. For instance if failures occur due to a competing process earlier, the failure time according to the studied process will never be determined. With entangled distributions, this can be very relevant to the data analysis.

There are alternative methods to analyze censored data sets, e.g., [15] and to use adjusted plotting positions (cf. Sect. 5.2.2 in [13]).

4.3 *Parameter Estimation*

Distributions have parameters and if these are known, various characteristics can be estimated such as the expected lifetime, confidence limits (e.g., a probability that the first out on n devices will fail before a certain time), etc. The parameters are usually not known, but are estimated based on a data set of n failure observations. Various methods exist to estimate parameters.

At least three aspects of parameter estimation are important: bias, efficiency, and consistency. Bias is the systematic error of an estimator. Preferably, the estimator should be unbiased, but if it is known at least, it may be corrected for with an unbiasing operation. The efficiency is related to the scatter or random error. It should be as small as possible and cannot be corrected for, it can be reduced by averaging the results from multiple experiments. A consistent estimator fulfills the requirement of becoming more accurate with increasing sample size such that bias and scatter approach zero.

Two classes of parameter estimators are maximum likelihood (ML) and linear regression using least squares (LS). The latter can be improved by applying weights (WLS). For large sample sizes n these classes are consistent and perform quite comparably. For small sample sizes n the bias and scatter can be significant. This effect should be considered when parameters are estimated with data sets that differ in size n. A comparison may be flawed if the bias depends on sample size.

Both classes optimize an expression to estimate parameters. For the ML method the so-called likelihood function L is maximized, which is a function of the

Table 14.3 A set of observed and censored data. Adjusted ranks are shown in the fourth row. The set is consistent with Table 14.2, i.e., after completion, the test could yield the same results

Rank i	1	2	3	4	5	6	7	8	9	10
Failure t_i (y)	13.9			20.5		22.1			28.6	31.1
Censored (y)		14.0	14.3		21.3		22.3	23.1		
Adj.rank I	1			2.25		3.71			6.14	8.57

parameters u_j ($j = 1,..,m$) to be estimated. For instance, u_1 could be the Weibull scale parameter estimator a of α and u_2 could be the shape parameter estimator b of β. The observations t_i ($i = 1,..,n$) are parameters. More convenient is to maximize $\ln(L)$:

$$\ln\left[L\left(u_1,;,\ldots,;,u_m,;,t_1,;,\ldots,;,t_n\right)\right] = \sum_{i=1}^{n} \ln f\left(t_i|,u_1|,\ldots|,u_m\right) \tag{14.32}$$

In case of $n-r$ censored data, $\ln(L)$ includes a term R^{n-r} that describes the surviving fraction at given $t = \tau$:

$$L\left(u_1,;,\ldots,;,u_m,;,n,;,r,;,\tau,;,t_1,;,\ldots,;,t_r\right) = \left(R(\tau)\right)^{n-r} \cdot \prod_{r}^{i=1} f\left(t_i|,u_1|,\ldots|,u_m\right)$$

The parameters can now be found by solving:

$$d\ln[L] = \sum_{i=1}^{n} \frac{\partial \ln[L]}{\partial u_i} du_i = 0 \tag{14.34}$$

In case of the Weibull scale parameter this yields (with τ_j the survival time of object j):

$$a^b = \frac{1}{r} \cdot \left(\sum_{i=1}^{r} t_i^{\,b} + \sum_{j=1}^{n-r} \tau_j^{\,b} \right) \tag{14.35}$$

And for the Weibull shape parameter:

$$\frac{\sum_{i=1}^{r} t_i^b \ln t_i + \sum_{j=1}^{n-r} \tau_j^b \ln \tau_j}{\sum_{i=1}^{r} t_i^b + \sum_{j=1}^{n-r} \tau_j^b} - \frac{1}{r} \cdot \sum_{i=1}^{r} \ln t_i = \frac{1}{b} \tag{14.36}$$

These expressions cannot be solved analytically but require a numerical method such as the Newton-Raphson method. The ML method can be extended to handle more complicated situations, such as accelerated aging testing or mixed distributions. Formulas for the bias and scatter exist [13]. There is also extensive literature on the subject and in various studies Eq. 14.36 is modified in order to even more decrease the bias and/or increase the efficiency of b, such as [16]. Bias formulas also conveniently enable to define unbiased estimators.

For (W)LS the (weighted) sum of residues is minimized to obtain estimators u_j. In the following weights w_i are included in the equations. The method is elaborated for the Weibull distribution. For ordinary LS (OLS) all weights w_i can be set to unity. The index 'i' can be replaced by the adjusted rank 'I' in case of censored data. The WLS weights w_i are defined as:

$$w_i = \frac{1}{\overline{Z_i^2} - \overline{Z_i}^2} \tag{14.37}$$

The weights also for non-integer adjusted ranking can be calculated conveniently with an approximation [19]. First two averages are assessed:

$$\overline{Z}_W = \frac{\sum_{i=1}^{n} w_i Z_i}{\sum_{i=1}^{n} w_i} \quad \text{and} \quad \overline{\log t}_w = \frac{\sum_{i=1}^{n} w_i \log t_i}{\sum_{i=1}^{n} w_i} \tag{14.38}$$

Next the Weibull shape parameter b is found as:

$$b = \frac{\sum_{i=1}^{n} w_i \left(\langle Z_i \rangle - \overline{\langle Z \rangle}_w \right)^2}{\sum_{i=1}^{n} w_i \left(\langle Z_i \rangle - \overline{\langle Z \rangle}_w \right) \cdot \left(\log\ t_i - \overline{\log\ t}_w \right)} \tag{14.39}$$

And the scale parameter a as:

$$a = 10^{\overline{\log\ t}_w - \frac{\overline{\langle Z \rangle}_w}{b_{WLS}}} \tag{14.40}$$

The WLS is a special case of the generalized LS (GLS) [2].

The estimators of both are biased except the (W)LS-estimator of $1/b$. As an advantage, (W)LS is consistent with plotting if the expected plotting position $\langle Z_i \rangle$ is used rather than the median plotting position.

The ML and (W)LS methods are widely used. The ML estimators are found by a numerical procedure, whereas the (W)LS method employs analytical expressions. As a consequence, error calculations can be performed for (W)LS in case the observations vary in accuracy. The ML and (W)LS estimators of the Weibull shape parameter β are biased. The same applies to the ML estimator $1/b$. Table 14.4 gives expressions for unbiasing factors and for the standard deviations of these estimators for complete data sets as a function of their sample size n. For example, the factor to unbias the ML estimator b is found to be $4/5.32 = 0.75$, $6/7.32 = 0.82$, and

Table 14.4 Unbiasing factors and standard deviations (st. dev.) of the ML, WLS, and LS estimators for the Weibull shape parameter, valid for complete data sets. The original estimators are b and $1/b$. The unbiased estimators are b_u and $(1/b)_u$. The true or theoretical parameters are β and 1/β. The formulas are discussed in Chap. 6 of [13]

Estimator b of Weibull shape parameter β			
Method	ML	WLS	LS
Bias	$b_U \approx b \cdot \frac{n-2}{n-0.68}$	$b_U \approx b \cdot \frac{n-2}{n-1.34}$	$b_U \approx b \cdot \frac{(n-2)^{0.9}}{(n-2)^{0.9}+0.68}$
St. dev.	$\sigma_{b_U} \approx \beta \cdot \frac{0.8}{\sqrt{n-3}}$	$\sigma_{b_U} \approx \beta \cdot \frac{0.8}{\sqrt{n-1.2}}$	$\sigma_{b_U} \approx \beta \cdot \frac{0.8}{(n-2.9)^{0.45}}$
Estimator 1/b of the inverse Weibull shape parameter 1/β			
Method	ML	WLS	LS
Bias	$\left(\frac{1}{b}\right)_U \approx \frac{1}{b} \cdot \frac{n-0.18}{n-0.94}$	$\left(\frac{1}{b}\right)_U \approx \frac{1}{b}$	$\left(\frac{1}{b}\right)_U \approx \frac{1}{b}$
St. dev.	$\sigma_{\left(\frac{1}{b}\right)_U} \approx \frac{1}{\beta} \cdot \frac{0.8}{\sqrt{n-1.2}}$	$\sigma_{\left(\frac{1}{b}\right)_U} \approx \frac{1}{\beta} \cdot \frac{0.8}{\sqrt{n-1.2}}$	$\sigma_{\left(\frac{1}{b}\right)_U} \approx \frac{1}{\beta} \cdot \frac{0.9}{(n-0.9)^{0.47}}$

8/9.32 = 0.86 for $n = 6$, 8 and 10 observations respectively. This signals that estimators with different sample sizes n require unbiasing to make a fair comparison possible. For censored data sets the bias and scatter are worse generally and often unbiasing expressions are as yet to be developed.

5 Stress-Related Lifetimes

Power electronics are designed to operate in a range of stresses like voltage, current, temperature, humidity, exposure to contamination, etc. Usage within a rated range should result in a satisfactory lifetime of the device. However, in the long run devices will age. If enhanced stresses are applied, then this will generally accelerate the deterioration of devices. In normal use the rated stresses should not be exceeded, but in testing this provides a means to assess the quality of devices in a relatively convenient short time. Accelerated aging in tests is very useful, but care must be taken to not enhance the stresses to such an extent that the deterioration becomes unrepresentative of normal operation. At that point a test loses its value.

The Acceleration Factor (AF) is defined as the ratio of the lifetime t_{B0} at a rated level B_0 (i.e., the assumed usage stress level) and the lifetime t_B at an enhanced stress level B. For instance, if a testing period of 2 h at an enhanced stress level is the equivalent of a service lifetime of 50 h, then the AF is 50/2 = 25. Put in other words, 1 h at the enhanced stress level is the equivalent of 25 h at rated stress level. In short:

$$AF = \frac{t_{B0}}{t_B} \tag{14.41}$$

The dependency of AF on the stresses varies with the underlying aging mechanism(s), as discussed in the following.

5.1 Power Law

Some stresses are known to influence the lifetime through a so-called power law. Again, if B is a given test stress, B_0 the rated stress, t_B the (expected, median) lifetime under stress B, and t_{B0} the (expected, median) lifetime under stress B_0, then the power law with power parameter p reads:

$$B^p \cdot t_B = B_0^p \cdot t_{B0} \tag{14.42}$$

The acceleration factor of the power law, AF_P, follows as:

$$AF_P = \frac{t_{B0}}{t_B} = \left(\frac{B}{B_0}\right)^p \tag{14.43}$$

Degradation types where the power law usually applies:

- Electric breakdown with electric field E versus rated field E_0.
- Humidity impact in e.g. breakdown with relative humidity RH.

For failure mechanisms that follow a Weibull distribution it can be shown that a power law acceleration only influences the scale parameter α and not the shape parameter β. This can be conveniently demonstrated with the cumulative hazard rate. The distributions at rated stress and test are the same except for the acceleration of time or put in other words the compression of the time scale. The cumulative hazard $H_{B0}(t_0)$ at stress B_0 and $H_B(t_B)$ at stress B relate as:

$$H_{B0}(t_{B0}) = H_B(t_B) \Leftrightarrow \left(\frac{t_{B0}}{\alpha_{B0}}\right)^{\beta_{B0}} = \left(\frac{t_B}{\alpha_B}\right)^{\beta_B} \tag{14.44}$$

$$\Rightarrow \ \beta_B = \beta_{B0} \ \wedge \ \alpha_B = \alpha_{B0} \cdot \left(\frac{B_0}{B}\right)^p \tag{14.45}$$

The fact that the shape parameter remains constant is apparent in a Weibull plot where data from different stresses have practically the same slope $1/\beta$, but are shifted along the time axis. Figure 14.7 shows a typical Weibull plot of failure data

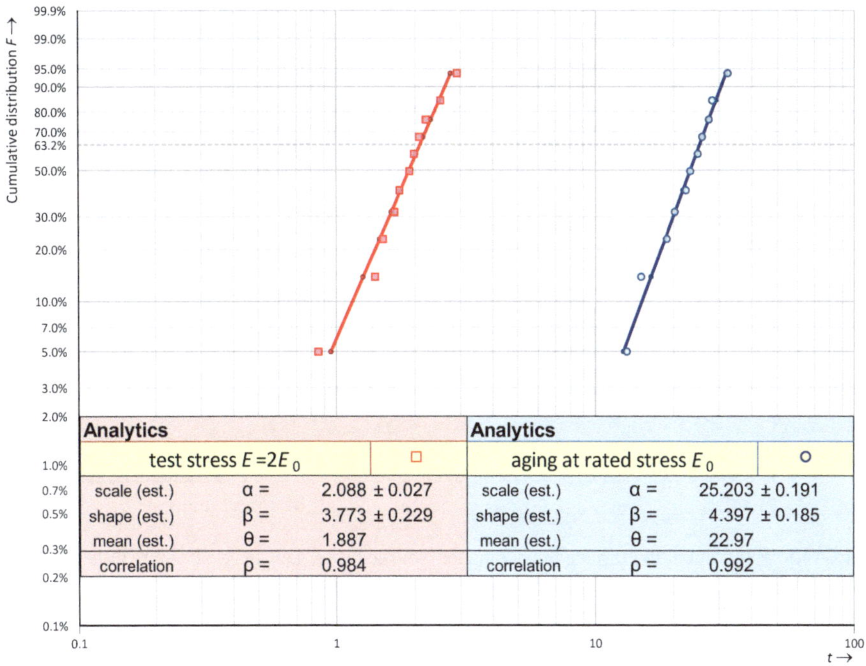

Fig. 14.7 Comparison of failure data at rated stress E_0 and enhanced test stress $E = 2E_0$. The power in the power law is about 3.6 in this case. The shape parameters of the series are similar

under rated stress E_0 and an accelerated aging test with stress $E = 2 \bullet E_0$. AF is 2^p in this example (Eq. 14.43).

If graphs of distributions at various stresses do not run more or less parallel in a Weibull plot, this may indicate that, e.g., the stress enhancement leads to unrepresentative degradation, multiple mechanisms play a role, the test objects are different, or that the sample sizes n of the test series significantly differ. The last situation may be corrected for with the equations in Table 14.4. Comparing the shape parameters from tests at different stresses is a good check whether the application of a power law is justified.

The purpose of accelerated testing is often to estimate the expected life t_{B0} under rated stress B_0 with Eq. 14.43 and given t_B and AF. This requires knowledge of p.

The parameter p may be known from experience. If not, it can be determined experimentally from tests at various fixed stresses B_i or from ramped stress tests. The latter involves a series of tests in which the stress B_i is linearly increased at a rate c_i with time. On a logarithmic scale the ramped stresses $B(t) = c \bullet t$ at various rates c show as parallel lines (cf. Fig. 14.8). The stresses $B(\tau_i)$ at mean breakdown times τ_i form the power law line for ramped stresses that runs parallel to the power law line for fixed stresses. The slope of these lines is $-1/p$ which provides an estimator for p.

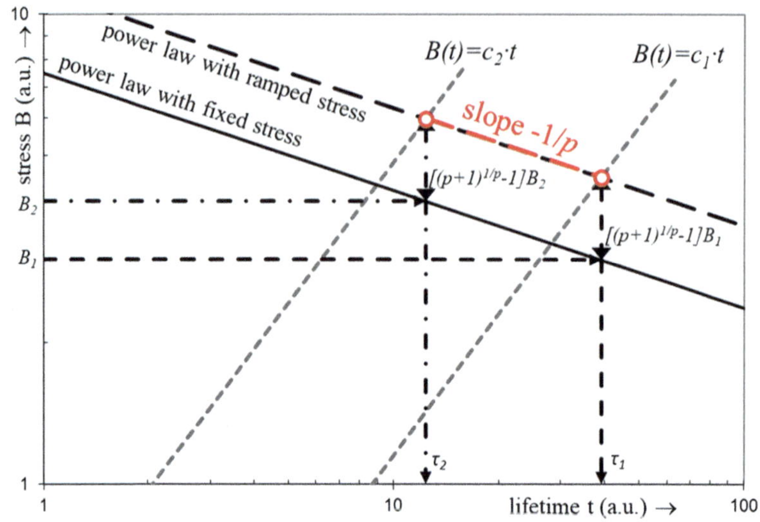

Fig. 14.8 Experiment to estimate the parameter p of the power law. In two or more tests the stress B is ramped up linearly in time. The combined mean or median breakdown times at the various ramped stress tests yield the power law line for ramped stress. The figure also shows the power law line for fixed stresses

5.2 *Arrhenius Law*

Some AFs have an activation energy relationship with the applied stress. Particularly temperature follows such an exponential law, that is called the Arrhenius law and is also referred to as the Eyring model. This type of AF is often encountered in thermal and chemical processes. According to the Arrhenius law the rate constant k of a chemical reaction depends on the absolute temperature T as:

$$k = A \cdot \exp\left(\frac{-E_a}{k_B T}\right) \tag{14.46}$$

Here A is the pre-exponential factor, k_B is the Boltzmann constant ($8.62 \bullet 10^{-5}$ eV/K) and E_a is a so-called activation energy (eV) that is related to an energy barrier to be overcome for the process to take place. With respect to reliability, such reactions can lead to degradation and failure. If applied to degradation of devices, the reaction rate is directly related to the failure rate in this model. When comparing aging at different temperatures, a reference or rated temperature T_0 (K) and a test temperature T (K) may be defined. The acceleration factor AF_T at T is then given by:

$$AF_T = \exp\left[\frac{E_a}{k_B} \cdot \left(\frac{1}{T_0} - \frac{1}{T}\right)\right] \tag{14.47}$$

For voltage stress a similar law is found to describe the physical damage on oxide layers. This is probably because of heating the layers by leakage currents. If the reference voltage is V_0 and the test voltage V, then the acceleration factor AF_V is found to be:

$$AF_V = \exp\left[\gamma \cdot (V - V_0)\right] \tag{14.48}$$

Similar to the power law, an enhanced temperature reduces the scale parameter and leaves the shape parameter untouched. Again, this may be employed to check whether an accelerated aging test remains representative. Transitions in material structure may have a large impact and limit the applicability of the Arrhenius law. For instance, relatively abrupt changes like the glass transition affect reaction rates (much) more than the Arrhenius model predicts [11].

The parameters of the Arrhenius law A and E_a are usually found by plotting the logarithmic reaction or hazard rate against the applicable aging parameter like temperature or voltage. The case of temperature follows from Eq. 14.46:

$$\ln k = \ln A - \frac{E_a}{k_B} \cdot \frac{1}{T} \tag{14.49}$$

Therefore, the slope of this plot is $-E_a/k_B$, from which E_a follows.

5.3 Combined Accelerated Aging Factors

Some processes are influenced in various ways. For instance aging of dielectric layers may be accelerated not only by increasing the electric field, but also by an increase in temperature and humidity. If these factors act independently, the combined acceleration factor AF_{combi} is the product of the separate factors AF_i:

$$AF_{\text{combi}} = \prod AF_i \tag{14.50}$$

As an example, an AF due to combined independent influences of an electric field, relative humidity, and temperature would yield:

$$AF_{E,RH,T} = \left(\frac{E}{E_0}\right)^{p_E} \cdot \left(\frac{RH}{RH_0}\right)^{p_{RH}} \cdot \exp\left[\frac{E_a}{k_B} \cdot \left(\frac{1}{T_0} - \frac{1}{T}\right)\right] = \frac{t_{E_0,RH_0,T_0}}{t_{E,RH,T}} \tag{14.51}$$

This can be extended with other factors. If the influences are not independent, then a tempering, but also often a synergistic effect may occur. E.g. a higher humidity may cause ions to be dissolved. The higher mobility or greater availability of ions may increase the rate of a reaction. As a second example, a higher electric field may help ions to overcome electrostatic barriers. The increased presence of ions in an insulating layer increases the hydrophilicity and changes the dielectric properties ε' and ε". That can have a synergistic effect where leakage currents play a role.

It should be noted that the action of combined stresses as discussed above applies to a single process. In case of competing processes and mixed subpopulations, the effects are evidently more complex.

6 System Lifetime

Many applications of power electronics concern a system or assembly of components. For instance, an HVDC converter may consist of various modules, one or more power supplies, filters, etc. Not only the component reliabilities, but also the exact configuration determine the system reliability R_S.

The basic building blocks for a system configuration are series and parallel connections. A block diagram visualizes the dependency of the system reliability on the component reliabilities. The diagram consists of a starting point (usually not indicated explicitly) on the left of the diagram. From this point a network of blocks and interconnecting lines or arrows develops towards an exit. If all elements between start and exit are functioning, then the system works, otherwise it does not. The former state is the working or up state, the latter the failed or down state.

Figure 14.9 shows the diagram of a series system. The system failure distribution follows the same set of rules as for the case of competing processes. In a series system the processes of all components compete. Some rules to calculate the distribution functions of a series system are:

$$R_S = \prod_i R_i \tag{14.52}$$

$$h_S = \sum_i h_i \tag{14.53}$$

$$H_S = \sum_i H_i \tag{14.54}$$

It can be seen that the hazard rate of a series system is always larger than that of each of its components. A general lesson is that longer series configurations tend to increase failure rates. This is why keeping systems as simple as possible may be a good strategy to make them more reliable. A chain is not as weak as its weakest link; it is weaker. This is because it may also fail due to another link.

Fig. 14.9 Block diagram of a series system

A special case of a series system is the situation where the system length is a variable. What if a connection would have M times the length of another connection with length L and otherwise the same properties? From Eq. 14.53 it follows that the hazard rate is increased with a factor M. It depends on the character of the distribution how this translates into lifetime. With an exponential distribution the mean lifetime is the inverse of the hazard rate and it would thus be reduced with a factor M. For a Weibull distribution the shape parameter would remain the same, but the scale parameter would be reduced:

$$\beta_{M \cdot L} = \beta_L \text{ and } \alpha_{M \cdot L} = \frac{\alpha_L}{M^{1/\beta}} \tag{14.55}$$

As β remains constant with varying M, the ratio of θ and α also remains constant (cf. Eq. 14.13). Therefore, θ changes with the same factor as α with varying M.

Figure 14.10 shows the diagram of a parallel system. It is down only if all parts fail. Some rules to calculate the distribution functions for a parallel system are:

$$F_S = \prod_i F_i \tag{14.56}$$

$$R_S = 1 - F_S = 1 - \prod_i (1 - R_i) \tag{14.57}$$

$$h_S = \frac{1}{R_S} \cdot \frac{\partial F_S}{\partial t} \tag{14.58}$$

The reliability and hazard rate are more complicated than in the series system case. If it comes to a parallel system of n identical components following an Exponential failure distribution with mean lifetime θ, then the mean system lifetime θ_S can be shown to be (Sect. 7.6 in [13]):

$$\theta_S = \theta \cdot \left(1 + \frac{1}{2} + .. + \frac{1}{n}\right) \tag{14.59}$$

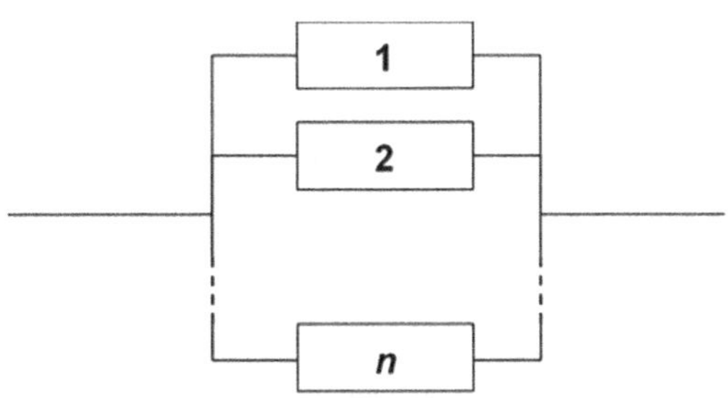

Fig. 14.10 Block diagram of a parallel system

This relates redundancy to mean lifetime. For instance, the electrical connection of TO-220 packaged devices to the top-side of the chip consists of two parallel Al wire-bonds [11]. Provided each wire is able to fulfill the function, this is a redundant connection with $n = 2$. There can be various processes that would disrupt this connection. However, if these would fail at random and independently then the mean lifetime of the wire pair is 150% of the mean lifetime of a single wire connection. In practice there are also other aspects to consider like thermal impact on degradation. In those cases, the Exponential distribution may not be applicable.

Systems may be more complicated. It may not be possible to analyze it in terms of pure series and parallel systems. Methods like the so-called conditional method, truth tables, and minimal path method can be applied to analyze the reliabilities (e.g. [12], Chap. 7 in [13]).

6.1 Reparability Considerations

Many power electronic modules are used with a run-to-failure maintenance strategy. This means the devices are operated until failure at which moment the system is down unless the system was equipped with redundancy. After failure the module is replaced to lift the state of failure. This type of maintenance is called corrective maintenance.

So far, corrective maintenance is quite standard on electronic devices and circuits because there is often little preventive maintenance that can be done, except for auxiliary systems such as the cooling. However, with Reliability 4.0 the ambition grows to monitor devices, and rather than to run until failure, the condition of circuits may be monitored as well. This would open the door to health index methodologies and timely replacement of modules to prevent system outages (Ross & Koopmans, Timely Detection of Non-compliance, [14]). This is particularly possible if redundancy is available and problematic modules can be exchanged without interrupting services. Such maintenance strategies are standard in infrastructures for energy and water, and may increasingly be applied in the domain of power electronics.

The reliability of components that cannot be repaired is usually described in terms of the survival function $R(t)$ and/or expected lifetime θ. Reparable systems are subject to an equilibrium between up and down states. For that purpose the availability is defined as the ratio between up time and total time (i.e., up time plus down time). Apart from the failure rate(s) also repair rate(s) are considered. Where reliability focuses on mean-time-until-failure, with availability mean-time-between-failures becomes important. The analysis of availability is often carried out with concepts as system states, Markov chains and Laplace transforms to solve the differential equations that describe the balance between failure and repair and the effect of absorbing states (i.e., unrepairable states).

With the upcoming smart systems, asset management strategies like condition-based and risk-based maintenance with concepts like health index and risk index may also become more important.

7 Disentangling Combined Distributions

So far, various aspects of reliability degradation of power electronics materials were discussed. If only a single process is active, the failure data analysis can be quite straightforward. However, the situation is more complicated if:

- Various processes can compete within each test object (cf. Figs. 14.3 and 14.4). The rules to calculate the resulting distributions are given in Eq. 14.15 through Eq. 14.19 for two processes. These can conveniently be extended to multiple processes starting from Eq. 14.15. An important example of competing processes is the conventional bath tub curve.
- Populations can be a mix of subpopulations with different processes (cf. Figs. 14.1 and 14.2). The rules to calculate the resulting distribution of a mixed population are given in Eq. 14.20 through Eq. 14.24. An important example is a subpopulation of defect items mixed into a batch of good products.
- When tests are terminated before all items fail and/or when multiple processes compete, there will be censored data (cf. Fig. 14.6). This means that the data sets for at least one (if not all) processes are incomplete. Times until which items were observed to survive, are relevant data and should be involved in the statistical analysis.
- Enhanced stresses speed up degradation (cf. Fig. 14.7). This acceleration depends on various circumstances such as the applied materials, stresses, types of degradation, ambient aspects, etc. In various cases the acceleration factors can be described by laws and characteristic parameters. Discussed are the power law and the Arrhenius law. The parameters associated with these laws differ from process to process.

Single processes are associated with single distributions. The discussed aspects can result in more complicated distributions. These can shift in time by enhanced stresses. However, since various processes can have a different acceleration factor, the appearance of the total distribution can change as well. We used the expression 'entangled distributions' for this combination.

If data are collected in such a complicated situation, the challenge is to recognize that situation and to disentangle the distributions. A general strategy is to firstly assume a model and secondly evaluate whether the observations match the model. For instance, one might assume that a bath tub model of competing processes applies. A basic bath tub model would consist of a child mortality process, a random process, and a wear-out process (cf. Fig. 14.3 without the weak subpopulation). This might be described with two Weibull-2 distributions and an Exponential distribution in the middle of the operational life. This would involve five parameters (α_1, β_1, θ,

α_2, β_2). Based on the observations, the parameter values must be estimated that best fit the assumed model.

Thirdly, another model might be proposed for which also the parameter values are estimated that fit that model. The more complicated the model, the more parameters will be involved. These are not only distribution parameters, but also the fractions p_i in case of subpopulations and parameters in acceleration factors if applicable. These parameter sets can grow rapidly. Therefore, models must be chosen wisely.

Instead of crunching the numbers to produce best-estimated parameters, a graphical approach might be used to explore models. Figs. 14.1 and 14.2 are examples of two mixed subpopulations. The Weibull plot of the total cumulative distribution with the original distributions is shown in Fig. 14.11. The observations are ideal in this illustrative example in the sense that the expected observations are used as data. Naturally these data are not obscured by scatter as practical data would. All figures indicate a mix of a child mortality type and a wear-out type of distribution. If confidence intervals were drawn, a part would probably exceed the boundaries. Such patterns can be recognized by eye or pattern recognition methods and would suggest to use a two-distribution model.

As a rule of thumb, an upward bend indicates a competition where a process at some moment in time starts to take over. In contrast, an zigzag shape indicates a mixed population of which a weak subpopulation is depleted (see also the example in Figs. 14.3 and 14.4). It is also possible that upper levelling off part of the zigzag shape is not yet visible in the observations (cf. Fig. 14.11). So, care must be taken when interpreting these shapes.

There are two optimizations required in this approach. Firstly, the best-estimated parameters require an objective method. With a single distribution ML and (W)LS methods are commonly used. For entangled distributions, it must be considered what is the most effective way to estimate parameters. Number crunching on the

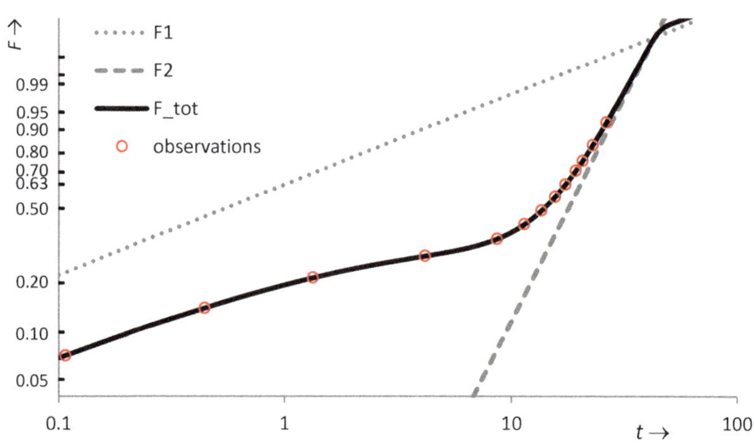

Fig. 14.11 Weibull plot of two entangled Weibull distributions and their combination due to a mixed population. See Figs. 14.1 and 14.2 for the corresponding hazard rate and distribution density. Note, the vertical scale is actually the Z-scale (cf. Eq. 14.30)

complete set is one way. Another way is to divide the data into subsets and subsequently estimate parameters for each data subset (acknowledging censored data). Examples employing the ML method are described in (Nelson, Applied Life Data Analysis, [9]).

Secondly, alternative models may be compared and the choice of model that matches the data best should also be based on objective grounds. Various measures may be used to quantitatively evaluate the best model and its parameters. Such measures include a range of goodness of fit tests [6] and measures for the difference between observed and expected data such as L2 (sum of squared residues) and the similarity index [17].

Acknowledgments This project has received funding from the ECSEL Joint Undertaking (JU) under grant agreement No 876659. The JU receives support from the European Union's Horizon 2020 research and innovation programme and Germany, Austria, Slovakia, Sweden, Finland, Belgium, Italy, Spain, Netherlands, Slovenia, Greece, France, and Turkey.

References

1. M. Abramowitz, I.A. Stegun, *Handbook of Mathematical Functions with Formulas, Graphs, and Mathematical Tables* (Dover Publications, New York, 1972)
2. A.C. Aitken, On least-squares and linear combinations of observations, in Proceedings of the Royal Society of Edinburgh, Vol. 55, pp. 42–48 (1936)
3. E.A. Amerasekera, F.N. Najim, *Failure Mechanisms in Semiconductor Devices*, 2nd edn. (Wiley, New York, 1997)
4. L.A. Dissado, J.C. Fothergill, *Electrical Degradation and Breakdown in Polymers* (Institution of Engineering & Technology, London, 1992)
5. IEC TC112, IEC 62539(E):2007 Guide for the statistical analysis of electrical insulation breakdown data. Geneva: International Electrotechnical Committee (2007–07)
6. M. Krit, *Goodness-of-fit tests in reliability: Weibull distribution and imperfect maintenance models* (General Mathematics [math.GM]/Université de Grenoble, Grenoble, 2014)
7. J.f. Lawless, *Statistical Models and Methods for Lifetime Data* (Wiley, New York, 2002)
8. W.Q. Meeker, L.A. Escobar, *Statistical Methods for Reliability Data* (Wiley, New York, 1998)
9. W. Nelson, *Applied Life Data Analysis* (Wiley, New York, 1982)
10. W. Nelson, *Accelerated Testing* (Wiley, New York, 1990)
11. A. Otto, R. Dudek, R. Döring, S. Rzepka, Investigating the Mold Compounds Influence on Power Cycling Lifetime of Discrete Power Devices. PCIM Europe 2019 (VDE Verlag GMBH/Offenbach, Nuremberg/Berlin, 2019), pp. 479–486
12. M. Rausand, A. Høyland, *System Reliability Theory – Models, Statistical Methods, and Applications* (Wiley, New York, 2004)
13. R. Ross, *Reliability Analysis for Asset Management of Electric Power Grids* (Wiley-IEEE Press, Hoboken, NJ, 2019)
14. R. Ross, G. Koopmans, Timely detection of non-compliance, in Cigré Belgium Conference Innovation for Secure and Efficient Transmission Grids (Cigré Belgium, Brussels, 2014), 8pp
15. W. Wang, Refined rank regression method with censors. Qual. Reliab. Eng. Int. **20**(7), 667–678 (2004, November). https://doi.org/10.1002/qre.575
16. Z. Yang, M. Xie, Efficient estimation of the Weibull shape parameter based on modified profile likelihood. J. Stat. Comput. Simul. **73**(2), 115–123 (2003). https://doi.org/10.1080/00949650215729

17. P.A.Ypma, R. Ross, Determining the similarity between observed and expected ageing behavior, in Proceeding ICEMPE (1st International Conference on Electrical Materials and Power Equipment), Xi'an, China, pp. 99–102 (2017)
18. M. Zelen, N.C. Severo, Probability functions, in *Handbook of Mathematical Functions*, ed. by M. Abramowitz, I. A. Stegun, 9th edn., (Dover Publications, New York, 1970), p. 932
19. R. Ross, P.A.C. Ypma, G. Koopmans, Weighted linear regression based data analytics for decision making after early failures, in Proceedings IEEE-ISGT 2021, Brisbane (2021)

Chapter 15
Degradation of Cure-Induced Stress Levels in Micro-electronics

Daoguo Yang and L. J. Ernst

1 Introduction

Thermo-mechanical reliability of electronic packages and assemblies is one of the major concerns in the electronics industry. Currently, thermo-mechanical-related failures account for about 65% of total failures in microelectronics [1]. This is expected to become even more critical in future products due to further miniaturization and function integration, which causes increased power dissipation density, higher interconnection density, and higher reliability demands.

Polymers are widely used in electronics industries. They are often classified into three groups according to their structure and mechanical properties: thermoplastic polymers (crystalline and non-crystalline), elastomers (rubbers), and thermosetting polymers. In thermosetting polymers, all macromolecular chains are connected by chemical bonds and form a large, three-dimensional network. A thermoset product therefore in fact consists of one, big molecule. The manufacturing of a thermoset product involves the polymerization process or curing process, in which chemical reaction called crosslinking occurs. The crosslinking reaction, which may occur either during or after the growth of the macromolecular chain segments, is also accompanied by a reduction in volume (shrinkage).

Thermosetting polymers are widely used in electronic manufacturing. They have been used in different forms and as different constituents, such as compounds, underfills, and adhesives, insulators, dielectrics, coatings, etc.

D. Yang (✉)
Guilin University of Electronic Technology, Guilin, China
e-mail: daoguo_yang@163.com

L. J. Ernst
Emeritus Professor in Mechanics of Materials of the Delft University of Technology, Delft, The Netherlands
e-mail: lje@ljernst.nl

W. D. van Driel, M. Yazdan Mehr (eds.), *Reliability of Organic Compounds in Microelectronics and Optoelectronics*,
https://doi.org/10.1007/978-3-030-81576-9_15

Generally speaking, no matter what the applications of the thermosetting polymers are, there are a few important issues, among others, need to be well understood:

- *The curing process*

The curing process is one of the major packaging processes for electronic packages. During the curing process, the polymer transforms from a liquid state into a viscoelastic solid with a relatively high glassy modulus, accompanied with chemical shrinkage. On the one hand, the material properties evolve in the process and the curing profile contributes significantly to the final properties. On the other hand, residual stress and warpage may be induced due to the polymerization shrinkage and subsequent cooling down from the curing temperature. Because of the cross-linking reaction and thermal shrinkage, stresses will arise in areas where shrinkage is (partly) prevented by geometrical constraints. Furthermore, subsequent cooling or thermal cycling will produce additional stress contributions and deformation due to the mismatches in thermal expansion among the materials in the package. The residual stresses may cause reliability issues and package dimensional instability. Interface delamination, cracking of silicon die, package cracking, and cure-induced warpage are often reported.

- *The time and temperature-dependent behavior*

It is well known that polymers and polymer-based composites show strong temperature and time-dependent behavior. Generally the behavior is viscoelastic, such that combined phenomena of creep and relaxation occur in packages at various levels. Such behavior has significant influence on the thermo-mechanical behavior of plastic IC packages.

- *The effect of filler*

Polymers are usually filled with fillers to improve mechanical, physical, optical, and electrical properties. It is very common in electronic packaging to use silica fillers in conjunction with polymers to increase the elastic modulus and thermal conductivity and decrease the coefficient of thermal expansion (CTE), and cure shrinkage to meet the requirements for electronic packaging. Also addition of silica fillers can reduce the cost. The prediction of mechanical properties based on the knowledge of filler and matrix properties gives great advantages for designers to select the material and to optimize the packaging structure and processing parameters.

In the design phase of electronic packaging, the influence of polymer applications on the thermo-mechanical reliability is often judged through thermal and mechanical simulations, under assumed processing and service conditions. For simplicity reasons and/or because of lacking insight into the mechanical processes during fabrication, the impact of the processing-induced residual stress fields is usually neglected by assuming a so-called "stress-free" temperature. For example, the curing temperature is often assumed to be the "stress-free" temperature. In that case, all residual stresses are assumed to initiate from the cooling step. Sometimes, this simplification can be partly justified by the fact that part of the cure-induced stress will relax because of the material's viscoelastic behavior. But in most cases, this assumption may not be appropriate.

The behavior of these polymer constituents determines the performance, such as functionality and reliability, of the final products. Therefore, the successful development of microelectronics depends to some extent, on the optimal design, selection, and processing of polymer materials. Due to the development trends of microelectronics, the development and application of polymers becomes one of the bottlenecks for the microelectronic industry. With the development and introduction of new packaging materials there are many new requirements to packaged device reliability. The development of accurate reliability models for microelectronic and/or optoelectronic devices requires an in-depth understanding of failure and degradation mechanisms related to the packaging materials. Therefore, understanding the thermo-mechanical behavior of packaging polymers is critical for the development of packages. In order to be able to conduct reliable curing process optimization and to predict thermo-mechanical reliability more accurately and further to provide a base for virtual thermo-mechanical prototyping of electronic packaging, a fundamental understanding of the process of stress build-up during cure is imperative.

2 Constitutive Modeling of Cure-Dependent Viscoelastic Properties

There has been limited work on modeling of the development of thermo-mechanical properties of polymers during cure. Generally, such modeling falls into two categories, i.e., elastic models and viscoelastic models. Among the elastic models, Plepys et al. [2] used an incremental linear elastic constitutive equation to calculate the residual stresses induced during cure. Several authors have tried to approximate the residual stress-state by assuming the stress relaxation to be much quicker than the evolution of chemical shrinkage [3–5]. Under this assumption an integral form of a rubbery elastic model was adopted using so-called "equilibrium moduli", i.e., the moduli just after transient relaxation.

Among the efforts on viscoelastic modeling, Lange et al. [6] used a single Maxwell model to account for relaxation behavior of shrinkage stresses. Cure-dependent relaxation times and (elastic) shear moduli were obtained through torsional DMA measurements. The Poisson's ratio was assumed to be constant. The model results did not give sufficient improvement compared to the earlier published incremental elastic model. The well-known time-temperature superposition principle has been widely used in modeling the effect of temperature on the viscoelastic behavior of polymers [7]. Lee and McKenna [8] used time-crosslinking superposition to describe the effect of crosslinking density on stress relaxation during physical aging in thermosetting networks. Tensile relaxation tests were carried out at temperatures below T_g. The capability of such superposition was observed. Since then, a few researchers have applied this principle to study the effect of the conversion on the viscoelastic behavior during cure. Kim and White [9] studied stress relaxation behavior of a diglycidyl ether of bisphenol A (DGEBA) resin cured with a multifunctional amine under various conversion levels. Relaxation measurements were carried out on partly cured specimens at different temperatures. The relaxation shear modulus was approximated with Prony series, in which both of the equilibrium modulus and Poisson's ratio were assumed to be constant with degree of conversion. The influence of ongoing cure was handled like long-term chemical aging by applying shift factors on the relaxation times. Complex theoretical considerations were invoked to predict the shift factors at lower degrees of cure. Simon et al. [10] proposed a model for the viscoelastic properties of an epoxy resin during cure after gelation. They also approximated the shear relaxation modulus with the generalized Maxwell model. The evolution of the rubbery shear modulus was modeled with the theory of rubber elasticity, in which the cure dependency was related through the effective concentration of network chains using the Miller and Macosko approach.

A research group in Sandia National Laboratories has conducted research on modeling the viscoelastic properties of epoxy resins during curing with the aim to establish a structure-property model for curing resins [11–13]. The shear modulus was considered to be a sum of the equilibrium shear modulus and the decaying portion. In the decaying portion, the relaxation behavior was represented by the characteristic relaxation time. The characteristic relaxation time was then treated similarly as the WLF temperature dependence, in which the reference temperature was taken to be the glass transition temperature. The glass transition temperature was described by a semi-empirical relationship, which depends on the rubbery shear modulus. A model based on scaling analysis was proposed to describe the evolution of the shear rubbery modulus after gelation.

The research group in TUDelft has made continuous efforts to establish models to describe the behavior of curing polymers. Kiasat [14] proposed a fully isotropic constitutive equation for the curing of polyester resin. Both the shear relaxation modulus and bulk relaxation modulus were represented with a Prony's series form. The relaxation behavior was described in such a way that the response at any loading time is only dependent on the material behavior at that time, and the effect of ongoing curing on the subsequent relaxation process is not taken into account. Ernst et al. [15–16] generalized the method and further it was applied for describing cure-dependent viscoelastic behavior of epoxy resin and for investigating the solder joint fatigue lifetime of a Flip Chip package [17]. Hof [18] proposed a fully cured dependent viscoelastic model, in which the history of state parameters is taken into account in the convolution integral expression. The model does not assume a master curve concept to account for the curing history effects. However, it is hard to be implemented into a FEM program. Jansen et al. [19, 20] and Yang [21, 22] applied time-cure superposition principle to model the mechanical behavior during isothermal cure. Epoxy resins were cured under isothermal temperature slightly below the ultimate T_g and DMA frequency sweeps were conducted during cure. This model was applied to predict the cure-induced warpage in the process of QFN packages. De Vreugd [23] conducted the thermo-viscoelastic characterization of curing compound and the effect of aging on the properties. Rezaie Adli [24] investigated shear stress building up in filling process by assuming a Newtonian flow with conversion and temperature-dependent viscosity, and predicted cure- and cooling-induced residual stresses using a simplified analytical model.

2.1 General Form of Cure-Dependent Stress-Strain Relation

General state-dependent viscoelastic equations both in relaxation mode and creep mode by using the convolution integral were proposed in [15]. The relaxation modulus function or creep compliance function can be considered to depend on state variables, such as stress level, temperature, humidity, or degree of cure. With concerns to the application in electronic packaging, the stress level in the packages is relatively low. Therefore, it is reasonable to assume linear viscoelasticity in such applications to reduce the complexity.

Accordingly, the state of stress at time t in a curing polymer system is given by:

$$\sigma_{\mathrm{ij}}(t)=\int_{-\infty}^{t} C_{\mathrm{ijkl}}\left[\alpha(\xi),T(\xi),(t-\xi)\right]\cdot\left(\frac{\partial\varepsilon_{\mathrm{kl}}}{\partial\xi}-\frac{\partial\varepsilon_{\mathrm{kl}}^{*}}{\partial\xi}\right)\mathrm{d}\xi \tag{15.1}$$

where ξ is the curing time or loading time, T the temperature, α the degree of conversion, and $\varepsilon_{\mathrm{kl}}$ strain. $\varepsilon_{\mathrm{kl}}^{*}$ is the initial strain, including thermal and polymerization strains. $(t-\xi)$ represents the time interval between load setup and current time. C_{ijkl} are the relaxation modulus functions or kernel functions, and here they are considered to be the function of curing state (degree of cure), temperature and the history between the curing time (or loading time) and the current time.

Assuming isotropy, the cure-dependent relaxation modulus functions can be decomposed into two independent parts, governed by the shear relaxation moduli and bulk relaxation moduli, respectively. The convolution integral for the stresses at current time t is thus simplified to:

$$\sigma_{\mathrm{ij}}(t)=\int_{-\infty}^{t}\left[2G[\alpha(\xi),T(\xi),(t-\xi)]\cdot\frac{\mathrm{d}\varepsilon_{\mathrm{ij}}^{d}}{\mathrm{d}\xi}+K\left[\alpha(\xi),T(\xi),(t-\xi)\right]\cdot\frac{\mathrm{d}\varepsilon_{v}^{\mathrm{eff}}}{\mathrm{d}\xi}\right]\mathrm{d}\xi \tag{15.2}$$

where $\varepsilon_{\mathrm{ij}}^{d}$ represents the deviatoric strains, with $\varepsilon_{\mathrm{ij}}^{d}=\varepsilon_{\mathrm{ij}}-\frac{1}{3}\varepsilon_{\mathrm{kk}}\delta_{\mathrm{ij}}$, and $\varepsilon_{v}^{\mathrm{eff}}$ is the effective volumetric strain.

Though it is possible that for some cases the relaxation modulus functions in Eq. (15.2) could be measured straightforwardly by experiments, it is not realistic for ongoing curing polymers. Models have to be established to describe the evolution of the relaxation moduli with the curing state.

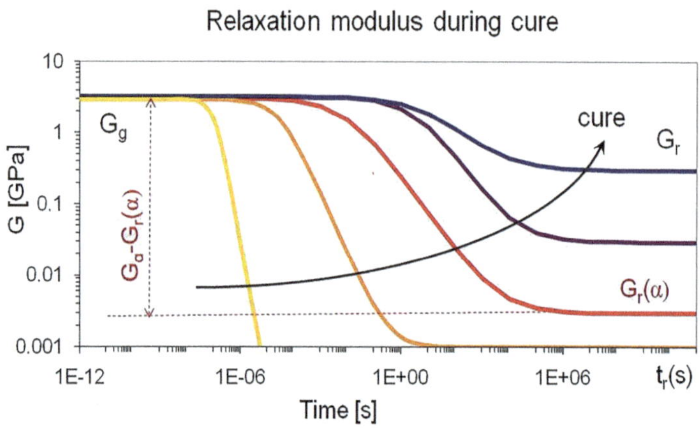

Fig. 15.1 Typical relaxation behavior of crosslinking polymer during cure

2.2 *Cure-Dependent Relaxation Moduli*

During the curing process, the crosslinking polymer transforms from a liquid state into a viscoelastic solid with a relatively high glassy modulus. Figure 15.1 illustrates the typical behavior of the shear relaxation modulus during cure. G_g is the glassy value (the glassy plateau). The other horizontal plateau is the equilibrium modulus (or rubbery modulus). From the observation, the following characteristics regarding relaxation behavior of curing polymers can be drawn:

- The glassy plateau is not strongly temperature- or conversion-dependent.
- The equilibrium modulus (or rubbery modulus) shows strong conversion-dependency. It increases from zero before gelation to a certain value at the fully cured state.
- The relaxation time increases as the curing process proceeds.
- The relaxation amplitude, which is the difference of G_g and $G_r(\alpha)$, i.e. $G_g - G_r(\alpha)$, changes from G_g before gelation to $G_g - G_r^f$ at the fully cured state.
- It is obvious that the relaxation curves at different degree of cure cannot be formed to be one single master curve by simple horizontal or vertical shifting. It means that the curing polymer shows thermo-rheologically complex behavior.

Thus in the modeling, the strong cure-dependency of the rubbery modulus must be considered. Here, the total relaxation modulus is split into two parts, i.e., the equilibrium part and the transient part:

$$G\left[\alpha(\xi),T(\xi),(t-\xi)\right]=G_r\left[\alpha(\xi),T(\xi)\right]+G_d\left[\alpha(\xi),T(\xi),(t-\xi)\right] \quad (15.3)$$

in which $G_r[\alpha(\xi), T(\xi)]$ is the equilibrium shear modulus, which is cure and temperature-dependent. $G_d[\alpha(\xi), T(\xi), (t - \xi)]$ is the transient part or the relaxation part of the total relaxation modulus, which is not only dependent on the cure state and temperature, but also on the history between the curing time or loading time ξ and current time t. For simplification, $\alpha(\xi)$ and $T(\xi)$ are denoted as α and T, respectively.

Because of the numerical efficiency and convenience, Prony series approximations are used to describe the relaxation part. Since the normalized relaxation master curves (the transient part) at different cured states can be eventually overlapped into one single by simply horizontal shift, it is implied that the transient relaxation modulus functions follow a thermo-rheologically simple behavior. Therefore, simplification can be made by taking the weighting factors as cure independent. In such way, the shear relaxation moduli can be described as:

$$G\left[\alpha,T,(t-\xi)\right]=G_r(\alpha,T)+\left[G_g-G_r(\alpha,T)\right]\cdot\sum_{n=1}^{N}g_n\cdot\exp\left[-\frac{t-\xi}{\tau_n(\alpha,T)}\right] \tag{15.4}$$

in which G_g is the glassy value of the relaxation modulus, which is considered to be not strongly temperature and cure dependent. g_n are the weighting factors, with $\sum_{n=1}^{N} g_n = 1$. $\tau_n(\alpha, T)$ is the relaxation time of the nth term.

In analogy to the modeling of the shear modulus, the relaxation bulk modulus is expressed in the same form:

$$K\left[\alpha,T,(t-\xi)\right]=K_r(\alpha,T)+\left[K_g-K_r(\alpha,T)\right]\cdot\sum_{n=1}^{N}k_n\cdot\exp\left[-\frac{t-\xi}{\tau'_n(\alpha,T)}\right] \tag{15.5}$$

where $K_r(\alpha, T)$ is the equilibrium bulk modulus, K_g is the glassy bulk modulus; k_n are the weighting factors, with $\sum k_n = 1$; $\tau'_n(\alpha, T)$ is the relaxation time of the nth term, which is not necessarily the same as the relaxation time in the shear modulus description.

Therefore, the stress-strain relation can be represented as:

$$\begin{aligned}\sigma_{ij}(t)=&\int_{-\infty}^{t}2\left\{G_r(\alpha,T)+\left[G_g-G_r(\alpha,T)\right]\cdot\sum_{n=1}^{N}g_n\cdot\exp\left[-\frac{t-\xi}{\tau_n(\alpha,T)}\right]\right\}\cdot\frac{d\varepsilon_{ij}^{d}}{d\xi}d\xi\\&+\int_{-\infty}^{t}\left\{K_r(\alpha,T)+\left[K_g-K_r(\alpha,T)\right]\cdot\sum_{n=1}^{N}k_n\cdot\exp\left[-\frac{t-\xi}{\tau'_n(\alpha,T)}\right]\right\}\cdot\frac{d\varepsilon_{v}^{eff}}{d\xi}d\xi\end{aligned} \tag{15.6}$$

As shown in Eq. (15.6), because the equilibrium moduli and the relaxation amplitudes in the transient parts are considered to be cure-dependent, the model actually describes a “quasi” thermo-rheologically simple behavior.

2.3 Equilibrium Moduli (Rubbery Moduli)

Basically, there are two approaches to establish models for describing the evolution of the equilibrium shear modulus during the curing process. One is based on the theory of rubber elasticity, in which the rubbery modulus is considered to be proportional to the effective concentration of network chains. Simon et al. [10] used this model to describe the effect of temperature and conversion on the rubbery shear modulus for an epoxy resin.

The other model is based on scaling analysis and was originally proposed by Adolf et al. [11]. Starting from the percolation analysis, they came up with a model relating the shear rubbery modulus with the conversion as follows:

$$G_r(\alpha,T) = G_r^f(T)\left[\frac{\alpha^2 - \alpha_{gel}^2}{1-\alpha_{gel}^2}\right]^{8/3} \tag{15.7}$$

where $\alpha(\xi)$ and α_{gel} are the degree of cure at the curing time ξ and at the gel point, respectively. G_r^f is the rubbery shear modulus at the fully cured state (α = *1.0*). Note that G_r^f is often temperature-dependent for most polymers. Yang et al. [21] applied the model to predict the evolution of the equilibrium shear modulus during the isothermal curing of an epoxy resin and also for the epoxy composites. A good agreement between the experimental measurement and the model prediction was observed.

The degree of conversion at the gel point, α_{gel}, can be determined either by experimental measurements or using theoretical calculation. Winter et al. [25] proposed a general criterion that can be used to identify the gel point. Ferry [7] proposed an equation to calculate the conversion at the gel point based on the fact that the conversion at gelation is determined by the (average) number of reaction sites of the monomer, and the mixing ratio of the two components.

In order to model the cure dependency of the rubbery bulk modulus, we propose an equation similar to Eq. (15.7):

$$K_r(\alpha,T) = K_r^l + \left[K_r^f(T) - K_r^l\right]\left[\frac{\alpha^2 - \alpha_{gel}^2}{1-\alpha_{gel}^2}\right]^{8/3} \tag{15.8}$$

in which K_r^l is the bulk modulus at liquid state, K_r^f is the bulk modulus at the fully cured state. The extra constant K_r^l in Eq. (15.8) is added to ensure a minimum value for the bulk modulus. A zero value for the bulk modulus would have been unacceptable from a physical point of view since this corresponds to a material with zero resistance to hydrostatic compression. The bulk modulus of liquid molding compounds is expected to be close to that of other liquids, i.e., in the range from 1 to 3 GPa.

2.4 The Effect of Temperature and Conversion on the Relaxation Behavior

The effect of temperature and cure on the relaxation behavior is usually described through the shift factor, which is either used for "reduced time" or "reduced frequency". For fully cured polymers, the time-temperature shift factor is often described by applying the WLF equation and or Arrhenius behavior or the general Vogel equation. While for curing polymers, the effect of conversion on the temperature dependence of the shift factors must be included. In the early work by Suzuki and Miyano [26], they found that the activation energy for the shift factors was independent of the degree of cure for an epoxy resin. Similarly, Plazek and Chay [27] found in their model epoxy system that the viscoelastic shift factors followed the same Vogel temperature dependence independent of the degree of cure when the reference temperature was taken as the glass transition temperature.

As indicated above, we assume that the transient part or the relaxation part follows a thermo-rheologically simple behavior. Under this assumption, then the effect of temperature and the degree of cure on the relaxation behavior can be established by applying the principles of time-temperature and time-cure superposition. The relaxation is expressed as [10]:

$$\tau_n(\alpha, T) = a_{T,\alpha} \cdot \tau_n(\alpha_{\mathrm{ref}}, T_{\mathrm{ref}}) \tag{15.9}$$

where $\tau_n(\alpha_{\mathrm{ref}}, T_{\mathrm{ref}})$ is the nth relaxation time corresponding to the reference temperature $T_{\mathrm{ref}}$and reference conversion α_{ref}. $a_{T,\alpha}$ is the shift factor that accounts for the temperature and cure dependence.

When the reference temperature is chosen as the glass transition temperature, according to the general Vogel equation, the shift factor for considering both the temperature and cure effects can be expressed as:

$$\log a_{T,\alpha} = \left(\frac{A}{T - T_\infty} - \frac{A}{T_{\mathrm{ref}} - T_\infty}\right) - \left(\frac{B}{T_{g\alpha} - T_\infty^{'}} - \frac{B}{T_g^{\mathrm{ref}} - T_\infty^{'}}\right) \tag{15.10}$$

where A and B are constants, T_{ref} is the reference temperature, T_∞ is a constant temperature, $T_{g\alpha}$is glass transition temperature, and $T_\infty^{'}$ is also a constant temperature.

3 Experimental Characterization of Cure-Dependent Viscoelastic Properties

Before implementation of the cure-dependent constitutive model into finite element simulations, experimental characterization has to be performed to investigate the material behavior and to obtain the cure-dependent parameters. As mentioned before, corresponding to the proposed model, the following cure-dependent behavior should be investigated experimentally:

- The cure kinetics and the Tg-α relationship;
- The evolution of relaxation moduli that are cure and temperature-dependent;
- Thermal and chemical shrinkage.

It is not an easy task to measure the material properties during the curing process, since the mechanical properties of the material undergo a huge change. During cure, polymers transfer from a liquid at the beginning to a crosslinked solid in the fully cured state, and simultaneously the stiffness is being built up. This imposes restrictions on the choice of possible sample geometries and measurement methods. Basically, people used two different approaches for curing DMA measurements. One is a DMA test on partly cured samples. Samples with different conversion levels are made using quenching to stop curing or using different chemical stoichiometric ratios [27]. The disadvantages of this method are that it is time-consuming and the variations in the sample making may have effect on the test results. The other approach is continuously measuring during a part or the whole curing process. In most cases, this approach has been used in the isothermal curing conditions. Eom et al. [28] used a torsional parallel plate rheometer for measuring the shear modulus during the isothermal curing of an epoxy-amine system. Yang et al. [21, 22] performed DMA measurements on an epoxy system during isothermal curing using a double simple shear test set-up. Due to the dramatic change of the material properties during cure, strict demands for the test machine and the test setup are required to get accurate results.

Hof [18] proposed a continuous DMA approach, "intermittent cure", in which the curing profile includes a number of cycles of isothermal cure, cooling down, and heating up. In between the cure intervals the material is cooled down to room temperature such that the curing is virtually stopped. DMA measurements are then performed during cooling down or heating up. Such a method enables people to obtain the relaxation behavior at different degree of cure using one single specimen.

In this part, the experimental characterization of the cure-dependent properties of the thermosetting polymers is presented.

3.1 *Cure Kinetics and* T_g - α *Relationship*

Epoxy compound for electronic packaging is usually comprised of epoxy resin, hardener, and fillers as well as other proprietary additives such as coupling agent, silicone modifiers, release agent, anti-oxidant, color pigment, flame retardant, etc. It has a complex chemical structure.

In order to focus on the polymer crosslinking behavior and exclude the influence of other additives, an epoxy resin with the basic chemical structure as in compound is chosen as the matrix material. As a model system, the epoxy resin is Araldite EPN 1180, a novolac epoxy. It has an epoxy/hydroxy equivalent of 175 ~ 182 (g/eq). The hardener is Bisphenol-A, with an epoxy/hydroxy equivalent of 114.1(g/eq). Triphenylphosphine (TPP) is used as an accelerator for the curing of the epoxy resin. In order to study the effect of filler on the cure-dependent viscoelastic properties of the composites and thermo-mechanical behavior of the packages, fused silica spheres are used as filler to make model EMCs with different filler loading. The filler has a median diameter of 15 μm, with a density of 2.20 g/cm^3. To make a filled composite, a certain amount of silica filler was added to the matrix resin. Three levels of filler loading were used. The final TPP concentrations and the filler loading percentage used in the experiments are shown in Table 15.1. The mixing was done in a special mixer (Kenwood, K. beater).

The chemical reaction of a thermosetting polymer results in the formation of a crosslinked network polymer structure. The physical and mechanical properties of the polymer are dependent on the degree of cure. Therefore, understanding the kinetic behavior is important not only for modeling of the cure-dependent viscoelastic behavior but also to optimize the process parameters and therefore to obtain and improve the mechanical quality of the end product.

The degree of cure, α, is often defined as:

$$\alpha(t) = \frac{H(t)}{H_u} \quad (15.11)$$

where $H(t)$ is the heat released by the reaction until time t and H_u is the ultimate heat of reaction.

Mechanistically, the curing kinetics of thermosetting polymers can be divided into two general categories: nth-order curing reaction kinetics and autocatalytic curing reaction kinetics. For the rate of conversion, Sourour and Kamal [29] proposed the following semi-empirical model to take into account both nth-order and autocatalytic curing reactions:

Table 15.1 Catalyst concentration and filler loading of the epoxy compounds

	F00	F40	F50	F65
TPP concentration [g/100 g epoxy]	0.85	1.40	1.60	2.35
Actual filler loading [wt %]	–	42.5%	50.7%	65.6%

$$\frac{d\alpha}{dt} = k_0 \exp\left[-\frac{Q}{RT}\right]\alpha^m (1-\alpha)^n \tag{15.12}$$

where k_0 is a constant, Q the activation energy, R the gas constant, T the absolute temperature, m and n represent the reaction orders.

It is well accepted that for many thermosetting polymers there is a unique relationship between the glass transition temperature and the degree of conversion, independent of the cure temperature and thermal history [30]. The modified empirical DeBenedetto equation is used to describe such a relationship, as:

$$\frac{T_g - T_{g0}}{T_{g\infty} - T_{g0}} = \frac{\lambda\alpha}{1-(1-\lambda)\alpha} \tag{15.13}$$

where T_{g0} is the T_g of the uncured resin ($\alpha = 0$), $T_{g\infty}$ is the T_g of the fully cured resin ($\alpha = 1$), and λ is an adjustable, structure-dependent parameter. When the conversion reaches a high level (e.g., above 95% conversion), a quantitative measurement of the residual heat is very difficult due to the small exotherm. In that case, measuring T_g is a more sensitive way to determine the degree of conversion.

DSC was used to obtain the curing process parameters, such as the degree and rate of chemical conversion, glass transition temperature (T_g). The DSC measurements are performed in a differential scanning calorimeter of TA Instruments (DSC2920). For isothermal cure, the samples were cured under the prescribed temperatures ranging from 90 to 150 °C for different time. After the isothermal run, the samples are cooled down quickly to −40 °C, and rescanned at a heating rate of 10 °C/min to 300 °C to determine the residual heat. For dynamic cure, six different constant rates as 1, 2, 5, 10, 15 and 20 °C/min are used. The samples are first cooled down to −50 °C, equilibrated for 2 min and then heated to 320 °C using the six different rates.

In order to determine the glass transition temperature (T_g) as a function of degree of conversion, a series of isothermal curing and subsequent heat scanning was performed for each material. The samples were cured in the DSC cell at the prescribed temperatures for various periods of time. After each isothermal run the partly cured samples were rapidly cooled down to −40 °C, equilibrated and rescanned at a heating rate of 10 °C/min to 300 °C to determine the T_g and the residual heat (ΔH_R). The corresponding residual heat was used to calculate the degree of conversion.

Figure 15.2 shows the relationship between T_g and conversion from DSC measurements. It can be seen that the experimental data for the three filled materials generally fall on an approximately single curve, and for the unfilled resin the other single curve could be seen. It can be concluded that there indeed exists a one-to-one relationship between T_g and conversion for all of the resins, which is independent of the cure temperature. The modified DiBenedetto equation (Eq. 15.13) is then used as a model to fit the experimental data, as shown in Fig. 15.2 in the solid lines.

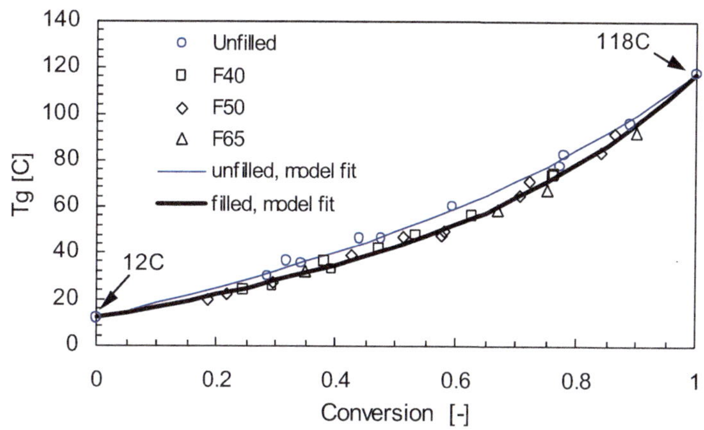

Fig. 15.2 Relationship between Tg and the degree of conversion for the unfilled and filled materials. The solid curves represent model fitting from the modified DiBenedetto Eq. (15.13). The symbols represent the DSC measured values

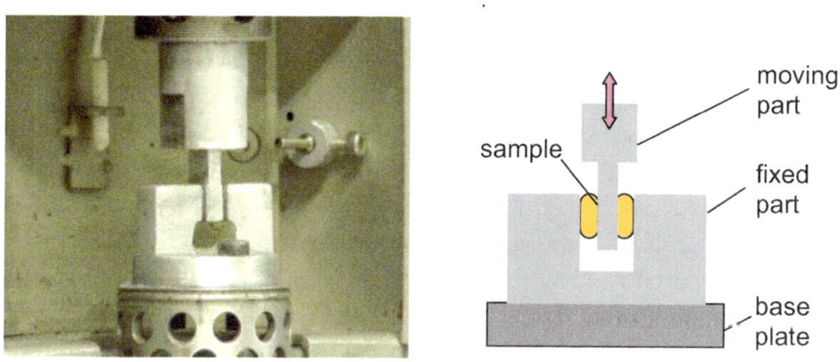

Fig. 15.3 Shear DMA setup with the schematic illustration (right)

3.2 *Cure-Dependent Shear Modulus*

The shear DMA measurements are performed in Metravib VA4000 Viscoanalyzer, as shown in Fig. 15.3. The equipment can operate either in stress-controlled or strain-controlled mode with frequencies up to 100 Hz and with a force range of 0.1–150 N. The displacement resolution is 0.1 μm. With the attached temperature chamber, test temperature up to 300 °C can be applied and controlled.

Due to the special requirement for holding the uncured resins, the commercial test fixtures are not suitable for DMA measurement of curing resins. New shear test setups were developed. In our measurements, the strain control mode is used.

3.2.1 Cure-Dependent Rubbery Shear Moduli

For fully cured material, DMA, relaxation, and creep tests can be used to measure the rubbery modulus. The tests should be conducted at a temperature well above the ultimate T_g. It is much more difficult to measure the rubbery modulus during the curing process. In that case relaxation and creep tests are not suitable due to the ongoing stiffness build-up as the curing proceeds. Therefore a novel method called Ultra Low Frequency Scan (ULFS) DMA is used to characterize the rubbery shear modulus during cure. In this method a displacement oscillation with a very low frequency (e.g., 0.001–0.01 Hz) is applied to the sample during isothermal cure at a temperature above the glass transition T_g and the reacting force is measured. Preliminary tests at both 0.005 and 0.01 Hz showed that measured modulus values for these epoxy systems were frequency independent, which shows that we really measure the rubbery plateau values. Figure 15.4 shows schematically the low-frequency DMA method for the rubbery measurement. Displacement oscillation with constant amplitude was applied to specimen and reacting force was recorded. As the curing conversion increases, passing through the gel point, the amplitude of the force increases. The rubbery modulus can be calculated either using the force amplitude within one cycle or using the increment of the displacement and force during a certain time increment.

The DMA tests during cure were performed in the viscoanalyzer using a dual-plane shear test setup as shown schematically in Fig. 15.3. Displacement was applied by the moving part and the base plate was connected to the force transducer. The measured displacement and force signals were recorded by a data acquisition system and used to calculate the stiffness increase with curing time.

The results of the Ultra Low Frequency Scan method and the applied temperature profile are shown in Fig. 15.5. The resin was cured in two steps: first at 130 °C for 5 h and then at 160 °C for 2 h. Afterward, the sample was cooled stepwise down to 120 °C to check the temperature dependency of the fully cured rubbery modulus.

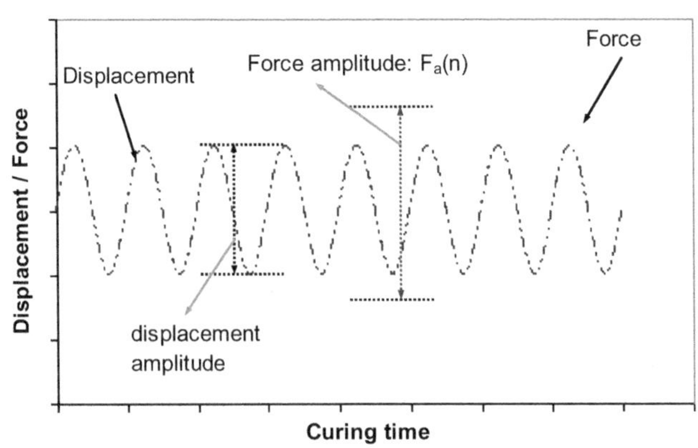

Fig. 15.4 Schematic illustration of an ultra-low frequency scan during cure

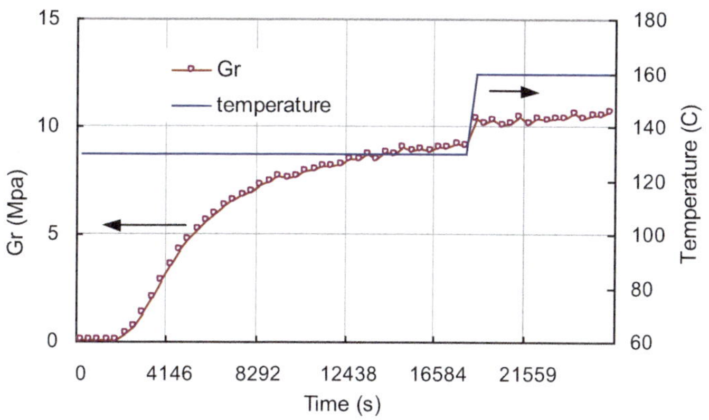

Fig. 15.5 Rubbery shear modulus versus curing time under two-step curing

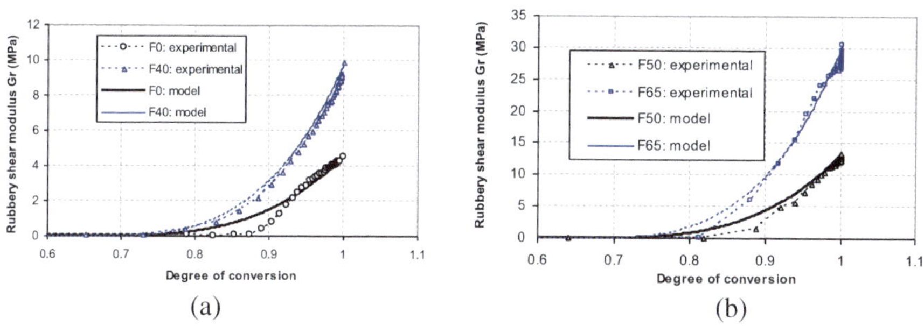

Fig. 15.6 Rubbery shear modulus evolution with conversion at 130 °C; (**a**): F00 and F40; (**b**): F50 and F60

The temperature for the first curing step is chosen such that an appropriate curing rate is obtained and the second curing step with higher temperature enables the resin to be in almost fully cured state. The frequency used is 0.01 Hz. It can be seen that the modulus is zero at liquid state and increases from certain point (the gel point).

Figure 15.6 shows the modulus build-up with conversion level for all materials at 130 °C. It can be seen that before 80% conversion, the modulus is very low. After that, it increases quickly during the final conversion period. Nearly two-thirds of the modulus is built up in the conversion range from 90 to 100%. It also indicates that the filler percentage has significant reinforcement on the rubbery modulus, especially at highly filled state. The 65% filled compound (denoted as F65) shows a sharp increase of rubbery modulus compared to the 50% filled material.

It also shows a comparison of the experimental measurement and the model prediction of the rubbery shear modulus for the materials with different filler loading during 130 °C isothermal cure. The evolution of the conversion level with curing time was calculated from Sourour and Kamal's kinetic model. The solid lines in the figures represent the prediction of Adolf's model. It is seen that the model predictions match well with the experimental data for F40, F50, and F65. For the unfilled resin, before 95% conversion a relatively large difference between the model prediction and the measured data is seen. In Adolf's model, the degree of cure at gel point α_{gel} is required. Here the gel point is obtained using DMA approach and the kinetics model, which may not match well with the real condition in the DMA measurements (such as sample handling and temperature ramping to the curing temperature). The deviation in prediction of α_{gel} could result in the difference between the model prediction and the measured data of the rubbery modulus, since Adolf's model is very sensitive to α_{gel}.

3.2.2 Isothermal Cure Approach

In the isothermal cure method, the sample is cured under isothermal condition and DMA frequency sweeps are applied continuously during the whole curing process. This measurement can be used for continuous probing of viscoelastic properties during cure. The cure temperature should be carefully chosen in order to be able to follow the changes in the full transition region. If the cure temperature is too high only the changes in the range close to the rubbery plateau are recorded. If the cure temperature is too low the final conversion level is low. After a few preliminary tests, the curing temperature is chosen as slightly below the ultimate T_g of the resins. To obtain higher conversion level, multi-isothermal cure steps can be utilized. In addition, if at the end of isothermal curing a cooling down with frequency sweeps is applied subsequently, then the glassy modulus at the end of cure and the viscoelastic properties of the fully cured resins can be determined.

Shear DMA measurements on the epoxy system are performed using the Viscoanalyzer VA4000. Continuous frequency sweeps during isothermal cure are used for continuous probing of viscoelastic properties during cure. Seven logarithmically spaced frequencies between 0.1 and 60 Hz are applied. Various cure temperatures are investigated.

To illustrate the procedure, the evolution of the storage and loss shear moduli under 105 °C isothermal cure (for various frequencies) with ongoing cure is presented in Fig. 15.7. To describe their dependency on the conversion level, the degree of cure is calculated using the Eq. (15.12). As discussed in Sect. 15.2.2, the influence of cure on the relaxation behavior of thermosetting polymers may show thermo-rheologically complex behavior, because the rubbery plateau increases with ongoing cure. When we examine the cure dependency, the effect of the evolving rubbery modulus should be taken into account.

In accordance with Eq. (15.4), the "transient parts" of the storage shear moduli are obtained by subtracting the rubbery modulus. Figure 15.8 illustrates the transient part of the storage shear modulus ($G' - G_r(\alpha, T)$) at different conversion levels as functions of the applied frequencies. Then the individual curves at different conversion can be shifted along the frequency axis to form an overlapping master curve and the cure-dependent shift factor can be obtained. The solid line is the transient part of the storage moduli at the fully cured state (T_{ref}=105 °C). The "cure shift

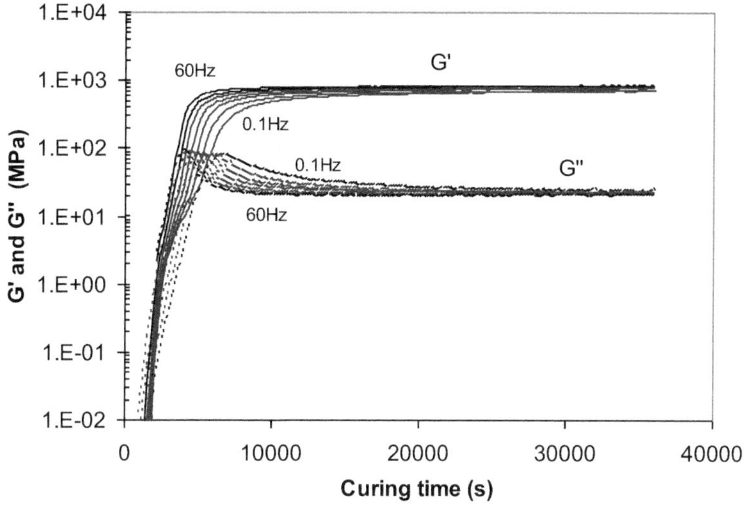

Fig. 15.7 Evolution of the shear storage and loss modulus of F65 during 105 °C isothermal curing

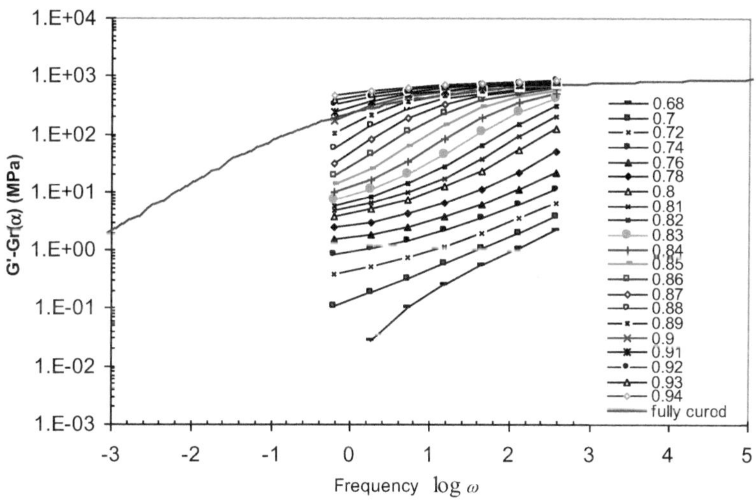

Fig. 15.8 Transient parts of storage moduli (G′ -Gr) versus frequency for various degree of cure + a "fully cured" master curve under 105 °C isothermal curing

factor" a_α is obtained after shifting the individual curves to match with the "fully cured" master curve. The dependency of the shift factor on the conversion is shown in Fig. 15.9a. In Fig. 15.9b, the shift factor is illustrated as a function of the glass transition temperature. It can be described quite well by the Arrhenius behavior as:

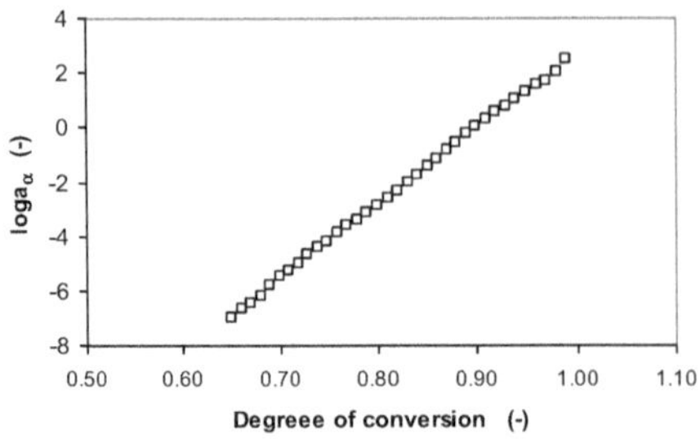

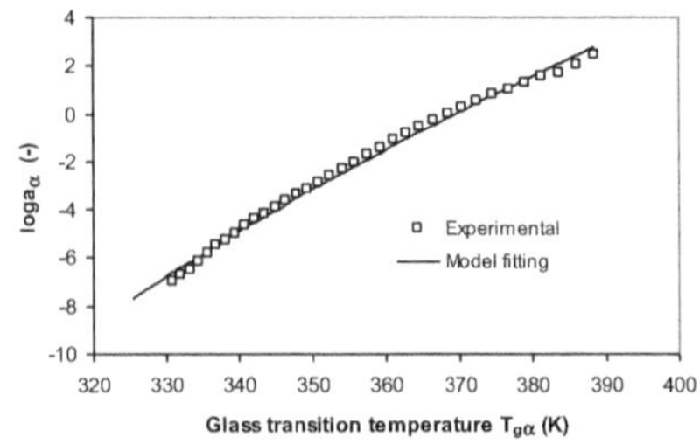

Fig. 15.9 Shift factor versus conversion (**a**) and versus glass transition temperature (**b**)

Table 15.2 Parameters for the shift factors

	F00	F40	F50	F65
ΔH_α	492.29	417.06	437.74	399.79
T_0	373.33	385.65	368.07	369.68

$$\log a_\alpha(\alpha) = -\frac{\Delta H_\alpha}{2.303R}\left(\frac{1}{T_{g\alpha}} - \frac{1}{T_0}\right) \tag{15.14}$$

where ΔH_α is the activation energy term corresponding to conversion. The procedure is repeated for various temperatures and for compounds with various filler loadings. The final master curves for the relaxation shear modulus G for various compounds (with different filler %) can be obtained. The fitting parameters for the conversion shift factors are presented in Table 15.2.

Though the isothermal approach is relatively simple, the temperature effect on the relaxation behavior at different curing states cannot be obtained by this method. Alternatively, multi-isothermal cure steps (to get fully cured) followed with a cooling down with frequency sweeps can be utilized to measure the temperature dependence of the fully cured material.

3.2.3 Intermittent Cure Approach

The intermittent cure approach was proposed by Hof [18] and was successfully applied to the unfilled resin and the filled resins in [20]. Here this approach is used to further investigate the effect of the curing state on the relaxation behavior. The basic idea of this method is that a temperature profile consisting of isothermal curing sections interrupted by several linear cooling and heating intervals is used and continuous frequency sweeps are applied to the sample during the temperature history. In this way a temperature scan is made at various stages during cure. Note that this method can only be used for slowly curing systems, since it requires scanning to temperatures above the Tg at each cure state without significant reaction. For many commercial resins the reaction already starts in the glassy state such that the viscoelastic range of the master curve cannot be obtained at a single conversion level.

Preliminary trial tests are done before the proper time intervals for each isothermal session are determined to get desired corresponding conversion levels. Figure 15.10 shows the temperature profile used in the intermittent cure. The ramp rate for the heating scanning is 2 °C/min and is identical for all the heating stages. DMA measurement data are obtained for each heat scanning, in which the material is in different degree of cure. Using the data processing procedure similar to that used for the fully cured material, the master curves of the storage and loss moduli for each of the curing states can be established. Figure 15.11 shows the master curves of the storage modulus at different degree of cure versus the reduced frequency, with an identical reference temperature of 120 °C. The master curves of the transient part of the storage modulus can also be established by subtracting the corresponding rubbery modulus. These master curves can be further shifted using a reference conversion to form a "general" master curve. Figure 15.12 shows the shifted master curves of the transient part of the storage moduli, with an identical reference temperature of 120 °C and using a reference conversion of 0.96 (corresponding to Tg of 107 °C). The shift factor related to conversion agrees well with the results of the isothermal cure approach. We can see that the master curves of the transient part at different curing states shows a good match by only horizontal shifting.

Figure 15.13 shows all the shift factor curves for establishing the master curves of the corresponding conversions, where 120 °C is used as the reference temperature. If the same data is re-plotted with T_g as a reference temperature, all shift factor curves are seen to merge into a single curve (see Fig. 15.14). Note that the deviations from this overall shift factor curve mainly occur at the temperature ranges well above $T_{g\alpha}$ or well below $T_{g\alpha}$ (i.e., outside the viscoelastic range) where the shift

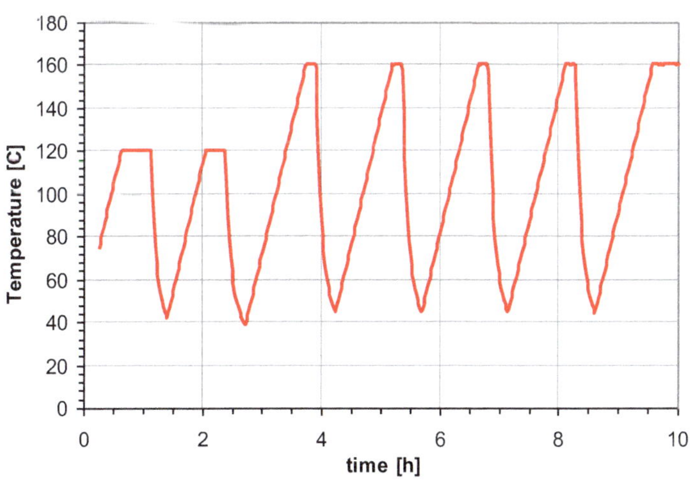

Fig. 15.10 Temperature profile used in the intermittent cure approach

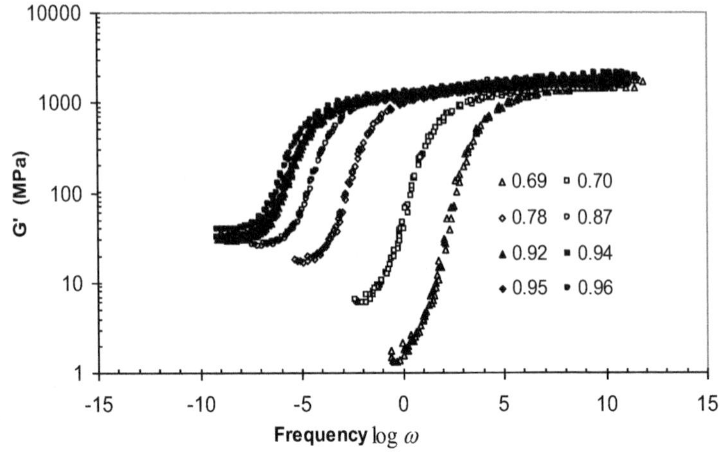

Fig. 15.11 Master curves of storage moduli at different conversion levels, established using reduced frequency. The reference temperature is 120 °C

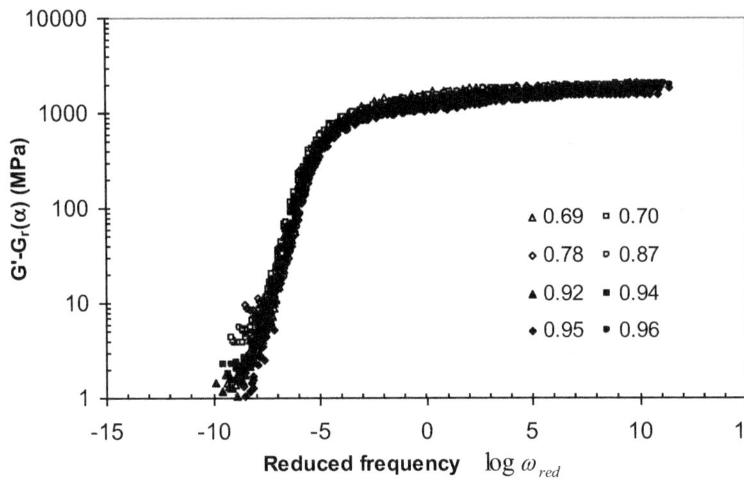

Fig. 15.12 Master curves of the transient parts of the storage moduli, derived by horizontal shifting using 0.96 as the reference conversion; the reference temperature is 120 °C

factor loses its usefulness (shifting is not needed in the glassy and rubbery elastic plateaus). This indicates that for the epoxy system in this study the temperature effect on the shift factors is independent of the degree of cure when the glass transition temperature is chosen as the reference temperature. As mentioned in Sect. 15.2.3, previous research work indicated that for many polymers temperature dependency of the shift factors is independent of the degree of cure when the reference temperature was taken as the glass transition temperature.

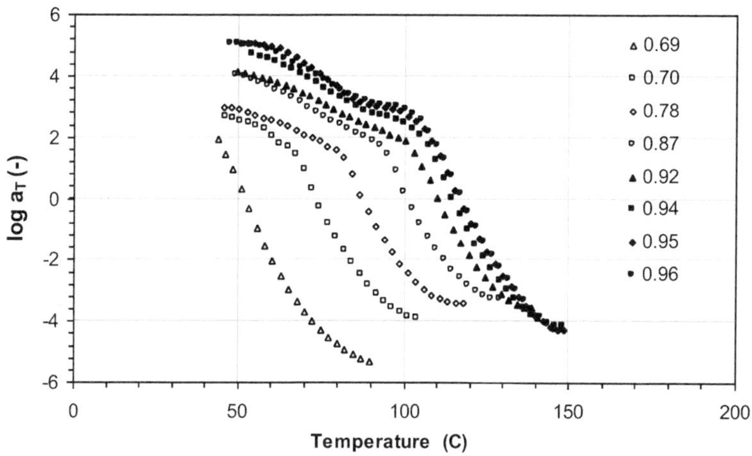

Fig. 15.13 Shift factors for obtaining the master curves of different curing states, using a reference temperature of 120 °C

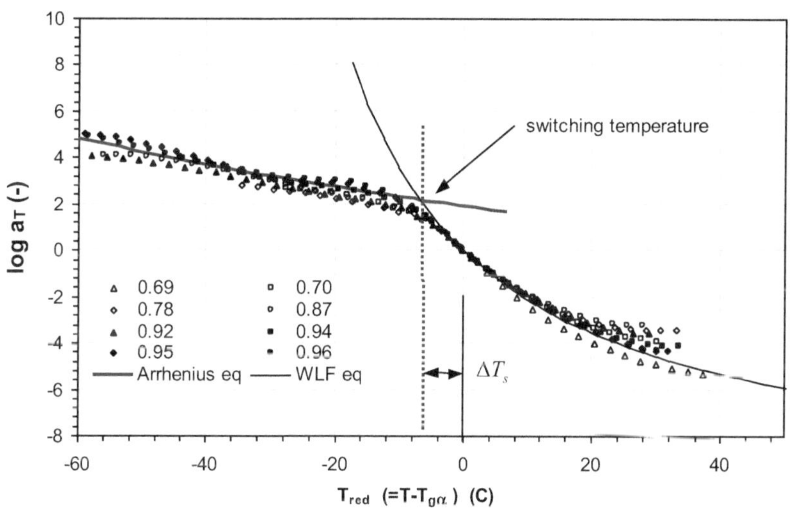

Fig. 15.14 Shift factors as a function of reduced temperature; reference temperature is the glass transition temperature $T_{g\alpha}$.

The shift factors for the temperature and conversion effects can then be expressed in an "integrated" way by using the new concept of "reduced temperature". The concept of reduced temperature is introduced, as $T_{\mathrm{red}} = T - T_{g\alpha}$. Thus, the two parts of the curve are fitted to the WLF equation and Arrhenius equation, respectively. The equation for the shift factors can then be written as:

$$\log a_{T,\alpha} = -\frac{C_1\left(T_{\text{red}} - T_{\text{ref}}^{\text{WLF}}\right)}{C_2 + T_{\text{red}} - T_{\text{ref}}^{\text{WLF}}}, \quad T \geq T_s \tag{15.15}$$

and

$$\log a_{T,\alpha} = \frac{\Delta H_T}{2.303R}\left(\frac{1}{T_{\text{red}}} - \frac{1}{T_0}\right) \quad T < T_s \tag{15.16}$$

in which $T_{\text{red}}$is the reduced temperature, with $T_{\text{red}} = T - T_{g\alpha}$. The switching temperature T_s is a few degrees lower than the glass transition temperature and the difference ΔT_s (=$T_s - T_{g\alpha}$) is almost the same for different curing states. For the resin of F65, $\Delta T_s$is about −6.55 °C.

The measured data shown in Fig. 15.14 can be fitted to Eqs. (15.15) and (15.16). The fitting parameters for each of the materials are given in Table 15.3. It is seen that the shift factor is independent of the degree of cure if the glass transition temperature is used as the reference temperature.

3.3 Cure-Dependent Bulk Modulus

The bulk modulus is one of the fundamental material properties, which relates the stress response to the applied volumetric change. Accurate measurement of the bulk modulus is extremely difficult, especially during cure. Even for fully cured polymers, no standard measurement tools for this property are available.

In principle the relaxation bulk modulus can be derived from the relaxation Young's modulus and the relaxation shear modulus by iteratively solving their Laplace interrelations [31]. Therefore, as our first attempt to establish the relaxation bulk modulus, longitudinal DMA measurements were performed on partly cured strip specimens. With DMA tests on the strip specimens the relaxation Young's modulus $E(t - \xi)$ is first established. With known relaxation Young's modulus and known relaxation shear modulus $G(t - \xi)$, the relaxation bulk modulus can be

Table 15.3 The parameters for the shift factors of Eqs. (15.15) and (15.16) using the reduced temperature

	F00	F40	F50	F65
C_1 [−]	10.7	10.7	10.7	10.7
C_2 [K]	40.7	40.7	40.7	40.7
$T_{\text{ref}}^{\text{WLF}}$ [K]	0	0	0	0
ΔH_T [KJ/Mol]	121.18	91.02	67.13	55.14
T_0 [K]	298.05	299.60	313.50	337.51
ΔT_s	−8.72	−5.2	−4.59	−6.55

obtained by solving the Laplace interrelation. However, difficulties were encountered in handling and clamping of the partly cured specimens. This method appeared to fail for specimens with a relatively low degree of cure or in the rubbery state. Besides, because of measurement inaccuracies combined with loss of accuracy during the iteration process it turned out that the obtained relaxation bulk modulus values showed a large scatter and were inadequate. Therefore, this method was discarded and new approaches had to be explored.

The pressure-deformation approach was used to obtain the cure-dependent behavior of the bulk modulus. An experimental setup with a pressure cell was built, as shown in Fig. 15.15a. This setup consists of a thick-walled steel cylinder filled with oil, a piston, a heating element, a thermocouple, and a pressure transducer. The sample specimens with embedded strain gages were placed in the cylinder and the deformations were measured with the strain gages. We first tried the normal strain gages with plastic substrate. It turned out that at higher temperatures the gage substrate becomes of comparable stiffness as that of the specimens, which makes the measurements in that region unreliable. Therefore a special kind of strain gage, called "free grid strain gage", was used to replace the normal strain gages. The free grid strain gage can be completely embedded into the materials. Figure 15.15b shows the specimen made of the unfilled resin, in which a free grid strain gage was embedded.

The set-up was installed in a test machine and a constant force was applied to the piston for a predetermined period of time. All signals were recorded with the data acquisition system. The bulk relaxation modulus is therefore measured as

$$K(t) \approx \frac{-p}{\varepsilon_V(t)}, \tag{15.17}$$

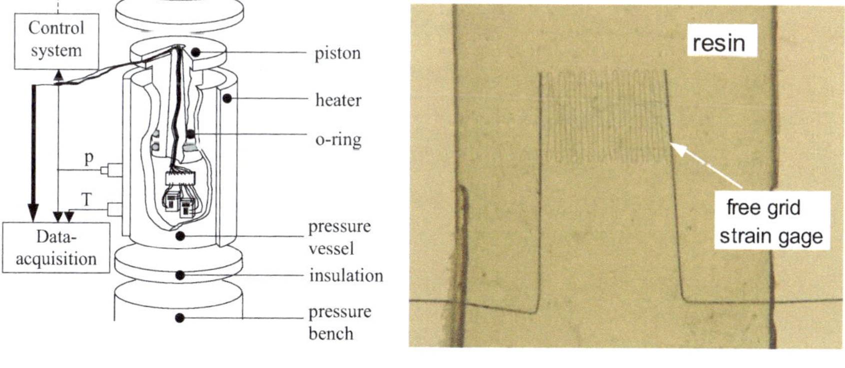

(a) (b)

Fig. 15.15 Pressure cell test setup for measuring the volumetric creep compliance of EMC, (**a**). schematic illustration of the experimental setup; (**b**) specimen with embed free grid strain gages in which p is the pressure and ε_v is the volumetric strain. Step pressure loadings up to 90 MPa (900 bar) were applied. The device was calibrated using calibration specimen with known bulk moduli (Quartz, Aluminum, and Polystyrene).

The bulk creep tests were performed on the fully cured materials using a step loading of 10 MPa. The temperatures used in the tests were 20 °C (ambient temperature), another four levels from 80 to 120 °C. Above about 100 °C these moduli dropped rapidly, as expected. The strain variations after step loading were recorded for 20 hours and plotted on a logarithmic time scale. Figure 15.16 shows typical creep curves obtained from the measurements, where the pressure history and the measured creep strains for F40 and F65 are presented. After having established the volumetric creep compliance for various temperatures, the corresponding volumetric (or bulk) relaxation modulus curves can be derived by solving their Laplace transform interrelation. The instantaneous values of the moduli at 20 °C were taken as the glassy moduli, which are listed in Table 15.4. The relaxation curves obtained at the above-mentioned temperatures can be used to establish a part of the master curves of relaxation bulk moduli (see Fig. 15.17).

From the pressure cell experiment it was found that only the glassy bulk modulus values at higher degree of cure could be obtained. For the rubbery values the disturbance of the samples by the free grid strain gages appeared to be non-negligible. The measurement data of the pressure cell test for the rubbery state (120 °C) is considered to be less reliable.

Then a disk-shaped specimen test was used to measure the rubbery bulk modulus. After measuring the axial stiffness of the specimen, with the known shear modulus the bulk modulus can be obtained. Compression DMA tests were performed using the disk-shaped specimen method. These measurements were performed in the Metravib Viscoanalyzer and in Zwick test machine, where different plate diameters were used. Then we tried to establish the master curves of relaxation bulk moduli based on the rubbery bulk modulus and the relaxation curves obtained at the above-mentioned temperatures. The relaxation curves obtained from pressure cell

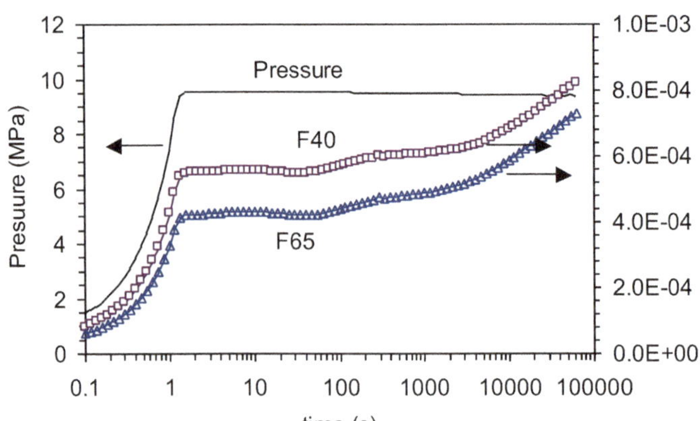

Fig. 15.16 Typical creep curves obtained from the measurements

Table 15.4 The glassy and rubbery bulk moduli

	F00	F40	F50	F65
K_g [MPa]	4770	6600	8330	8960
K_r [MPa]	210	258.71	350.72	524.56

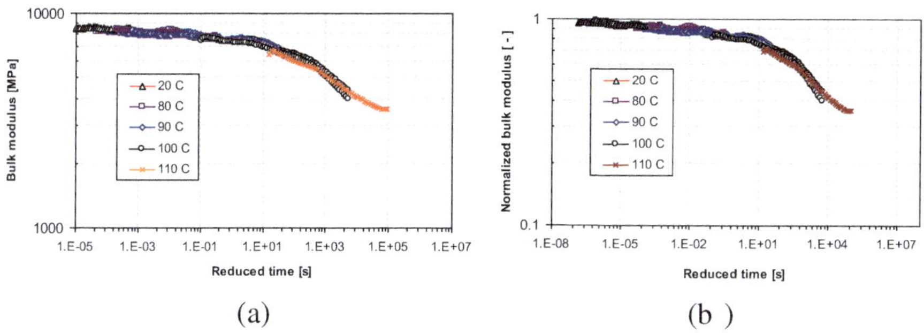

Fig. 15.17 (**a**) The master curve of relaxation bulk modulus of F65 and (**b**) the normalized master curve of F65

tests can be shifted along the time axis to obtain the bulk modulus master curves. Because these tests only provided the curves at certain temperatures, only a part of the bulk modulus master curve can be acquired. Figure 15.18 shows the master curve and the normalized master curve of F65. Then a Prony series approximation of the normalized bulk modulus curve can be made and the coefficients are presented in Table 15.5. Since no full bulk relaxation curves of the other materials were available, we assume that these materials have the same bulk relaxation kernel as F65, in analogy to the relaxation kernel for shear relaxation modulus. This means that the same Prony series data apply in combination with their own glassy bulk moduli and rubbery moduli given in Table 15.4.

3.4 The Curing Shrinkage Measurement

A specially designed online density measuring setup based on the Archimedes principle was used to measure the chemical shrinkage of the resins during the curing process. The setup consists of an accurate weighing device, a Teflon cup, an oil bath with controlled heating device, and a data acquisition system. The uncured sample of about 20 ~ 30 g was carefully weighted and put in the Teflon cup. Then the cup and the sample were put into silicon oil, which has been kept at the desired temperature. The evolution of the sample's density was obtained by measuring the weight change.

The curing shrinkage measurements were carried out under isothermal curing condition at 90 °C, 120 °C, and 150 °C for all of the resins. The kinetics model described in Sect. 15.3.1 is used to calculate the degree of conversion and a plot of

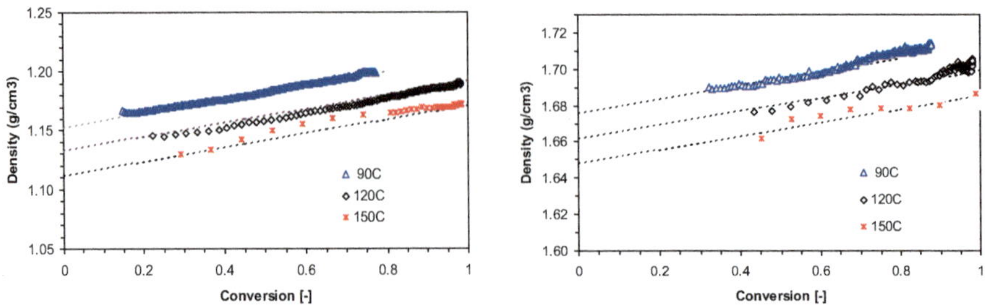

(a) (b)

Fig. 15.18 Density evolution versus conversion (**a**): F00, (**b**): F65

Table 15.5 Relaxation times and Prony coefficients for the relaxation bulk modulus of F65

Relaxation times τ_n [s]	Prony coefficients k_n
1.0E-05	0.0389
1.0E-04	0.0287
1.0E-03	0.0509
1.0E-02	0.0274
1.0E-01	0.0872
1.0E+00	0.0266
1.0E+01	0.160
1.0E+02	0.153
1.0E+03	0.261
1.0E+04	0.187
1.0E+05	0.00688

the density change versus conversion can be constructed. Figure 15.18 presents the density change with the conversion. It can be seen that the density increases almost linearly with conversion for the pure and filled resins, which also shows no curing temperature dependency. Such linear relationship was also observed by other researchers, e.g., Zarrelli, et al. [32].

Based on the measurement data, we assume a linear shrinkage model for the epoxy resin [20], as:

$$\rho^{c+T}(T,\alpha) = \rho_{\text{ref}}\left[1 - 3\beta_L^{g,r}(T - T_{\text{ref}}) + 3\gamma_L(\alpha - \alpha_{\text{ref}})\right] \quad (15.18)$$

where T_{ref} is the reference temperature, α_{ref} is the reference conversion, and ρ_{ref} is the density at T_{ref} and α_{ref}. β_L is the linear coefficient of thermal expansion as shown in Table 15.6 and γ_L denotes the linear cure shrinkage parameter. The density parameters can be obtained by fitting equation to measurement data. Figure 15.18 presents the density evolution as a function of degree of cure for all the materials at the tested

Table 15.6 Parameter for the density equation

	F00	F40	F50	F65
α_{ref} [−]	1.00	1.00	1.00	1.00
ρ_{ref} [g/cm³]	1.193	1.463	1.560	1.700
T_{ref} [C]	118	118	118	118
γ_L	0.0167	0.0133	0.010	0.0073

temperatures, together with the model prediction. It shows a good agreement with the experimental values.

4 Case Study: Warpage of QFN Packages Induced during the Process

Warpage due to processing-induced residual stresses has long been recognized as an important issue for package reliability. Excessive warpage not only causes device failure issues such as die cracking and interface delamination, but also may cause assembly problems in the subsequent processes, such as dimension instability, non-coplanarity, singulation problems, etc.

Many efforts have been made to predict and/or measure process-induced warpage of packages. Kelly [33] measured the warpage of a plastic power package and indicated that the package was significantly deformed at the temperature, which is attributed to chemical shrinkage of the compound. In [34], the impact of processing conditions on warpage prediction of a plastic quad flat package (PQFP) was investigated. It was suggested that low temperature and longer molding time or high temperature and shorter time would result in less warpage. They also showed that the viscoelastic model predicted the warpage more accurately than the thermoelastic model. In [35], a coupled cure-thermal-stress analysis approach was proposed to predict warpage and stresses during sequential fabrication of electronic packaging structures. They concluded that neglecting the effect of the viscoelastic behavior of polymers and the influence of the process history on the FE modeling could lead to incorrect predictions. However, due to fast and huge change of the material properties during the curing process, it is still an open and challenging research field to accurately predict the process-induced stress and warpage, especially under complex curing temperature profiles.

As a case study, an investigation on the process-induced warpage during the array-molding process of QFN packages is conducted both experimentally and numerically. The Philips business carrier HVQFN package is chosen as a demonstrator. The package is composed of an IC die attached to an exposed pad and in a later stage encapsulated with an epoxy compound. A series of experiments for QFN matrix strips with the model molding compounds are performed. In the molding experiments, different processing parameters, filler loading, and die thickness are used. The warpage is measured by a contact probe coordinate measuring system.

The cure-dependent constitutive model is applied to describe the behavior of the compound during the curing process. The cure-dependent material parameters are incorporated into a finite element code Marc and numerical simulations were carried out for the QFN array-molding process. The effects of filler concentration, curing temperature profiles, and the die thickness on the warpage are studied.

The curing process of EMC and subsequent cooling down will introduce warpage due to polymerization shrinkage and CTE difference of the materials. Warpage is a critical issue for the map-mold technology. Figure 15.19 shows a warped molded strip after the molding process. It is expected that with increased density of QFN units, warpage induced in the manufacturing process will be more serious. Such warpage in the map mold will obstruct the delicate sawing process. Furthermore, after sawing, the residual warpage in the QFN package can cause assembly problems and reliability issues. Therefore, it is very important that, first, a process-dependent material model should be established and the warpage should be predicted and verified with experimental data. And then an optimized design for package, material parameters, and process conditions can be obtained through virtual prototyping.

4.1 Experiments

The filled resins (F40, F50, and F65) described and characterized in session 3 were used as the molding compounds. A series of molding experiments with various combinations of processing conditions and package parameters was conducted and warpage after molding and post-cure was measured.

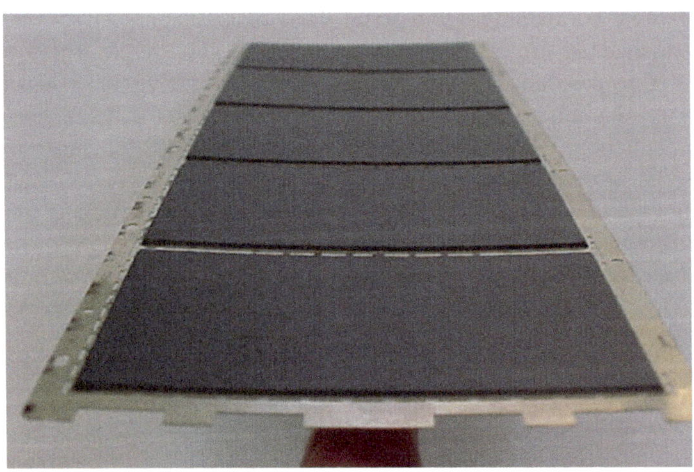

Fig. 15.19 Warpage induced during the process

The products were cured in the mold under the pressure for a prescribed time and then de-molded and cooled down freely to room temperature. Then, the warpage along the chosen longitudinal and lateral cross-sections was measured by using a contact probe coordinate measuring system. Meanwhile, a DSC measurement was conducted to determine the glass transition temperature, which is a measure for characterizing the de-molding conversion level. Afterward, the products were post-cured in the oven for 1 h at 200 °C. After cooling down, the product shape was re-measured. Figure 15.20 shows schematically the temperature profile for the molding and post-curing.

The filler loading, die thickness, molding temperature, and the de-molding conversion were taken as parameters in the molding experiments. Each parameter has three levels, as listed in Table 15.7. In order to limit the number of experiments we chose to vary the parameter level only one at a time. As standard conditions were chosen: 65 wt% filled molding compound, 250 μm die thickness, 5 min curing in a mold at 140 °C, resulting in a conversion level at mold opening of about 80%. Note that the curing times for the temperature history study were chosen such as to result in the standard de-mold cure levels (80%). The de-molding conversion level was varied by changing the curing times while keeping the mold temperature at 140 °C. Experiments with lower de-mold cure levels than 75% were unsuccessful, since at these levels the molding compound after mold opening is not stiff enough for the product to be extracted from the mold. After full determination of the molding conditions from the preliminary molding experiments, for each experimental condition two or three products were molded.

The amount of warpage was determined from measurements of the surface shape using the setup with a programmable precision XYZ-table, which is equipped with 1 μm accurate stepper motors. The vertical coordinate was measured with an LVDT sensor (Fig. 15.21).

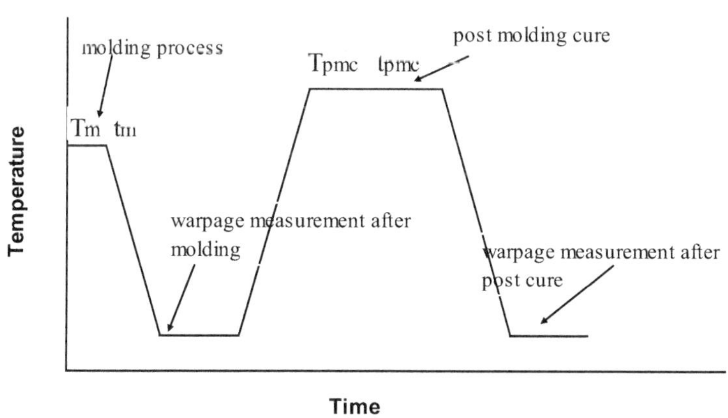

Fig. 15.20 Illustration of temperature history of molding process (*Tm* molding temperature, *tm* in-mold curing time, *Tpmc* post-molding curing temperature, *tpmc* post-molding curing time)

Table 15.7 Parameter lists in the molding experiments

Parameter	Level 1	Level 2	Level 3
Filler loading (%wt)	40	50	65
Die thickness (μm)	180	250	380
Molding temperature T_m (°C)	120	140	160
Conversion after molding	0.75	0.80	0.90

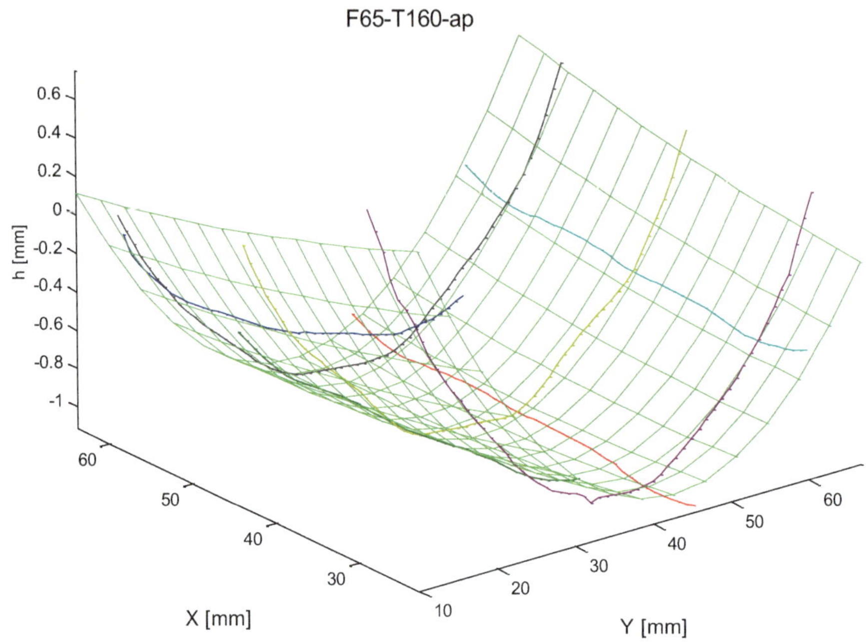

Fig. 15.21 Shape of the map mold after post-curing, 65 wt% filled molding compound, 250 mm die thickness, 5 min curing in a mold at 140 °C. The full lines are measured coordinates and dashed lines correspond to curved plane fit

4.2 Finite Element Modeling

The package consists of silicon die, EMC, die-attach, leadframe, die-pad. Figure 15.22a shows the bottom view of a single map. The HVQFN map mold used in both modeling and measurements has six 10 × 10 mm dies and the following thickness dimensions:

- total package thickness: 0.85 mm,
- lead thickness: 200 μm,
- glue thickness: 25 μm,
- die thickness: 180/250/380 μm.

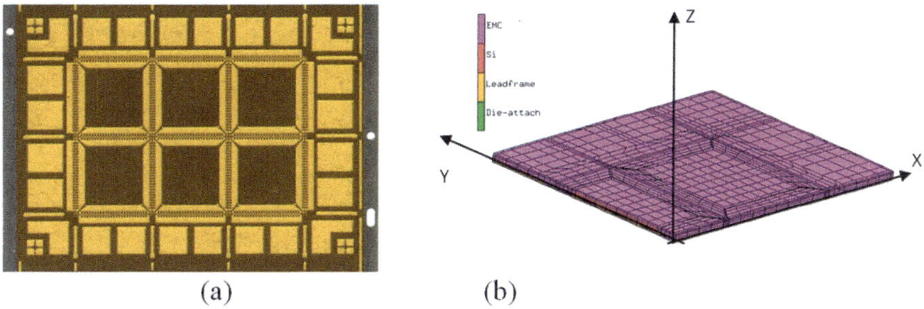

Fig. 15.22 Geometry and FE mesh; (**a**). bottom view of a single map mold; (**b**). 3D mesh used in the FE model of a quarter of the HVQFN map mold

Table 15.8 Material properties

Materials	E (GPa)	ν [−]	σ_y (MPa)	CTE (ppm/°C)
Silicon die	169	0.26	–	3
Leadframe & die pad	123	0.35	300	16.3
Die-attach	See fig. 6.11	0.3	–	81, T < 38 °C 181, T > 38 °C

Parametric 3-D FE models are built for the package with variation of the die thickness. Figure 15.22b shows a typical finite mesh. The cure-temperature-dependent viscoelastic model proposed above is applied here to describe the material behavior of EMC's during the curing process and cooling down phase.

The silicon die is assumed elastic and isotropic and the leadframe and diepad are considered elastic and ideally plastic. The material properties used in the simulation are shown in Table 15.8. The die-attach is a silver-filled conductive adhesive and it has a glass transition temperature of about 38 °C. Because it is rather thin compared to the compound, it is expected its viscoelastic behavior has insignificant influence on the overall stress and strain state in the package. Therefore, in our simulations the die-attach is considered to be temperature-dependent elastic. To measure the temperature-dependent elongation modulus, DMA measurement in single cantilever mode under 1 Hz frequency was performed in IMO-Backend, Philips. Figure 15.23 shows the storage Young's modulus of the die-attach from the DMA measurement, which is used for the temperature-dependent modulus in the simulation. Note that at rubbery state a slight increase of the modulus with temperature can be seen, showing that the rubbery modulus has slightly temperature dependency. A constant Poisson's ratio is assumed for the die-attach.

The thermal history used in the simulation is taken as the same as in the experiments, as already shown in Fig. 15.20. The stress-free state is assumed at the starting of the curing process for all the constituents.

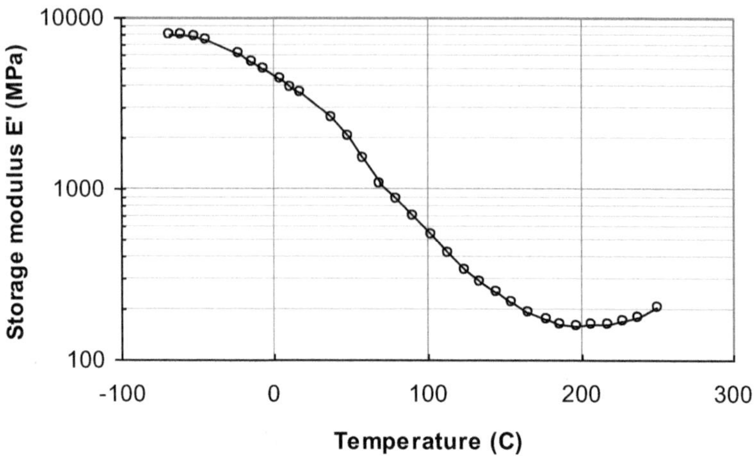

Fig. 15.23 Storage Young's modulus of the die-attach from 1 Hz DMA measurement

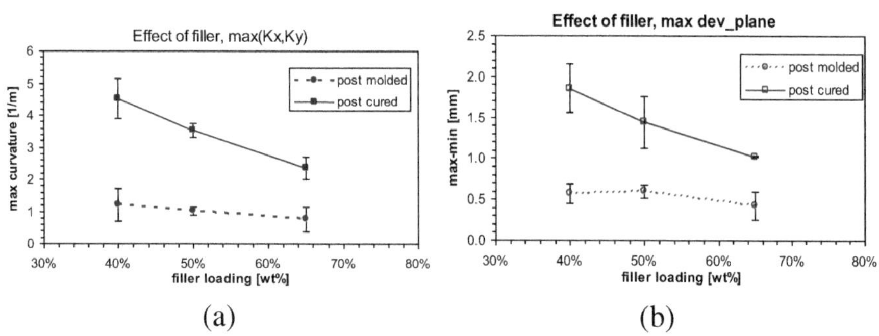

Fig. 15.24 Effect of filler loading on warpage, (**a**) maximum curvature; (**b**). maximum plane deviation

4.3 *Results and Discussion*

The warpage results from the molding experiments are shown in Figs. 15.24, 15.25, and 15.26. On the left side are all warpage graphs according to the maximum curvature warpage criterion and on the right are the graphs where the deviations from the average plane are used as a warpage measure. It can be seen that both measures give similar trends in all cases. The curvature directly after de-molding (dashed lines) is always much lower than after post-curing (a factor of 3–3.5). The difference in warpage between after de-molding and after post-curing is possibly due to the T_g difference and also the faster stress relaxation in the partly cured product.

Figure 15.24 presents the warpage measurement versus the filler loading after de-molding and after post-curing. It can be seen that the filler loading has a strong effect on the warpage level. The curvature reduces by a factor 2 if the filler content

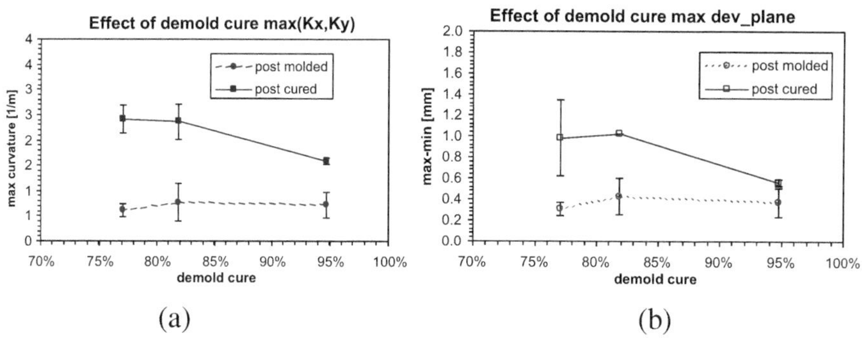

Fig. 15.25 Effect of de-molding conversion on warpage: (**a**) maximum curvature; (**b**) maximum plane deviation

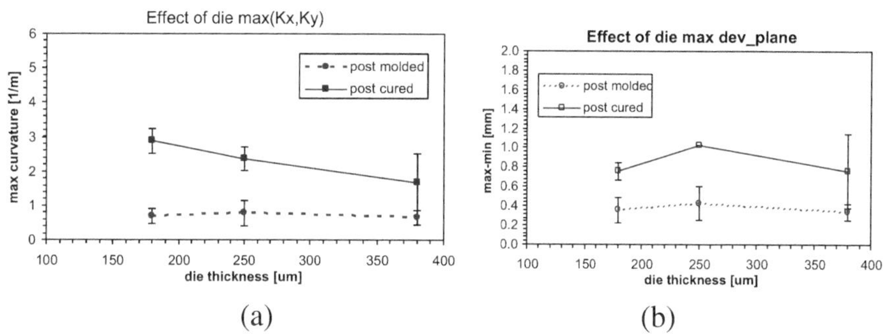

Fig. 15.26 Effect of die thickness on warpage: (**a**) maximum curvature; (**b**) maximum plane deviation

is increased from 40 to 65 wt%. The decrease is approximately linear. With the increase of the filler loading, both the CTE and the polymerization shrinkage will decrease, which reduces the process-induced warpage.

Figure 15.25 shows the effect of conversion level at de molding on the warpage. It is indicated that high de-molding conversion tends to reduce warpage. It is perhaps due to the constrained curing in the mold that partly reduces the total warpage. There appears to be no significant effect of the molding conditions on the warpage after post-curing. Unfortunately, the interpretation of these measurement series is somewhat obscured by the relatively large uncertainty in the low-temperature curing experiment (20 min at 120 °C). Note that the de-molding conversion levels were kept equal in these experiments.

High de-molding conversions (>90%) tend to reduce warpage. The reduction is 30–40% going from 82% de-mold cure to 95%.

The effect of die thickness on warpage is shown in Fig. 15.26. It can be seen that increasing the die thickness reduces the final warpage. It is due to the fact that the thicker die makes the matrix stiffer. Die thickness only has a slight effect on post-molded warpage.

Figures 15.27a and b shows the vertical displacement patterns after molding and after post-curing, respectively. Fig. 15.29a and b shows the out-of-plane displacement along the x-direction sawing lines (as defined in Fig. 15.28) with a comparison to measured data for the packages molded with 65% filled EMC with 250 μm and 380 μm die thickness, respectively. The curing profile is the standard molding conditions, i.e., in mold cure at 140 °C for 20 min and the degree of cure at de-molding 80%. Because only a quarter of the one map mold is used in the FE simulation, symmetry of the simulation results along the symmetric axis is used for convenient comparison with the measurements (Fig. 15.29).

A reasonably good agreement between measured and simulated displacements can be observed. For these reference cases the simulated and measured "maximum mean curvature" deviates less than 10%.

The other reason may be in the cooling rate. Preliminary cyclic thermal experiments were performed, where fully cured packages were heated above T_g and subsequently cooled down at several cooling rates. Observed warpage appeared to be dependent from the cooling rate. This could not be explained with viscoelasticity only. A strong indication that aging phenomena are responsible for the deviations remains. Future inclusion of aging phenomena in the descriptions should be explored.

Thus, the results confirm that the present model is capable of giving reliable predictions for real molding process and show the engineering applicability of the present method for reliability investigation and optimization including the curing effects.

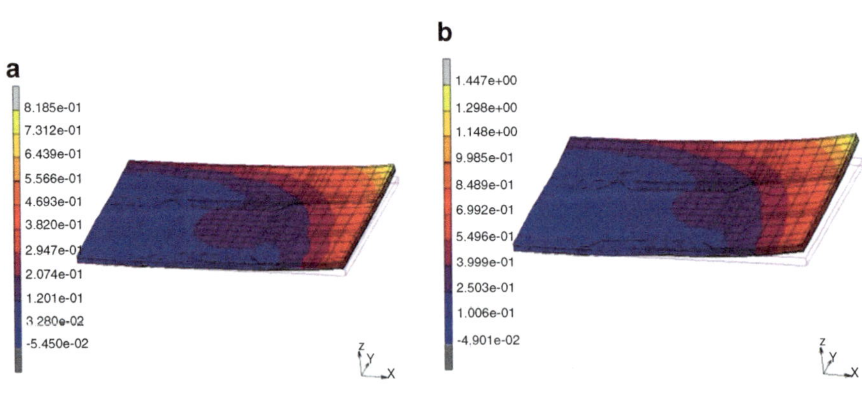

Fig. 15.27 Displacement contour after cooling down to room temperature (**a**) from molding and (**b**) from post-cure (65% filled EMC, 380 μm die)

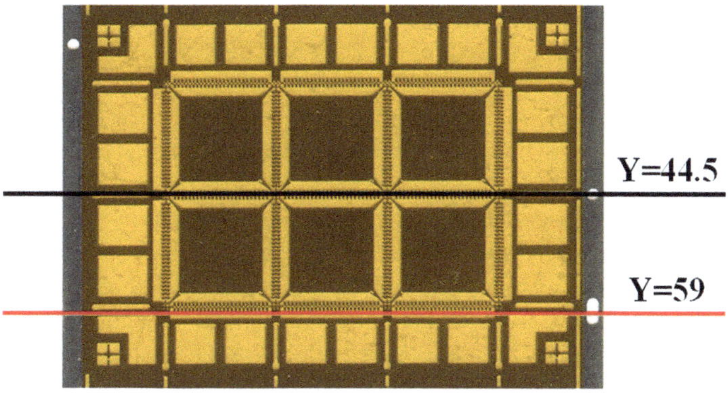

Fig. 15.28 Illustration of x-direction sawing lines

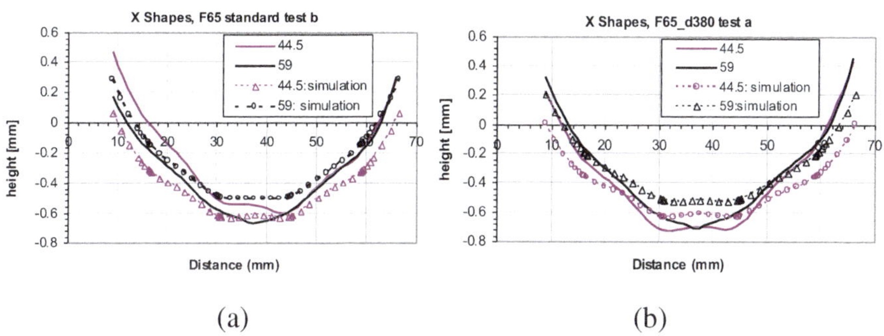

Fig. 15.29 Vertical displacement along the x-direction sawing lines (65% filled). Solid lines are measurements and symbols refer to simulation results. (**a**). 250 μm die and standard molding conditions. (**b**). 380 μm die and standard molding conditions

5 Conclusions

The curing process of thermosetting polymers will introduce residual stress and warpage in the microelectronic devices, due to the polymerization shrinkage and subsequent cooling down from the curing temperature. In this chapter, a cure-dependent viscoelastic model is developed to describe the mechanical behavior of thermosetting polymers. The model is able to describe a "quasi" thermo-rheologically simple behavior but has the advantage of convenient implementation into a general FEM program. The experimental characterization of the cure-dependent properties including DMA measurements, DSC analysis, and the measurement of the curing shrinkage is presented. Shear DMA measurements using both isothermal and intermittent curing approaches are performed to characterize the cure-dependent relaxation behavior. It is indicated that the temperature effect on the shift factors is independent of degree of cure when the glass transition temperature is chosen as the reference temperature. The temperature effect is described well with the WLF equation if the temperature is above a switching temperature, while below the switching temperature the Arrhenius equation gives a good prediction. A new concept of "reduced temperature" is introduced. The equation for the shift factors is modified using the reduced temperature. In this way, the effect of the temperature and the degree of cure on the relation behavior can then be expressed in an 'integrated' way.

As an application case, the process-induced warpage during the array-molding process of QFN packages is investigated both experimentally and numerically. The simulated warpage results using the developed model were compared to the experimentally observed warpage and a reasonable agreement is obtained, showing that the model is capable of giving reliable predictions for real molding process.

Acknowledgments The author D.G. Yang would like to cordially thank Prof. Kouchi Zhang and Dr. Kaspar Jansen for their guidance and support for the research work. He also likes to acknowledge the support of the Natural Science Foundation of Guangxi Province (Grant No. 2017GXNSFDA198006) and The Key Research and Development Program of Guangxi Province (Grant No. GuiKe AB20159038).

References

1. G.Q. Zhang, W.D. van Driel, X.J. Fan, *Mechanics of Microelectronics* (Springer, 2006)
2. A. Plepys, M.S. Vratsanos, R.J. Farris, Determination of residual stresses using incremental linear elasticity. Compos. Struct. **27**, 51–56 (1993)
3. R.S. Chambers et al., A rubbery cure shrinkage model for analyzing encapsulated assemblies. J. Electron. Packag. **117**(3), 249–254 (1995)
4. J. Lange, S. Toll, J. Manson, A. Hult, Residual stress build-up in thermoset films cured above their ultimate glass transition temperature. Polymer **36**(16), 3135–3141 (1995)
5. V.N. Korotkov, Y.A. Chekanov, Y.N. Smirnov, I.D. Zenkov, Shrinkage stresses during cure of constraint polymer matrix, in ECCM-7: Seventh European Conference on Composite Materials. Realizing their Commercial Potential, Vol. 1, pp. 141–146 (1996)
6. J. Lange, S. Toll, J.E. Manson, A. Hult, Residual stress build-up in thermoset films cured below their ultimate glass transition temperature. Polymer **38**(4), 809–815 (1997)
7. D.J. Ferry, *Viscoelastic Properties of Polymers* (Wiley, New York, 1980)
8. A. Lee, G.B. McKenna, Effect of crosslink density on physical ageing of epoxy networks. Polymer **29**, 1812–1817 (1988)
9. Y.K. Kim, S.R. White, Stress relaxation behavior of 3501-6 epoxy resin during cure. Polym. Eng. Sci. **36**(23), 2852–2862 (1996)
10. S.L. Simon et al., Modeling the evolution of the dynamic mechanical properties of a commercial epoxy during cure after gelation. J. Appl. Polym. Sci. **76**(4), 495–508 (2000)
11. D. Adolf, J.E. Martin, Time-cure superposition during cross-linking. Macromolecules **23**(15), 3700–3704 (1990)
12. D. Adolf, J.E. Martin, Calculation of stresses in crosslinking polymers. J. Compos. Mater. **30**(1), 13–34 (1996)
13. D.B. Adolf, J.E. Martin, R.S. Chambers, S.N. Burchett, Stresses during thermoset cure. J. Mater. Res. **13**(3), 530–550 (1998)
14. M.S. Kiasat, Curing Shrinkage and Residual Stresses in Viscoelastic thermosetting resins and composites, PhD thesis, Delft University of Technology (2000)

15. Ernst, L.J., C. van't Hof, D.G. Yang, M.S. Kiasat, G.Q. Zhang, H.J.L. Bressers, J.F.J. Caers, A.W.J. den Boer, J. Janssen, Mechanical characterization and simulation of curing packaging polymers, in Proceedings of the 3rd International Conference on Micro Materials (2000), pp 162–167
16. L.J. Ernst, C. van't Hof, D.G. Yang, M.S. Kiasat, G.Q. Zhang, H.J.L. Bressers, J.F.J. Caers, A. den Boer, J. Janssen, Mechanical modeling and characterization of the curing process of underfill materials. J. Electron. Packag. **124**(2), 97–105 (2002)
17. D.G. Yang, G.Q. Zhang, L.J. Ernst, C. van't Hof, J. Caers, H.J.L. Bressers, J. Janssen, Investigation on flip chip solder joint fatigue with cure-dependent underfill properties. IEEE Trans. Compon. Packag. Technol. **26**(2), 388–398 (2003)
18. C. van't Hof, Mechanical Characterization and Modeling of Curing Thermosets, PhD thesis, Delft University of Technology 2006
19. Jansen, K.M.B., L. Wang, D.G. Yang, C. van't Hof, L.J. Ernst, H.J.L. Bressers and G.Q. Zhang, Constitutive Modeling of Compounds, 2004 ECTC (2004), pp. 890–894
20. K.M.B. Jansen, Recommended Test Methods and Material Data, Mevipro Report (D038), Consortium Confidential2004
21. D.G. Yang, K.M.B. Jansen, L.J. Ernst, G.Q. Zhang, W.D. van Driel, H.J.L. Bressers, J.H.J. Janssen, Numerical modelling of warpage induced in QFN array molding process. Microelectron. Reliab. **47**(2–3), 310–318 (2007)
22. D.G. Yang, Cure-Dependent Viscoelastic Behaviour of Electronic Packaging Polymers, PhD thesis, 2007, Delft University of Technology
23. Jan De Vreugd, The effect of aging on compound properties, PhD thesis, 2011, Delft University of Technology
24. Ali R. Rezaie Adli, Numerical Analysis and Experimental Verification of Stresses Building up in Microelectronics Packaging, PhD thesis, 2017, Delft University of Technology
25. H.H. Winter, Can the gel point of a cross-linking polymer be detected by the G′ – G″ crossover? Polymer Eng. Sci. **27**, 1698–1702 (1987)
26. K. Suzuki, Y. Miyano, Change of viscoelastic properties of epoxy resin in the curing process. J. Appl. Polymer Sci. **21**, 3367–3379 (1977)
27. D.J. Plazek, I.C. Chay, The evolution of the viscoelastic retardation Spectrum during the development of an epoxy resin network. J. Polymer Sci. Part B:Polymer Phys. **29**, 17–29 (1991)
28. Y. Eom, L. Boogh, V. Michaud, et al., Time cure-temperature superposition for the prediction of instantaneous viscoelastic properties during cure. Polymer Eng. Sci. **40**(6), 1281–1292 (2000)
29. S. Sourour, M.R. Kamal, Thermochmica ACTA **14**, 41 (1976)
30. E.A. Turi, *Thermal Characterization of Polymeric Materials*, vol 2, 2nd edn. (Academic Press, 1997)
31. N.W. Tschoegl, *The Phenomenological Theory of Linear Viscoelastic Behavior; an Introduction* (Springer, Berlin, 1989)
32. M. Zarrelli, A.A. Skordos, I.K. Partidge, Investigation of cure induced shrinkage in unreinfored epoxy resin. Plast. Rubber Compos. **31**(9), 377–384 (2002)
33. G. Kelly, Accurate prediction of PQFP Warpage, in proceedings of 44th ECTC, 1994, pp. 102–106
34. D.T.S. Yeung, M.M.F. Yuen, Warpage of plastic IC packages as a function of processing conditions. J. Electron. Packag. **123**(3), 265–272 (2001)
35. R.C. Dunne, S.K. Sitaraman, An integrated process modeling methodology and module for sequential multilayered substrate abrication using a coupled cure-thermal-stress analysis approach. IEEE Trans. Elec. Pack. Manufact. **25**(4), 326–334 (2002)

Chapter 16
Manufacturing for Reliability of Panel-Level Fan-out Packages

T. Braun and O. Hölck

1 Introduction

Fan-out Panel Level Packaging (FOPLP) is one of the latest trends in microelectronics packing. Manufacturing is currently done on wafer level up to 12″/300 mm and 330 mm respectively. For higher productivity and therewith lower costs larger form factors are introduced. Instead of following the wafer-level roadmaps to 450 mm, Fan-out Panel-Level Packaging (FOPLP) might be the next big step [1, 2].

However, an easy upscaling of technology when moving from wafer to panel level is not possible. Materials, equipment, and processes have to be further developed or at least adapted [3, 4].

A crucial step in the process of FOPLP is the encapsulation of the die assembly on a temporary carrier with an epoxy molding compound and the following debond from that carrier, regardless of the process flow (see next paragraph). The flow during spreading and compression of the encapsulation material and the chemical reaction in the course of curing during a multi-step temperature profile lead to the later physical and chemical structure of the material which in turn gives rise to potential reliability concerns for the Fan-Out Package:

- Cure shrinkage of the epoxy molding compound (EMC) leads to shifts in die position leading to layout incompatibility to the redistribution layers or stresses in the panel and later package.
- Coefficient of thermal expansion (CTE) mismatch between EMC, temporary carrier, and assembly components leads to stresses in the package and to panel warpage affecting the downstream processes and reliability.

T. Braun · O. Hölck (✉)
IZM Fraunhofer, Fraunhofer Institut Reliability and Microintegration, Berlin, Germany
e-mail: Ole.Hoelck@izm.fraunhofer.de

W. D. van Driel, M. Yazdan Mehr (eds.), *Reliability of Organic Compounds in Microelectronics and Optoelectronics*,
https://doi.org/10.1007/978-3-030-81576-9_16

- Particle disintegration in the EMC leads to flow marks which can mark soft spots for chemical attack and inhomogeneous CTE.

By controlling the process and observation of key characteristics that are influenced by the material properties and interaction, the basis for later understanding of the degradation of material integrity and package reliability is laid.

Two of the main observable characteristics in the encapsulation process are warpage and die shift. In the recent past, a number of authors have published investigations on warpage and die shift in fan-out (FO) wafer- and panel-level packaging. However, before discussing the literature, it is advantageous to consider the process of encapsulation and carrier debonding in FO packaging and to discuss the origin of warpage and die shift as key indicators for the appraisal of material performance and interaction.

2 Process Description

For Fan-Out Wafer and Panel-Level Packaging two basic process flows are encountered: the "Mold first" and the "RDL first" approach (see Fig. 16.3). Where for the "Mold first" process meanwhile a face-down and a face-up option exists. Both variants are already in mass production. Common to all options is the use of a temporary carrier with a release layer for either thermal or laser debond, and an epoxy molding compound that is applied in a compression mold process under high pressures and temperatures. The molded wafer or panel needs to undergo a Post Mold Cure step (PMC) to fully cure the EMC. This chapter will focus on a "Mold first" face down approach, but most of the challenges described may be also a topic for "Mold first" face down or "RDL first".

The process starts with the choice of carrier material considering not only the thermo-mechanical behavior but also properties as, e.g., weight or stability. After laminating the thermal release tape, the chips and components are positioned by pick and place assembly on the carrier posing a first bottleneck for high throughput. Here, high speed but also high accuracy assembly are required. In the next step, typically compression molding is used for chip embedding and to form the reconfigured panel. Liquid, granular, and sheet type encapsulants are available as EMC material.

They are applied onto the assembly by dispensing in patterns (see Fig. 16.1), by evenly distributing or by exact positioning, respectively. Upon EMC distribution, the assembled panel is subjected to the compression mold step. The chamber containing the carrier is evacuated and the EMC compressed inside the cavity at typical pressures of 5 MPa to keep voids to a minimum (Fig. 16.2).

Under high temperatures the EMC is cured within the mold (in mold cure, IMC) until stable enough to remove from the mold, while curing to a higher degree of conversion is performed in a batch oven (PMC).

For application of the redistribution layer (RDL) in the next process step, the reconfigured panel needs to be released from the carrier material in a thermal

Fig. 16.1 Liquid EMC dispensed in a point pattern, granular EMC distribution, and EMC sheet positioned on assembled carrier (left to right)

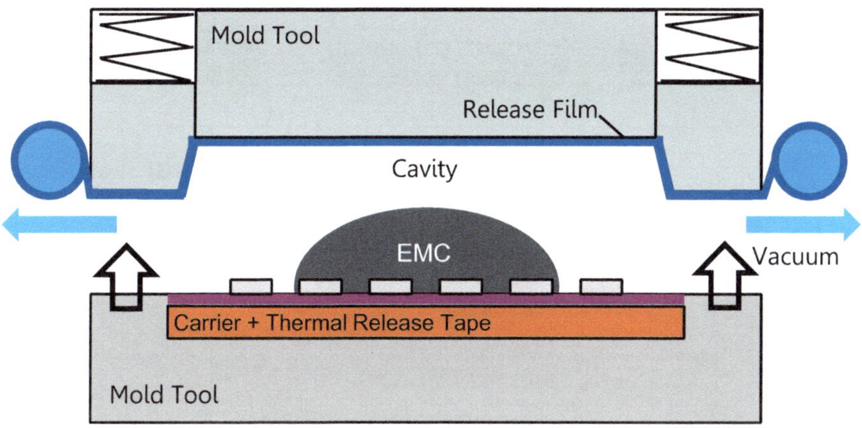

Fig. 16.2 Large area compression mold principle: A carrier with an assembly and EMC is compressed at high pressure and temperature, the EMC is cured within the cavity until stable enough to take out

debonding step, requiring temperatures above the PMC. For RDL layer formation a large variety of lithography tools and dielectric material options exist. As dielectrics photosensitive as well as non photosensitive or liquid versus dry-film materials can be considered. Mask-based lithography as, e.g., stepper technology is just as maskless-based tools as laser direct imaging (LDI) available for panel sizes. Both offering different capabilities and strategies to overcome challenges from die placement accuracy and die shift after molding (Fig. 16.3).

The individual process steps of the mold first face down approach for the manufacturing of a reconfigured panel that will be discussed here are as follows and the corresponding temperature profile is depicted in Fig. 16.4:

1. Component assembly on carrier at room temperature.
2. Compression molding including in-mold cure (IMC).
3. Post mold cure (PMC).
4. Carrier release (debonding).
5. Cooling down to room temperature.

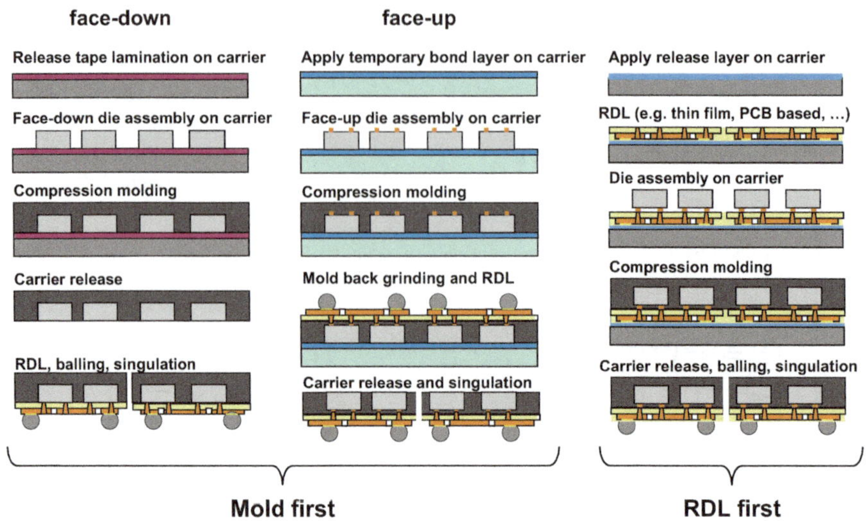

Fig. 16.3 Fan-out process flow options

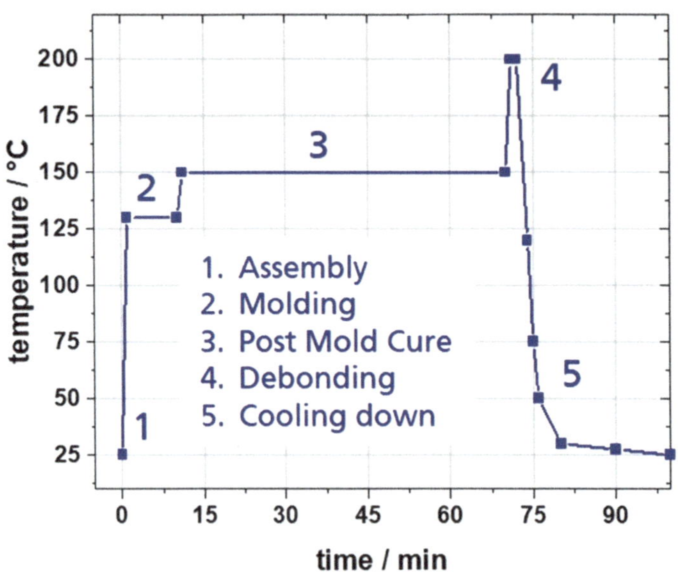

Fig. 16.4 Typical temperature profile of the discussed process steps

3 Technical Challenges

Warpage and die shift are, among others, challenges for FO packaging which have to be taken into account especially when scaling technology from wafer to panel level and discussing material and equipment options. The root cause of the effects has not yet been sufficiently enough described to be able to make predictions on either warpage or die shift, e.g., using FE simulations. While this challenge is not new to wafer-level packaging, with panel-level packaging the processed area becomes larger at similar thicknesses, making the effects more pronounced or inviting side effects like gravity. Thus, solutions have to be found to more thoroughly understand the phenomena.

4 Involved Materials and Main Properties

The process as detailed above contains several temperature steps (Fig. 16.4), combining to a profile that moreover contains complex changes in material properties (EMC cure) or boundary conditions (carrier release). Responsible for the technical challenges is to a large part the interplay between the involved materials and their properties, especially the thermal expansion coefficients compiled in Table 16.1.

There are several carrier materials available resulting in a wide range of CTEs and elastic properties. While for considerations of handling and reuse steel carriers are an obvious choice, the CTE of glass types can be adapted to match that of the EMC or assembly better. Besides the effect of carrier thickness on its stiffness (e.g., deformation during PMC), the heat transfer during heating to IMC temperature or debond temperature may be of relevance.

Materials used for mold embedding are with SiO_2 highly filled (up to 90%wt.) epoxy molding compounds (EMC). They are available as liquid, granular, or sheet compound, resulting in different handling concepts. Besides the chemical composition and CTE, elastic properties influence the warpage result as well as the chemical shrinkage.

The thermal release tapes are typically double-sided adhesive tapes. One side is facing the carrier and contains a thermally activated mechanism which releases the carrier from the adhesive. The other side is pressure sensitive and holds the

Table 16.1 CTE ranges of different materials used in FO packaging material

Material	CTE range
Die (silicon)	2.3 ppm/K
Carrier (metal, glass, polymer)	3.3–24 ppm/K
EMC	6–10 ppm/K

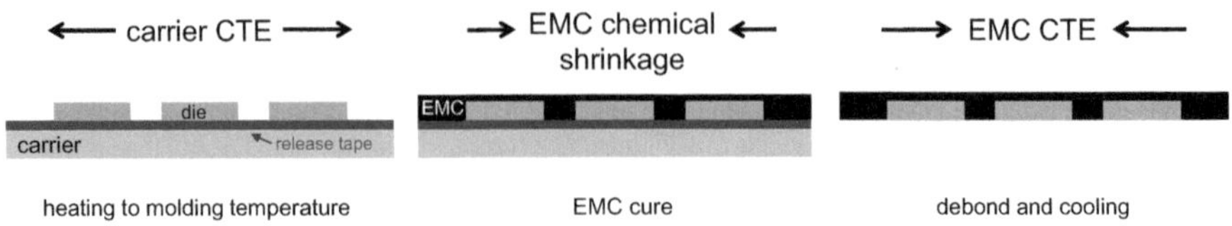

Fig. 16.5 Sketch depicting the interaction between materials of different CTEs leading to stresses in a molded panel

assembled dies in place. Its main impact to warpage and die shift is indirect: the release mechanism by necessity activates at a temperature above the post-mold cure, adding to the temperature profile. Further but smaller influence may come from the elastic properties of the core layer and undefined adhesive layers.

The assembly typically consists of silicon dies and possibly other components as, e.g., passives. Silicon typically has the lowest thermal expansion coefficient of the materials involved (2–3 ppm/K). In a reconfigured wafer or panel, depending on panel and package layout, the area occupation ranges as high as 60%. With die thicknesses in the range for example of 200 μm and overmold thicknesses of 100 μm, the panel resembles a layered material with different CTEs, resulting in stresses and deformation with temperature changes. The principle is depicted in Fig. 16.5.

5 Die Shift

Die shift describes the effect of chip movement after placement during compression molding, debonding, and cooling of the reconfigured panel. Dies are placed at room temperature on the temporary carrier with laminated thermal release tape (process step 1). Upon introducing the assembled carrier with distributed uncured EMC into the heated molding machine, it expands according to the thermal expansion coefficient (CTE) of the carrier. Accordingly, the die positions are shifted outward. During compression molding, the EMC is spread under pressure to fill the cavity and gaps between the assembled dies and cures to a state beyond gelation, where it is stable enough to be released from the molding tool. During the curing reaction, the EMC exhibits typically a decrease in volume called chemical shrinkage, which is related to the forming of covalent bonds and a three-dimensional molecular network. This cure shrinkage exerts stresses on the dies and carrier, and it acts towards the inside of the panel area. However, only the shrinkage beyond a certain point after gelation is relevant to mechanical stresses, because while still a viscous fluid, these stresses are relaxed immediately by the material. During PMC (step 3), this shrinkage continues until the EMC is fully cured and acquires its final properties of the solid material. In step 4, the panel is shortly heated to activate the release mechanism of the release tape, yet expanding the carrier/assembly/EMC stack, and the carrier is removed, and the release tape peeled off the panel. Stresses between carrier and panel induced by the cure shrinkage can now relax. During cool down, regardless of any thermal treatments on the way, the carrier CTE is irrelevant, whereas the thermal properties of the EMC now dominate contraction, shifting the assembly to their final positions.

It is not ruled out that dies my slide on the thermal release tape due to pressure of the mold flow. However, since no correlating shifting patterns could be observed in panels molded with point patterns of liquid EMC (see Fig. 16.1), both effects must be very small and are neglected in the further discussion. Figure 16.6 shows the die shift as the difference of positions before and after molding for different wafer and panel sizes. It can be observed that the shift is isotropic leading to a concentric shift and independent of substrate size. Since the steel carrier material exhibits a larger CTE (~12 ppm/K) than the EMC (~7 ppm/K), the shift is toward the outside and overcompensates effects of the shrinkage.

Figure 16.7 shows results for different carrier CTE acting in concert with EMC CTE. The large difference in die shift confirms the importance of CTE mismatch between carrier and EMC in contrast to the contribution of shrinkage effects. However, it also illustrates the significance of the compensation strategy needed to reliably hit contact pads when applying the RDL layers.

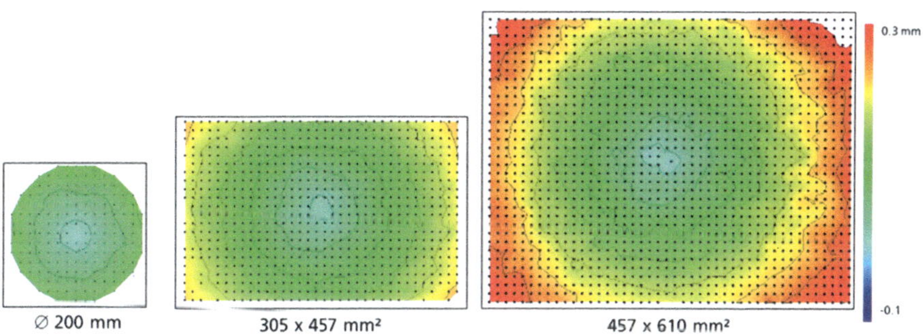

Fig. 16.6 Differences of measured die positions before and after the molding process show a concentric die shift (color scale) that is independent of the size (wafer, small rectangular panel, large rectangular panel)

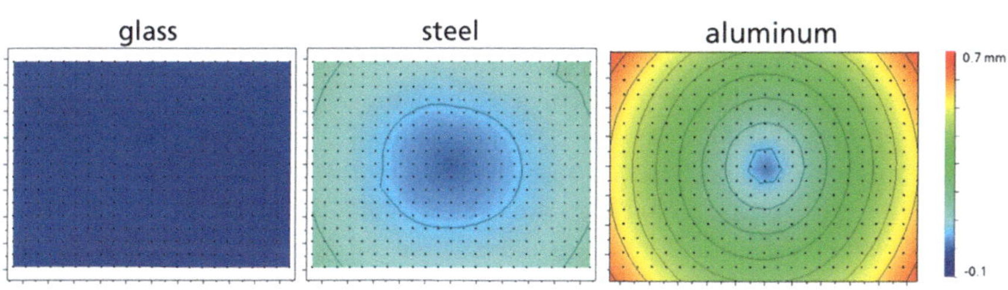

Fig. 16.7 Die shift on 457 × 610 mm² panel for different carrier materials show the effect of carrier CTE (left: 9 ppm/K; center: 12 ppm/K; right: 22 ppm/K)

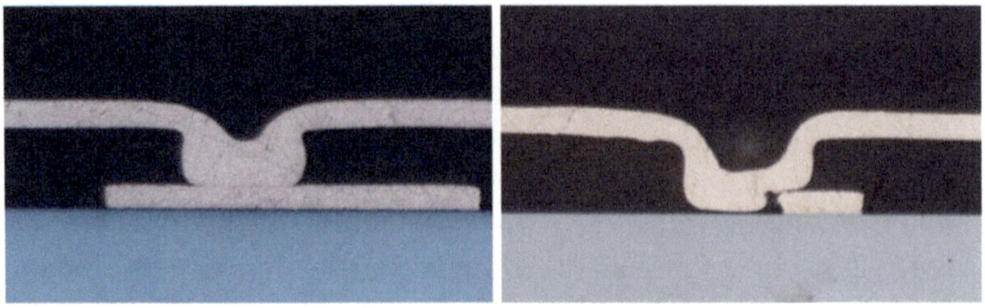

Fig. 16.8 Via and contact pad on chip in cross sections illustrating the need for precise compensation of die shifts. Left: RDL perfectly aligned to contact pad and via well centered on target; right: insufficient alignment of via to contact pad causing reliability issues

Figure 16.8 illustrates the importance of die shift compensation with respect to aligning the RDL layers to the contact pads on chips. Depending on via and pad size, the margin for errors in alignment needs to be considered in order to hit the target. In the extreme cases shorts between pads or non-contacted pads result, however, when misaligned as in the figure on the right side, failure may not immediately result, but the reliability is sure to be severely compromised.

6 Warpage

Warpage is discussed in this chapter as the deformed state the panel adopts after debonding of the carrier and cooling down.

Warpage is also mainly based on the CTE differences of the carrier, EMC, and the embedded die and the chemical shrinkage of epoxy mold compound as described above for the die shift. However, material stiffness (mainly of the EMC) and geometries like overmold thickness play important roles as well. During compression molding (also: in mold cure, step 2), the EMC is cured under high pressure and the panel can be regarded as flat (no deformation). The post-mold cure (step 3) is typically not performed under pressure but still on the carrier, and hence the deformation depends on the geometry and stiffness of the carrier as well as on the curing state and hence dynamic material properties of the EMC (thermomechanical and shrinkage). During the debonding process (step 4) the panel is again held flat by vacuum chucks while increasing temperature to near or above glass transition. The debond releases stresses induced by the carrier/panel CTE mismatch. Depending on the procedure, the cooling down to room temperature (final state) may involve temperature profiles to reduce stresses ("warpage adjust"), but below glass transition temperature it is assumed that no further relaxation of stresses occurs. The final deformation after release from the cooling chuck is induced by the stresses between assembly (silicon dies) and EMC including CTE mismatch of the materials and processing history of the EMC.

Figure 16.9 exemplarily shows the warpage of three different sizes (round wafer and rectangular panel). In all cases a tunnel shape is observed. Considering the

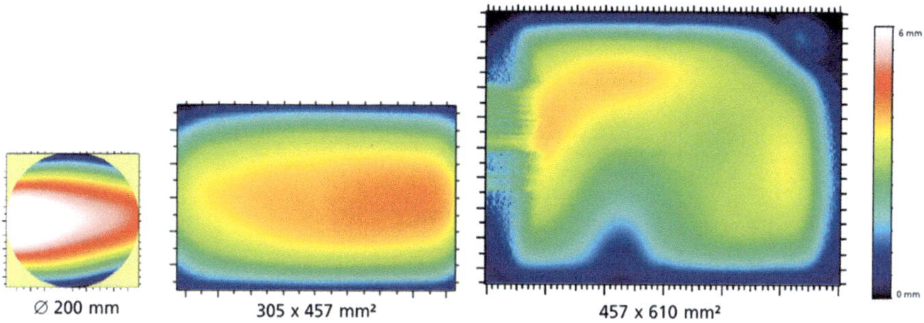

Fig. 16.9 Warpage of wafer and panel of different area and otherwise same process flow. The large rectangular panel shows considerable influence of gravity

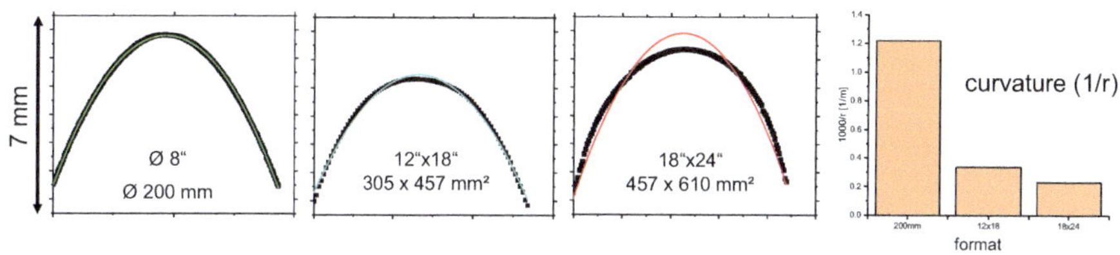

Fig. 16.10 Evaluation of warpage curvature shows decrease with format size (left to right)

small differences in assembly area utilization, the absolute values of wafer and small rectangular panel are similar, whereas the larger rectangular panel demonstrates the influence of gravity effects due to the large aspect ratio of panel dimensions. In contrast to the constant effects on die shift, the curvature, taking into account the deformation height and the dimensions, decreases with larger size (Fig. 16.10).

Since warpage is mainly rooted in thermomechanical interaction between materials; the simulation of the effects may help interpret these results [5]. However, results of such simulations highly depend on the quality of the input parameters, some of which are rather difficult to determine. As Fig. 16.11 shows, even though the simulation can reproduce the experimental result at higher thickness, the lower end is not sufficiently represented.

7 Flow Marks

Another challenge going from small wafer to large panel sizes is the development of flow marks. When using liquid molding compound and point-patterns of its distribution, the EMC needs to flow over long distances in order to be spread evenly, leading to flow marks as depicted in Fig. 16.12.

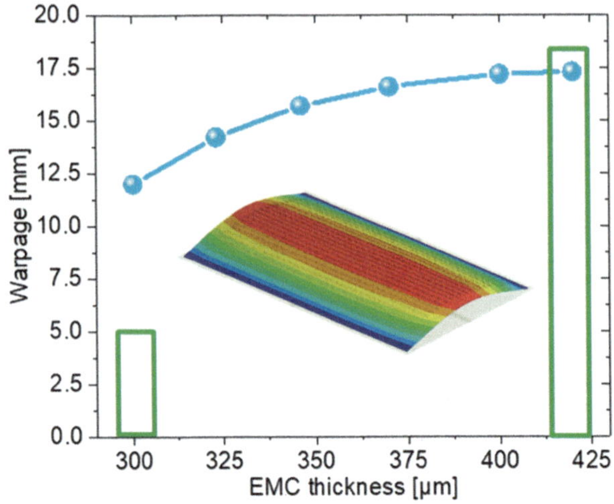

Fig. 16.11 Warpage simulation showing the high degree of dependence on thickness (connected curve) and large deviations from experimental values (green columns) at low thickness

Fig. 16.12 Large panel showing flow marks in areas of long flow lengths (late times in closing compression mold tool)

Cross sectioning shows the white marks to be filler particles at the surface (see Fig. 16.13).

Thermo-gravimetrical analysis (TGA) (see Fig. 16.14, left) reveals a larger filler content at these marks, which is confirmed by the CTE difference in a TMA analysis (right).

The exact origin of flow marks is still disputed. The observations suggest that this kind of flow marks form when a combination of low confined space (in z-direction while closing the cavity), fast flowing (to reach filling of the cavity) and a large filler cut (in case of Figure: 25% of final cavity height) is met. While larger particles tend to agglomerate and clog the flow with longer flow path and higher degree of cure and therewith also higher viscosity, streams of lesser particle content must form due to the compression and thus a segregation of particles results.

The thermo-mechanical analysis (TMA) also shows that this phenomenon is not only an optical issue, but the difference in thermomechanical properties leads to inhomogeneous behavior during follow-up processes and reliability issues of the product in thermal cycles. Figure 16.15 shows a combined picture of flow

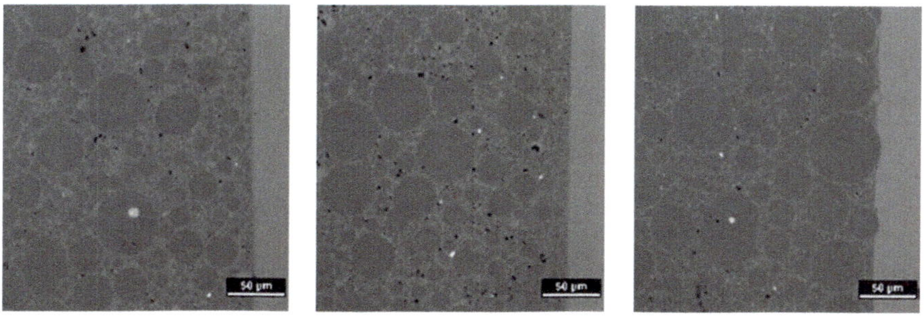

Fig. 16.13 Cross section of a panel from an area without flow marks (left), in between flow marks (center) and at a position of a flow mark. The protrusion of particles through the surface can be observed on the right side

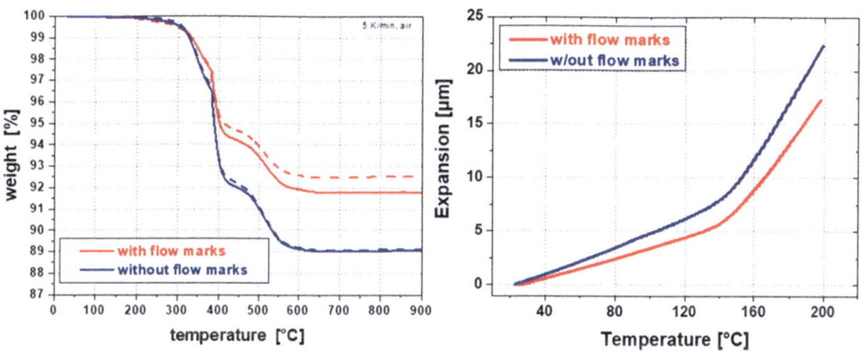

Fig. 16.14 TGA measurements (left) reveal higher filler particle content in flow marks while in TMA measurements a lower CTE is observed (right)

simulation, highlighting tracer particles which shows the path of the EMC flow. The simulation is reproducing the experimental flow pattern and allows to virtually optimize the distribution patterns with respect to flow mark patterns (see Fig. 16.12).

8 Approaches to Manufacture for Reliability

In order to manufacture for reliability, it is crucial to understand materials and their properties and interactions as well as their processing. The quality of reliability predictions highly depends on the quality of the input data and a deep understanding of the process in terms of the aspects that need to be considered. The view on the process chain of FOPLP shows that the processing of the involved material already imprints a history into the materials and thus into the packaged products. As the processed area becomes larger from 12″ wafers to larger rectangular packages, inhomogeneity, e.g., in mold flow, that are to date rarely investigated, may well lead

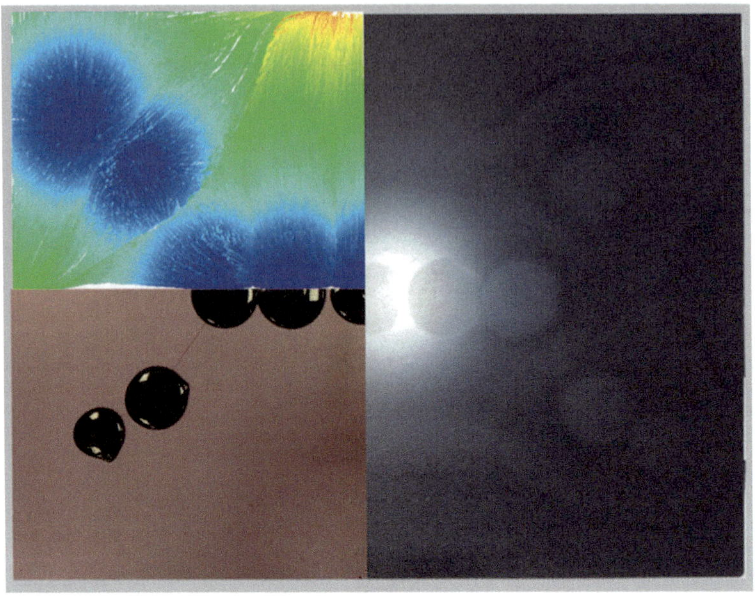

Fig. 16.15 Combination of dispense pattern (bottom left), molded panel (right) and mold flow simulation (top left) showing the trace of compression flow

to inhomogeneity in thermal properties, as was shown above for flow marks. In the literature (see next paragraph) most authors agree on the main impacts to warpage and die shift, which are also indicators for stress levels within the processed panels: CTE mismatch, EMC shrinkage, and (visco-) elastic properties. However, these properties are established during the process and thus the transition between liquid EMC and the cured moldcap needs to be taken into account. Other than the use of two simulation tools for mold flow on the one hand and thermo mechanics on the other hand suggests, there is no clear point of handover defined because the material properties may not even be final at the end of the process. The following uncertainties remain:

- Which stresses during chemical shrinkage cannot be relaxed by the solidifying EMC (mechanically relevant shrinkage).
- The degree of cure is dependent on time but also on curing temperature, leaving the potential for change in CTE and Tg and shrinkage in post-curing or chemical aging.
- How homogeneously does the mold flow and cure, also considering the out-of-plane axis, with implications to thermal properties and shrinkage stresses.

Reviewing the literature, it becomes clear that there is an understanding of the importance of these phenomena; however, it is extremely difficult to systematically relate published data. Different materials, formats of panels, layouts of packages, die sizes, and thicknesses are discussed, depending on equipment and project needs.

Varying emphasis on material properties, as well as process conduct and parameters, further complicate the comparison of results found in the literature, where warpage and die shift has been investigated on wafer and panel level in FO packaging for over 10 years now [6–9]. Most authors make use of both FE simulation and experimental warpage and die shift analysis, albeit to a different extent. While most earlier works lay the foundation for warpage and die shift analysis investigating 12″ wafer formats, more recently also large area panels of different formats are investigated [9–19].

There are five components to the investigations found in the literature that relate to the challenges as stated above.

a. A thorough material characterization to correlate results to physical properties or to generate input parameters for simulations.
b. A systematic record of process details and parameters and an assessment of their impact to the results.
c. The consideration of panel and package layout and geometries.
d. Details of the simulation methodology and the necessary reduction of complexity.
e. Validation of results in terms of comparison between experiment and simulation or transfer to different package specifications.

In most cases, the focus of investigation lies only on one or two of these aspects, whereas a systematic and comprehensive approach showing which details may be disregarded and under which circumstances has yet to be published.

Material Characterization While carrier and assembly materials are typically considered homogenous and elastic, the most important material and at the same time the most complex is the epoxy molding compound. Characterization for mold flow simulations needs to include rheology and cure kinetics [6, 12, 18, 20–22]. Some authors characterize the chemical shrinkage [7, 18, 23], while others use it as a calibration parameter (e.g., [24]). The results describe the properties of the EMC during cure until gelation of the solidifying resin. The thermomechanical properties are measured by thermomechanical analysis (CTE and Tg) and dynamic mechanical analysis (Young's modulus) to varying extents. Ideally, a set of CTEs is used to describe thermal expansion with temperature, and viscoelastic properties are determined to describe time- and temperature-dependent relaxations [21, 22, 24, 25].

Process Details In the previous paragraph, we have described one of the FOPLP processes to some extent until debonding of the carrier and cooling down (steps 4 and 5). Overviews of similar processes are also given in [19, 20, 26]. Bu et al. investigate the influence of dispense patterns of liquid EMC and compression mold parameters on the die shift [12, 13]. The investigation of Lin et al. [15, 27] describes an RDL first wafer and panel mold process, detailing much of the RDL process, but coming short on details on the compression mold process and input data for the simulations. Sanchez et al. [28] report on the importance of the debonding process including thermal treatments to the warpage effect.

Package Details Warpage and die shift result from the mismatch in properties between the involved materials (a) during process (b), where stresses build up and deform the panel or wafer. The magnitude obviously also depends on geometric parameters such as thicknesses of die and moldcap, layout of dies in the package, and of packages on the wafer or panel. In the majority of the investigations, a single geometric setup is used which typically consists of a single die. While few authors use a more fundamental approach with EMC on a bare silicon wafer [8, 10], Gadhiya et al. combine the two approaches [24]. Other authors investigate a package of a large and a small die [20, 21, 26, 29] or vary the die size [30].

Simulation Only very few investigators do not employ simulation tools for their studies. But there is no universal agreement on methodology (e.g., coupling of mold flow and thermo mechanics) and the significance of the integration of material properties such as shrinkage or viscoelasticity. A thorough simulation study was presented, e.g., by Gadhiya et al. [24] and by Che et al. [14, 25]. The latter team investigate large area panel warpage in an RDL first approach process step by step. The simulation DOE on large area panels comprises three different EMC properties, glass carrier CTE and thickness, die and overmold thickness and RDL variants. Another quite comprehensive study on die shift in 12" FOWLP is presented by Cheng et al. [25], covering material models of curing, viscosity, shrinkage, and thermomechanical properties of the EMC. An unusual method to predict warpage from input parameter variations of geometric specifications of an FOPLP is presented by Liu et al. [17]. The method is based on a trained AI algorithm, using a basic set of FE-models as database. However, it can be debated if such algorithms contribute to knowledge-based predictions rather than optimization by data mining.

Validation The final aspect to be considered is the validation of results. Larger DOEs are typically put up for simulations only, whereas the database for experimental results is scarce. Some investigations rely on simulations only, like the transfer to reconfigured wafers in [24] or the parametric die shift study in [25]. Others test one configuration which is valuable but does not allow a transfer to general use. A validated methodology showing control of the technical issues is still missing in the literature.

9 Outlook

To manufacture microelectronic Fan-Out Wafer and Panel -Level Packages for reliability, it is a key task to bring together the understanding of materials and materials interaction not only along the lifetime of the product. It is important to take into account material aging and degradation under conditions of manufacturing and use in reliability assessment. However, material properties and interfaces are established step by step along the process chain, imprinting a thermomechanical history into the package as it is formed. For future production, it will be useful to develop a comprehensive knowledge base that takes into account all aspects of the process. As a first step toward this goal, the authors of this chapter have equipped a compression mold tool with sensors, similar to efforts that have been done in the transfer mold process [31, 32]. The sensors will allow to collect in-situ data on pressure, temperature, and cure (conversion as well as shrinkage). The aim is to systematically explore the compression mold process on a medium area-sized panel in spatial resolution and correlate data to warpage and die shift as indicators of stresses and process history. In the long term, control of these indicators can be achieved by a detailed description of the process. The identification of contributions like out-of-plane inhomogeneity in network forming, e.g., due to temperature or pressure distribution, as well as the role of filler particles also need to be solved when working toward establishing a digital twin process of FO packaging.

Acknowledgments The authors would like to acknowledge partners of the Fan-Out Panel-Level Consortium for fruitful discussions and supply of materials. Results that were achieved by publicly funded projects and that have been published elsewhere [1]–[5] also contributed to this work.

References

1. T. Braun et al., Large area compression molding for fan-out panel level packing, in 2015 IEEE 65th Electronic Components and Technology Conference (ECTC), San Diego, CA, May 2015, pp. 1077–1083. https://doi.org/10.1109/ECTC.2015.7159728
2. T. Braun et al., Recent developments in panel level packaging, in 2018 International Wafer Level Packaging Conference (IWLPC), San Jose, CA, USA, October 2018, pp. 1–7. https://doi.org/10.23919/IWLPC.2018.8573277
3. T. Braun et al., Panel level packaging – a view along the process chain, in 2018 IEEE 68th Electronic Components and Technology Conference (ECTC), San Diego, CA, May 2018, pp. 70–78. https://doi.org/10.1109/ECTC.2018.00019
4. T. Braun et al., Fan-Out Wafer And Panel Level Technology For Advanced Led Packaging, presented at the IWLPC, San Jose, CA, USA, 2017
5. M. van Dijk et al., Simulation challenges of warpage for wafer- and panel level packaging, in 2020 21st International Conference on Thermal, Mechanical and Multi-Physics Simulation and Experiments in Microelectronics and Microsystems (EuroSimE), Cracow, Poland, July 2020, pp. 1–6. https://doi.org/10.1109/EuroSimE48426.2020.9152797
6. S.-S. Deng, S.-J. Hwang, H.-H. Lee, D.-Y. Huang, G.-S. Shen, Warpage simulations with P-V-T-C equation and experiments of fan-out wafer level package after encapsulation process, in 2010 5th International Microsystems Packaging Assembly and Circuits Technology Conference, Taipei, Taiwan, October 2010, pp. 1–4. https://doi.org/10.1109/IMPACT.2010.5699590
7. T.-C. Chiu, C.-L. Gung, H.-W. Huang, Y.-S. Lai, Effects of curing and chemical aging on Warpage – Characterization and simulation. IEEE Trans. Device Mater. Reliab. **11**(2), 339–348 (June 2011). https://doi.org/10.1109/TDMR.2011.2135860
8. R. Farrugia, Warpage issues in large area mould embedding technologies. XJENZA Line **2**, 37–44 (October 2014). https://doi.org/10.7423/XJENZA.2014.2.06
9. H.-D. Chang et al., Development and characterization of new generation panel fan-out (P-FO) packaging technology, in 2014 IEEE 64th Electronic Components and Technology Conference (ECTC), Orlando, FL, USA, May 2014, pp. 947–951. https://doi.org/10.1109/ECTC.2014.6897402

10. E. Kuah, J. Y. Hao, Y. Bin, W. L. Chan, W. Kai, S. C. Ho, Large format encapsulation, in 2015 IEEE 17th Electronics Packaging and Technology Conference (EPTC), Singapore, December 2015, pp. 1–5. https://doi.org/10.1109/EPTC.2015.7412419
11. E. Kuah et al., Large format packaging – An alternative format for discrete packaging: Its challenges and solutions, in 2016 IEEE 18th Electronics Packaging Technology Conference (EPTC), Singapore, Singapore, November 2016, pp. 411–416. https://doi.org/10.1109/EPTC.2016.7861513
12. L. Bu, F. X. Che, V. S. Rao, X. Zhang, Mold Flow Simulation for Fan-out Panel-Level Packaging (FOPLP), in 2018 IEEE 20th Electronics Packaging Technology Conference (EPTC), Singapore, Singapore, December 2018, pp. 691–694, https://doi.org/10.1109/EPTC.2018.8654341
13. L. Bu, F. X. Che, V. S. Rao, X. Zhang, Mechanism of Moldable Underfill (MUF) process for RDL-1st Fan-Out Panel Level Packaging (FOPLP), in 2019 IEEE 69th Electronic Components and Technology Conference (ECTC), Las Vegas, NV, USA, May 2019, pp. 1152–1158. https://doi.org/10.1109/ECTC.2019.00178
14. F. X. Che, K. Yamamoto, V. S. Rao, V. N. Sekhar, Study on warpage of Fan-Out Panel Level Packaging (FO-PLP) Using Gen-3 Panel, in 2019 IEEE 69th Electronic Components and Technology Conference (ECTC), Las Vegas, NV, USA, May 2019, pp. 842–849. https://doi.org/10.1109/ECTC.2019.00132
15. Y.-M. Lin et al., An RDL-First Fan-Out Panel-Level Package for Heterogeneous Integration Applications, in *2019* IEEE 69th Electronic Components and Technology Conference (ECTC), May 2019, pp. 1463–1469. https://doi.org/10.1109/ECTC.2019.00225
16. F. X. Che, Panel Warpage and Die Shift Simulation and Characterization of Fan-Out Panel-Level Packaging, p. 8, 2020
17. S. W. Liu, N. T. H. University, S. K. Panigrahy, N. T. H. University, K. N. Chiang, N. T. H. University, Prediction of Fan-out Panel Level Warpage using Neural Network Model with Edge Detection Enhancement, p. 6
18. C.H. Tsai, S.W. Liu, K.N. Chiang, Warpage analysis of fan-out panel-level packaging using equivalent CTE. IEEE Trans. Device Mater. Reliab. **20**(1), 51–57 (March 2020). https://doi.org/10.1109/TDMR.2019.2956049
19. V. Srinivasa Rao, C. Ser Choong, K. Yamamoto, Development of Mold-First Fan-Out Panel-Level Packaging (FO-PLP) on 600mm x 600mm Size Panel, in 2020 IEEE 70th Electronic Components and Technology Conference (ECTC), Orlando, FL, USA, June 2020, pp. 1154–1160. https://doi.org/10.1109/ECTC32862.2020.00185
20. Y. Han, M. Z. Ding, B. Lin, C. S. Choong, Comprehensive investigation of Die shift in compression molding process for 12 inch fan-out wafer level packaging, in 2016 IEEE 66th Electronic Components and Technology Conference (ECTC), Las Vegas, NV, USA, May 2016, pp. 1605–1610. https://doi.org/10.1109/ECTC.2016.46
21. T.-C. Chiu, E.-Y. Yeh, Warpage simulation for the reconstituted wafer used in fan-out wafer level packaging. Microelectron. Reliab. **80**, 14–23 (Jan. 2018). https://doi.org/10.1016/j.microrel.2017.11.008
22. J. Chao et al., Low Warpage Liquid Compression Molding (LCM) Material for High Density Fan-out and Wafer Level Packaging Applications, p. 7, 2020
23. S.-S. Deng, S.-J. Hwang, H.-H. Lee, Warpage prediction and experiments of fan-out wafer-level package during encapsulation process. IEEE Trans. Compon. Packag. Manuf. Technol. **3**(3), 452–458 (March 2013). https://doi.org/10.1109/TCPMT.2012.2228005
24. G. Gadhiya, S. Rzepka, S. Kersjes, and F. Fernandes, The Systematic Study of Fan-Out Wafer Warpage Using Analytical, Numerical and Experimental Methods, p. 9, 2020
25. H.-C. Cheng, C.-H. Chung, W.-H. Chen, Die shift assessment of reconstituted wafer for fan-out wafer-level packaging. IEEE Trans. Device Mater. Reliab. **20**(1), 136–145 (March 2020). https://doi.org/10.1109/TDMR.2020.2965000
26. M. Z. Ding, S. C. Chong, D. S. W. Ho, S. P. S. Lim, Molding process development for high density I/Os Fan-Out Wafer Level Package (FOWLP) with fine pitch RDL, *in 2016 IEEE*

18th Electronics Packaging Technology Conference (EPTC), Singapore, Singapore, November 2016, pp. 13–18. https://doi.org/10.1109/EPTC.2016.7861433
27. Y.-M. Lin et al., An RDL-First fan-out wafer level package for heterogeneous integration applications, in 2018 IEEE 68th Electronic Components and Technology Conference (ECTC), May 2018, pp. 349–354. https://doi.org/10.1109/ECTC.2018.00060
28. D. C. Sanchez, K. Reitinger, S. Oldeide, K. Zhao, W. Song, I. Khwaja, A unique and robust technique to eliminate warpage for FOWLP and FOPLP during the termal debonding process, in 2019 International Wafer Level Packaging Conference (IWLPC), San Jose, CA, USA, October 2019, pp. 1–6. https://doi.org/10.23919/IWLPC.2019.8914129
29. T.-C. Chiu, W.-J. Yin, E.-Y. Yeh, Y.-T. Yang, D.-L. Chen, Y.-H. Tseng, Physical aging of epoxy molding compound and its influences on the warpage of reconstituted wafer, in 2018 IEEE 68th Electronic Components and Technology Conference (ECTC), San Diego, CA, USA, May 2018, pp. 1848–1855. https://doi.org/10.1109/ECTC.2018.00277
30. S. S. B. Lim, S. C. Chong, L. P. S. Sharon, W. W. Seit, X. Zhang, Comprehensive study on die shift and die protrusion issues during molding process of Mold-1st FOWLP, in 2018 IEEE 20th Electronics Packaging Technology Conference (EPTC), Singapore, Singapore, December 2018, pp. 201–205. https://doi.org/10.1109/EPTC.2018.8654399
31. R. Kahle, K.-F. Becker, J. Bauer, T. Braun, T. Thomas, K.-D. Lang, Development of a sensor concept to in situ measure process data in a transfer mold process, presented at the Smart Systems Integration Conference (SSI), Munich, 2016
32. B. Kaya, Concept development and implementation of online monitoring methods in the transfer molding process for electronic packages, Doctoral thesis, Technische Universität Berlin, Berlin, 2018

Chapter 17
Outlook: From Physics of Failure to Physics of Degradation

W. D. van Driel, M. Yazdan Mehr, X. J. Fan, and G. Q. Zhang

The history of reliability as we know it now goes back to the 1950s, when electronics started to play a major role for the first time. Now, seven decades later, with million times more complex electronic systems, the industry is facing a continuous increase of early and wear-out failures with accompanying consequences. Nowadays, products with high failure rates may come under public scrutiny due to negative customer feedback publicly shared on websites, eventually building bad reputation for a company [1, 2]. To cover the increasing demands in product reliability performance, three distinct waves can be noted [3]:

- Wave 1: Stress-based

The first wave was characterized with the establishment of a test-to-failure approach based on standardized stress-based tests. Examples are thermal cycling, moisture testing and operational tests under combined conditions. Each of these tests got standardized in the semiconductors industry by dedicated bodies, like, e.g. JEDEC, IEEE or IEC [4, 5], to enable smooth comparison between suppliers and test houses. Understanding of possible failure modes gradually increased in several industries using semiconductor devices, but the use of prediction models was still limited.

W. D. van Driel (✉)
Signfy, HTC7, Eindhoven, The Netherlands

Delft University of Technology, Delft, The Netherlands
e-mail: willem.van.driel@signify.com

M. Y. Mehr · G. Q. Zhang
Delft University of Technology, Delft, The Netherlands
e-mail: m.yazdanmehr@tudelft.nl

X. J. Fan
Department of Mechanical Engineering, Lamar University, Beaumont, TX, USA

W. D. van Driel, M. Yazdan Mehr (eds.), *Reliability of Organic Compounds in Microelectronics and Optoelectronics*,
https://doi.org/10.1007/978-3-030-81576-9_17

- Wave 2: Knowledge-based

The second wave continued from all the test results obtained over a period of 30–40 years in the first wave. Companies started to understand the physics that caused failure modes in their products. Test schemes changed to test-to-failure instead of test-to-pass. Still standardized tests are used under the condition of similarity: if a previous product that differed slightly from a new one, no new testing was required. This wave is characterized as knowledge-based qualification [6]. Models became commonly used in this wave both analytical and/or numerical (using finite element methods) ones.

- Wave 3: Application-based

In the third wave, application conditions are considered. All industries performed a substantial amount of application studies in which with dedicated sensor are used to measure the actual loading, in terms of temperatures, vibrations and/or external forces. Here measure, in some cases, means monitoring so that the data is logged continuously and send to an online database. Although standards are not yet available, some bodies did publish guidelines [7].

Per today, most industries are in the transfer coming from wave 2 towards wave 3. Wave 3 goes hand in hand with the current development of machine learning, digital twin driven diagnostics or prognostics and health monitoring [8]. These technologies are needed to move to *wave 4: degradation-based.* (Fig. 17.1)

Degradation is apparent in all things and is fundamental to both manufactured and natural objects. It is often described by the second law of thermodynamics, where entropy, a measure of disorder, tends to increase with time in a closed system. Simply said things age. The natural ageing and degradation of materials have been a subject of study by engineers and scientists for many, many years. But with the demands placed on new engineered materials and devices for electronics,

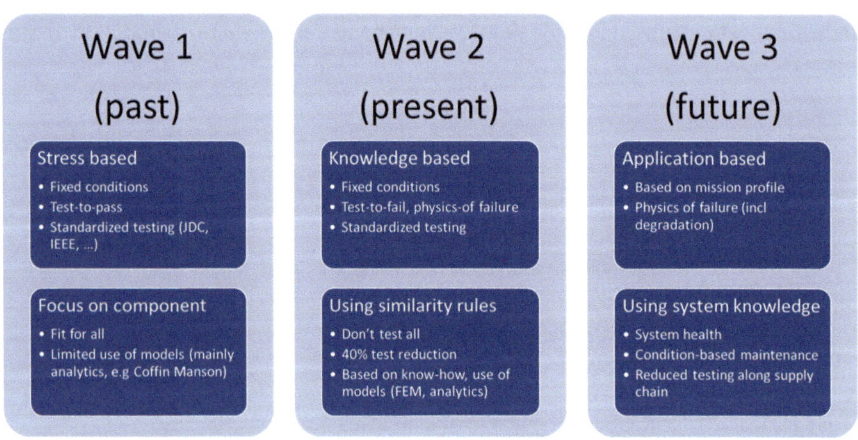

Fig. 17.1 Waves in reliability

computing, aerospace and biomedical applications, the reliability of such over time has become more and more crucial [9, 10].

Degradation is apparent in naturally occurring materials and structures as well as human-engineered materials and devices. In everyday experience, it is the ever-present phenomena of spontaneous loss of some quality, functionality and order. Work must be done from outside the system of interest to maintain that functionality or that order. The second law of thermodynamics formally captures this idea with the concept of entropy or disorder, which states [10]:

> *During real processes, the entropy of an isolated system always increases. In the state of equilibrium, the entropy attains its maximum value.*

This loss of order or degradation has many terms or phrases to label the phenomena, such as ageing, deterioration, devolution and wear-out. It is this degradation in electronics that this book has explored.

The mechanisms of degradation for a variety of materials and structures cover a wide range of discipline categories such as thermal, mechanical, chemical, biological and so on. All associated degradation mechanisms require the knowledge and understanding of natural processes and thus are grouped together as the physics of degradation. As mentioned above, the fundamental underlying principle is entropy and the second law of thermodynamics.

Regardless how carefully crafted, devices are made of materials that generally exist in metastable states. A state is referred to as being metastable if it is only apparently stable and susceptible to change/degradation. The driving force for materials degradation is a lower Gibbs Potential. When we apply a generalized stress ξ to a material, it tends to increase (not lower) the Gibbs Potential. Therefore, a stressed material is even more unstable and even more susceptible to degradation. Since electronic devices are fabricated from materials (and materials degrade with time), then these devices will degrade with time. Engineers are confronted with the very difficult situation—they must manage the degradation rate in order to prevent failure.

Here, health monitoring [8] and/or digital twin technologies [11, 12] may support the engineers to understand, master and forecast the physics of degradation. The concept of digital twin is relatively new. It was conceptualized during the early years of the twenty-first century and has gained traction mainly during the last decade. The primary reason behind it is the further digitalization of the electronic industry, which has been accelerated by the newly emerging IT technologies. Digital twin enables system optimization, monitoring, diagnostics and prognostics using integration of artificial intelligence, machine learning and big data analytics. It can be used for predicting failures and estimating lifetime of electronic components, which then allows for scheduling preventive maintenance. Launching a preventive maintenance program like this allows company to save time and costs and avoid customer dissatisfaction as well as unwanted lawsuits. This will be embraced in the fourth wave of reliability, physics of degradation, which will also reduce the amount (and cost) of product release testing.

Acknowledgements This project has received funding from the ECSEL Joint Undertaking (JU) under grant agreement No 876659. The JU receives support from the European Union's Horizon 2020 research and innovation programme and Germany, Austria, Slovakia, Sweden, Finland, Belgium, Italy, Spain, Netherlands, Slovenia, Greece, France and Turkey.

References

1. T. de Groot, T. Vos, R.J.M.J. Vogels, W.D. van Driel, Quality and reliability in solid state lighting, In: W.D. Van Driel, X.J. Fan (editors), Solid State Lighting Reliability: Components to System. 01/2013; ISBN 978-1-4614-3067-4 Springer New York
2. Warrantee Week, Warranty claims & accruals in financial statements, September 16, 2011., http://www.warrantyweek.com/
3. R. Rongen, On the relevance of physics of fail and mission profiles for microelectronics in automotive applications, presentation at the seminar Trends and challenges in Reliability: from components to systems, Eindhoven, May 2019
4. EIA/JEDEC Standard EIA/JESD47, Stress-Test-Driven Qualification of Integrated Circuits, Electronic Industries Association, July 1995
5. JEDEC Standard JESD34, Failure-Mechanism-Driven Reliability Qualification of Silicon Devices, Electronic Industries Association, March 1993
6. International SEMATECH Publication #00053958A-XFR, Knowledge-Based Reliability Qualification Testing of Silicon Devices, 2000
7. International SEMATECH Publication #99083810A-XFR, Use Condition Based Reliability Evaluation of New Semiconductor Technologies, 1999
8. M.S. Ibrahim, J. Fan, W.K.C. Yung, A. Prisacaru, W. van Driel, X. Fan, G. Zhang, Machine learning and digital twin driven diagnostics and prognostics of light-emitting diodes. Laser Photonics Rev, 2000254 (2020). https://doi.org/10.1002/lpor.202000254
9. J. Swingler, *The Physics of Degradation in Engineered Materials and Devices: Fundamentals and Principles* (Momentum Press, New York, 2014). https://doi.org/10.5643/9781606504680
10. J.W. McPherson, Physics of degradation, in *Reliability physics and engineering: Time-to-failure modeling*, (Springer International Publishing, ISBN: 978-3-319-93683-3, 2019). https://doi.org/10.1007/978-3-319-93683-3_2
11. F. Tao, M. Zhang, Y. Liu, A. Nee, Digital twin driven prognostics and health management for complex equipment. CIRP Ann. **67**(1), 169–172 (2018)
12. A.M. Madni, C.C. Madni, S.D. Lucero, Leveraging digital twin technology in model-based systems engineering. Systems **7**(1), 7 (2019)

Index

W. D. van Driel, M. Yazdan Mehr (eds.), *Reliability of Organic Compounds in Microelectronics and Optoelectronics*,
https://doi.org/10.1007/978-3-030-81576-9

F

Q

Printed by Printforce, the Netherlands